METHODS OF BIOCHEMICAL ANALYSIS

Volume 21

METHODS OF
BIOCHEMICAL ANALYSIS

Edited by **DAVID GLICK**

Stanford University Medical School
Stanford, California

VOLUME 21

An Interscience® Publication

JOHN WILEY & SONS, New York • London • Sydney • Toronto

An Interscience ® Publication

Copyright © 1973, by John Wiley & Sons, Inc.

All rights reserved.   Published simultaneously in Canada.

No part of this book may be reproduced by any means, nor transmitted, nor translated into a machine language without the written permission of the publisher.

Library of Congress Catalogue Card Number: 54-7232

ISBN 0-471-30751-3

Printed in the United States of America.

10 9 8 7 6 5 4 3 2

PREFACE

Annual review volumes dealing with many different fields of science have proved their value repeatedly and are now widely used and well established. These reviews have been concerned primarily with the results of the developing fields, rather than with the techniques and methods employed, and they have served to keep the ever-expanding scene within the view of the investigator, the applier, the teacher, and the student.

It is particularly important that review services of this nature should now be extended to cover methods and techniques, because it is becoming increasingly difficult to keep abreast of the manifold experimental innovations and improvements which constitute the limiting factor in many cases for the growth of the experimental sciences. Concepts and vision of creative scientists far outrun that which can actually be attained in present practice. Therefore an emphasis on methodology and instrumentation is a fundamental need in order for material achievement to keep in sight of the advance of useful ideas.

The volumes in this series are designed to try to meet the need in the field of biochemical analysis. The topics to be included are chemical, physical, microbiological, and if necessary, animal assays, as well as basic techniques and instrumentation for the determination of enzymes, vitamins, hormones, lipids, carbohydrates, proteins and their products, minerals, antimetabolites, etc.

Certain chapters will deal with well-established methods or techniques which have undergone sufficient improvement to merit recapitulation, reappraisal, and new recommendations. Other chapters will be concerned with essentially new approaches which bear promise of great usefulness. Relatively few subjects can be included in any single volume, but as they accumulate these volumes should comprise a self-modernizing encyclopedia of methods of biochemical analysis. By judicious selection of topics it is planned that most subjects of current importance will receive treatment in these volumes.

The general plan followed in the organization of the individual chapters

v

is a discussion of the background and previous work, a critical evaluation of the various approaches, and a presentation of the procedural details of the method or methods recommended by the author. The presentation of the experimental details is to be given in a manner that will furnish the laboratory worker with the complete information required to carry out the analysis.

Within this comprehensive scheme the reader may note that the treatments vary widely with respect to taste, style, and point of view. It is the Editor's policy to encourage individual expression in these presentations because it is stifling to originality and justifiably annoying to many authors to submerge themselves in a standard mold. Scientific writing need not be as dull and uniform as it too often is. In certain technical details, a consistent pattern is followed for the sake of convenience, as in the form used for reference citations and indexing.

The success of the treatment of any topic will depend primarily on the experience, critical ability, and capacity to communicate of the author. Those invited to prepare the respective chapters are scientists who either have originated the methods they discuss or have had intimate personal experience with them.

It is the wish of the Advisory Borad and the Editor to make this series of volumes as useful as possible and to this end suggestions will always be welcome.

DAVID GLICK

CONTENTS

Techniques for the Characterization of UDP Glucuronyltransferase, Glucose-6-phosphatase, and Other Tightly-Bound Microsomal Enzymes. *By David Zakim, Molecular Biology Division and Department of Medicine, Veterans Administration Hospital, and Department of Medicine, University of California Medical Center, San Francisco, California, and Donald A. Vessy, Molecular Biology Division, Veterans Administration Hospital, Department of Biochemistry and Biophysics, University of California Medical Center, San Francisco, California*   1

Determination of Selenium in Biological Materials. *By O. E. Olson, I. S. Palmer, and E. I. Whitehead, Experiment Station Biochemistry Department, South Dakota State University, Brookings, South Dakota* .   39

High-Performance Ion-Exchange Chromatography with Narrow-Bore Columns: Rapid Analysis of Nucleic Acid Constituents at the Subnanomole Level. *By Csaba Horvath, Yale University, New Haven, Connecticut*   79

Newer Developments in Enzymic Determination of D-Glucose and Its Anomers. *By Jun Okuda and Ichitomo Miwa, Faculty of Pharmaceutical Science, Meijo University, Nagoya, Japan*   155

Radiometric Methods of Enzyme Assay. *By K. G. Oldham, The Radiochemical Centre, Amersham, Buckinghamshire, England*   191

Polarography and Voltammetry of Nucleosides and Nucleotides and Their Parent Bases as an Analytical and Investigative Tool. *By Philip J. Elving, James E. O'Reilly, and Conrad O. Schmakel, The University of Michigan, Ann Arbor, Michigan*   287

Integrated Ion-Current (IIC) Technique of Quantitative Mass Spectrometric Analysis: Chemical and Biological Applications. *By John R. Majer, Department of Chemistry, University of Birmingham, Birmingham, England, and Alan A. Boulton, Psychiatric Research Unit, University Hospital, Saskatoon, Saskatchewan, Canada*   467

Author Index.   515

Subject Index.   543

Cumulative Author Index, Volumes 1–21 and Supplemental
    Volume.   553

Cumulative Subject Index, Volumes 1–21 and Supplemental
    Volume.   562

METHODS OF BIOCHEMICAL ANALYSIS

Volume 21

Techniques for the Characterization of UDP-Glucuronyltransferase, Glucose-6-Phosphatase, and Other Tightly-Bound Microsomal Enzymes

DAVID ZAKIM, *Molecular Biology Division and Department of Medicine, Veterans Administration Hospital, and Department of Medicine, University of California Medical Center, San Francisco, California*

and

DONALD A. VESSEY, *Molecular Biology Division, Veterans Administration Hospital, Department of Biochemistry and Biophysics, University of California Medical Center, San Francisco, California*

I.	Introduction	2
II.	Specific Assays of Microsomal Enyzmes	3
	1. UDP Glucuronyltransferase	3
	A. Background	3
	B. Choice of Aglycone for Assay of UDP Glucuronyltransferase	4
	a. *p*-Nitrophenol	4
	b. *o*-Aminophenol	9
	c. *p*-Aminophenol	11
	d. *o*-Aminobenzoate	11
	e. Bilirubin	12
	f. Phenolphthalein	14
	g. Assays Using ¹⁴C-labeled Glucuronyl Acceptors	14
	C. Assays of UDP Glucuronyltransferase Based on the Reverse Reaction	15
	2. Glucose-6-phosphatase	16
	A. Background	16
	B. Assay of Phosphohydrolase activity	16
	a. Hydrolysis of Glucose-6-Phosphate	16
	b. Hydrolysis of Pyrophosphate	17
	C. PPi-Glucose Phosphotransferase activity	18
III.	Investigations of Phospholipid-Protein Interactions	19
	1. Background	19
	2. Phospholipase A	20
	A. Purification of Phospholipase A from *Naja naja* Venom	20
	B. Purification of Phospholipase A from *Crotalus adamanteus* Venom	20
	C. Properties of Phospholipase A	21

1

3. Phospholipase C: Preparation and Properties 21
4. Phospholipase D: Preparation and Properties 22
5. Treatment of UDP Glucuronyltransferase and Glucose-6-Phosphatase with Phospholipases 22
6. Properties of Phospholipase A-Treated Enzymes 23
7. Activation by Detergents 26
IV. Preparation of Microsomes 26
1. Choice of Homogenization Medium 26
2. Homogenization, Isolation, and Storage 27
3. Large-Scale Preparation 28
4. Subfractionation of Microsomes 28
5. Variability in the Properties of Microsomal Enzymes 30
V. Interfering Enzymes and Substrate Forms 31
1. Background 31
2. Metabolism of Substrates by Two or More Enzymes 32
A. Determination of Initial Rates 32
B. Suppression of the Activity of Competing Enzymes by Modification of Assay pH 33
C. Purification 33
D. Use of Inhibitors 34
E. Substrate Forms 34
Acknowledgments 35
References 35

I. INTRODUCTION

Since the isolation and identification of microsomes (1,2) relatively little progress has been made in characterizing most of the enzymatic systems contained within these structures. Problems of multiplicity, substrate specificity, mechanism of action, and dynamics of regulation remain unresolved. Although the classical technique for careful study of an enzyme begins with purification, the problem of removing microsomal enzymes from their attachment to membranes has contributed to the difficulties in working with these enzymes. It has also become apparent recently that the catalytic properties of many tightly bound microsomal enzymes depend on interactions with their microsomal environments (3–10). Hence the proper study of microsomal enzymes actually requires that they be characterized in experiments with intact microsomes. Investigators therefore face problems of experimental design which do not arise or are avoided easily when working with unbound cytoplasmic enzymes. For example, microsomal enzymes exist in a heterogeneous particle containing enzymes which may metabolize products and substrates in pathways other than the one due to the enzyme of interest. There are also problems in the preparation and storage of microsomes since physical or chemical agents which alter the microsomal lipids, such as endogenous phospholipase A, can modify the properties of membrane-bound enzymes. The dietary history, hormonal balance, and age of animals also may influence the kinetic parameters of microsomal enzymes. In addition, substrates for many microsomal enzymes have limited solubility in H_2O, or are amphipathic and activate or inactivate microsomal enzymes because of nonspecific effects on the microsomal membrane.

It is not the purpose of this review to cover currently available techniques for the assay of microsomal enzymes in an encyclopedic way. Rather, emphasis is placed on the problems encountered in characterizing the properties of tightly bound enzymes in liver microsomes by assaying in the presence of a complex mixture of other microsomal enzymes, and on the ways in which most of these difficulties can be dealt with and meaningful assays for many microsomal enzymes developed. The assays of UDP glucuronyltransferase and glucose-6-phosphatase are presented in more detail, since these two tightly bound microsomal enzymes have been studied most extensively from the points of view of the number of separate species of protein needed to account for homologous reactions with different substates, the regulatory importance of protein-phospholipid interactions, and the determination of exact kinetic constants and kinetic mechanisms. The detailed descriptions of assay techniques for UDP glucuronyltransferase and glucose-6-phosphatase have general applicability to the problems likely to be encountered in examining the properties of other tightly bound microsomal enzymes. In addition to assay techniques, the closely related problems of preparation, storage, and subfractionation of microsomes, the techniques for studying the effects of treatment with phospholipases and detergents on the properties of tightly bound microsomal enzymes, and substrate forms are discussed in detail.

II. SPECIFIC ASSAYS OF MICROSOMAL ENZYMES

1. UPD-Glucuronyltransferase

A. BACKGROUND

Excretion of exogenous compounds as sugar conjugates was observed more than 100 years ago (11) in studies which led eventually to the elucidation of glucuronic acid as the conjugated sugar derivative. Work with intact organs, tissue slices, and homogenates, as well as the availability of ^{14}C-labeled sugars, established that glucose was the precursor of the glucuronic acid and that an "active factor," later shown to be UDP-glucuronic acid, was required for glucuronide synthesis in liver homogenates. It is now known that a variety of compounds are metabolized according to [1]:

$$UDP\text{-}GA + RH = R\text{-}GA + UDP \qquad [1]$$

where RH is an organic acid, a phenol, or an amine. Reaction [1], catalyzed by UDP glucuronyltransferase, is important for the detoxification of pharmacologic agents and several endogenously produced compounds in the microsomal fraction of the cell; activity is highest in liver but is present also in skin, kidney, intestinal mucosa, and some endocrine organs (12).

B. CHOICE OF AGLYCONE FOR ASSAY OF UDP GLUCURONYLTRANSFERASE

How many species of UDP glucuronyltransferase exist in liver microsomes is not known, but it is certain that all O-glucuronides are not synthesized by a single enzyme. Evidence obtained in this laboratory from kinetic studies of UDP glucuronyltransferase (13) and the properties of the —SH groups of this enzyme indicate that o-aminophenol and p-nitrophenol are glucuronidated by different enzymes, and it is clear that o-aminobenzoate does not share a common aglycone binding site with either of the other substrates. Therefore, assays of UDP glucuronyltransferase conducted with these glucuronyl acceptors, and probably bilirubin as well, do not measure the activity of the same enzyme. On the other hand, it is not known at this time what other substrates, if any, are glucuronidated by the p-nitrophenol, o-aminophenol, and o-aminobenzoate metabolizing forms of UDP glucuronyltransferase, and in general how many substrate-specific forms of UDP glucuronyltransferase exist. Although a discussion of the technical aspects of the problem of multiplicity is beyond the scope of this review, it should be stressed that with aglycones other than those listed above one cannot be certain what UDP glucuronyltransferase enzyme is being assayed.

a. p-Nitrophenol. p-Nitrophenol at alkaline pH has an absorption maximum at 400 nm which is lost on formation of the glucuronide; the assay with this substrate is based on the disappearance of p-nitrophenol as measured by the decrease in optical density at 400 nm.

REAGENTS. 1. *Sodium phosphate buffer*, 0.25M, pH 7.1.
           2. *UDP-glucuronic acid*, ammonium salt, 0.05M, pH 7.1.
           3. *p-Nitrophenol*, 0.002M.
           4. *Trichloroacetic acid*, 0.1M.
           5. *Potassium hydroxide*, 10N.

Procedure. For determination of activity at a single set of substrate concentrations the following final concentrations of reagents are convenient: $2 \times 10^{-4}M$ p-nitrophenol (0.05 ml), $5 \times 10^{-3}M$ UDP-glucuronic acid (0.05 ml), 0.05M phosphate buffer (0.10 ml), and 0.5 to 1.0 mg of microsomal protein in a final volume of 0.5 ml. Tubes are warmed to 37°, and the reaction is started by the addition of microsomes. After rapid mixing, a 0.1-ml aliquot of the reacion mixture is removed immediately and added to 2.0 ml of $0.1M$ trichloroacetic acid (TCA). This sample is the blank and should be determined separately for each assay. Serial aliquots of 0.10 ml are removed 4, 8, and 12 min after the addition of enzyme and similarly deproteinized by addition to 2.0 ml of TCA. After brief centrifugation to remove denatured protein, the supernatants are decanted into tubes containing 0.05 ml of $10N$ KOH, which raises the pH to >10.0, and the optical density is determined at 400 nm. The extinction coefficient for p-nitrophenol at pH >10 is 1.81×10^4 cm²/mole. Because the rate of disappearance of substrate is used to follow the course of the reaction, accurate pipetting is essential if quantitatively good data are to be obtained. Hence, the 0.1-ml aliquots should be removed from the reaction mixture with micropipettes. Also, we have found that the use of "Repipettes" (Labindustries, Berkeley, Calif.) provides the most convenient and reproducible method of accurately dispensing 2.0 ml of TCA.

The timing of the removal of serial aliquots from the reaction mixture can be adjusted according to the activity and amount of enzyme added; with untreated guinea pig microsomes and the concentrations of substrate specified above, rates of optical density change of about 0.020 per 4 min are observed, and the assay is linear with time. Hence, a single-point assay can be used. However, at lower concentrations of p-nitrophenol or UDP-glucuronic acid, assays are not linear with time, and serial time points must be used to estimate initial rates of activity. With rat liver microsomes linearity is not maintained even at relatively high concentrations of substrates, since these microsomes contain a highly active nucleotide pyrophosphatase which consumes UDP-glucuronic acid at a rapid rate, depleting the substrate concentration in the UDP glucuronyltransferase reaction. With rat microsomes, therefore, several time points always must be used to estimate initial rates of activity. The reaction rate varies little in the pH range of 7.0–7.8. At pH 8.0 and above, UDP glucuronyltransferase is activated irreversibly, activation being maximal at pH 10.5 (5). At pH values below 7.0 the assay cannot be used since in the presence of low concentrations of p-nitrophenol a chromophore, not precipitated by TCA and absorbing at 400 nm, is released from the microsomes. This chromophore is released also in the pH range 7.0–8.0, but not to a significant extent, at p-nitrophenol concentrations less than 0.6mM. Assay at pH less than 7 is also complicated by anomalous kinetics, since UDP glucuronlytransferase is activated by the phenolate form of p-nitrophenol (14). This complication also restricts the upper limit of concentrations of p-nitrophenol to 0.6mM in the pH range 7.0–8.0.

Increasing the concentration of phosphate to greater than $0.10M$ inhibits UDP glucuronyltransferase assayed with p-nitrophenol. This effect may be a general action of salts since the enzyme is inhibited also by NaCl at concentrations greater than $0.2M$.

Effect of Mg^{2+}. Addition of Mg^{2+} enhances the activity of UDP glucuronyl-transferase, primarily by increasing the activity at V_{max}, but glucuronidation of *p*-nitrophenol does not require Mg^{2+}. Probably EDTA complexes endogenous heavy metals in the microsomes, since it decreases the activity with *p*-nitrophenol as glucuronyl acceptor. If activity is measured in the presence of Mg^{2+}, EDTA should be added to the TCA tubes to give a final concentration of $5 \times 10^{-3}M$ in order to prevent precipitation of $Mg(OH)_2$ when the pH is raised to > 10.0. Unless the specific effects of Mg^{2+} on the activity of UDP glucuronyltransferase are to be studied with *p*-nitrophenol as aglycone, it is best not to include Mg^{2+} in the assays. In the presence of Mg^{2+}, primary double reciprocal plots of $1/v$ versus [UDP-glucuronic acid] are not linear over a range of concentrations of UDP-glucuronic acid of 2.5 to $40 \times 10^{-3}M$, whereas they are linear in the absence of Mg^{2+}.

Interpretation of Data. Although the method outlined above is given for a single set of substrate concentrations, it should be made clear that the rates as measured are far from those prevailing at saturating concentrations of substrates. These values can be obtained only graphically, since it is not practical to use saturating concentrations of UDP-glucuronic acid or *p*-nitrophenol. With *p*-nitrophenol as aglycone, K_{UDPGA} is $1.2 \times 10^{-2}M$ with guinea pig liver microsomes as the source of the enzyme. Also, at concentrations of UDP-glucuronic acid greater than $6 \times 10^{-2}M$ there is substrate inhibition. As mentioned above, *p*-nitrophenol at relatively high concentrations has several nonspecific effects on the properties of UDP-glucuronyltransferase and the microsomes.

The data in Figures 1 and 2 illustrate the method for measuring activity at V_{max} for UDP glucuronyltransferase with *p*-nitrophenol as aglycone. Initial rates of activity are determined as a function of the concentration of UDP-glucuronic acid at several different fixed concentrations of *p*-nitrophenol. The intercepts on the $1/v$ axis of the primary double reciprocal plots (Figure 1) are replotted versus the reciprocals of the concentration of the fixed substrate (Figure 2). The intercept on the $1/v$ axis of the secondary plot is $1/V_{max}$.

The data presented in Figures 1 and 2 were obtained with concentrations of UDP-glucuronic acid greater than 2.5 mM. At concentrations below this level, plots of $1/v$ versus 1/[UDP-glucuronic acid] are not linear but bend concave downward. Thus, rates of glucuronidation are greater at low concentrations of UDP-glucuronic acid than would be anticipated by extrapolation of the rate data obtained at high concentrations of UDP-glucuronic acid. Careful analyses of the data indicate that the most likely explanation for non-linearity in double reciprocal plots for v as a function of the concentration of UDP-glucuronic acid is negative cooperativity in the sequential

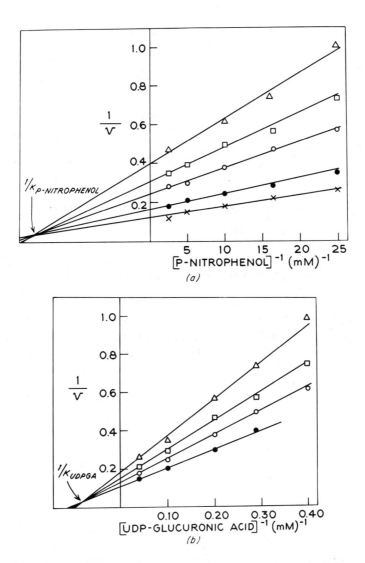

Figure 1. Determination of kinetic parameters of UDP-glucuronyltransferase. Initial rates of UDP-glucuronyltransferase were determined and plotted in double reciprocal form. (a) Rate as a function of the concentration of p-nitrophenol at several fixed concentrations of UDP-glucuronic acid: 2.5mM (△); 3.5mM (□); 5mM (○); 10mM (●); 25mM (×). (b) Rate as a function of the concentration of UDP-glucuronic acid at fixed concentrations of p-nitrophenol: 0.04mM (△); 0.06mM (□); 0.1mM (○); 0.2mM (●).

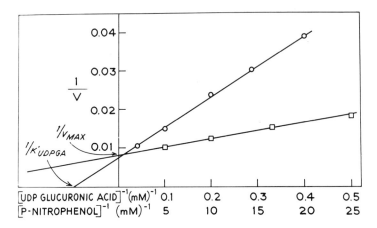

Figure 2. The intercepts on the $1/v$ axis in Figure 1 are replotted against $1/$[UDP-glu-curonic acid] ($\bigcirc$) and $1/$[p-nitrophenol] ($\square$) in order to obtain $1/V_{\max}$.

binding of UDP-glucuronic acid to UDP glucuronyltransferase (14a, 14b). In the presence of added Mg^{2+} double reciprocal plots are non-linear even at high concentrations of UDP-glucuronic acid probably for the same reason.

For an enzyme fulfilling the criteria of a Michaelis-Menten kinetic model, the intersection of the family of primary double reciprocal plots is the K_m for the variable substrate when it is the first substrate bound to the enzyme. The intercept on the $1/$[s] axis of the secondary plot is the K_m when this substrate is bound second. More precise estimates of K_m for binding to free enzyme can be obtained by determining the ratio of secondary replots of the slopes and intercepts of the data in Figure 1 (14c). Obviously this type of straight forward interpretation of secondary plots is not possible with UDP glucuronyltransferase because of apparent homotropic cooperativity in substrate binding. Nevertheless, the K's determined above do reflect a real property of the enzyme. The exact physical meaning of these constants depends on the kinetic mechanism of the enzyme under study. For UDP glucuronyltransferase, which has a rapid-equilibrium, random-order mechanism (14), K_m is the enzyme-substrate dissociation constant for the binding of substrate to the n*th* subunit of the enzyme. It is especially important to emphasize that studies of activity during induction of the enzyme or development of p-nitrophenol conjugating activity in fetal animals should be based on measurements of activity at $V_{\max}$; changes in activity which are based on rates of reaction at a single set of substrate concentrations cannot delineate differences in the binding of substrates or in the catalytic rate constant of the enzyme.

b. *o*-Aminophenol. Assays with this aglycone are based on the fact that *o*-aminophenylglucuronic acid can be diazotized selectively in the presence of unreacted *o*-aminophenol by careful control of the pH of the diazotization reaction, the conditions for which were established by Levvy and Storey (15). The diazotized *o*-aminophenylglucuronide is then complexed with *N*-(-naphthyl)ethylenediamine dihydrochloride. The product of this reaction is measured at its absorption maximum; the extinction coefficient for the coupled product is $2.9 \times 10^4 \text{ cm}^2/\text{mole}$ at 555 nm.

REAGENTS.
1. *Sodium phosphate* buffer, $0.25M$, pH 7.6.
2. *UDP-glucuronic acid*, ammonium salt, $0.05M$, pH 7.6.
3. *o-Aminophenylglucuronide* (Koch-Light Laboratories, Colnbrook, England).
4. *o-Aminophenol*, $0.002M$, containing 2 mg/ml ascorbate, pH 7.0. The *o*-aminophenol solution should be prepared fresh each week by sublimation and stored at $-20°$.
5. *Ascorbate*, 2 mg/ml, pH 7.0.
6. *Trichloroacetic acid-sodium phosphate*, $1M$, pH 2.0, mixed daily from solutions of TCA, $2M$, pH 2.0, and sodium phosphate, $2M$, pH 2.0.
7. *Sodium nitrite*, 0.05% (w/v).
8. *Ammonium sulfamate*, 0.5% (w/v).
9. *N-(1-naphthyl)ethylenediamine dihydrochloride.* Add 47.6 ml water to 55 mg in preweighed vials (Sigma).

Procedure. For assays at a single set of substrate concentrations pipette 0.2 ml *o*-aminophenol (final concentration $2 \times 10^{-4}M$), 0.2 ml UDP-glucuronic acid (final concentration $5 \times 10^{-3}M$), 0.4 ml phosphate buffer, and enough H_2O to produce a final volume of 2.0 ml. Allow the mixture to come to thermal equilibrium at 37°, and start the reaction by adding 1 to 2 mg of microsomal protein. At 5-min intervals during the course of the reaction transfer 0.5-ml aliquots to 0.5 ml of the TCA-sodium phosphate reagent. A single zero-time sample can serve as blank for a series of assays.

After removal of precipitated protein from the mixtures by centrifugation and decantation, add 0.1 ml sodium nitrite to each tube; shake and allow to stand at least 5 min. Add 0.1 ml ammonium sulfamate and, after 5 min, 0.1 ml *N*-(1-naphthyl)ethylenediamine dihydrochloride. Incubate the tubes in the dark at 25° for 2 hr; then read the optical density at 555 nm. Because of variability in the color yield from day to day, standards of *o*-aminophenylglucuronide should be run with each set of assays. With guinea pig liver microsomes the reaction is linear with time under these conditions. When microsomes contain nucleotide pyrophosphatase activity, or when the concentrations of substrates are reduced, the initial rates of activity must be estimated by extrapolation to zero time of serial estimates of the o-amino-phenylglucuronic acid synthesized. We have observed that in some situations the time for maximal color development may be variable, and should be determined if the assay system specified above is modified. The final pH of the diazotization mixture must be between 2.1 and 2.3 (16). At pH values less than 2.0 the aglycone, as well as the glucuronide, will be diazotized, giving spuriously high rates of glucuronidation. At pH levels greater than 2.3, diazotization of the glucuronide will be inhibited with consequent falsely low reaction rates. The pH of the TCA-phosphate reagent and the final pH of the diazotization mixture should be checked daily, and adjustments made in the pH of the stock TCA-phosphate in order to maintain the final pH of the mixture, after addition of the assay aliquot, in the desired range.

Substitution of amine-containing buffers for phosphate is not recommended, since we have found that even small amounts of these substances interfere with the diazotization. The effect of high concentrations of salt on the glucuronidation of o-aminophenol has not been investigated. If the amount of o-aminophenol added to assay tubes is less than that specified above, additional ascorbate is needed to maintain a constant concentration of this compound, since ascorbic acid is added to prevent oxidation of the o-aminophenol.

Effect of Mg^{2+}. As with p-nitrophenol, the UDP glucuronyltransferase responsible for the glucuronidation of o-aminophenol is enhanced by Mg^{2+}, though there is no absolute dependence on Mg^{2+}. The effect of Mg^{2+} is on activity at V_{max}. The addition of EDTA decreases the rate of glucuronidation of o-aminophenol.

Interpretation of Data and Limitations of the Method. Rates of o-aminophenol glucuronidation measured at a single set of substrate concentrations do not reflect maximum rates, and comparisons of activities under different experimental conditions should be based on determinations at V_{max} so that effects on the amount, catalytic constants, and binding affinity for substrates can be resolved. There are, however, more limitations on the estimate of V_{max} with o-aminophenol than with p-nitrophenol. Not only does o-aminophenol activate UDP glucuronyltransferase at high concentrations (14) but also plots of $1/v$ versus $1/$[UDP-glucuronic acid] are nonlinear; and relatively high concentrations of UDP-glucuronic acid (greater than $15 \times 10^{-3}M$) are inhibitory. A detailed consideration of the causes of the anomalous kinetic behavior, is beyond the scope of this review, but it again seems to reflect negative cooperativity in the binding of UDP-glucuronic acid. It is possible to obtain good estimates of maximal activity in the same manner as with p-nitrophenol, with careful selection of substrate concentrations. The concentration of o-aminophenol should be kept below $2 \times 10^{-4}M$, and the concentration of UDP-glucuronic acid in the range of $3-15 \times 10^{-3}M$.

c. *p*-Aminophenol. The glucuronidation of this compound can be studied in the same way as that of aminophenol, with only minor modifications. The maximum absorbance of the coupled glucuronide occurs at 540 nm for *p*-aminophenylglucuronide; as with *o*-aminophenylglucuronide, the final pH during diazotization must be kept between 2.1 and 2.3. Maximum color development at 25° requires 1.5 hr.

d. *o*-Aminobenzoate. This assay differs from the others in that, after the reaction is completed, unreacted aglycone is extracted into an organic phase. The glucuronide remains in the aqueous phase and is measured directly by diazotization and coupling with *N*-(1-naphthyl)ethylenediamine dihydrochloride.

REAGENTS. 1. *Sodium phosphate*, 0.25*M*, pH 7.6.
2. *UDP-glucuronic acid*, ammonium salt, 0.02*M*, pH 7.6.
3. *o-Aminobenzoate*, 0.002*M*, pH 7.6.
4. *Trichloroacetic acid*, 0.22*M*.
5. *Sodium phosphate*, 0.5*M*, pH 3.9.
6. *Sodium phosphate*, 1.0*M*, pH 1.0.
7. *Anhydrous ether*.
8. *Sodium nitrite*, 0.05% (w/v).
9. *Ammonium sulfamate*, 0.1% (w/v).
10. *N-(1-naphthyl)ethylenediamine dihydrochloride;* 55 mg in a pre-weighed vial (Sigma) is dissolved in 47.6 ml water.

Procedure. The reaction mixture contains 0.1 ml sodium phosphate, pH 7.6, 0.2 ml UDP-glucuronic acid (final concentration $5 \times 10^{-3}M$), 0.1 ml *o*-aminobenzoate (final concentration $5 \times 10^{-4}M$), and 2 mg microsomal protein in a final volume of 0.8 ml. After equilibration at 37°, the reaction is started by addition of microsomes. Aliquots of 0.2 ml are removed 5, 10, and 15 min after initiation of the reaction, and added to 0.3 ml 0.22*M* TCA. A single zero-time sample (blank) is sufficient. After centrifugation to remove precipitated protein, the supernatants are decanted into 0.3 ml sodium phosphate, pH 3.9. This mixture is extracted 3 times with anhydrous ether, and the ether layer discarded. Sodium phosphate, 0.1 ml, pH 1.0, is added to the remaining aqueous phase, and the glucuronide is diazotized and coupled with *N*-(1-naphthyl)ethylenediamine dihydrochloride, as for *o*-aminophenol. Color yield is maximal in 1 hr, and optical density is determined at 555 nm. Previous workers hydrolyzed the glucuronide in acid after separation of unreacted aglycone before diazotization, but we have found this step to be unnecessary.

The properties of the glucuronidation reaction have not been investigated as extensively with *o*-aminobenzoate as with *p*-nitro- and *o*-aminophenols. It is known, however, that Mg²⁺ and EDTA have qualitatively identical effects on the rate of glucuronidation of all three substrates. On the other hand, o-aminobenzoate differs from the two phenolic substrates in that at high concentrations it has no nonspecific activating effect on the rate of its own metabolism. As with the phenols, however, reaction rates at any single set of substrate concentrations are not a good measure of activity at V_{max}, and graphical methods should be used for the determination of this value and other kinetic constants.

e. Bilirubin. Under suitable conditions bilirubin glucuronide can be diazotized selectively in the presence of bilirubin. The procedure described below for diazotization is basically that of Van Roy and Heirwegh (16).

REAGENTS. 1. *Potassium phosphate*, 0.1M, pH 7.5.
2. *Potassium phosphate*, 1.0M, pH 2.6.
3. *UDP-glucuronic acid*, 0.05M, pH 7.5.
4. *Bilirubin*, 2.5 × 10^{-3}M in 0.01 N sodium hydroxide containing 0.002M EDTA. This reagent is prepared in the dark just before use.
5. *HCl*, 0.15M.
6. *Sodium nitrite*, 0.5% (w/v).
7. *Ammonium sulfamate*, 0.7% (w/v).
8. *Ethylanthranilate*.
9. *2-Pentanone*.
10. *Butyl acetate*.
11. *Ascorbate*, 40 mg/ml, prepared just before use.

Procedure. Working in the dark, add 0.25 ml potassium phosphate, pH 7.5, 0.03 ml bilirubin (final concentration 3 × 10^{-5}M), 10 mg of microsomal protein, and sufficient water so that the final volume will be 2.5 ml. After equilibration at 37°, start the reaction by addition of 0.25 ml UDP-glucuronic acid (final concentration 5 × 10^{-3}M). Immediately remove a 0.5-ml aliquot as the zero-time sample (blank), and add to a glass-stoppered centrifuge tube on ice containing 0.3 ml potassium phosphate, pH 2.6. Remove additional 0.5-ml aliquots 5, 10, and 15 min later and treat them similarly. Place the potassium phosphate-treated samples in a 25° bath for 10 min.

During this time the diazo reagent should be prepared by adding 0.1 ml of ethylanthranilate to 10 ml of 0.15M HCl, followed by 0.30 ml of the sodium nitrite solution; 0.1 ml of 0.7% ammonium sulfamate is added, and the diazo reagent allowed to stand for 3 min more. After preparation, 0.5 ml of diazo reagent is added to each potassium phosphate tube and the samples are warmed for 20 min at 25°. At the end of this incubation, the tubes are removed from the water bath, and 4 ml of ascorbate is added to each, followed by 3 ml of a mixture of 2-pentanone and butyl acetate (17/3, v/v). The tubes are then shaken vigorously, and the phases separated by centrifugation. The optical density of a portion of the upper, organic phase, which contains the coupled bilirubin glucuronide, is measured at 530 nm. The extinction coefficient for the coupled bilirubin glucuronide is 44.4×10^3 cm^2/mole. Note that all operations should be carried out in the dark, and optical density should be determined promptly.

The bilirubin assay is complicated by the limited solubility of the glucuronyl acceptor, a problem which does not arise with the phenols or o-aminobenzoate. In several previously described assay methods bilirubin was added as a bilirubin-albumin complex in order to make the bilirubin more "soluble." No direct measurements were made, however, of the concentration of bilirubin in true solution as a function of different ratios of bilirubin to albumin. Studies in this laboratory indicate in fact that at concentrations of bilirubin less than its limit of solubility the addition of albumin decreases the rate of synthesis of bilirubin glucuronide. Also, if data are based on measurement of initial rates of activity, the rate of formation of bilirubin glucuronide is independent of the amount of bilirubin-albumin complex when the ratio of bilirubin to albumin is constant.

With the conditions given above, the concentration of bilirubin is at the limit of its solubility, $3 \times 10^{-5}M$. Solubility can be increased to a limited extent by increasing the pH or the concentration of salt (17). As long as the concentration of bilirubin added to the assays is less than the limit of its solubility, the enzyme shows typical Michaelis-Menten kinetics for dependence of the rate of glucuronidation on the concentration of bilirubin

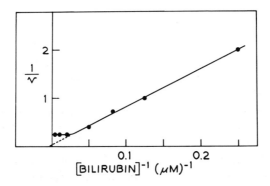

Figure 3. The rate of UDP glucuronyltransferase as a function of variable concentrations of bilirubin. Initial rates of synthesis of bilirubin glucuronide were determined with the indicated amounts of bilirubin and 5.0mM UDP-glucuronic acid. Rates are expressed as optical density change per minute per milligram protein.

(Figure 3). At bilirubin concentrations greater than saturating there is, as expected, no dependence of rate on the concentration of bilirubin.

The addition of Mg^{2+} has no direct effect on the enzyme catalyzing the synthesis of bilirubin glucuronide. At greater than saturating concentrations of bilirubin, Mg^{2+} increases the rate of glucuronidation by increasing the solubility of the bilirubin. Potassium phosphate also enhances the solubility of bilirubin, but in addition, at relatively high concentrations, directly increases the rate of synthesis of bilirubin glucuronide via what appears to be an independent effect on the enzyme. For this reason, the concentration of potassium phosphate in the assay system should be kept constant and preferably below $0.2M$.

As with the other substrates, maximal rates of synthesis of bilirubin glucuronide cannot be estimated from assays at a single set of substrate concentrations. Values can be estimated by extrapolation to V_{max}, as described in the previous discussion, although this determination is inherently less accurate because of the limited solubility of bilirubin.

f. Phenolphthalein. In addition to the methods outlined above, other substrates have been utilized for measuring UDP glucuronyltransferase. Experience in this laboratory with one commonly used compound, phenolphthalein, has shown it to be a less suitable substrate than the others discussed. The reason for this is extensive binding of the substrate by the microsomal membrane. This binding not only complicates the problem of kinetic studies but may be associated also with changes in the properties of the microsome. Unless it can be shown that phenolphthalein is glucuronidated by a separate UDP glucuronyltransferase, for which there is no other known substrate, there appears to be no need for using this compound as a glucuronyl acceptor.

g. Assays Using ^{14}C-Labeled Glucuronyl Acceptors. Lucier et al. (18) have described a simple technique which should be useful in the study of the glucuronidation of a number of compounds. This technique is based on the difference in the solubilities of relatively apolar aglycones and their more polar glucuronides in aqueous and organic solvents. The method has been used with $^{14}C_4$-testosterone, ^{14}C-1-naphthol, and ^{14}C-dieldrin as aglycones, but should have general application for any aglycone which can be separated quantitatively from its glucuronide via solvent extraction.

REAGENTS. 1. *$^{14}C_4$-testosterone* dissolved in benzene.
2. *^{14}C-1-naphthol* dissolved in benzene.
3. *UDP-glucuronic acid.*
4. *Tris-HCl*, pH 7.4, $0.05M$.

Procedure. One of the labeled substrates is added to a liquid scintillation counting vial, and the solvents are evaporated under a stream of nitrogen. Tris (1.0 ml, 0.05M) and UDP-glucuronic acid to give a final concentration of 5 $\times$ $10^{-3}M$ are then added. The vial is warmed to 37°, and the reaction started by the addition of 0.4 to 2.0 mg of microsomal protein. After a 6-min incubation under nitrogen, the reaction is stopped by adding 10 ml of a nonaqueous scintillation fluid (18). Unreacted aglycone is partitioned into the scintillation fluid and is measured by liquid scintillation counting after removal of the aqueous phase. More accurate results probably could be obtained by direct counting of the glucuronide products in the aqueous phase, although this would be more time consuming. Any glucuronyl acceptor could be used in this type of assay if appropriate solvent systems could be found for complete partition of aglycone into the organic phase while retaining the glucuronide in the aqueous phase.

C. ASSAYS OF UDP GLUCURONYLTRANSFERASE BASED ON THE REVERSE REACTION

REAGENTS. 1. *UDP, 0.05M, pH 7.1.*
2. *Sodium phosphate buffer, 0.25M, pH 7.1.*
3. *Saccharic acid–1,4-lactone, 0.05M.*
4. *p-Nitrophenylglucuronide, 0.05M, pH 7.1.*
5. *Trichloroacetic acid, 0.1M.*
6. *Potassium hydroxide, 10N.*

Procedure. Add 0.05 ml of UDP (final concentration 5 $\times$ $10^{-3}M$), 0.05 ml saccharic acid–1,4-lactone (final concentration 5 $\times$ $10^{-3}M$), 0.1 ml p-nitrophenylglucuronide (final concentration 0.01M), 0.1 ml sodium phosphate buffer, and sufficient H_2O to give a final volume of 0.5 ml. Allow the mixture to come to 37°, and start the reaction by adding 1 mg of microsomal protein. Immediately, and 2, 4, and 6 min later, remove 0.1-ml aliquots with a micropipette and add to centrifuge tubes containing 2.0 ml of TCA. After centrifugation to remove precipitated protein, decant into tubes containing 0.05 ml of 10N potassium hydroxide, mix, and determine the optical density at 400 nm.

Measurement of reaction rates in the reverse direction at a single set of substrate concentrations yields no better estimates of activity at V_{max} than those obtained in the forward direction, since K_{UDP} and K_{p-NPGA} are very large, and in some species, such as the guinea pig, the decline in reaction rate with time is quite fast because of rapid destruction of UDP by an interfering enzyme. On the other hand, large concentrations of glucuronide do not alter the properties of the microsome; and the rate of the reverse reaction in rat and beef liver microsomes, which destroy UDP-glucuronic acid via an alternative reaction, but not UDP, is faster than that of the forward reaction. An additional advantage of the back reaction is that low reaction rates can be measured more accurately because product formation is measured directly. Saccharic acid-1,4-lactone is added to the assay in order to inhibit β-glucuronidase, which interferes with estimates of the rate of the reverse reaction by hydrolyzing O-glucuronides. The lactone has no direct effect on the activity of UDP glucuronyltransferase. With 0.01M p-nitrophenylglucuronic acid and 5×10^{-3} M saccharic acid-1,4-lactone there is no production of p-nitrophenol in the absence of UDP. Thus far, the reverse reaction has been found useful only with p-nitrophenylglucuronide as substrate; we have not been able to demonstrate UDP-dependent hydrolysis of o-aminophenylglucuronide or phenolphthalein glucuronide.

2. Glucose-6-Phosphatase

A. BACKGROUND

It has been shown recently that glucose-6-phosphatase is a multifunctional enzyme which acts as a phosphohydrolase and phosphotransferase. Although the enzyme has not been purified, a large body of kinetic evidence suggests that the two types of reaction are catalyzed by a single protein (19–21). The mechanism of reaction involves the formation of a covalently bound enzyme-P_i intermediate which reacts with a variety of P_i acceptors, such as water (phosphohydrolase function) or glucose (phosphotransferase function).

$$\text{Glucose-6-P} + \text{E} \rightarrow \text{E-P}_i + \text{glucose} \qquad [2]$$

$$\text{PP}_i + \text{E} \rightarrow \text{E-P}_i + \text{P}_i \qquad [3]$$

$$\text{E-P}_i + \text{H}_2\text{O} \rightarrow \text{E} + \text{P}_i \qquad [4]$$

$$\text{E-P}_i + \text{glucose} \rightarrow \text{E} + \text{glucose-6-P} \qquad [5]$$

A detailed review of the properties of the glucose-6-phosphatase catalyzed reaction is now available (19). In the methods described below, procedures are given for estimating the rate of the phosphohydrolase reaction with glucose-6-P and PP_i as substrates and the phosphotransferase reaction with PP_i and glucose as substrates.

B. ASSAY OF PHOSPHOHYDROLASE ACTIVITY

a. Hydrolysis of Glucose-6-Phosphate

REAGENTS. 1. *Sodium acetate* buffer, 1.0M, pH 5.75.
        2. *Glucose-6-P*, 0.4M, pH 5.75.
        3. *Trichloroacetic acid*, 10% (w/v).

4. *Ammonium molybdate*, 1.6% (w/v) in $1N$ sulfuric acid.
5. *Ferrous sulfate*, 2.5 g per 25 ml $0.15N$ sulfuric acid, prepared fresh each day.

Procedure. Pipette into a test tube 0.1 ml sodium acetate buffer, 0.2 ml glucose-6-P (final concentration $8 \times 10^{-2}M$), and enough water so that the final volume will be 1.0 ml. After equilibrium at 37°, start the reaction by adding 1 mg of microsomal protein. Stop the reaction, usually after a 5-min incubation, by adding 0.5 ml TCA. Remove precipitated protein by centrifugation, and add 0.5 ml of supernatant to 5.0 ml of ammonium molybdate solution. Then add 0.8 ml of ferrous sulfate solution, and read the optical density at 660 nm after shaking.

With this procedure for the determination of phosphorus, maximum color intensity develops almost instantaneously, is stable for at least 1 to 2 hr, and is linear with phosphorus concentrations up to an optical density of 1.00. The rate of glucose-6-P hydrolysis is linear with time for at least 20 min with untreated microsomes. Assays at concentrations of glucose-6-P of $0.08M$ are nearly equal to activity at V_{max}, since K_{G-6-P} is approximately $5 \times 10^{-3}M$.

With liver preparations we have found it sufficient to run only a reagent blank as control. In some tissues, however (e.g., intestinal mucosa), there is appreciable hydrolysis of glucose-6-P because of the action of nonspecific phosphatases. In this instance the rate of phosphorus release by the nonspecific phosphatase can be estimated by utilizing β-glycerophosphate as substrate, or by measuring the rate of hydrolysis of glucose-6-P after destruction of glucose-6-phosphatase by heating microsomes for 30 min at 37° (22). Standard curves for the phosphorus assay are determined as above, using 1.3613 g of KH_2PO_4 per liter, which gives a phosphorus concentration of 10 μmoles/ml.

In untreated microsomes glucose-6-phosphatase has a broad pH optimum between 5.5 and 6.5. Although the shape of the pH-activity curve shifts after treatment of the microsomes with NH_4OH and detergents (23,24), the pH optimum remains close to 6.0 and the curve is flat in this region of the pH scale. Hence pH 5.75 is suitable for the assay of glucose-6-phosphatase in all preparations.

b. Hydrolysis of Pyrophosphate

REAGENTS. 1. *Sodium acetate* buffer, $1.0M$, pH 5.75.
2. *Sodium pyrophosphate*, $0.2M$, pH 5.75.
3. *Trichloroacetic acid*, 10% (w/v).
4. *Sodium acetate* buffer ($0.1M$ acetic acid, $0.025M$ sodium acetate), pH 4.0.
5. *Ascorbate*, 0.5% (w/v).
6. *Ammonium molybdate*, 7.6% (w/v) in $0.05N$ sulfuric acid.

Procedure. Pipette into a test tube 0.1 ml sodium acetate buffer, pH 5.75, 0.2 ml sodium pyrophosphate (final concentration $0.04M$), and enough H_2O for a final volume of 1.0 ml. After equilibration at 37°, start the reaction by adding 1 mg microsomal protein. Stop the reaction 5 min later by the addition of 0.5 ml TCA, and determine the inorganic phosphorus on a 0.5-ml portion of the protein-free supernatant by the method of Lowry and Lopez (25).

The aliquot from the reaction mixture is added to 5 ml of sodium acetate buffer, pH 4.0, making certain that the final pH of the resulting mixture is between 3.5 and 4.2. After addition of 2.5 ml of H_2O, 0.5 ml of ascorbate and then 0.5 ml of ammonium molybdate are added. Optical density is determined at 660 nm, against a reagent blank, 10 min after the addition of ammonium molybdate. In comparison with the original method of Lowry and Lopez, it is necessary to add larger amounts of ammonium molybdate because of the high concentration of pyrophosphate. The phosphorus assay is linear to at least 2.0 μmole phosphate per assay. Pyrophosphatase activity is equal to one half of the amount of phosphate produced. Standard curves for phosphate should be determined, using the above conditions in the presence of pyrophosphate.

As with the glucose-6-phosphatase activity, there is a relatively broad pH optimum for the cleavage of pyrophosphate in untreated rat liver microsomes with an optimum at pH 5.75 (23). This peak becomes sharp and is shifted to pH 5.0 after treatment of microsomes with NH_4OH or with detergents (24). With untreated beef liver microsomes the pH optimum is 5.25 and shifts to 5.75 after the microsomes are treated with phospholipase A. The pH for assays of pyrophosphatase activity thus depends on the source of the enzyme and its previous treatment.

C. PP$_i$-GLUCOSE PHOSPHOTRANSFERASE ACTIVITY

REAGENTS. 1. *Sodium acetate* buffer, $1.0M$, pH 5.50.
2. *Sodium pyrophosphate*, $0.2M$. pH 5.50.
3. *Glucose*, $1.0M$.
4. *NADP*, 12.5 mg/ml.
5. *Glucose-6-phosphate dehydrogenase*, 5 mg/ml (Sigma, Type VII from baker's yeast. Very low to absent contamination with 6-phosphogluconate dehydrogenase).
6. *Tris-HCl* buffer, $1.0M$, pH 8.0.

Procedure. Sodium acetate (0.1 ml), sodium pyrophosphate (0.4 ml), glucose (0.4 ml), and H_2O to a final volume of 1.0 ml are brought to 37°. The reaction is started by adding 1 mg of microsomal protein and is stopped 5 min later by heating the tubes for 3 min in a boiling-water bath. Precipitated protein is removed by centrifugation, and the amount of glucose-6-P synthesized measured enzymatically by adding 0.025- to 0.1-ml aliquots of supernatant to a quartz cuvette containing 0.1 ml Tris, pH 8.0, 0.025 ml NADP, 0.002 ml glucose-6-phosphate dehydrogenase, and H_2O to give a final volume of 1.0 ml. The increase in optical density at 340 nm due to the reduction of NADP to NADPH is a direct measure of the glucose-6-P in the sample, which is calculated using an extinction coefficient for NADPH of $6.22 \times 10^6/cm^2/mole$. Zero-time samples, after addition of microsomes, serve as the blank. Carbamyl-P, a variety of acyl-P compounds, CTP, CDP, deoxy-CTP, ATP, ADP, GTP, GDP, and IDP can also serve as P_i donor in the phosphotransferase reaction (19). Assays can be conducted with these compounds by substituting 30 μM of each for the PP_i. The pH of each phosphate compound should be brought to the pH of the final reaction mixture before addition to the assay.

The nucleotide-sugar transferase reaction is severely constrained in untreated microsomes. Measurement of this reaction requires that the microsomes be treated with detergent, or that detergent be added to the assay tubes. About 1 mg of cholate or deoxycholate should be added per 4 mg of microsomal protein. The amount needed for maximum rates of reaction will vary with the preparation and can be determined by titration of the microsomes with detergent.

The problem of the selection of pH for assays of the phosphotransferase activity is greater than for the other activities of glucose-6-phosphatase in that the phosphotransferase has a sharp pH optimum at 5.50 in untreated beef liver microsomes (4), but a broad pH optimum between 4.80 and 5.20 in rat liver microsomes (21,23). In both species the pH optimum is shifted to more alkaline values by treatment of the microsomes with activating agents such as phospholipases (4,24).

III. INVESTIGATIONS OF PHOSPHOLIPID-PROTEIN INTERACTIONS

1. Background

As mentioned in the Introduction (Section I), the activities of several microsomal enzymes are constrained in some manner in the native microsome by the relationship between the membrane phospholipids and the enzyme protein. The evidence for this is that treatment with phospholipases or detergents can increase enzyme activity. Another sort of enzyme-phospholipid interaction is phospholipase-induced inactivation of some microsomal enzymes, which appears to be reversible on subsequent treatment of the phospholipase-treated form of the enzyme with phospholipid micelles. It is thus apparent that studies of the effects of treatment with phospholipases, detergents, and phospholipids on the properties of microsomal enzymes are important for full characterization of some of these enzymes.

2. Phospholipase A

A. PURIFICATION OF PHOSPHOLIPASE A FROM NAJA NAJA VENOM (28)

Digestion of microsomes by proteolytic enzymes has been reported both to activate (26) and to inactivate (27) microsomal enzymes. Ideally, phospholipases should be free of proteolytic enzyme activities.

Procedure. Adjust the pH of a 10% (w/v) solution of crude *Naja naja* venom (Miami Serpentarium, Miami, Florida) to 3.5 to 3.7 with $1N$ sulfuric acid, and heat in a boiling-water bath for 5 to 10 min. After cooling the venom to room temperature, add $1M$ potassium phosphate to yield a final concentration of $0.05M$, and adjust the pH to 7.6 with $3M$ NH_4OH. Remove precipitated protein by centrifugation at 0°, and chromatograph the supernatant on a column of Sephadex G-75 equilibrated with $0.05M$ potassium phosphate, pH 7.6. The V_0 of the column should be at least 30% greater than the volume of enzyme put on the column. Elute the column with $0.05M$ potassium phosphate, pH 7.6, and collect fractions equivalent to one fourth of the V_0.

Phospholipase A is eluted in the last portion of the excluded volume, and in the first few tubes of the included volume. The excluded portion contains the highest specific activity for phospholipase A as measured by the rate of release of fatty acids from an artificial substrate (29). The enzyme is stable almost indefinitely when frozen. Pure isoenzyme fractions can be obtained by isoelectric focusing of the purified phospholipase A preparation (29). Since there are differences in the substrate specificities of the isoenzymes of *Naja naja* venom (30), detailed studies with single isoenzyme forms of phospholipase A may offer advantages. There are, however, no data on this point as yet.

B. PURIFICATION OF PHOSPHOLIPASE A FROM *Crotalus adamanteus* VENOM (31)

Procedure. Adjust the pH of a 1% (w/v) solution of crude dried *Crotalus adamanteus* venom (Miami Serpentarium) to 9.0 with $0.1N$ potassium hydroxide, and remove the precipitated protein by centrifugation at $12,800g$ for 10 min at 4°. Adjust the pH of the supernatant to 7.0, and dialyze at 4° for at least 30 hr against 300 vols of $10^{-3}M$ EDTA and then 300 vols of $10^{-4}M$ EDTA for an additional 40 hr. Adjust the enzyme solution to pH 3.0, and heat at 90° for 5 min while stirring. Cool the solution in ice-cold water and adjust the pH to 7.4. Remove precipitated protein by centrifugation at

12,800g for 10 min at 4°. Dialyze the supernatant against 300 vols of potassium phosphate, 0.005M, pH 7.4, containing $10^{-3}M$ EDTA. Equilibrate DEAE-cellulose with the same buffer and pack into a column 1.8 × 37 cm. After passing buffer through the column overnight at 4°, apply the protein and collect fractions of about 25 to 30 ml each, at a flow rate of 50 ml/hr. The fractions are monitored by their absorption at 280 nm. When the breakthrough peak is reached and absorption at 280 nm returns to the base line, begin a linear gradient of 0.005 to 0.1M potassium phosphate, pH 7.4, containing $10^{-3}M$ EDTA. Two peaks with phospholipase A activity are resolved.

C. PROPERTIES OF PHOSPHOLIPASE A

Although we have observed no differences in the effects of phospholipase A from *Naja naja* and *Crotalus adamanteus* venoms on the catalytic properties of glucose-6-phosphatase and UDP glucuronyltransferase, this result may not apply to all microsomal enzymes, since the two phospholipases have differing specificities with synthetic phospholipid substrates and also have differing effects on the solubilization of mitochondrial NADH dehydrogenase (30). Phospholipase A from *Naja naja* venom has considerably greater activity with microsomes than that from *Crotalus adamanteus* on a milligram basis.

The pH activity curve of partially purified phospholipase A from *Crotalus adamanteus*, which is a mixture of two isoenzymes, is essentially flat from pH 4.5 to 9.5 (31). The pH optimum of phospholipase A from *Naja naja* venom is relatively sharp, for most of the isoenzymes studied, with a peak at approximately 7.8 (30). With a mixture of isoenzymes from *Naja naja* venom, activity at pH 7.0 or 8.5 is 25% of that at the optimum pH. Since treatment of microsomes with phospholipases can alter the stability of microsomal enzymes as a function of pH, selection of the phospholipase to be used on the basis of its pH-activity curve may be an important consideration in some experiments.

With synthetic substrates all phospholipases require an exogenous source of Ca^{2+} for activity. This is true also for the action of *Crotalus adamanteus* phospholipase A on microsomes, but the *Naja naja* variety of phospholipase A has full activity with liver microsomes in the absence of added Ca^{2+}. The hydrolysis of phospholipids by both enzymes is inhibited completely by 5.0 × $10^{-3}M$ EDTA (5).

3. Phospholipase C: Preparations and Properties

Crude preparations of phospholipase C from filtrates of *Clostridium welchii* have generally been used without further purification. The enzymatic activity of this material has not been investigated in detail, but the filtrates contain "toxins" which do not have "lecithinase" activity (32). All but the "lecithinase" activity can be destroyed, however, by heating for 10 min at 100°, pH 7.6, in a sealed ampoule. Approximately 50% of the phospholipase C is also lost during heating.

The effects of pure preparations of phospholipase C on the properties of microsomal membranes have not been studied. Two isoenzymes of phospholipase C can be purified from extracts of *Clostridium welchii* (33,34); sphingomyelin is the preferred substrate for one isoenzyme, and lecithin for the other (34). The pH optimum for the sphingomyelin-hydrolyzing isoenzyme is 7.8 to 8.8, and this form of phospholipase C is activated by Mg^{2+} but inhibited almost completely by $10^{-3}M$ Ca^{2+}. The rate of hydrolysis of lecithin is 9% of that for sphingomyelin in the presence of $10^{-3}M$ $MgCl_2$. The isoenzyme for which lecithin is the preferred substrate has a pH optimum at 7 and shows nearly equal activity with sphingomyelin or lecithin as substrate in the absence of Ca^{2+}. In the presence of $10^{-3}M$ Ca^{2+}, however, there is no activity with sphingomyelin. The addition of Mg^{2+} has no effect on the activity of this isoenzyme. Since all studies of microsomal enzymes reported to date have utilized phospholipase C treatment of microsomes in the presence of Ca^{2+}, the effects of phospholipase C on the properties of microsomal enzymes almost certainly reflect the effects of hydrolysis of microsomal lecithins.

4. Phospholipase D: Preparations and Properties

This enzyme, which has been found only in plants, is activated by Ca^{2+} and anionic detergents (35). The pH optimum is 5 to 6, with little activity below 2 or above 8.0. To date few studies of protein-lipid interactions in microsomal enzymes have been made with phospholipase D.

5. Treatment of UDP Glucuronyltransferase and Glucose-6-Phosphatase with Phospholipases

REAGENTS. 1. Partially purified *phospholipase A*, 10 mg/ml, from *Naja naja* or *Crotalus adamanteus* venom. *Phospholipase C* from *Clostridium welchii*, 10 mg/ml, prepared fresh each day.
2. *Calcium chloride*, 0.1*M*.
3. *Tris-HCl*, 0.5*M*, pH 8.0.
4. *Microsomes*, 20 mg/ml in 0.25*M* sucrose.
5. *EDTA*, 0.25*M*.

Procedure. Microsomes (0.5 ml), Tris-HCl (0.1 ml), and H_2O (0.45 ml) are placed in the reaction tube; 0.025 ml $CaCl_2$ is added if phospholipase A from *Crotalus adamanteus* or phospholipase C is to be used. After equilibration at

25°, the reaction is started by addition of 0.005 ml phospholipase A or 0.1 ml phospholipase C. Aliquots are removed from the reaction mixture at $\frac{1}{2}$, 1, 2, 5, and 10 min after addition of phospholipase, and placed in appropriate assay tubes containing EDTA at a final concentration of $0.005M$ to inhibit further action of phospholipase.

There is no correlation between the rate of hydrolysis of microsomal phospholipids and phospholipase-induced modifications in the properties of glucose-6-phosphatase (4), and the time course for these effects are variable for different enzymes and in different species. The times used in the above method are satisfactory when working with glucose-6-phosphatase and UDP glucuronyltransferase, but it is essential to establish carefully for each enzyme and species the time course for the effects of treatment with phospholipases. If effects (especially, e.g., inactivation) appear to be immediate, the ratio of phospholipases to microsomal protein should be reduced in order to be certain that a rapid activation is not being misssed.

6. Properties of Phospholipase A-Treated Enzymes

It should be appreciated that treatment with phospholipase A may have mixed effects on the properties of microsomal enzymes. The data in Figure 4 are illustrative of this point. Microsomes were treated with phospholipase A and assayed for UDP glucuronyltransferase in the presence of EDTA at single time points under conditions which gave linear kinetics for untreated microsomes. The data suggest that treatment with phospholipase A has a biphasic effect on the activity of UDP glucuronyltransferase. That this is an incorrect conclusion is clear from the experiment presented in Figure 5, in which the initial rates of activity were determined at 23°. When this was done, treatment with phospholipase A is seen to have only an activating effect on UDP glucuronyltransferase. The phospholipase A-activated form

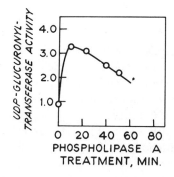

Figure 4. The effect of treatment with phospholipase A on the apparent activity of UDP-glucuronyltransferase. Microsomes from beef liver were treated with partially purified phospholipase A from *Naja naja* venom at 23° in $0.05M$ Tris, pH 8.0, at a microsomal phospholipase A protein ratio of 60:1. Assays were carried out at a *p*-nitrophenol concentration of 0.6mM and 3.0mM UDP-glucuronic acid. Assay time was 10 min. Rate is expressed as millimicromoles *p*-nitrophenol conjugated per minute per milligram protein.

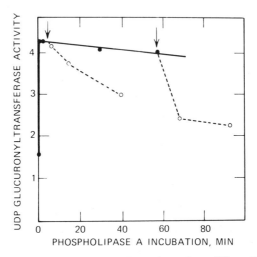

Figure 5. Effect of treatment with phospholipase A on the stability of UDP-glucuronyltransferase. Microsomes from beef liver were treated with phospholipase A as in Figure 4 at a microsome phospholipase A protein ratio of 25:1. (●). At the indicated times (arrows) aliquots were removed, made 5mM in EDTA in order to inhibit the further action of phospholipase A, and incubated at 37° (○). Reproduced from Vessey and Zakim (5).

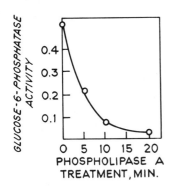

Figure 6. Effect of treatment with phospholipase A, as in Figure 4, on the activity of glucose-6-phosphatase in beef liver microsomes. Activity is expressed as micromoles phosphorus produced per minute per milligram microsomal protein.

of the enzyme, however, is unstable at 37° (dotted line in Figure 5). The effect appears to be biphasic in Figure 4 because conditions which gave zero-order kinetics for the untreated enzyme yield nonlinear kinetics for the phospholipase A-treated form of UDP glucuronyltransferase.

The importance of carefully designed studies of the effect of phospholipases and any other activators on stability is illustrated also in Figures 6 and 7. Treatment of beef liver microsomes with phospholipase A appears to inactivate glucose-6-phosphatase directly (Figure 6). As seen in Figure 7,

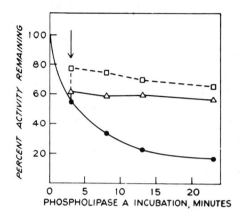

Figure 7. Effect of EDTA (●), EDTA plus albumin (□), or EDTA plus asolectin (△) on the activity of glucose -6-phosphatase in phospholipase A-treated beef liver microsomes. Microsomes were treated with phospholipase A, as in Figure 4, at a microsomal phospholipase A protein ratio of 50:1 At the arrow 5mM EDTA (final concentration), EDTA plus albumin, or EDTA plus asolectin was added to portions of the phospholipase A-treated microsomes and the incubations were continued at 23°. Reproduced from Zakim (4).

however, treatment with phospholipase A has no direct effect on the catalytic properties of glucose-6-phosphatase but only makes the enzyme unstable. In the absence of independent studies on stability, it might appear from selection of data in Figure 7 that treatment with phospholipase A inactivates glucose-6-phosphatase and that this effect is reversed by subsequent treatment with phospholipids. This conclusion, which has proved incorrect, was reached by Duttera et al. (35a) in studies which did not investigate the stability of phospholipase A-treated glucose-6-phosphatase (4).

Care should be taken also in interpreting the effects of phospholipase C on the properties of microsomal enzymes. The reason for this, as pointed out by Cater and Hallinan (36), is the presence of an endogenous microsomal acylhydrolase. It has been shown that the fatty acids released by the combined action of phospholipase C and endogenous diglyceride acylhydrolase make the phospholipase C-treated form of glucose-6-phosphatase unstable at 37°. To what extent presumed actions of phospholipase C on the properties of glucose-6-phosphatase and other microsomal enzymes result from these combined effects is unknown, since it is not possible to inhibit selectively the endogenous microsomal acylhydrolase.

With enzymes like UDP glucuronyltransferase, which is assayed at less than saturating concentration of substrates, activation or inactivation could reflect changes in the binding affinity of substrates and/or the catalytic rate constant. For this reason the kinetic parameters of the activated and un-activated UDP glucuronyltransferases should be studied as in Section II, in order to determine the kinetic mechanism for changes in activity after treatment with phospholipases or other activating agents. Another important reason for determining the kinetic constants of microsomal enzymes after treatment with activating agents is that each agent may produce a different form of the enzyme. In the case of UDP glucuronyltransferase, activations by phospholipase A, Triton X-100, and alkaline pH appear to yield identical results if activity is determined at only one set of substrate concentrations; but the kinetic constants of the enzyme are different for each type of treatment. This is less of a problem with glucose-6-phosphatase since it is not too difficult to measure activity at close to V_{max} by using very high concentrations of glucose-6-P.

7. Activation by Detergents

The kinds of precautions important in investigating the effects of phospholipases on the properties of microsomal enzymes apply also to studies with detergents. Each enzyme should be titrated with a wide range of detergent concentrations, and the design of experiments should be such that changes in the stability of the enzyme are monitored at each stage of the study. Especially important in using detergents is careful control of the temperature, since treatment with detergents makes microsomal enzymes extremely labile.

IV. PREPARATION OF MICROSOMES

1. Choice of Homogenization Medium

Experience in this laboratory with liver microsomes has shown that unbuffered $0.25M$ sucrose is the most suitable homogenization medium. The pH of an unbuffered sucrose homogenate of microsomes is approximately 6. Buffering of this homogenate at pH values above 7.0 is to be avoided, since microsomes prepared in this manner can clump after freezing and thawing. High ionic strength or the addition of divalent cations to homogenizing media has a similar effect. In addition, Na^+ is reported to activate the endogenous phospholipase A of microsomes (37). The choice of homogenizing media depends also on the tissue studied. For example, optimal conditions for the isolation of sarcoplasmic reticulum are different from those for the preparation of liver microsomes (7,38).

2. Homogenization, Isolation, and Storage

For small-scale preparations we have homogenized in loose-fitting Potter-Elvehjem tubes. Preparations of good quality, biochemically and morphologically, can also be obtained after brief homogenization in a Waring blender, which we have used for large-scale preparations. Microsomes are then isolated by differential centrifugation. These microsomes are not completely stable on storage, however, in that at 0°, or even prolonged storage at $-20°$, the activity of glucose-6-phosphatase and UDP glucuronyltransferase increases slowly (39). The exact cause for the increasing activity of these two enzymes on storage is not completely clear, but changes observed in the kinetic properties of UDP glucuronyltransferase in beef microsomes on aging are similar to the effects of treatment with phospholipase A (5). Thus, after either addition of exogenous purified phospholipase A from *Naja naja* venom or storage, $K_{p\text{-nitrophenol}}$ increases as the activity increases. The theory that the increase in activity on storage results from the action of phospholipases is further supported by the fact that $5 \times 10^{-3}M$ EDTA partially inhibits spontaneous activation.

Elevation of the pH of the microsomal homogenate to 7.0 slows down the rate of spontaneous activation; and since lysosomal phospholipases have a pH optimum in the acid range, it is likely that the source of phospholipase which leads to activation is contaminating lysosomes, rather than the endogenous phospholipase A of the microsomes. Although addition of EDTA or adjustment of the pH to the range of 7–8 tends to increase the stability of glucose-6-phosphatase and UDP glucuronyltransferase on storage, these methods are unsuitable for the reasons cited above. Alternatively, we have attempted to diminish the contamination of microsomes with lysosomes by including an extra centrifugation step. Thus the postmitochondrial supernatant is centrifuged at 10,000g for 20 min before the collection of microsomes at 100,000g. The microsomal pellet collected by centrifugation at 100,000g for 60 min is resuspended and centrifuged again at 10,000g for 20 min before preparation of the final pellet. The rate of spontaneous activation of UDP glucuronyltransferases is slower in this type of microsomal preparation than in microsomes collected after a single centrifugation of mitochondrial supernatant at 100,000g, and these preparations have a useful age of at about 2 weeks when stored at $-20°$. Once thawed, microsomal preparations should not be refrozen and used again since freezing and thawing alone alters their properties.

Although our experience is limited to studies of the properties of glucose-6-phosphatase and UDP glucuronyltransferase, it should be pointed out that the activities of several other microsomal enzymes can be modified by treatment *in vitro* with phospholipases, detergents, and organic solvents.

It is likely, therefore, that spontaneous activation of other microsomal enzymes will be observed when looked for.

The problems of spontaneous activation of tightly bound microsomal enzymes are not limited to difficulties in comparing experiments done on different days. Spontaneous activation can alter the stability of enzymes and, more important, the effects of experimental treatments on the properties of the enzymes. For example, Triton X-100 is a potent activator of UDP glucuronyltransferase assayed with p-nitrophenol as aglycone (5,39). After spontaneous activation, however, treatment with Triton X-100 becomes inhibitory (39). In fact, lack of attention to the potential difficulties caused by spontaneous activation of UDP glucuronyltransferase may have contributed to conflicting reports in the literature on the effects of treatment with phospholipase A on the activity of this enzyme (5,40).

3. Large-Scale Preparation

One of the difficulties encountered in attempts to purify microsomal enzymes is the length of time needed for the preparation of starting material. Although we have not investigated its value for all microsomal enzymes, a preparation of microsomes with high specific activity for glucose-6-phosphatase in high yield can be made with a Sharples centrifuge. Chunks of beef liver are homogenized in 4 vols of 0.25 sucrose for 40 sec in a Waring blender. The homogenate is strained through cheesecloth, and centrifuged in a Sharples centrifuge with a separation bowl producing a force of 65,000g, at a flow rate of 200 ml/min. The supernatant from this process is equivalent to that obtained after centrifugation for 10 min at 10,000g. The postmitochondrial supernatant is diluted by the addition of $\frac{1}{3}$ vol of 0.25M sucrose and recentrifuged in the Sharples with a flow rate of 80 ml/min. The sedimented microsomes obtained in this manner have 70% of the glucose-6-phosphatase activity present in the postmitochondrial supernatant. Examination of this fraction with an electron microscope shows that the preparation contains smooth and rough microsomes with relatively little contamination by other cellular components.

4. Subfractionation of Microsomes

The distribution of enzymes within microsomal membranes is not random. For example, liver microsomal UDP N-acetylglucosaminyl-, galactosyl-, and sialyltransferases, which account for the terminal trisaccharide unit of a variety of plasma glycoproteins, appear to be concentrated in the Golgi apparatus (41). Fouts and his coworkers have found that several NADPH-dependent drug-metabolizing enzymes are concentrated in smooth microsomes (42). On the other hand, UDP glucuronyltransferase activities with

p-nitro- and o-aminophenols as aglycones were found predominantly in rough microsomes from rabbit liver (43).

Not only is there an uneven distribution of enzyme activities within the microsomal membranes, but also the properties of tightly bound microsomal enzymes are not the same in all subfractions. The specific activity of glucose-6-phosphatase is greater in rough than in smooth microsomes, but after pretreatment of microsomes at alkaline pH or with deoxycholate the specific activities in rough and smooth microsomes increase and become identical (45). Apparently, the environment of the smooth microsomes imposes a greater constraint on the maximum potential activity of glucose-6-phosphatase than does that of the rough microsomes. Although the situation is more complex with UDP glucuronyltransferase, constraint on the activity of this enzyme is also different in rough and smooth microsomes when p-nitrophenol is used as aglycone.*

Rough and smooth microsomes may be separated in three different ways. The first technique, described by Moule et al. (45), makes use of a series of differential centrifugations in $0.88M$ sucrose. As pointed out by Dallner and Ernster (46), this method does not yield highly purified subfractions. Rothchild's method (52) separates rough and smooth microsomes on a discontinuous sucrose gradient, but the 8-hr centrifugation required introduces problems related to stability and spontaneous activation of microsomal enzymes. The third method, which is the simplest, was developed by Dallner and his coworkers (46,48); it is based on selective aggregation of the rough microsomal fraction by monovalent cations. The Cs^+ cation is used most often, though other members of group IA of the periodic table have a similar effect. Aggregation seems to depend on neutralization of the net negative surface charge of the rough fraction of microsomes and is reversible in all cases (46).

Cesium chloride to a final concentration of $0.015M$ is added to postmitochondrial supernatant (4.5 ml) in $0.25M$ sucrose and layered over 2 ml of $1.3M$ sucrose containing $0.015M$ CsCl. After centrifugation for 90 min at 40,000 rpm in a Spinco No. 40.2 rotor, the rough microsomes sediment as a pellet and the smooth collect at the interface. The CsCl is removed by repelleting each fraction in $0.25M$ sucrose. Rotors other than the 40.2 can be used with the limitation that the tube-to-rotor angle be at least 34°; the method will not yield good separation of fractions with a Spinco No. 30 or 40 rotor. The conditions described are suitable only for liver and must be modified when other tissues are studied (46).

Smooth and rough microsomes can be separated further into subfractions. The smooth microsomes can be fractionated into two subgroups as a result of

* D. Zakim, D. A. Vessey, and J. Goldenberg, unpublished data.

selective aggregation of a portion of the smooth microsomes with $MgCl_2$ (48). The fluffy interfacial layer obtained after centrifugation in $0.015M$ CsCl is made $0.007M$ in $MgCl_2$ and layered over $1.15M$ sucrose-$0.007M$ $MgCl_2$. Centrifugation for 45 min at $102,000g$ in a No. 40.2 rotor yields one fraction of smooth microsomes as a pellet and another at the interface. The specific activity of glucose-6-phosphatase is different in these two subfractions of smooth microsomes (49). Unfortunately, however, the aggregating effect of $MgCl_2$ cannot be reversed (46), and Mg^{2+} alters the kinetic properties of some microsomal enzymes (see Section II).

The rough microsomal fraction has been fractionated further by short periods of centrifugation on continuous sucrose density gradients in the range 1.17–$1.25M$ (46). The activities of NADPH and NADH cytochrome c reductase and glucose-6-phosphatase are not distributed evenly in these subfractions.

Morre et al. (50) have prepared highly enriched fractions of Golgi apparatus by gentle homogenization of minced liver for 30 to 60 sec in a Polytron (Brinckmann Instruments, Westbury, N.Y.), which releases subcellular structures intact. Inclusion of a relatively high concentration of sucrose and 1% dextran in the homogenization medium seems to be required for good results. After centrifugation of whole homogenates at $2000g$ for 20 min, the Golgi are isolated on a discontinuous sucrose density gradient. Electron microscopy of purified fractions shows morphologically intact Golgi apparatus. Golgi membranes have been prepared also after homogenization of liver in a Potter-Elvejhem tube by zonal centrifugation or by single-step gradient centrifugation (51,52). It has been suggested that UDP galactose-N-acetylglucosamine galactosyltransferase, which is 100-fold concentrated in Golgi as compared to whole homogenates of liver, is a useful marker enzyme for this organelle (52).

5. Variability in the Properties of Microsomal Enzymes

An additional problem encountered in working with microsomal enzymes is a lack of constancy in their molecular properties during the life of an animal. It has been observed that the apparent $K_{aniline}$ and $K_{ethylmorphine}$ of microsomal aniline hydroxylase and ethylmorphine demethylase, respectively, change continuously in male rats between 1 and 12 weeks of life (53). Whether these developmental changes result from a heterogeneity of aniline hydroxylases and ethylmorphine demethylases or from interactions between the enzyme and a changing microsomal environment is unclear. Nevertheless, care must be taken in comparing the molecular properties of microsomal enzymes in animals of different ages. Unfortunately, no useful data are available on changes in the kinetic characteristics of other microsomal enzymes.

Changes in the properties of microsomal enzymes are not limited to developmentally induced alterations. Thus alloxan-diabetes changes the K_{G-6-P} of glucose-6-phosphatase (54, 55), and fasting selectively increases the phosphohydrolase, as compared with the phosphotransferase, activity of the enzyme (56). In this instance the alteration appears to be related to changes in enzyme-lipid interactions, since after treatment of microsomes with detergents the kinetic properties of glucose-6-phosphatase become identical in normally fed, diabetic, and fasted animals (56,57).

The influence of specific modification of dietary fats on the properties of microsomal enzymes has been examined in only a limited way. Although the activities of a variety of microsomal enzymes can be modified *in vitro* by perturbation of the microsomal phospholipids, the effects of variation in the amounts of polyunsaturated and saturated fatty acids and of cholesterol in the diet on the properties of microsomal enzymes have not been studied. It would seem a wise precaution, however, to maintain animals being used as a source of liver microsomes on as constant a diet as possible in order to avoid potential problems with dietary-induced alterations in the properties of a single enzyme.

V. INTERFERING ENZYMES AND SUBSTRATE FORMS

1. Background

Since the microsomal fraction of the cell is enzymatically heterogeneous, it is important to consider that substrates may be metabolized by more than a single enzyme, and that the products of the reaction under study may be substrates for another microsomal enzyme. The importance of this has already been pointed out with regard to assays of UDP glucuronyltransferase, which indicate in a general way that considerable care should be given to the design of systems for determining the activity of microsomal enzymes, irrespective of the problems posed by the specific enzyme to be studied. Our experience with UDP glucuronyltransferase has shown that the most critical aspect of the study of microsomal enzymes has been to define experimental conditions which allow for measurement of their true kinetic parameters. There are a great many instances in the literature, however, in which the potential complications of competing enzymes have been neglected. It is not possible to state with certainty the number of enzymes which will compete for substrates in standard enzyme assays.

The potential problem caused by interfering enzymes is twofold, in that they may make it difficult to control the concentration of substates and/or to follow the course of the reaction. It is necessary, therefore, to determine for each substrate and product whether changes in its concentrations are due only to the reaction being studied, and also to establish that initial rates of enzyme activity are being measured by the assay technique. The only manner in which these difficulties can be resolved is to incubate separately each product and substrate with microsomes in order to measure its rate of removal by alternate enzyme pathways. In this way it is possible to determine which product or substrate may be used as an indicator of the rate of the reaction being studied, and to assess the rates at which other products and substrates are being consumed by interfering enzymes.

2. Metabolism of Substrates by Two or More Enzymes

A. DETERMINATION OF INITIAL RATES

One difficulty resulting from the metabolism of a substrate via an alternate pathway is that reaction rates will not be linear with time if the concentration of substrate falls significantly during the course of the assay. This problem can be avoided if the reaction is studied at "saturating" concentrations of substrates and the rate of the competing reaction is relatively low, in which case a single-point assay may be useful for measuring rates of reaction. This technique, in fact, has been popular for a variety of microsomal enzymes. The importance of verifying that substrates are actually present at saturating concentrations is obvious, but usually it has not been shown that the substrate concentrations used in these systems are truly saturating. With UDP glucuronyltransferase in rat liver microsomes, for example, it is not possible with a single-point assay technique to study activation and inactivation, since inhibitors of nucleotide pyrophosphatase will appear to be activators of UDP glucuronyltransferase, and activators of the pyrophosphatase will appear to be inhibitors of glucuronyltransferase. If the rate of the competing reaction is large in comparison to that of the enzyme of interest, good estimates of initial rates of activity obviously cannot be obtained. For example, it is possible to measure accurately the rate of glucuronyltransferase reactions in rat liver microsomes at high concentrations of UDP-glucuronic acid and to study activation and inhibition of the enzyme in a direct manner. It is not possible, however, to measure rates over a wide range of substrate concentrations, even with a multiple sampling technique, and other methods must be used.

Reaction rates which are linear with time are no assurance that a competing enzyme is not removing substrates for the reaction being studied and thereby influencing the interpretation of rate data. A side reaction (E_2) which is fast in comparison to the primary reaction (E_1), is fully reversible, and has an equilibrium point not too favorable to product formation would establish an equilibrium between substrate (A) and alternate product (P_2):

$$A + B \xleftrightarrow{E_1} P_1 \qquad [6]$$

$$A \xleftrightarrow{E_2} P_2 \tag{7}$$

Although the concentration of A may be constant during the assay in this situation, the true concentration of A will be different from its apparent concentration as calculated from amounts added to the assay system, and inaccurate estimates of K_A will be obtained. In addition, compounds which alter the rate only of the side reaction will appear to act as effectors of the reaction being studied.

B. SUPPRESSION OF THE ACTIVITY OF COMPETING ENZYMES BY MODIFICATION OF ASSAY pH.

If preliminary experiments indicate that substrates and/or products are metabolized at rapid rates by alternate pathways which are too active for accurate measurements of initial rates of activity, then variation of the pH of the assay media may allow for selective depression of the activities of the competing enzymes. This method has not been used generally; it finds applications, however, in the assay of UDP glucuronyltransferase where there may be hydrolysis of products of this reacion by β-glucuronidase. Although the latter enzyme is considered to be primarily lysosomal, cyto-chemical (58,59) and tissue fractionation studies (60) indicate its presence in microsomes. Its pH optimum is 4.5 in all tissue fractions examined (61), so that hydrolysis of the glucuronide product can be suppressed by keeping the pH of the glucuronyltransferase reaction at a relatively higher level. At pH 7.1 and higher there is no detectable β-glucuronidase activity with the level of glucuronides produced by the glucuronyltransferase reaction.

C. PURIFICATION

It is theoretically possible to obviate the problem of interfering enzymes by removing them from the microsome, but all currently available techniques for removing selectively either the interfering enzymes or those of direct experimental interest may alter the properties of the enzyme being studied. The only separation method likely to be useful at the present time is the preparation of subfractions of microsomes. A possible alternative is to use different species or organs. For example, liver microsomes from guinea pigs do not possess nucleotide sugar pyrophosphatase activity and constitute an excellent material for studying not only the properties of UDP glucuronyl-transferase but also a variety of UDP-sugar donating reactions. Also, in comparison with the rat, we have found that the activity of the microsomal nucleotide pyrophosphatase is lower in rabbit, mouse, and human liver. The activity of another potentially troublesome enzyme, acyl-CoA ligase, which interferes in the study of other acyl-CoA utilizing enzymes, is high in rat liver microsomes but lower in pig, guinea pig, and beef liver microsomes (62,63). The extremely high ATPase activity of brain microsomes and sarcoplasmic reticulum makes it difficult to study the kinetic parameters of ATP-utilizing enzymes in these preparations, whereas ATPase activity is relatively low in liver microsomes.

D. USE OF INHIBITORS

The compounds NADH and NADPH are reoxidized by two distinct microsomal flavoproteins, NADH-cytochrome b_5 reductase and NADPH-cytochrome c reductase, which funnel electrons to O_2 via cytochrome P-450 and also perhaps directly from cytochrome b_5 in the case of NADH. The presence of these enzymes in high concentration makes it difficult to couple assays of microsomal enzymes to oxidation-reduction reactions. A solution to the problem of reoxidation of reduced nucleotides as a competing reaction seems to be to study reactions requiring NADH or NADPH anaerobically, since in the microsomes O_2 is the only terminal electron acceptor. Care should be exercised in the use of known inhibitors of interfering enzymes, and the effects of the inhibitor on the activity of the enzyme of primary interest should be susceptible to investigation in an independent manner. For example, EDTA and sugar-nucleotides, other than UDP-glucuronic acid, are effective inhibitors of the microsomal nucleotide pyrophosphatase (64); but EDTA inhibits the activity of UDP glucuronyltransferase, and UDP-N-acetylglucosameine activates this enzyme in guinea pig livers. Addition of these inhibitors to assay systems containing rat liver microsomes, therefore, will yield mixed effects on the rate of synthesis of glucuronides, making clear interpretations of the data difficult.

E. SUBSTRATE FORMS

Substrates for several tightly bound microsomal enzymes are nonpolar or amphipathic, and as a result have a strong affinity for the hydrophobic portions of the membrane. Phenols (Section II.1.B) produce extensive alterations in microsomal structure, as indicated by turbidity changes, release of a TCA-soluble chromophore from the membrane, and activation of UDP glucuronyltransferase. Substrate-induced inactivation occurs in reactions utilizing acyl-CoA or lysophosphatides as substrates. The problem in this instance is related to the formation of micelles, which behave as detergents (65). Another source of difficulty is nonspecific binding of substrates to microsomes, even in the absence of modification of the properties of microsomal enzymes. An example of this has been noted in assays of UDP glucuronyltransferase with phenolphthalein as aglycone (Section II.1.B). An especially difficult experimental problem involves the kinetics of acyl-CoA-acyltransferase reactions, since significant amounts of the long-chain acyl-CoA compounds formed remain bound to the microsomal membrane (66,67) and contributions of free and bound acyl-CoA to the total reaction rate are uncertain. If the bound pool makes a significant contribution to this rate, one has to consider not only that kinetics will be anomolous, but also that perturbation of the microsomal lipids could affect enzyme activity because of effects on the nonspecific binding of substrates.

Some substrates may be so insoluble in aqueous solution that measurable reaction rates cannot be obtained. What has been done in many such instances is to add substrates attached to a carrier (e.g., albumin), or to decrease the polarity of the aqueous phase by adding the substrate in ethanol or polyethylene glycol. The problems thereby raised are uncertainty as to the availability of carrier-bound substrate and the effects of organic solvents on microsomal structure and on the properties of the enzyme being studied. For substrates with limited solubilities or those with relatively low critical micelle concentrations, modification of pH, salt concentrations, or buffer strength may appear to activate or inhibit the enzyme under investigation, not because of direct effects, but through alterations of the solubility of the substrate and/or shifts in the equilibrium between carrier-bound and free substrate. This is true for the apparent Mg^{2+}-induced activation of the glucuronidation of bilirubin (Section II.1.B) and also for the acylation of 1-acylglycerol-3-P (65).

Acknowledgments

This work was supported by grants from the United States Public Health Service (HE 10027) and the National Science Foundation (8248) to Dr. Thomas P. Singer. The authors thank Dr. Singer for his continuing interest and support of this work and Dr. Edna B. Kearney for her critical review of the manuscript.

References

1. A. Claude, *Harvey Lectures Ser.*, *43*, 121 (1947–48).
2. W. C. Schneider, and G. H. Hogeboom, *Cancer Res.*, *11*, 1 (1951).
3. H. Beaufay and C. DeDuve, *Bull. Soc. Chim. Biol.*, *36*, 1551 (1954).
4. D. Zakim, *J. Biol. Chem.*, *245*, 4953 (1970).
5. D. A. Vessey and D. Zakim, *J. Biol. Chem.*, *246*, 4649 (1971).
6. H. M. Abou-Issa and W. W. Cleland, *Biochim. Biophys. Acta*, *176*, 692 (1969).
7. A. Martonosi, J. Donley, and R. A. Halpin, *J. Biol. Chem.*, *243*, 61, (1968).
8. J. S. Ellingson, E. E. Hill, and W. E. Lands, *Biochim. Biophys. Acta*, *196*, 176 (1962).
9. R. C. Nordlie and W. J. Arion, *J. Biol. Chem.*, *239*, 1680 (1964).
10. M. R. Stetten, *J. Biol. Chem.*, *239*, 3576 (1964).
11. A. Baeyer, *Ann. Chem.*, *155*, 257 (1870).

12. G. J. Dutton, in *Glucuronic Acid, Free and Combined*, G. J. Dutton, Ed., Academic Press, New York, 1966, p. 185.
13. D. Zakim and D. A. Vessey, *Federation Proc.*, *31*, 4044 (1972).
14. D. A. Vessey and D. Zakim, *J. Biol. Chem.*, *247*, 3023 (1972).
14a. A. Levitzki and D. E. Koshland, Jr., *Proc. Nat'l Acad. Sci.*, *62*, 1121 (1969).
14b. D. A. Vessey, J. Goldenberg, and D. Zakim, *Biochim. Biophys. Acta*, **309** 58 (1973).
14c. W. W. Cleland, in *The Enzymes*, Vol. II, 3rd Ed., P. D. Boyer, Ed., Academic Press, New York, 1970, p. 1.
15. G. A. Levvy, and I. D. E. Storey, *Biochem. J.*, *44*, 295 (1949).
16. F. P. Van Roy and K. P. M. Heirwegh, *Biochem. J.*, *107*, 507 (1968).
17. R. C. Burnstine and R. Schmid, *Proc. Soc. Exptl. Biol. Med.*, *109*, 356 (1962).
18. G. W. Lucier, O. S. McDaniel, and H. B. Matthews, *Arch. Biochem. Biophys.*, *145*, 520 (1971).
19. R. C. Nordlie, in *The Enzymes*, Vol. IV, 3rd Ed., P. D. Boyer, Ed., Academic Press, New York, 1971, p. 543.
20. R. C. Nordlie and W. J. Arion, *J. Biol. Chem.*, *239*, 1680 (1964).
21. R. C. Nordlie and W. J. Arion, *J. Biol. Chem.*, *240*, 2155 (1965).
22. H. Beaufay, H. G. Hers, J. Berthet, and C. DeDuve, *Bull. Soc. Chim. Biol.*, *36*, 1539 (1954).
23. M. R. Stetten and H. L. Taft, *J. Biol. Chem.*, *239*, 4041 (1964).
24. M. R. Stetten and F. R. Burnett, *Biochim. Biophys. Acta*, *139*, 138 (1967).
25. O. H. Lowry and J. A. Lopez, *J. Biol. Chem.*, *162*, 421 (1946).
26. M. Gorlich and E. Heise, *Nature*, *197*, 698 (1963).
27. A. Ito and R. Sato, *J. Cell Biol.* 40, 179 (1969).
28. T. Cremona and E. B. Kearney, *J. Biol. Chem.*, *239*, 2328 (1964).
29. J. I. Salach, P. Turini, R. Seng, J. Hauber, and T. P. Singer, *J. Biol. Chem.*, *246*, 331 (1971).
30. J. I. Salach, R. Seng, H. Tisdale, and T. P. Singer, *J. Biol. Chem.*, *246*, 340 (1971).
31. K. Saito and D. J. Hanahan, *Biochemistry*, *1*, 521 (1962).
32. M. G. MacFarlane, *Biochem. J.*, *42*, 587 (1948).
33. M. W. Slein and G. F. Logan, *J. Bacteriol.*, *90*, 69 (1965).
34. I. Pastan, V. Macchia, and R. Katzen, *J. Biol. Chem.*, *243*, 3750 (1968).
35. M. Kates, *Can. J. Biochem. Physiol.*, *34*, 967 (1956).
35a. S. M. Duttera, W. L. Byrne, and M. C. Ganoza, *J. Biol. Chem.*, *243*, 2216 (1963).
36. B. R. Cater and T. Hallinan, *FEBS Letters*, *16*, 137 (1971).
37. M. Paysant, R. Wald, and J. Polonovski, *Compt. Rend.*, *163*, 2100 (1969).
38. G. Meissner and S. Fleischer, *Biochim. Biophys. Acta*, *241*, 356 (1971).
39. K. K. Lueders and E. L. Kuff, *Arch. Biochem. Biophys.*, *120*, 198 (1967).
40. D. Attwood, A. B. Graham, and G. C. Wood, *Biochem. J.*, *123*, 875 (1971).
41. H. Schachter, I. Jabbal, R. L. Hudgin, L. Pinteric, E. J. McGuire, and S. Roseman, *J. Biol. Chem.*, *245*, 1090 (1970).
42. J. R. Fouts, L. A. Rogers, and T. E. Gram, *Exp. Mol. Pathol.*, *5*, 475 (1966).
43. T. E. Gram, A. B. Hansen, and J. R. Fouts, *Biochem.*, *J.*, *106*, 587 (1968).
44. M. R. Stetten and S. B. Ghosh, *Biochim. Biophys. Acta*, *233*, 163 (1971).
45. Y. Moule, C. Rouiller, and J. Chauveau, *J. Biophys. Biochem. Cytol.*, *1*, 547 (1960).
46. G. Dallner and L. Ernster, *J. Histochem. Cytochem.*, *16*, 611 (1968).
47. J. Rothschild, *Biochem. Soc., Symp.*, *22*, 4 (1963).
48. G. Dallner, P. Siekevitz, and G. E. Palade, *J. Cell Biol.*, *30*, 73 (1966).
49. A. Bergstrand and G. Dallner, *Anal. Biochem.*, *29*, 351 (1969).

50. D. J. Morre, R. L. Hamilton, H. H. Mollenhauer, R. W. Mahley, W. P. Cunningham, R. D. Cheetham, and V. S. LeQuire, *J. Cell Biol.*, *44*, 484 (1970).
51. B. Fleischer, S. Fleischer, and H. Ozawa, *J. Cell. Biol.*, *43*, 59 (1969).
52. B. Fleischer and S. Fleischer, *Biochim. Biophys. Acta*, *219*, 301 (1970).
53. T. E. Gram, A. M. Guarino, D. H. Schroeder, and J. R. Gillette, *Biochem. J.*, *113*, 681 (1969).
54. H. L. Segal and M. E. Washko, *J. Biol. Chem.*, *234*, 1937 (1959).
55. R. A. Freedland and J. K. Barnes, *Proc. Soc. Exptl. Biol. Med.*, *112*, 995 (1963).
56. R. C. Nordlie, W. J. Arion, T. L. Hanson, J. R. Gilsdorf, and R. N. Horne, *J. Biol. Chem.*, *243*, 1140 (1968).
57. T. L. Hanson and R. C. Nordlie, *Biochim. Biophys. Acta*, *198*, 66 (1970).
58. W. H. Fishman, S. S. Goldman, and R. Delellis, *Nature*, *213*, 457 (1967).
59. R. A. Delellis and W. H. Fishman, *Histochemie*, *13*, 4 (1968).
60. K. Paigen, *Exptl. Cell Res.*, *25*, 286 (1961).
61. G. W. Lucier and O. S. McDaniel, *Biochim. Biophys. Acta*, *261*, 168 (1972).
62. R. C. Reitz, M. El-Sheikh, W. E. M. Lands, I. A. Ismail, and F. D. Gunstone, *Biochim. Biophys. Acta*, *176*, 480 (1969).
63. W. E. M. Lands and P. Hart, *J. Biol. Chem.*, *240*, 1905 (1965).
64. H. Ogawa, M. Sawada, and M. Kawada, *J. Biochem.*, *59*, 126 (1966).
65. W. L. Zahler and W. W. Cleland, *Biochim. Biophys. Acta*, *176*, 699 (1969).
66. P. B. Garland, D. Shepherd, and D. W. Yates, *Biochem. J.*, *97*, 587 (1965).
67. N. Baker and F. Lynen, *European J. Biochem.*, *19*, 200 (1971).

Determination of Selenium in Biological Materials

O. E. Olson, I. S. Palmer, and E. I. Whitehead, *Experiment Station Biochemistry Department, South Dakota State University, Brookings, South Dakota*

I. Introduction. 40
II. Sample Preparation and Storage. 40
III. Destructive Analysis. 41
   1. Destruction of Organic Matter. 42
     A. Ashing. 42
     B. Closed-System Combustion. 42
     C. Wet Digestion. 43
     D. Hydrogenolysis. 46
     E. Comments. 46
   2. Isolation of the Selenium. 47
     A. Distillation. 47
     B. Toluene-3,4-dithiol Complex. 47
     C. Arsenic Coprecipitation. 48
     D. Ion-Exchange Treatment. 48
   3. Measurement of the Selenium. 48
     A. Colorimetry. 49
     B. Spectrophotometry. 50
     C. Fluorometry. 52
     D. Atomic Absorption Spectrophotometry. 54
     E. Gas Chromatography. 55
     F. Polarography. 55
     G. Titration. 56
     H. Gravimetric. 56
     I. Spark Source Mass Spectrometry. 57
IV. Nondestructive Analysis. 58
   1. Neutron Activation Analysis. 58
     A. Methods of Measurement. 58
     B. Modes of Activation. 59
     C. Sample Preparation. 61
     D. Factors Affecting Detection Limits. 63
   2. X-Ray Fluorescence Analysis. 66
V. Methods. 67
   1. Method A. 67
   2. Method B. 70
   3. Method C. 72
   References. 74

I. INTRODUCTION

Selenium occurs naturally in biological materials at levels that vary from a few parts per billion to a few per cent. Methods for its determination in these materials must, therefore, cover a very wide range of concentrations. As a result there have been developed a number of procedures for selenium analysis, many of which are discussed briefly in a review by Watkinson (1).

The methods generally used for analyzing biological materials for selenium are based on the destruction of the organic constituents with the concomitant oxidation of the element to Se^{4+} or Se^{6+} and its subsequent determination by one of a variety of procedures, either with or without its separation from other elements. Measurement of the element by nondestructive techniques, especially neutron activation analysis, is, however, becoming increasingly popular. In the discussion that follows, the principles and techniques involved in using these two general types of analysis are reviewed, and some of the most commonly used methods are outlined in detail.

No attempt is made to discuss methods of analysis for different chemical forms of the element or for radioactive selenium used as a tracer in biological studies.

II. SAMPLE PREPARATION AND STORAGE

Some of the chemical forms of selenium in biological materials are volatile, and this must be considered in preparing and storing samples for analysis. Early in their experience with "alkali disease," farmers thought they observed a decrease in the toxicity of their grains on storage (2). Later, it was found that on heating these grains to temperatures of 160°C or more their toxicity decreased and a significant part of the selenium was lost (3). It was further reported that, when highly seleniferous grains were stored for from 3 to 5 years without temperature control, much of the selenium was lost (4).

Beath et al. (5) found large losses of selenium during the air drying of *Astragalus bisulcatus* plants. They later (6) confirmed this in other highly seleniferous "indicator" plants, but stated that there was no significant loss of the element on air-drying ordinary farm crops such as grains and grasses. Asher et al. (7) found that only 0.5 to 3.0% of the selenium was lost when alfalfa was dried at 70°C for 48 hr, and Ehlig et al. (8) reported losses of less than 5.0% of the element on drying a number of species of green plants at 70°C for 30 hr. These two groups of workers used plants labeled with ^{75}Se, counting samples before and after drying. The results of Gissel-Nielson (9) suggest, however, that because of errors in counting, due to self-absorption by the water in the undried samples, the actual losses are probably greater than those determined by this method.

Whereas the volatilization of selenium by animals and molds has been quite well investigated (see ref. 10, pp. 175–176) there has been little study of the loss of the element on drying animal tissues or fluids. Vacuum-oven drying or lyophilization has been used when methods of analysis have required starting with a dried sample.

Allaway and Cary (11) point to the need for care in the grinding of plant samples for analysis. Leaves reduce to fine particles much more readily than do stems, and there is the possibility of separation of these two plant parts in the ground sample. Since leaves generally contain the higher concentrations of selenium, important sampling errors can result unless care is taken to ensure good mixing, especially when very small amounts are used for analysis. Fine grinding followed by careful mixing and the use of larger samples reduce this problem.

Although dried tissues may be stored at room temperatures for short periods, for long storage they should be kept in tight containers at temperatures near or below freezing, especially if they contain volatile selenium compounds. Wet tissues should be stored frozen to reduce the likelihood of the enzymatic formation of volatile compounds (12). Faulkner et al. (13) have reported that selenium is lost from urine under several storage conditions. The loss was prevented by making the urine $1M$ in nitric acid or was reduced by freezing.

Analyzing samples without drying should give the most reliable results. Dry matter contents can be determined on separate samples when it is necessary to have the results expressed on a moisture-free basis. To reduce sampling error when wet digestion is used, predigesting large (2 to 20 g) samples by heating with 100 ml of concentrated nitric acid to bring about their solution is helpful. After adjusting or recording the volume, a suitable aliquot can be used for completing the digestion and analysis.

For grasses, grains, and many other plants that contain little if any volatile selenium, drying at 70°C to constant weight is satisfactory for most studies. Animal tissues or fluids that require drying before analysis are probably best lyophilized to prevent the possible formation of volatile selenium compounds as the result of biological activity (12).

III. DESTRUCTIVE ANALYSIS

The determination of selenium in biological materials by chemical means requires the destruction of organic matter. After this step a number of methods may be used for measuring the element, some of which require its isolation from interfering substances.

1. Destruction of Organic Matter

A. ASHING

Shortly after 1930, when selenium was found to cause the "alkali disease" of livestock (3), Robinson (14) described a method for the determination of the element in which wheat was dry ashed in the presence of calcium acetate. He reported that oxidation with hydrogen peroxide, nitric acid, or perchloric acid resulted in excessive losses of selenium and that fusion with sodium peroxide was unsatisfactory because too much of the reagent was required. Later, Robinson et al. (15) described a method of analysis of plant or animal tissues based on dry ashing in the presence of magnesium nitrate. The large losses of selenium found to occur on open ashing at 700°C without added calcium or magnesium were probably not effectively prevented even when these reagents were present, since this method for the destruction of organic matter was soon abandoned for wet digestion procedures.

Gleit and Holland (16) investigated a low-temperature, dry-ashing process for decomposing organic matter in samples for trace element analysis. They used a high-frequency electromagnetic field to excite a stream of oxygen and greatly increase its reactivity. The excited oxygen was then passed over the sample heated by induction to temperatures of less than 100°C for 90 min, during which time the organic matter was destroyed. A sample of alfalfa grown on soil enriched with ^{75}Se was ashed by the process, and 99% of the radioactivity was recovered in the ash. Later, Mulford (17) used this method and reported that selenium added as selenite to powdered cellulose which also contained added mercury, arsenic, and copper was recovered after ashing at low power. In the presence of copper alone, however, large selenium losses occurred during ashing of the cellulose, and these losses increased as power input increased. At present, this process is not being used in selenium determinations on biological materials.

B. CLOSED-SYSTEM COMBUSTION

In 1961, Gutenmann and Lisk (18) described a method for selenium analysis using the closed-system oxygen flask of the Schöniger type for the dry combustion of pelletized samples of oats. They obtained good recoveries of added selenium, and their results compared well with those obtained by the method of Klein (19). Dye et al. (20) studied the use of the Schöniger flask further and also used the Parr bomb at a pressure of 30 atm of oxygen for combusting samples. Fluids and wet tissues were dried by lyophilizing before burning. The oxidized selenium was allowed to absorb in water added to the flask before combustion. The use of alkali or acid instead of water offered no advantages. Excellent recoveries of ^{75}Se from alfalfa grown on this isotope and good agreement with results obtained by neutron activation analysis attested to the efficiency of oxidation and retention of the selenium by either Schöniger flask or Parr bomb combustion. Although slightly better precision was obtained by using the oxygen flask, the authors found little to choose from between the two methods.

Allaway and Cary (11) also developed a procedure in which the oxygen flask was used for combustion. They too obtained good results as compared with neutron activation analysis. They found that adding magnesium nitrate to certain plant tissues before pelleting for the combustion step yielded a brighter colored residue. This procedure was not used routinely, however, since it resulted in the formation of a slowly soluble bead of magnesium nitrate. Sodium peroxide and periodate were also tried as oxidizers but gave too violent a reaction.

Cukor et al. (21) and Taussky et al. (22) have also described methods in which the oxygen flask served for the combustion of biological samples. Watkinson (23) compared it with wet digestion using nitric and perchloric acids and found excellent agreement of results when the insoluble residue from the combustion was digested with perchloric acid. The Parr bomb at a pressure of 30 atm of oxygen was also used by Johnson (24) for combusting municipal solid waste, compost, and paper for selenium analysis.

C. WET DIGESTION

In 1934, Robinson et al. (15) used the ordinary Kjeldahl procedure to destroy organic matter in biological samples before analysis for selenium. The digestion was conducted in an all-glass system in which the escaping gases were passed through bromine water to trap any selenium. Since the bromine was continuously reduced, especially during the first stages of digestion, repeated addition of this element was required, making control of bromine fumes in the laboratory difficult. Shortly thereafter, Williams and Lakin (25) used open digestion with a mixture of sulfuric and nitric acids and temperatures not exceeding 120°C for plant tissues. After some study of this method of digestion and modifications to include the addition of mercury as a selenium "fixative" (26–28), it was adopted for use in the analysis of foods and plants by the Association of Official Agricultural Chemists (29). Klein (19) used a similar mixture and found it satisfactory even at much higher temperatures. Grant et al. (30) obtained good recoveries of added selenium from nitric-sulfuric acid digests of biological samples, and Olson (31) found good agreement between the results of the analysis of plant samples oxidized with the nitric-sulfuric acid mixture and those obtained by several other methods. However, others (32–34) have reported losses of the element during digestion with this mixture. Apparently, when a loss occurs it is the result of excessive charring after the nitric acid has evaporated. When charring is avoided, good recoveries can be obtained (32).

Down and Gorsuch (35) investigated the use of sulfuric acid-hydrogen peroxide digestion of organic materials for analysis for a number of elements. Recoveries of selenium were poor, however, and this method cannot be considered as satisfactory for its determination.

A number of methods based on digestion of samples with mixtures containing perchloric acid have been described. These vary as follows: nitric acid and perchloric acids (23,31,36–40); nitric, sulfuric, and perchloric acids (41–44); nitric and perchloric acids plus ammonium metavanadate (45); sulfuric and perchloric acids plus sodium molybdate and water (46–48); and nitric and perchloric acids plus hydrogen peroxide (49). There is a considerable body of evidence that little, if any, selenium is lost during the digestion of biological materials with these mixtures. A number of investigators (23,32,36,39,41–43,45) have, on using them, reported excellent recoveries of inorganic selenium added to biological materials. Others have obtained good comparisons between the results on using such mixtures and the results with the closed-system oxygen combustion (23,31) or neutron activation (23,31,32,37) methods. Furthermore, recoveries of ^{75}Se from the blood of sheep dosed with this radioactive element (see ref. 1, p. 99), or from the blood and liver of rats previously administered the isotope, have been excellent after digestion with mixtures containing perchloric acid. On the other hand, Schwarz (see ref. 1, p. 112) has observed severe and variable losses of selenium during nitric-perchloric acid digestion of biological materials, especially urine. Moreover, Hall and Gupta (49) found low and variable recoveries of added selenium when plant samples were digested with a mixture of nitric and perchloric acids. To further investigate this matter, the present authors compared the radioactivity, before and after digestion with a nitric-perchloric acid mixture, of wheat grown on $Na_2^{75}SeO_4$, of the urine, blood, and tissues of a rat fed this wheat, and of *A. racemosus* leaves from cuttings treated with $Na_2^{75}SeO_3$. The results of this study are summarized in Table I. The data confirm the previous reports of excellent recoveries of selenium after wet digestion in the presence of perchloric acid.

Gorsuch (32) found that with nitric-sulfuric acid mixtures heavy charring occurs after the nitric acid has distilled off and that when this happens some selenium is lost. Perchloric acid helps prevent this loss by maintaining oxidizing conditions. Under any circumstance, wet digestion should be carried out to avoid excessive charring, either by using a slow rate of heating, by using a condenser to slow the distillation of nitric acid, or by stopping the digestion and adding nitric acid whenever the solution becomes excessively brown. When such precautions were taken, neither an increase in the rate of

TABLE I

Recovery of ^{75}Se from Digests of Wheat, of *Astragalus racemosus*, and of
Rat Urine, Blood, and Tissues

Material digested	Size of sample	[75]Se content before digestion (counts/min/ml or g)	[75]Se recovered after digestion[a] (%)
Wheat (seed)[b]	0.50 g	138,600	99.9
A. racemosus (leaves)[c]	0.20 g	1,961,000	98.4
Rat urine (0–24 hr)[d]	2.0 ml	89,285	99.6
Rat urine (24–72 hr)	2.0 ml	11,590	102.2
Rat spleen	0.26 g	11,630	102.1
Rat blood	1.08 g	5,890	98.3
Rat kidney	1.11 g	39,800	97.4
Rat heart	1.17 g	7,787	98.8
Rat liver	0.48 g	25,200	98.0
Rat muscle	1.30 g	2,770	100.4
Average recovery			99.5

[a] The samples were counted with a Packard Auto-Gamma spectrometer with a 2 in. × 2 in. well-type NaI(Tl) detector. They were then digested with a nitric-perchloric acid mixture (31). The digest was transferred to a 5.0-ml volumetric flask, washing to volume with small amounts of water, and 2.0-ml aliquots were counted. The data are averages of determinations on duplicate samples.
[b] The wheat was grown on a sand-soil mixture to which $Na_2{}^{75}SeO_4$ had been added, as described in ref. 49a. Its selenium content was 3.8 ppm.
[c] The cut ends of the upper portions of *Astragalus racemosus* plants in the blossom stage of growth were dipped into a $Na_2{}^{75}SeO_3$ solution for 1 hr and then in water under lights for 18 hr. After air drying, the leaves were finely ground with about 4 parts of leaves from untreated plants. The selenium content of the mixture was 1200 ppm.
[d] One albino rat weighing about 350 g was fed the wheat containing ^{75}Se for 24 hr, during which time urine was collected. The wheat was then replaced by a stock diet to which some selenium had been added, and the urine was collected for an additional 48 hr. The rat was then decapitated, and blood and tissue collections were made.

digestion nor an extension in the time of heating beyond the appearance of perchloric acid fumes caused a decrease in the selenium found in plant material (31).

Of the many mixtures containing perchloric acid that have been used to destroy organic matter in biological materials for selenium analysis, none has any decided advantages. The nitric-perchloric acid combination offers simplicity and possibly a lower blank than the others (1), but it may also require longer digestion periods than some. Mixtures containing perchloric acid seem to have decided advantages over those not containing it, and when properly used they offer no serious explosion hazard.

During digestion with perchloric acid, selenium is converted to selenite and selenate, and for many methods of determination the reagents used are specific for selenite. Treatment with hydrochloric acid in strongly acid solution readily accomplishes the reduction of any selenate to this form (23,40).

D. HYDROGENOLYSIS

Raney nickel has been used in the hydrogenolysis of organic sulfur compounds (50). Wiseman and Gould (51) used the process to remove selenium from various organic compounds, and Throop (52) adapted it to the determination of selenium in organic compounds occurring as contaminants in steroids. The selenium was removed as nickel selenide by treatment with Raney nickel, separated by filtration, oxidized and solubilized with nitric acid, and then determined colorimetrically. The method was designed for samples containing at least 20 μg of selenium and to date has not been applied to the analysis of biological tissues or fluids.

E. COMMENTS

In the analysis of biological materials for selenium, the method used for the destruction of the organic matter will depend on a number of factors. For instance, digestion with nitric-sulfuric acid mixture containing added mercury may serve as well as other types of oxidation for samples of "indicator" plants containing high levels of the element. Normally, however, mixtures containing perchloric acid will be the oxidants of choice. Though Parr bomb or oxygen flask combustion is also very satisfactory, digestion with a mixture containing perchloric acid seems to be used most often. It has the advantages of requiring simpler equipment, of being more readily adaptable to large numbers of samples and to larger samples, and of facilitating the handling of liquids or wet samples. Although closed-system oxidation with the Parr bomb or Schöniger flask would seem better suited to prevent loss of the element, apparently little, if any, selenium is lost if excessive charring is avoided during wet digestion. Nitric acid has been used to trap the volatile selenium in the breath of rats (53); therefore it seems probable that the selenium which occurs in the tissues will not be lost on wet digestion. Indeed, there is no evidence to date for the occurrence of any volatile selenium compound in natural materials that would escape on solution in nitric acid.

When oxygen combustion is used, care should be taken to ensure destruction of the organic matter, at least to the point that what remains will not interfere in the subsequent measurement of the selenium. This is especially important when the selenium is not separated from the mixture before its determination.

2. Isolation of the Selenium

Some methods of analysis require isolation of the selenium, or at least its separation from other elements or compounds that interfere in its determination. Usually this has been accomplished by (*a*) distillation as the tetrabromide with or without subsequent precipitation as the element, (*b*) extraction as the toluene-3,4-dithiol complex, (*c*) coprecipitation with arsenic, or (*d*) removal of cationic interference by ion-exchange treatment.

A. DISTILLATION

Rosenfeld and Beath (10) have reviewed the theory and procedures of separating selenium as the volatile tetrabromide and precipitating it as the insoluble element. Digests are heated with a bromine-hydrobromic acid mixture in a closed system, and the distillate is collected in a cooled receiving flask containing water or other suitable trapping agent. The purpose of the bromine is to ensure that all of the selenium is oxidized to at least the quadravalent form. The excess bromine distills at a low temperature, and the hydrobromic acid reduces any hexavalent selenium that may be present to the quadravalent form, which then forms the tetrabromide and distills. In addition to selenium, arsenic, tin, germanium, antimony, and tellurium also distill, at least in part. The free bromine and the selenium can then be reduced, the latter to the elemental form, which is insoluble and can be filtered off and separated from the elements that distill with it. The most commonly used procedure for this reduction is treatment with sulfur dioxide to remove the bromine and saturate the solution, followed by heating with hydroxylamine. This method of isolating the element has been used for its gravimetric determination (3) on samples such as the highly seleniferous indicator plants and on organic selenium compounds. It has also been incorporated with various modifications into a number of other methods.

Instead of precipitating the selenium, Handley and Johnson (42) treated the distillate with phenol to remove the bromine and completed the determination of the selenite colorimetrically. On the other hand, Bonhorst and Mattice (54) used sulfur dioxide and hydroxylamine to precipitate the element directly from a digestion mixture without distillation as the tetrabromide. By filtering, they isolated the selenium for colorimetric determination.

B. TOLUENE-3,4-DITHIOL COMPLEX

In a series of studies on the use of *o*-dithiols in elemental analysis, Clark (55) found that in hydrochloric acid solutions $10N$ or stronger selenium(IV) reacted with toluene-3,4-dithiol to form an insoluble complex which could be extracted with ethylene chloride. The only other elements that reacted and precipitated at this high acidity were palladium(II), tellurium(IV), gold, tungsten(VI), and rhenium(VII), although a number of others precipitated at lower concentrations of acid. Since toluene-3,4-dithiol is rather unstable on storage, the use of its zinc complex has been suggested (56).

Watkinson (39) used the reaction of selenium(IV) with toluene-3,4-dithiol and the extraction of the complex formed with 1:1 ethylene chloride-carbon tetrachloride to separate the element from interfering substances before its fluorometric determination. The carbon tetrachloride made the organic solvent phase heavier than the water phase, but did not prevent the extraction of the complex by the ethylene chloride. After the extraction step, the solvent was evaporated in the presence of a little nitric-perchloric acid mixture, which destroyed the organic residue and dissolved the selenium.

C. ARSENIC COPRECIPITATION

Luke (57) used the coprecipitation of selenium with arsenic to separate the selenium from metals in copper and lead samples. Arsenic did not interfere with the subsequent analysis, and neither did the small amounts of lead, bismuth, silver, antimony, and tin that accompanied the precipitate. Interference by small amounts of copper was prevented by complexing with EDTA. Hypophosphorus acid was used to reduce the selenium and arsenic to their insoluble forms. Cousins (36) adapted the procedure to biological materials, and several others have also used it (11,37,48).

D. ION-EXCHANGE TREATMENT

Schumann and Koelling (58) used a cation-exchange resin to remove interfering metals before the precipitation of selenium in its determination in sludges and dusts of the sulfuric acid industry. Lott et al. (59) employed this procedure in the analysis of copper samples for selenium, but Cukor et al. (21) reported that, in the analysis of plant samples, ion-exchange resins were unsatisfactory for removing interfering cations. The resin was carried over in trace amounts, which could not be filtered out, and its fluoroscence gave high results. The possibility that ion-exchange treatment can be used in the analysis of highly mineralized diets needs examination.

3. Measurement of the Selenium

Nearly all methods of quantification for selenium, with the exceptions of neutron activation and X-ray fluorescence, require its oxidation to the 4+ or 6+ valence state in a medium relatively free of organic material. In general, certain methods then require the reduction of selenium to the insoluble elemental state with subsequent separation, while others permit its determination in solution. For the latter, the bulk of the literature concerns colorimetric, spectrophotometric, and fluorometric procedures. These appear to be the most widely used techniques and are quite rapid if the element is determined directly in solution.

A. COLORIMETRY

Colorimetry was used in some of the earliest procedures for selenium analysis. In 1924, Cousen (60) determined selenium in glass by reducing it to elemental form with phenylhydrazine hydrochloride in the presence of gum arabic to form a stable sol, which was then visually compared with standards. A similar method was described by the American Public Health Association (61) for the determination of selenium in water and sewage, except that sulfur dioxide and hydroxylamine served as reducing agents. The method was further modified by Fogg and Wilkinson (34) in that ascorbic acid was used as a reducing reagent.

Franke et al. (62) reduced selenium and filtered it onto barium sulfate mats, which in turn were compared with previously prepared standards. A more recent attempt to directly estimate selenium involves filtering the reduced element through millipore filters (38) and then comparing the pink disks with previously prepared standards. The lowest detection limit was stated as 0.2 μg, but the method appears to lack accuracy.

Other workers have attempted to use ring colorimetry based on the ring oven technique of Weisz (63). This method is basically a special kind of spot test, carried out on paper, in which the substances to be detected are deposited in a well-defined ring. The ring is then treated with a color-producing reagent and compared visually with prepared standards. Biswas and Dey (64) have employed this method to determine selenium in water, and West and Cimerman (65) have proposed its use for air pollution studies. In the former instance, color was developed with thiourea, but extensive separation of selenium was required and relatively high levels of the element were necessary (0.23 ppm in water). In the latter instance, the color rings were developed with 3,3'-diaminobenzidine; however, interference was caused by several ions which would certainly be present in biological samples, and provision was not made for all of them. At best, such techniques seem semiquantitative (see ref. 63).

Horn (66) reported a qualitative test in which inorganic selenium forms a blue complex with codeine. Klein (19) studied the potential of the codeine method in quantitative analysis, as well as the applicability of the colorimetric evaluation of the carbon disulfide extract of elemental selenium. In both instances, comparisons with standards were made visually; although

the codeine method showed promise, it lacked the sensitivity and reproducibility of the spectrophotometric methods subsequently developed.

B. SPECTROPHOTOMETRY

The most studied methods of selenium analysis are based on the reaction
of selenious acid with o-diamines. Hoste (67) and Hoste and Gilles (69) were
the first to report such a method for the detection of selenium, based on the
measurement of the yellow compound formed on the reaction of 3,3'-diamino
benzidine (DAB) with Se^{4+}. They assumed the colored material to be a
dipiazselenol (see compound 1 in [1]). Parker and Harvey (68) have shown
that, in fact, the colored product is a monopiazselenol (2 in [2]). This is

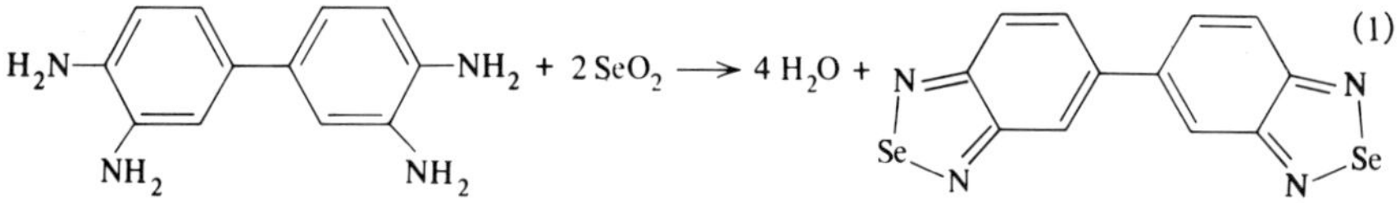

(1)

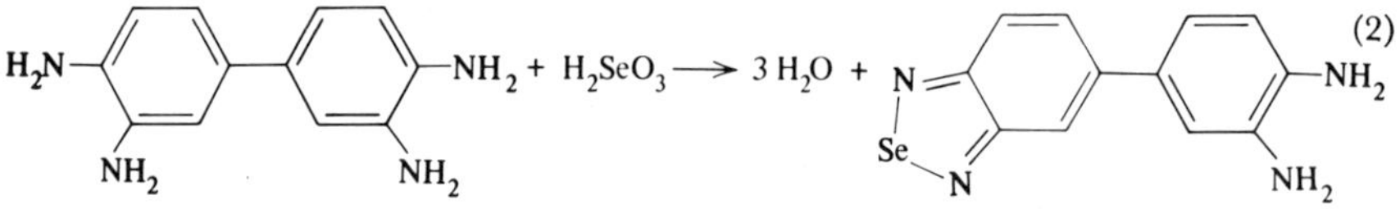

(2)

supported by the effect of pH on the spectrum and extractability of the
compound. According to Hoste (67) and Hoste and Gillis (69), the reaction
was specific for selenium except that oxidizing agents interfered, and
iron(III), copper(II), and vanadium reacted with the reagent. Interference
by iron was prevented by the addition of fluoride or phosphoric acid; copper
interference, by oxalic acid. No method was given for the prevention of
vanadium interference.

Cheng (70) developed several modifications of the DAB method which
made it more specific for selenium. He introduced the use of EDTA to mask
the effect of many of the divalent ions. He also extracted the piazselenol into
toluene, which negated the effects of vanadium and several other elements,
since their DAB derivatives were insoluble in toluene. However, the procedure involved pH adjustment, since the formation of the piazselenol
required acidic conditions, whereas the DAB-Se complex was quantitatively
soluble in toluene only in neutral or slightly alkaline solution. The yellow
piazselenol was measured at 420 nm, and a blank was required because

unreacted DAB had a weak absorbance at this wavelength. The limit of sensitivity was stated to be 50 ppb with a 1-cm absorption cell, but later work by others shows this appraisal to be rather optimistic. The method appeared to have good precision (3.73 C.V. on 10 μg of Se in a 1-g sample).

Several workers have applied the method of Cheng to the analysis of selenium in inorganic materials (57,71–74). In 1959, Handley and Johnson (42) and Bonhorst and Mattice (54) applied the DAB method to biological materials. The method was found suitable for the determination of about 0.2 μg of Se or more and appeared to be reliable.

Kelleher and Johnson (43) attempted to increase the sensitivity of the Cheng method by application of ultramicrotechniques. Although this appeared to improve the sensitivity, considerable losses of selenium occurred in certain of the preparative steps, and it was necessary, therefore, to use a combined spectrophotometric and isotope dilution method. The method appears somewhat cumbersome and has not been popular, especially since the more sensitive fluorometric procedures have been developed.

More recently, Cummins et al. (46,47) have developed modifications of the original Cheng procedure, primarily concerned with sample preparation, which allow the analysis of biological samples. They have compared their results with those obtained by neutron activation techniques and have found excellent agreement. They stated the lower limit of sensitivity to be 1.0 μg. The method is rapid, with no loss of selenium, and should be sufficient for studies of selenium toxicosis.

Parker and Harvey (75) investigated the reaction of selenious acid with a series of aromatic o-diamines and observed that 2,3-diaminonaphthalene (DAN) was a more sensitive reagent than DAB for the detection of selenium. Subsequently, Lott et al. (59) developed a procedure for the use of DNA which is nearly identical with that for DAB, except that pH adjustment is not necessary before extracting the Se-DAN complex into the organic solvent. At the same time DAN was found to be a very sensitive reagent in the fluorometric analysis for selenium, and it has had much wider application for this purpose.

Several other aromatic o-diamines have been utilized in the spectrophotometric analysis of selenium. These include o-phenylenediamine (52,76), 1,2-diaminonaphthalene (52), 4-methylthio-1,2-phenyldiamine (77,78), 4-dimethylamino-1,2-phenyldiamine (77), 4,5-diamino-6-thiopyrimidine (79), and 1-methyl-4-chloro- and 4,5-dichloro-o-phenylenediamine (21).

Another type of spectrophotometric method involves the formation of complexes between selenium and various sulfur ligands. 2-Mercaptobenzothiazole (80) and 2-mercaptobenzoic acid (81) have both been investigated and apparently give sensitivity and selectivity comparable to those obtained with DAB.

A few spectrophotometric methods which involve the catalytic effect of selenium on the reduction of various compounds have also been studied. Kawashima and Tanaka (82) studied the possible application of the catalytic effect of selenium on the reduction of 1,4,6,11-tetraazanaphthacene in the presence of hypophosphoric acid and glyoxal. The reduced compound, 1,6-dihydro-1,4,6,11-tetraazanaphthacene, has a blue color which is measured at 600 nm. Several ions, including tellurium, interfere with the method, and data were not given on the effectiveness of extraction procedures in the removal of the interferences. Taking advantage of the catalytic effect of selenium on the reduction of methylene blue by sodium sulfide, West and Ramakrishna (83) have proposed a method which they claim is suitable in the range 0.1–1.0 μg. Iron(III) potentiated the reaction, and copper, which was found to be the only ion not masked by EDTA, was removed by ion exchange. Since visual color comparisons were made, the precision was poor at certain concentrations. Mesman and Doppelmayr (83a) have developed a simple photometric device to increase the efficiency of the technique. Neither of the above catalytic methods was tried on biological samples.

Yet another class of colorimetric procedures involves the reaction of selenium with a reagent to produce an intermediate, which in turn is reacted with a color-producing reagent. Kirkbright and Yoe (84) oxidized phenyl-hydrazine-p-sulfonic acid with selenious acid and then measured the resulting diazonium product by coupling with 1-naphthylamine. In a similar method (85) selenious acid was reacted with hydroxylamine to produce nitrite. The nitrite was then used to convert sulfonamide to its corresponding diazonium derivative, which in turn was coupled with N-(1-naphthyl)-ethylenediaminedihydrochloride. Naidu and Rao (86) oxidized selenium directly with permanganate and then measured the excess permanganate at 530 nm. Cresser and West (81) have investigated the spectrophotometric determination of selenium by formation of the ketone-chloro complex with cyclohexanone in $7M$ HCl. This method is subject to interference by various ions which must be removed.

Of the colorimetric and spectrophotometric methods, it appears that those utilizing DAB and DAN are the most suitable for biological samples. Although some of the others show promise, most have not been used on biological samples, and usually they have not been compared to other methods.

C. FLUOROMETRY

The colorimetric and spectrophotometric methods are suitable for determining toxic concentrations of selenium, but they are generally not sensitive enough for studies of trace amounts of the element. In the search for more sensitive methods, Cousins (36) took advantage of the fluorescence of the Se-DAB complex in hydrocarbon solvents. After the selenium was coprecipitated with arsenic, the Se-DAB complex was formed under the same conditions as used in the colorimetric procedures. A filter system giving maximum transmission at 450 nm was used for excitation, and the fluorescent emission was measured at 580 nm. The fluorescence in the blank was relatively low, and the method was 20 times more sensitive than the DAB spectrophotometric procedures. Watkinson (39) further extended the fluorometric procedure by extracting the oxidized selenium with toluene-3,4-dithiol as a purification step; however, this procedure required digestion periods before and after the extraction. The method was applied to the determination of selenium in arsenic (68) and in various biological materials (30).

While investigating the relative luminescence of piazselenols formed from selenious acid and various o-diamines, Parker and Harvey (75) found the product formed with 2,3-diaminonaphthalene (DAN) to be strongly fluorescent and considerably more sensitive than the Se-DAB complex. Lott et al. (59) applied the reagent to the determination of selenium in samples containing several foreign ions and found it more sensitive than DAB but also more subject to interference. Consequently, the method was coupled with isotope dilution analysis (21). Allaway and Cary (11) used DAN to determine selenium, but they separated the element by coprecipitation with arsenic. They also added hydroxylamine to decrease the oxidation of the DAN during the formation of the Se-DAN complex. Comparing the method to the previously developed spectrophotometric procedure which utilized DAB and to neutron activation analysis, they found good agreement.

Watkinson (23) conducted a thorough study on the use of DAN to determine selenium without previous separation of the element. The sample was digested in a nitric-perchloric acid mixture, and the digest was heated with HCl to convert all selenium to the 4+ state. After the pH was adjusted to 1, the selenium was complexed with DAN in a water bath at 50° for 20 min. The complex was extracted with cyclohexane, and the fluorescence was measured with a 606-nm filter.

In 1969, Olson (31) conducted a collaborative study comparing a slightly modified version of the Watkinson method with several spectrophotometric and fluorometric procedures and with neutron activation analysis. The results from the modified method compared well with those from the other techniques. Several other workers have developed further procedural modifications, most of which concern the conditions of oxidation and separation of the selenium before color development (24,37,44,48,49,87,88).

Concerning the extraction of the piazselenol, studies on the fluorometric methods (20) showed that mesitylene was most efficient in extracting the piazselenol of DAB, but toluene was chosen primarily because of cost. In fluorometric studies with Se-DAN complex, extraction with toluene gave extracts that were not stable (75). Cyclohexane was shown by Parker and Harvey (75) to be a good solvent, but it is quite volatile; therefore they recommended using decalin. However, Watkinson (23) has recommended cyclohexane, since it is more easily purified and because the disadvantages of high volatility can be readily controlled. The Watkinson method (23) and several of the modified versions seem well suited for the rapid analysis of large numbers of samples containing more than 0.02 μg of selenium.

D. ATOMIC ABSORPTION SPECTROPHOTOMETRY

Atomic absorption spectrophotometry is based on the absorption of light by the unexcited atoms of an element which have been sprayed into a flame. Consequently, it is limited to elements which are dissociated into atoms in the flame. It was recognized by Allan (89) that selenium was such an element. However, the various forms of the element still had to be converted to the inorganic state before measurement, and the early hollow cathode lamps used for measuring the absorption by selenium were rather insensitive (requiring 1.5 to 5 ppm Se). Although Allan recognized that the most sensitive absorption line for selenium was at 196 nm, this was beyond the limit of his equipment.

Rann and Hambly (90) measured the absorption of selenium at 196 nm in an air-acetylene flame and achieved a sensitivity of about 1 ppm. When the method was applied to the analysis of biological samples, good agreement was found with the Klein (19) procedure. Several others have attained a similar sensitivity (90a,90b). Chakrabati (91) attempted to increase the sensitivity by extracting the selenium-diethyldithiocarbamate complex into an organic solvent and then measuring the absorption due to selenium in an air-acetylene flame.

Although the present methods involving atomic absorption spectrophotometry lack the sensitivity for submicrogram quantities of selenium, it appears that more refined instrumentation is possible (1). Improvement in sensitivity has been obtained by use of an argon-hydrogen flame (92), the "sampling boat" technique (93), the Delves sampling cup technique (93a), or the high sensitivity system for selenium, which converts selenium to a gaseous hydride before sweeping it into the flame (93b). The recent development of the heated graphite atomizer (93c) appears to present possibility for increasing the sensitivity of the method for selenium.

An indirect atomic absorption procedure has been developed by Lau and Lott (93d) which appears to be quite sensitive. It involves extracting the product from the reaction of selenium with DAN and then further reacting this product with palladium chloride. The amount of selenium-containing product is then determined by measuring the palladium with normal atomic absorption procedures.

E. GAS CHROMATOGRAPHY

Very little has been done on the application of gas chromatography to selenium analysis; however, Nakashima and Toei (94) have formed the 5-chloropiazselenol under the same conditions as described in Section III.3.B. The 5-chloropiazselenol was extracted into toluene and analyzed by gas chromatography on a column packed with 15% SE-30 on 60–80 mesh Chromosorb W. An electron capture detector was used to measure the chloropiazselenol. The minimum amount detectable was stated to be 0.04 μg Se; however, an internal standard had to be run along with the sample since the sensitivity of the electron capture detector varied during operation. Others (95) have used 4-nitro-o-phenylenediamine in the determination, since the nitro group is more sensitive to electron capture detection than the chlorine group. The same conditions were used, and the sensitivity was somewhat greater. No attempts have been made to apply the method to biological samples.

F. POLAROGRAPHY

Another means of assaying selenium is by polarography, which seems to be similar to spectrophotometry in sensitivity and rapidity. The first polarography of selenium was reported by Schwaer and Suchy (96), but Faulkner et al. (13) pioneered in applying the technique to the analysis of selenium in biological material. In a series of papers, Christian et al. (41,97,98) further clarified the polarographic characteristics of selenium. Copper interferes but may be removed by extraction with dithizone in chloroform (97). Christian et al. (41) found that for samples containing less than 4 μg of Se it was best to separate the element by precipitation or extraction of the DAB complex before polarography. The Se-DAB complex was extracted into a mixture of chloroform and ethylene dichloride and then back-extracted into perchloric acid to obtain the polarogram. With prior separation, as little as 0.2 μg of Se could be determined in 1- to 2-g samples. The technique has been applied to water (99) for the detection of acutely toxic levels of various elements, including selenium. Griffin (100) has reported that, by using single-sweep polarography instead of conventional polarography to determine the selenium-DAB complex, sensitivity equal to or greater than that obtainable with the spectrofluorometric method can be achieved (1 ppb). Although the method has been applied to biological samples, the author states that it involves considerable unreliability and that further development is required.

G. TITRATION

One of the oldest and most often used procedures for selenium analysis before the development of the spectrophotometric methods is based on titration with thiosulfate. The original work on this procedure was done by Norris and Fay (101) and depends on the following reaction:

$$4Na_2S_2O_3 + H_2SeO_3 + 4HBr \rightarrow Na_2SeS_4O_6 + Na_2S_4O_6 + 4NaBr + 3H_2O$$

An excess of thiosulfate is required to drive the reaction to completion, and this excess is titrated with standard iodine in the presence of a starch indicator. The method was studied further by Coleman and McCrosky (102), by Curl and Osborn (26), and in some detail by Klein (19). The Klein method has been used most extensively. This method requires a wet digestion and separation of the selenium by distillation with Br_2-HBr and precipitation in elemental form. After filtration the selenium is dissolved in Br_2-HBr and the excess Br_2 removed with phenol. The thiosulfate-iodine-starch titration is carried out in a special vessel. The method has been widely used and is generally acceptable for samples containing more than a few micrograms of selenium.

Another volumetric method, described by Hillebrand and Lundell (103) and studied further by Klein (19), involves the reduction of selenious acid to elemental selenium by iodide. The resulting I_2 is then titrated with thiosulfate according to the following reactions:

$$H_2SeO_3 + 4HBr + 4KI \rightarrow Se + 2I_2 + 4KBr + 3H_2O$$

$$2I_2 + Na_2S_2O_3 \rightarrow 2Na_2S_4O_6 + 4NaI$$

Although the method is accurate, the presence of elemental selenium interferes with observing the starch-iodine end point.

There have been other reports of the oxidation of selenite to selenate with potassium permanganate as the titrant (104,105). Hypobromite, hypochlorite (106), and potassium ferricyanide (107) have also been used as oxidimetric titrants. The methods involving oxidation are subject to inaccuracies and to interference by other oxidizable materials. They have not been tried on biological materials and have not received wide usage.

H. GRAVIMETRIC

Many procedures for selenium analysis require the separation of the selenium in elemental form. If milligram quantities are present, the element may be filtered on an asbestos pad and weighed. Hydroxylamine in 12% HBr (14) and a combination of sulfur dioxide followed by hydroxylamine

(3,15) have been used to precipitate the element for its gravimetric determination. Other reducing agents such as ascorbic acid have also been employed (34,38). For biological samples containing high levels of selenium or for pure seleno compounds the gravimetric method is reliable if adequate care is taken that impurities are not occluded in the precipitated element.

Lott et al. (59) have reported that organic precipitants can also be used for the gravimetric determination of selenium if milligram amounts of the element are present. Selenium in milligram amounts reacts with DAN to form a red precipitate. The precipitate is washed with 1.5N HCl to remove other ions which coprecipitate. The same masking reagents that are employed in spectrophotometric procedures can be used. There will be some error at lower concentrations of selenium, since microgram quantities of Se-DAN complex are soluble.

I. SPARK SOURCE MASS SPECTROMETRY

In 1951, Gorman et al. (108) described a method for the analysis of solids with the mass spectrometer, applying it to the analysis of steel for chromium and nickel. They used a system now referred to as spark source mass spectrometry (SSMS), the principles of which have been discussed by Roboz (109). Sasaki and Watanabe (110), Wolstenholme (111), and Brown et al. (112) applied the technique to the semiquantitative or qualitative analysis of some biological samples for trace elements, and Evans and Morrison (113) improved its accuracy, reporting on the analysis of lung tissue for 30 elements, including selenium. Yurachek et al. (114) analyzed human hair for 23 elements, including selenium, and Harrison et al. (115) have discussed the forensic applications of SSMS, reporting on the analysis of hair and fingernails for a number of elements, including selenium.

The SSMS technique has excellent sensitivity, responds to nonmetals as well as to metals, and offers considerable promise as a tool for the simultaneous analysis of biological material for a number of elements. Its accuracy should be improved when it is coupled with the isotope dilution technique (116). It requires destruction of the sample by dry or wet ashing (the latter in the case of selenium), a disadvantage when the method is compared with nondestructive neutron activation analysis. As is the case also with neutron activation analysis, its greatest disadvantage is probably the cost of the equipment required. At present, at least, this cost greatly restricts the general use of this method. When selenium is the only element of interest in an analytical program, SSMS analysis may seldom be used; but when samples must be surveyed for a number of trace elements, the technique appears to have considerable promise.

IV. NONDESTRUCTIVE ANALYSIS

1. Neutron Activation Analysis

The theory and principles of neutron activation analysis (NAA) have been discussed by several authors (117–122). It has been used, both with and without destruction of the sample, to determine trace levels of selenium in a number of biological materials. Examples of this application have been reported for lung tissue (123), various muscle tissues (30,123,124), blood (125–127), kidney and liver (30,124,128,129), pancreas and aorta (124), hair (130), eye lens (131), dental enamel (132), feces (133), torula yeast (134), tobacco and tobacco products (135), forages (133,136), various feeds (125,136–138), enzymes and proteins (138), nicotinic acid (139), cystine (140), and organoselenium compounds (141,142).

A. METHODS OF MEASUREMENT

After neutron activation, the amount of an element in the sample can be determined either by the direct (or absolute) method or by the comparator technique. With the direct method, the following basic NAA equation (143) is used:

$$W = \frac{A_0 M}{(6.02 \times 10^{23})\theta\sigma f S}$$

where W = the weight of the element (g); M = the atomic weight of the element; θ = the isotopic abundance of the target nuclide; σ = the activation cross section of the target nuclide for neutron capture (cm²); f = the neutron flux to which the sample has been exposed (neutrons/cm²/sec); S = the saturation factor, $1 - e^{-0.693}(t_0/T)$, where t_0 is the duration of the irradiation period and T is the half-life of the product radionuclide; and A_0 = the activity (disintegrations per second) of the product radionuclide at t_0. All of the foregoing factors, except W, are known or can be determined. Activity, A, is usually measured at a time, t, after termination of the irradiation, and A_0 is determined from the relationship $A = A_0 e^{-0.693}(t/T)$, knowing the efficiency of the detector counting system for a given counting geometry.

The direct method has not been extensively used in trace element analysis, however, since it is considered semiquantitative with systematic errors approaching 20%. Activation analysis for selenium has been accomplished in most instances by use of the comparator technique. The unknown sample and a known weight, w_s, of the pure element being determined are irradiated together, so that each is exposed to the same neutron flux at the same time. After irradiation, the standard and the unknown sample are counted under conditions in which, ideally, the induced specific activities (disintregration rate of the product radionuclide/unit weight of the element) in both standard and unknown would be the same. The weight of the element in the unknown, w_u, can be determined quite simply by means of the equation:

$$w_u = \frac{w_s R_u}{R_s}$$

where R_u = the count rate (counts per second, corrected) for the unknown sample, and R_s = the count rate for a given weight, w_s, of the pure element.

The comparator technique is designed to obviate many of the uncertainties of the direct method of determination. The standards used, however, should meet certain criteria (120, p. 10), including highest purity and resistance to thermal decomposition. To improve the handling of microgram quantities, the standard may be mixed with a larger amount of inert material, such as paraffin (144), water (133,145), water dispersed on filter paper (125,146), or purified silicon dioxide (147).

Leliaert et al. (148) have described a method of internal standardization in which samples are irradiated both with and without the homogeneous admixture of a known amount of the element being determined, and this can be applied to selenium. If the sample contains another activatable element which can be used as a monitor of the neutron flux, these irradiations need not be performed simultaneously.

B. MODES OF ACTIVATION

Neutrons are produced in a broad range of energies as by-products of the fission process in nuclear reactors. This is illustrated for one reactor for which neutron flux data were presented by Yule (149):

Neutron energy	Flux (neutrons/cm^2/sec)
Thermal	4.3×10^{12}
>10 keV	3.5×10^{12}
>1.35 MeV	7.5×10^{11}
>3.68 MeV	1.2×10^{11}
>6.1 MeV	1.9×10^{10}

Moderating materials such as hydrogen-containing compounds and graphite serve to slow down neutrons, and those with energies reduced to 0.025 eV (2200 m/sec) are designated as thermal. The ease with which thermal neutrons are captured by the nuclei of most nuclides leads to loss in the thermal neutron flux, which is maintained, however, through replacement by the processes degrading the energies of the ever-present fast neutrons.

The principal activation reaction resulting from thermal neutron capture is the (n, γ) reaction. The captured neutron increases the mass of the nuclide by 1, and nuclear rearrangement leads to γ-ray emission. Fast neutrons present in the reactor, however, may give rise to interfering activities.

Thermal neutron fluxes of 10^{11} to 10^{13} neutrons/cm²/sec are available in the various reactors. A constant flux is desirable and is generally used in thermal neutron irradiations (which may be extended for weeks if necessary). There are variations of the method, one of which is the pulsing technique by which the reactor is caused to produce intense neutron fluxes ($\sim 10^{17}$ neutrons/cm²/sec) of very brief duration (several milliseconds). It has found application in determining certain elements having radionuclides of short half-lives (150, p. 72). Greater than 70-fold improvement in sensitivity can be achieved for radionuclides with half-lives <1 sec. The pulsing technique offers no advantage, however, in selenium analysis.

An analysis for selenium using prompt γ-ray emission has been reported (151). Production of prompt γ-rays (representing the binding energy of the neutron to the nucleus in response to neutron capture by a given nuclide) is not dependent on the half-life of the product radionuclide or on the formation of a radioactive product. A comparison of the sensitivities obtained, counting the 0.402-MeV γ-photon of ^{75}Se (undergoing radioactive decay) or the 6.58-MeV prompt photon of ^{76}Se, favors the former by a factor of 10^5. For activation of selenium isotopes, this method does not presently compete with conventional neutron activation.

Neutron activations can also be performed using the intense, fast neutron flux of the reactor, if special techniques are employed to absorb thermal neutrons. Usually, activation with fast neutrons is accomplished using generators which supply a single beam of neutrons of certain defined energies in a flux that may reach 10^{10} neutrons/cm²/sec (152). By bombarding a tritiated titanium target with accelerated deuterons, 14-MeV neutrons are produced, the reaction being ^{3}H(d, n)^{4}He. Neutrons of this energy are suitable for (n, p), (n, α), (n, 2n), (n, n') and (n, f) reactions. Generally, the activation cross sections for 14-MeV neutrons are lower than those for the thermal (n, γ) reactions with the same target nuclides, but for certain elements better yields of a specific radionuclide are obtained through fast neutron reactions. For the same flux level, the detection limit for selenium is about the same using either fast or thermal neutron activation (150, pp. 59, 61). Since the limited life of target materials restricts fast neutron generators to intermittent operation, this activation technique is better suited to the production of short-lived radionuclides.

γ-Photon (18 to 27 MeV) activation has important applications, particularly for the activation of nuclides of low atomic number (153). The photon-induced reactions are of the (γ, γ'), (γ, p), and (γ, n) forms. The reactions forming nuclear isomers (γ, γ') has been used in selenium activation, in one instance (154) employing the high-energy bremsstrahlung from platinum or lead targets and, in another trial (155), the gamma radiation from a 100-kCi ^{60}Co source. A sensitivity approaching 10^{-5} g Se was reported for the ^{77}Se$(\gamma, \gamma')^{77m}$Se reaction with each of these activation conditions.

Activation can also be achieved through reactions involving high-energy particles. Debrun et al. (155a) have measured the short-lived radioisotopes produced by the ^{80}Se$(p, 2n)^{79m}$Br and ^{77}Se$(p, p')^{77m}$Se reactions. The usefulness of this mode of activation, or for that matter of γ-photon activation, in the determination of selenium in biological samples remains to be demonstrated.

C. SAMPLE PREPARATION

The preparation of samples for irradiation is dictated in some degree by the product isotope which has been selected for measurement. Generally, the photopeaks of short-lived isotopes (e.g., ^{77m}Se) will of necessity be measured at the irradiation facility, whereas samples to be analyzed for ^{75}Se (and possibly ^{81m}Se or ^{81}Se, depending on the distance-time factor) can be returned to the sender's laboratory for this determination. At the expense of the nondestructive advantage of NAA, some pretreatment of the sample may be necessary. Also, removal of interfering activities after irradiation is frequently recommended in order to attain the greatest sensitivity possible with the method.

Girardi (153) and Haller et al. (137) have defined some of the problems associated with sample preparation. Excessive manipulation is to be avoided, and it may be necessary to employ glove-box or clean-room techniques to minimize dust contamination. Any inhomogeneity of the sample must be considered in its preparation. Also, care should be taken not to contaminate the sample with trace elements in any of the reagents used (these may be present in greater amounts in the reagent than in the sample) or to lose the trace elements of interest by exchange or adsorption on container surfaces. High-moisture materials can be sampled by means of Lucite knives or borers; leaf tissue can be freeze-dried and homogenized using all plastic equipment.

Specific examples of pretreatments used in selenium NAA may be mentioned. Tissue and blood samples were dialyzed (126,145) to remove sodium before neutron activation. Wet or liquid samples may be irradiated under certain conditions, but drying samples before irradiation is usually recommended. In addition to reducing possible pressure buildup in the capsule during irradiation, the removal of water should reduce ^{19}O interference (145). Blood and tissue samples have been oven-dried at 80 to 90°C (145), tobacco products at 90°C for 12 hr (135), and kale powder at 90°C for 20 hr (147).

Pear, apple, raisin, and seed samples have been dehydrated by freeze drying for 2 days (137). Veveris et al. (156) powdered chicken tissue, after treatment with liquid N_2, and dried it to constant weight at 50°C with a loss of volatile selenium estimated at 7%.

For short-period (1-hr) irradiations in a thermal neutron flux of about 3×10^{13} neutrons/cm^2/sec (150, p. 62) or for irradiations of liquid samples in a flux not exceeding 5×10^{12} neutrons/cm^2/sec for a few days (150, p. 169), high-purity polyethylene vials are satisfactory as sample containers. Polyethylene containers have been frequently used in selenium neutron activation studies, and in some irradiations they have been further cleaned (distilled water or nitric acid) and protected against surface contamination (aluminum foil) during handling (137). The small amount of radioactive contamination in the vials after irradiation can be eliminated by transferring the irradiated sample to a fresh vial before photopeak measurement.

At high reactor temperatures (e.g., ≮80°C) or at higher flux conditions (10^{14} neutrons/cm^2/sec) organic materials may undergo appreciable radiation decomposition, even though dehydrated before irradiation (150, pp. 63, 170). High-purity quartz vials or aluminum containers are recommended for these irradiation conditions in order to contain the gases released by decomposition. In such cases it may be advisable to transfer the irradiated samples to fresh containers after irradiation. Heat production, leading to decomposition of the sample, is ascribed to large neutron capture cross sections in (n, γ) reactions or to (n, α) reactions (if the sample contains appreciable amounts of lithium or boron), the kinetic energy being almost entirely absorbed by the sample (150, p. 63). γ-Ray radiolysis due to the intense γ-ray flux in the reactor is also a contributing factor (150, pp. 64, 170). When sample decomposition has been found to be severe, low-temperature ashing, using a current of atomic oxygen (16), has been recommended as a part of sample preparation (see Section I.1.A).

To achieve greater sensitivity in measuring the radioselenium isotopes, several postirradiation procedures have been used. Bowen and Cawse (125), counting the β^- activity of ^{81}Se, used (a) a wet-ashing procedure with added carrier, followed by (b) distillation, (c) precipitation of elemental selenium with SO_2, (d) centrifugation, and (e) dissolution of the selenium precipitate, with repetition of steps c and d; the entire proecdure required 18 min. Maziere et al. (127) digested plasma and serum samples in a HNO_3-H_2SO_4 mixture, removed interfering ions on a column of Dowex 2, eluted selenium with $6N$ HCl, formed the diethyldithiocarbamate-selenium complex, and then extracted the complex with CCl_4. Samsahl (157) also used a wet digestion for irradiated animal tissue samples, and followed this step with a distillation of selenium and its separation on an anion-exchange column. Dry distillation of selenium in a Leco furnace has been recommended by Conrad and Kenna (158); the radioselenium is collected in a quartz tube packed with glass wool. When separation of the selenium isotopes is not performed, a period of decay is allowed for the disappearance of shorter-lived, interfering activities and the emergence of the radioselenium photo-peak(s) against the background. A decay period of 20 to 50 days has been used in ^{75}Se measurements (123,137,147,159). In tobacco and tobacco product irradiations, Nadkarni and Ehmann (135) cleaned the exterior of the quartz ampoules in aqua regia after activation and before ^{75}Se measurement; to prevent the escape of any volatile selenium products, the ampoules were not opened.

D. FACTORS AFFECTING DETECTION LIMITS

Detection limit values (μg Se) have been reported as follows: 2400 (144) for ^{73m}Se or ^{73}Se; $\sim 10^{-3}$ (127), 0.01 (159), 0.02 (138,145) and 0.41 (149) for ^{75}Se; 0.005 (131), 0.013 (149), 0.04 (129), 0.05 (156), 0.10 (145,160), 14 (152), and 50 (163) for ^{77m}Se; 0.07 (161) for ^{77}Se; 10 (162) for ^{79m}Se; 0.021 (149) and 190 (144) for ^{81m}Se; $\sim 10^{-2}$ (145), 0.05 (163), and 0.23 (149) for ^{81}Se; and 2 (163) for ^{83}Se. It should be noted that the high values reported by Perdijon (144) were normalized for a flux of 10^9 14-MeV neutrons/cm^2/sec. The wide range of these values for a given selenium radionuclide apparently reflects the sample size, the neutron energies and flux used to achieve activation, the irradiation period, and the fact that chemical separations were or were not employed to remove interferences. When an appreciable amount of time is involved in transportation and/or chemical separations after irradiation, the sensitivities achieved with the short-lived radioisotopes can be markedly affected.

The sensitivity of the nondestructive method in analysis for selenium (as well as for other elements) is dependent on the abundance and the activation cross section of the nuclide producing the radioisotope whose photopeak(s) has been chosen for measurement after activation. The natural abundance of the selenium isotopes, their half-lives, and their activation cross sections, as recently revised (164), are shown in Table II. Cross-section values have been considered a major source of error in calculating radionuclide activity induced through neutron activation (153). Also, a certain amount of error can be introduced if it is assumed that all of the neutrons in the reactor flux are thermal. Values for nuclide abundance, on the other hand, are known with good precision (165), and values for half-lives also are generally well established.

TABLE II[a]

Isotope	Natural abundance (%)	Half-life	Thermal neutron cross section (barns)
^{70}Se		44 min	
^{71}Se		5 min	
^{72}Se		8.4 days	
^{73}Se		7.1 hr	
^{74}Se	0.87		55 ± 5
^{75}Se		120.4 days	
^{76}Se	9.02		21 ± 2
^{77m}Se		17.5 sec	
^{77}Se	7.58		42 ± 4
^{78}Se	23.52		0.33 ± 0.04 ^{79m}Se
			0.20 ± 0.106 ^{79}Se
^{79m}Se		3.9 min	
^{79}Se		6.5×10^4 years	
^{80}Se	49.82		0.08 ± 0.01 ^{81m}Se
			0.23 ^{81}Se
^{81m}Se		57 min	
^{81}Se		18.6 min	
^{82}Se	9.19		0.040 ± 0.010 ^{83m}Se
			0.006 ± 0.001 ^{83}Se
^{83m}Se		70 sec	
^{83}Se		23 min	
^{84}Se		3.3 min	
^{85}Se		39 sec	

[a] The cross-section value for the reaction ^{80}Se (n, γ)^{81}Se was taken from Leddicotte and Reynolds (163).

As discussed earlier, reactions induced by 14-MeV neutrons are more varied than those induced by thermal neutrons, but thermal neutron activation seems to offer greater advantages in terms of sensitivity, freedom from interference, and simplicity of use (153). Since the induced radioactivity is proportional to the flux (120, p. 8), it would seem advisable to use a high neutron flux. A very high flux (10^{14} neutrons/cm²/sec), however, can introduce errors associated with second-order interference (166). In the presence of interfering reactions the choice of irradiation time appears to be one way of achieving greater selectivity for the radionuclide of immediate interest. Isenhour and Morrison (167) have provided a computer program to optimize time of irradiation (and decay) for analysis of any element in a complex mixture. Notwithstanding the consideration of interferences, selenium nuclides, except for ^{74}Se, can conveniently be irradiated in the reactor for the period of time required to achieve saturation. The half-life of ^{75}Se makes its saturation measurement in biological materials impractical, and generally irradiation periods of hours to several days are used for this radioisotope. The limited life of targets used in producing a beam of 14-MeV neutrons makes saturation activation of the other selenium nuclides, except for those with half-lives of a few minutes or less, infeasible.

A number of interfering nuclear reactions are possible, depending on the matrix composition of the material being irradiated. Primary, secondary, second-order, and competing reactions have been discussed by Koch (117, p. 9) and Schulze (150, p. 11). Second-order interferences have been further detailed by Ricci and Dyer (166). The following matrix elements have been cited for possible interference contributions in selenium determinations through radioactivation: ^{24}Na (128,135,168); ^{42}K (136,147); ^{41}Ar (123,149); ^{32}P, the bremsstrahlung of its 1.7-MeV β^--particle being most intense at energies <500 keV and providing a high background throughout the energy range of many of the gamma photopeaks of the selenium radio-nuclides (119,137,156); ^{198}Au (147) and ^{203}Hg (137,147,159) for their respective interferences in measuring the 401-keV and 260- to 280-keV photopeaks of ^{75}Se; ^{38}Cl or ^{13}N (128) and ^{185}W or ^{197}Au (154) for interferences in measurements of ^{77m}Se; ^{75}Se (149) and ^{180}Ta (144) for interferences in measurements of ^{81m}Se; and ^{124}Sb, ^{131}Te, ^{38}Cl, or ^{80}Br for interference in measurements of ^{81}Se (125). It is fortunate, considering the nature of biological samples, that carbon, hydrogen, and nitrogen are not, for all practical purposes, activated by thermal neutrons and that oxygen is not appreciably activated (169), although the production of ^{19}O is thought to limit the accuracy and sensitivity of ^{77m}Se measurement (145).

Criteria for selection of the photopeak to be used in computations have been provided by Yule (149). Representative photopeak selections which have been used in selenium NAA measurements are the following (in keV, except as otherwise noted): ^{75m}Se or ^{73}Se—*510* (144); ^{75}Se—*136* (138,158, 159), *260* (146), *264* (159,170), *265* (149), *265* to *280* (137,158), *401* (135), *402* (151), and *410* (127); ^{76}Se—*6.58* MeV prompt γ (151); ^{77m}Se—*130* to *220* (154), *150* to *170* (171), *160* (128,144,149,156,172,173), and *162* (152); ^{81m}Se—*103* (144) and *104* (149); and ^{81}Se—*103* (158,173), *280* (149), and *1.60* MeV β^- (125). The improved resolution of γ-ray photopeaks obtained with Ge(Li) semiconductor detectors has been a great aid in trace element analysis in biological materials, particularly when neutron activation is not followed by radiochemical separations (129,147,174,175).

Koch (117, p. 14) has discussed the analytical and nuclear sources of error in neutron activation measurements. Included in the nuclear sources of error are self-shielding for thermal (or resonance neutrons) in the sample or in the comparator or monitor samples, and flux gradients in the irradiation sample. Anders (172) has pointed out the need to prepare standards that are of approximately the same weight and geometry as the unknowns, particularly when the element of interest has a high cross section; otherwise, self-shielding can introduce errors estimated to be $> 100\%$. The attenuation process occurring with fast neutrons and common to all fluxes has been studied in some detail (176–178).

Flux monitors are frequently used to compare the neutron flux during experimental conditions with that prevailing under reference conditions. Cited here are examples using copper foils (137), titanium foil (159), and cobalt wire (179), the composition of the latter being 1% Co, 9% Mg, and 90% Al. Variations in thermal flux over the sample of about 5% have been noted (137,146,159). To compensate for flux variations, irradiations have been performed using rotating specimen racks. Examples of this technique have been reported by Yule (180) for thermal neutron activations, and by Lundgren and Nargolwalla (181), who developed a dual-sample, biaxial rotating assembly for use in 14-MeV NAA.

Selenium assays by neutron activation have an error estimated to be 10% by Rancitelli et al. (123) and 6% by Maziere et al. (127). In a comparison of activation values with colorimetric assay values a difference of 12% was reported by Lambert et al. (128). The analysis of samples of standard kale showed poor interlaboratory precision for selenium (150, p. 106), with results ranging from 0.046 to 0.62 ppm (average 0.15). Fluorometric analysis of the same material gave a value of 0.14 ppm Se. In a separate study with standard kale, Nadkarni and Ehmann (147) obtained selenium values ranging from 0.14 to 0.17 ppm (average, 0.15) for five runs.

2. X-Ray Fluorescence Analysis

X-ray fluorescence analysis can be used nondestructively to determine selenium in biological samples. At present, however, this method does not offer the sensitivity of measurement associated with NAA, fluorescence spectrophotometry, or certain other chemical methods which have been discussed in this chapter. Handley (182) determined selenium in plant material with an estimated probable error of $< 10\%$ for samples containing 10 ppm Se (collecting a minimum of 10,000 counts). He dried the plant samples for 24 hr at 60°C, ground the dried material to pass a 40-mesh screen, and then pressed it into a disk (1 in. in diameter), using a packing pressure of 2000 to 16,000 psi. Counts for the Se $K\alpha$ peak were corrected for scattered primary radiation due to the matrix, as well as for internal shielding effects.

Lisanti (183), Olsen and Shell (184), and Goldman and Beckman (185), respectively, have also determined selenium in plant samples, organic compounds, and tissue samples of medical interest, using the X-ray fluorescence technique.

V. METHODS

The preceding discussion has pointed out the many methods now available for selenium analysis. Although neutron activation analysis is becoming increasingly popular, its use is restricted by the requirement for sophisticated equipment. At present, therefore, methods based on fluorometric measurement of the piazselenol of 2,3-diaminonaphthalene are most commonly used. These are relatively simple, sensitive, and accurate. They have been applied with a number of modifications. Two of these are described below, one using a wet digestion without separation of the selenium before its reaction with 2,3-diaminonaphthalene, the other employing an oxygen flask combustion and coprecipitation of the selenium with arsenic before its measurement. A method for the gravimetric analysis of samples of very high selenium content is also described.

1. Method A

The following is the method of Watkinson (23) with minor modifications. It has been accepted, essentially as described, as Official First Action for selenium in plants by the Association of Official Analytical Chemists (186), and it has been found satisfactory by the authors for other biological samples. As described, its lower limit of reliability is about 0.02 μg Se (0.02 ppm in a 1-g sample), but this can be somewhat improved by using larger samples and by making certain modifications, provided appropriate reagent blank determinations are also made.

REAGENTS. 1. *EDTA–NH₂OH solution.* Slowly add approximately 5N NH₄OH with mixing to 4 g ethylenediaminetetraacetic acid, acid form (EDTA), suspended in about 20 ml of water until the EDTA dissolves. Some excess of NH₄OH is not harmful. Add about 100 ml water and 12 g NH₂OH·HCl, stir to dissolve, and make to 500 ml with water.

2. *Cresol red indicator.* Add 1 drop of 5% NaOH to 20 mg cresol red (o-cresolsulfonphthalein), and mix to wet. Make to 100 ml with water.

3. *Standard selenite solution.* Add about 10 ml of concentrated nitric acid to 40.0 mg of elemental selenium (99+% pure), and warm to dissolve. Cool to room temperature, make to 100 ml with water, and mix. Transfer 1.0 ml to a 30-ml

micro Kjeldahl flask, and add 1 glass bead and 2 ml of 70% $HClO_4$. Boil gently until $HClO_4$ fumes and then for an additional 2 to 3 min. Cool, add 1 ml of water and 1 ml of $1 + 4$ $HCl:H_2O$ solution, and heat in a boiling-water bath for 30 min. After cooling, make to 1000 ml in approximately $1N$ HCl. This solution is stable at room temperature for several months when stored in all-glass containers. However, contamination of the solution by dust or traces of organic material can result in a reduction of the selenite, and hence the preparation of a fresh solution every 10 to 14 days is advisable.

4. *DAN solution. (Prepare in a semidarkened room or in a room equipped with a yellow light.)* Add a few drops of approximately $0.1N$ HCl to 100 mg of 2,3-diaminonaphthalene (DAN), and mix with a stirring rod to wet the reagent. Add 100 ml of the approximately $0.1N$ HCl, and place in a 50°C water bath for 15 min. After cooling to room temperature, extract twice with 10 ml of cyclohexane, discarding the extracts. Filter through coarse filter paper saturated with water, and use immediately. *The solution must be prepared fresh for each set of determinations.*

All reagents used should be analytical grade.

APPARATUS

Digestion unit. A micro Kjeldahl digestion unit with a glass fume duct attached to a glass aspirator. Use in fume hood.

Fluorometer. Should be capable of illuminating sample at about 370 nm and measuring fluoresced light at about 525 nm. A Turner model 110 fluorometer equipped with a GE F4T4 lamp, a No. 7-60 primary filter, and a No. 58 secondary filter has been found very satisfactory.

Water bath. A 50°C bath, covered or in a dark room.

Procedure. *The following procedure is written for samples containing less than about 2 ppm of selenium. Modifications to provide for the measurement of larger amounts are discussed under "Comments" below, as are also provisions for overcoming some difficulties in obtaining representative samples.*

Weigh not more than 1 g of dry tissue or 2.5 g of wet tissue or measure not more than 3 ml of blood or other fluid into a 30 ml borosilicate micro Kjeldahl flask. Add 10 ml of concentrated HNO_3 and one glass bead. At this point, allowing the mixture to stand for 3 or 4 hr or overnight reduces the possibility of foaming during the early stages of the digestion, but it is not necessary. Place the flask on the digestion rack and heat at low flame until the initial reaction and danger of foaming are past. Then increase the heat until the HNO_3 condenses in the lower part of the neck of the flask and continue heating for 10 to 15 min. Turn off the burners, cool for a few minutes, and then add 2 ml of 70% $HClO_4$, washing down the sides of the flask. Continue heating until $HClO_4$ fumes appear, drawing off the fumes in the fume duct. At this point, care should be taken to prevent excessive charring of the mixture. A dark, straw-brown colored digest is acceptable, but if the mixture blackens the heat should be turned off, the digest cooled, and a few milliliters of HNO_3 added. The digestion should be continued at a gentle boil for an additional 15 min to ensure maximum destruction of the organic matter and the elimination of HNO_3. Turn off the heat, cool, add 1 ml of water and 1 ml of 1 + 4 $HCl:H_2O$ solution, and heat in a boiling-water bath for 30 min. Remove from the bath and allow to cool.

Prepare a blank containing 2 ml of $HClO_4$ and a standard containing 1.0 ml of the standard selenite solution plus 2 ml of $HClO_4$. To the blank, standard, and unknowns add 5 ml of the $EDTA-NH_2OH$ solution and 2 drops of the cresol red indicator. Then add $5N$ NH_4OH with mixing to a yellow color and 1 + 4 $HCl:H_2O$ to an orange-pink color. *From this point on, perform all manipulations in a semidarkened room or a room with a yellow light only.*

Prepare the DAN solution, add 5 ml to each flask, and mix. Then add approximately $0.1N$ HCl up to the neck of each flask, mix by swirling, and place the flasks in the 50° water bath for 25 min, mixing again at about 10 min. Remove from the bath and cool to room temperature in a pan of cold water.

Transfer the solutions into 125-ml separatory funnels containing 10.0 ml of cyclohexane (avoid pouring the glass bead into the funnel), using 5 to 10 ml of water to assist in the transfer. Shake the separatory funnels vigorously for at least 30 sec, allow to stand for a few minutes, draw off the aqueous layer, and discard. Wash the cyclohexane layer twice by shaking with 25-ml portions of approximately $0.1N$ HCl, discarding the washings. (A VirTis Extractomatic shaker with 100-ml separators may be used in place of the separatory funnels. In this case, extract with cyclohexane for 5 min and wash the extract twice by shaking for 1 min with $1.0N$ HCl.)

Transfer the cyclohexane layers to 12-ml centrifuge tubes, and centrifuge at a moderate speed for about 1 min. Then decant the cyclohexane into fluorometer tubes, and read all tubes against the blank set at zero. Calculate the selenium content as follows:

$$\text{ppm Se} = \frac{0.4 \times \text{sample reading}}{\text{standard reading} \times \text{sample weight (g)}}$$

Comments. With some types of materials, it may be necessary to start with more than a few grams in order to obtain a representative sample. In this circumstance, a larger portion can be heated on a hot plate with 5 vols of concentrated HNO_3 until the volume is reduced by about one half; after cooling and allowing the fat to rise to the surface, the solution is made to volume (at the bottom of the fat layer) with HNO_3. A suitable aliquot is then removed for digestion and completion of the analysis.

Samples containing more than 2 ppm Se may be digested in the usual manner, reduced with HCl, and then made to an appropriate volume. A suitable aliquot with 2 ml of 70% $HClO_4$ added is then treated with the $EDTA$–NH_2OH solution, and the analysis is completed as described above.

Decalin (decahydronaphthalene) may be used in place of cyclohexane to extract the piazselenol. It has the advantage of giving some increased sensitivity and is also less volatile. It is, however, a more expensive reagent, and it produces a somewhat higher and more variable blank.

When samples of very low selenium content are being analyzed, the analysis should be carried out from the beginning with reagents only to obtain a reagent blank. Experience to date indicates that with analytical-grade reagents this blank is negligible; nevertheless it should be determined occasionally.

This method has been used successfully on a wide variety of biological samples. It has also been found satisfactory for the analysis of animal diets, with the exception that highly mineralized supplements occasionally give problems because of the precipitation of iron during the neutralization procedure. Coprecipitation of the selenium with arsenic, as described below, should be used in these cases.

The accuracy of the method might be slightly improved by using standards containing about the same amounts of selenium as those found in the unknowns. However, the intensity of fluorescence is essentially proportional to the concentration of selenium present, up to about 0.5 μg of the element, so that a single standard is sufficient for most work.

2. Method B

The following is a modification of the method of Allaway and Cary (11). It uses oxygen flask combustion to oxidize the selenium and to destroy the organic matter, and it separates the selenium from interfering materials by arsenic coprecipitation. Its sensitivity, precision, and accuracy are about the same as those of method A.

REAGENTS. 1. *EDTA–NH₂OH solution*. See method A.
2. *Cresol red indicator*. See method A.
3. *Standard selenite solution*. See method A.
4. *DAN solution*. See method A.

5. *Arsenic solution.* Dissolve 0.25 g As_2O_3 (oxidimetric standard) and 10 pellets of reagent-grade NaOH in water and dilute to 200 ml.

6. *Reducing solution.* Mix 2 parts of concentrated HCl and 1 part of 30% hypophosphorus acid, preparing daily as needed.

APPARATUS

Pelletizer. A press with smooth metal dies for pelletizing dry, powdered material into pellets $\frac{1}{2}$ in. in diameter.

Combustion flask. A 5-liter-capacity Schöniger flask as modified by Thomas-Gutenmann-Lisk (Arthur H. Thomas Co., Philadelphia) is used. Balloons, used as safety valves, are rinsed with 1:1 HNO_3, 1:10 HCl, and water and then stored submerged in water. They should be tested for defects with compressed air before use.

Fluorometer. See method A.

Water bath. See method A.

Magnetic stirrer.

Procedure. Weigh out and pelletize 1 g (or less if selenium content is over 0.5 ppm) of finely ground dried material (see "Comments" below for suggestions for wet samples). Place the pellet on a filter paper with a $\frac{1}{8}$-in. fuse $2\frac{1}{2}$ in. long in the platinum basket of the combustion flask, allowing the fuse to project upward and slightly out from the basket.

Measure 25 ml of 0.5N HCl into the flask after fitting it with a balloon. Place a 1-in. Teflon-coated stirrer in the flask. Holding the stopper with its attached platinum basket in readiness, direct a fairly rapid stream of oxygen into the flask so as to replace all of the air. Immediately ignite the fuse; when the flame is about $1\frac{1}{2}$ in. from the pellet, insert the sample, basket, and stopper into the flask. While the sample is burning, the balloon will expand to about the size of a grapefruit. On completion of the combustion, stir the contents rapidly with the magnetic stirrer until the atmosphere in the flask is clear.

Transfer the contents of the flask into a 250-ml beaker, washing it and the balloon with a minimum of water. If a cinder forms in the platinum basket, it should be crushed with a stirring rod and washed into the beaker. Filter the solution (S & S white ribbon No. 589 paper) into a second beaker with minimal washing, and evaporate without boiling to about 50 ml.

Add 5 ml of the arsenic solution to the filtrate, mix, and then add 60 ml of the reducing solution. Boil gently with occasional stirring for about 5 min. A black precipitate of elemental selenium and arsenic forms. Draw off the liquid with an immersion tube with a fritted disk, washing the precipitate with a small amount of water. Disconnect the filter, add 7 ml of 1:1 HNO_3–H_2O, and dissolve the precipitate. Using an atomizer bulb, force any liquid in the stem of the filter back into the beaker, add 2 ml of 1:1 HNO_3–H_2O into the stem of the filter, and force it back through into the beaker, repeating this twice. Force 5 ml of water back through the filter into the beaker, and then wash the filter with a small amount of water. After covering the beaker with a watch glass, boil gently for about 2 min, without allowing the volume to go below 10 ml. Cool the solution and transfer it to a 50-ml volumetric flask, washing it in 5 ml of 0.5N HCl and 5 ml of water. Add 2 ml of 70% $HClO_4$ and proceed as in method A, starting with "Prepare a blank containing . . .," substituting 50 ml volumetric flasks for the micro Kjeldahl flasks.

Comments. For samples of very low selenium content, greater accuracy is obtained by burning successive 1-g portions over the same charge of absorbing solution, provided the atmosphere in the flask is clear after each combustion.

Milk, blood, and urine may be analyzed without drying by pipetting up to 0.5 ml onto a pellet of filter paper. Larger samples may be dried *in vacuo* on filter paper.

A reagent blank should be determined by burning an appropriate size of filter paper and completing the analysis as described.

3. Method C

When large amounts of selenium are present in samples or when pure selenium-containing compounds are being analyzed, it may be more convenient to use a gravimetric procedure. The method described here is a modified version of the one developed by Robinson et al. (15).

REAGENTS. 1. H_2SO_4 *solution.* Dilute 2 parts of concentrated H_2SO_4 with 1 part of water.
2. Br_2–*HBr solution.* Add 10 ml of Br_2 to 1 liter of HBr (40 to 48%).
3. $NH_2OH \cdot HCl$ *solution*, 10% (w/v).

APPARATUS
Distilling apparatus. See Figure 1.

Procedure. Weigh out material sufficient to contain at least 0.5 mg Se and transfer it to a digestion flask. Digest the material as described under method A. If more than about 1 g of sample is required, a 100-ml Kjeldahl flask, and additional amounts of the acids, should be used. Transfer the digest to the distillation apparatus shown in Figure 1, completing the transfer by rinsing with 75 ml of the sulfuric acid solution. Add 25 ml of the Br_2–HBr solution through the thistle tube, and distill the mixture into a receiving

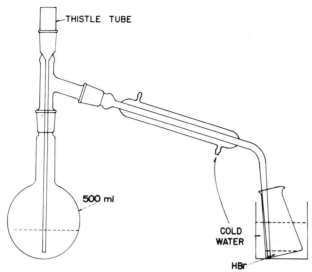

Figure 1. Apparatus for selenium tetrabromide distillation.

flask containing 5 ml of 40 to 48% HBr, collecting about 75 ml of distillate. Some free bromine should distill in the beginning, indicating an excess of this reagent.

Remove the receiving flask, washing the liquid on the tip of the condenser into it with a few milliliters of water. Bubble sulfur dioxide through the solution until it is colorless, add 1 ml of the $NH_2OH \cdot HCl$ reagent, cover the flask with a watch glass, and heat it on the steam bath for 30 min. After cooling the solution, collect the precipitated selenium on a Gooch crucible having about a 0.5-cm pad of acid-washed asbestos fiber. The crucible should have been previously dried and weighed, and the asbestos pad on it should not lose weight when washed with concentrated hydrobromic acid. (Millipore filters may also be used to collect the precipitate.) Complete the transfer with the aid of a rubber policeman, washing with water until the filtrate is free of bromide. Dry the filter and its contents to constant weight at 110°C.

Comments. Elemental selenium has a tendency to stick to glass and to creep back onto the surface of the liquid during the transfer. These tendencies introduce some error into the method, but with care it can give very satisfactory results. With amounts of selenium over about 20 mg, too rapid a reduction with the sulfur dioxide may result in the formation of clumps of the element, which may occlude impurities. The errors due to this have not, however, been found to be serious.

References

1. J. H. Watkinson, in *Selenium in Biomedicine*, O. H. Muth, J. E. Oldfield, and P. H. Weswig, Eds., AVI, Westport, Conn., 1967, p. 97.
2. K. W. Franke and E. P. Painter, *Cereal Chem.*, *15*, 1 (1938).
3. A. L. Moxon, *S. Dakota State Coll. Agr. Expt. Sta. Bull.*, *311*, 1936.
4. A. L. Moxon and M. Rhian, *Proc. S. Dakota Acad. Sci.*, *18*, 20 (1938).
5. O. A. Beath, H. F. Eppson, and C. S. Gilbert, *Wyoming Agr. Expt. Sta. Bull.*, *206*, 1935.
6. O. A. Beath, H. F. Eppson, and C. S. Gilbert, *J. Am. Pharm. Assoc*, *26*, 394 (1937).
7. C. J. Asher, C. S. Evans, and C. M. Johnson, *Australian J. Biol. Sci.*, *20*, 737 (1967).
8. C. F. Ehlig, W. H. Allaway, E. E. Cary, and J. Kubota, *Agron. J.*, *60*, 43 (1968).
9. G. Gissel-Nielson, *Plant Soil*, *32*, 242 (1970).
10. I. Rosenfeld and O. A. Beath, *Selenium*, Academic Press, New York, 1964.
11. W. H. Allaway and E. E. Cary, *Anal. Chem.*, *36*, 1359 (1964).
12. H. E. Ganther, *Biochemistry*, *5*, 1089 (1966).
13. A. G. Faulkner, E. C. Knoblock, and W. C. Purdy, *Clin. Chem.*, *7*, 22 (1961).
14. W. O. Robinson, *J. Assoc. Offic. Agr. Chemists*, *16*, 423 (1933).
15. W. O. Robinson, H. C. Dudley, K. T. Williams, and H. G. Byers, *Ind. Eng. Chem., Anal. Ed.*, *6*, 274 (1934).
16. C. E. Gleit and W. D. Holland, *Anal. Chem.*, *34*, 1454 (1962).
17. C. E. Mulford, *At. Absorption Newsletter*, *5*, 135 (1966).
18. W. H. Gutenmann and D. J. Lisk, *J. Agr. Food Chem.*, *9*, 488 (1961).
19. A. K. Klein, *J. Assoc. Offic. Agr. Chemists*, *24*, 363 (1941).
20. W. B. Dye, E. Bretthauer, H. J. Seim, and C. Blincoe, *Anal. Chem.*, *35*, 1687 (1963).
21. P. Cukor, J. Walzcyk, and P. F. Lott, *Anal. Chim. Acta*, *30*, 473 (1964).
22. H. H. Taussky, A. Washington, E. Zubillaga, and A. T. Milhorat, *Microchem. J.*, *10*, 470 (1966).
23. J. H. Watkinson, *Anal. Chem.*, *38*, 92 (1966).
24. H. Johnson, *Environ. Sci. Technol.*, *4*, 850 (1970).
25. K. T. Williams and H. W. Lakin, *Ind. Eng. Chem., Anal. Ed.*, *7*, 409 (1935).
26. A. L. Curl and R. A. Osborn, *J. Assoc. Offic. Agr. Chemists*, *21*, 228 (1938).
27. R. A. Osborn, *J. Assoc. Offic. Agr. Chemists*, *19*, 236 (1936).
28. R. A. Osborn, *J. Assoc. Offic. Agr. Chemists*, *22*, 346 (1939).
29. Association of Official Agricultural Chemists, *Official and Tentative Methods of Analysis*, 5th ed., Washington, D. C., 1940.
30. C. A. Grant, B. Thafvelin, and R. Christella, *Acta Pharmacol. Toxicol.*, *18*, 285 (1961).
31. O. E. Olson, *J. Assoc. Offic. Anal. Chemists*, *52*, 627 (1969).
32. T. T. Gorsuch, *Analyst*, *84*, 135 (1959).
33. I. Rosenfeld and H. F. Eppson, *Wyoming Agr. Expt. Sta. Bull. 414*, 1964, p. 53.
34. D. N. Fogg and N. T. Wilkinson, *Analyst*, *81*, 525 (1956).
35. J. L. Down and T. T. Gorsuch, *Analyst*, *92*, 398 (1967).
36. F. B. Cousins, *Australian J. Exptl. Biol. Med. Sci.*, *38*, 11 (1960).
37. J. C. Lane, *Irish J. Agr. Res.*, *5*, 177 (1966).
38. C. L. Newberry and G. D. Christian, *J. Assoc. Offic. Agr. Chemists*, *48*, 322 (1965).
39. J. H. Watkinson, *Anal. Chem.*, *32*, 981 (1960).
40. A. B. Grant, *New Zealand J. Sci.*, *6*, 577 (1963).
41. G. D. Christian, E. C. Knoblock, and W. C. Purdy, *J. Assoc. Offic. Agr. Chemists*, *48*, 877 (1965).
42. R. Handley and C. M. Johnson, *Anal. Chem.*, *31*, 2105 (1959).

43. W. J. Kelleher and M. J. Johnson, *Anal. Chem.*, *33*, 1429 (1961).
44. I. Hoffman, R. J. Westerby, and M. Hidiroglou, *J. Assoc. Offic. Anal. Chemists, 51*, 1039 (1968).
45. J. S. McNulty, *Anal. Chem.*, *19*, 809 (1947).
46. L. M. Cummins, J. L. Martin, and D. D. Maag, *Anal. Chem.*, *37*, 430 (1965).
47. L. M. Cummins, J. L. Martin, G. W. Maag, and D. D. Maag, *Anal. Chem.*, *36*, 382 (1964).
48. R. C. Ewan, C. A. Baumann, and A. L. Pope, *J. Agr. Food Chem.*, *16*, 212 (1968).
49. R. J. Hall and P. L. Gupta, *Analyst*, *94*, 292 (1969).
49a. O. E. Olson, E. J. Novacek, E. I. Whitehead and I. S. Palmer, *Phytochemistry, 9*, 1181 (1970).
50. R. Mozingo, D. E. Wolf, S. A. Harvis, and K. Falkers, *J. Am. Chem. Soc.*, *65*, 1013 (1943).
51. G. E. Wiseman and E. S. Gould, *J. Am. Chem. Soc.*, *76*, 1706 (1954).
52. L. J. Throop, *Anal. Chem.*, *32*, 1807 (1960).
53. O. E. Olson, B. M. Schulte, E. I. Whitehead, and A. W. Halverson, *J. Agr. Food Chem.*, *11*, 531 (1963).
54. C. W. Bonhorst and J. J. Mattice, *Anal. Chem.*, *31*, 2106 (1959).
55. R. E. D. Clark, *Analyst*, *83*, 396 (1958).
56. R. E. D. Clark, *Analyst*, *82*, 182 (1957).
57. C. L. Luke, *Anal. Chem.*, *31*, 572 (1959).
58. H. Schumann and W. Koelling, *Z. Chem.*, *1*, 371 (1961); from *Chem. Abstr.*, *57*, 3076c (1962).
59. P. F. Lott, P. Cukor, G. Moriber, and J. Solga, *Anal. Chem.*, *35*, 1159 (1963).
60. A. Cousen, *J. Soc. Glass Technol.*, *7*, 303 (1923); from *Chem. Abstr.*, *18*, 1735 (1924).
61. American Public Health Association, *Standard Methods for the Examination of Water, Sewage, and Industrial Wastes*, 10th ed., New York, 1955, p. 179.
62. K. W. Franke, R. Burris, and R. S. Hutton, *Ind. Eng. Chem.*, *Anal. Ed.*, *8*, 435 (1936).
63. H. Weisz, *Microanalysis by the Ring-Oven Technqiue*, 2nd ed., Pergamon Press, Oxford, 1961.
64. S. D. Biswas and A. K. Dey, *Analyst*, *90*, 56 (1965).
65. P. W. West and C. Cimerman, *Anal. Chem.*, *36*, 2013 (1964).
66. M. J. Horn, *Ind. Eng. Chem.*, *Anal. Ed.*, *6*, 34 (1934).
67. J. Hoste, *Anal. Chim. Acta*, *2*, 402 (1948).
68. C. A. Parker and L. G. Harvey, *Analyst*, *86*, 54 (1961).
69. J. Hoste and J. Gillis, *Anal. Chim. Acta*, *12*, 158 (1955).
70. K. L. Cheng, *Anal. Chem.*, *28*, 1738 (1956).
71. T. Danzuka and K. Ueno, *Anal. Chem.*, *30*, 1370 (1958).
72. J. B. Magin, Jr., L. L. Thatcher, S. Rettig, and H. Levine, *J. Am. Water Works Assoc.*, *52*, 1199 (1960).
73. B. G. Russell, W. V. Lubbe, A. Wilson, E. Jones, J. D. Taylor, and T. W. Steele, *Talanta*, *14*, 957 (1967).
74. R. E. Stanton and A. J. McDonald, *Analyst*, *90*, 497 (1965).
75. C. A. Parker and L. G. Harvey, *Analyst*, *87*, 558 (1962).
76. H. Ariyoshi, M. Kiniwa, and K. Toei, *Talanta*, *5*, 112 (1960).
77. D. Demeyere and J. Hoste, *Anal. Chim. Acta*, *27*, 288 (1962).
78. E. Sawicki, *Anal. Chem.*, *29*, 1376 (1957).
79. F. L. Chan, *Talanta*, *11*, 1019 (1964).
80. B. C. Bera and M. M. Chakrabartty, *Analyst*, *93*, 50 (1968).
81. M. S. Cresser and T. S. West, *Analyst*, *93*, 595 (1968).

82. T. Kawashima and M. Tanaka, *Anal. Chim. Acta*, *40*, 137 (1968).
83. P. W. West and T. V. Ramakrishna, *Anal. Chem.*, *40*, 966 (1968).
83a. B. D. Mesman and H. A. Doppelmayr, *Anal. Chem.*, *43*, 1346 (1971).
84. G. F. Kirkbright and J. H. Yoe, *Anal. Chem.*, *35*, 808 (1963).
85. R. L. Osburn, A. D. Shendrikar, and P. W. West, *Anal. Chem.*, *43*, 594 (1971).
86. P. P. Naidu and G. G. Rao, *Talanta*, *17*, 817 (1970).
87. R. F. Burk, Jr., W. N. Pearson, R. P. Wood, and F. Viteri, *Am. J. Clin. Nutrition*, *20*, 723 (1967).
88. J. P. Molloy, *Irish J. Agr. Res.*, *6*, 133 (1967).
89. J. E. Allan, *Spectrochim. Acta*, *18*, 259 (1962).
90. C. S. Rann and A. N. Hambly, *Anal. Chim. Acta*, *32*, 346 (1965).
90a. G. F. Kirkbright, M. Sargent, and T. S. West, *At. Absorption Newsletter*, *8*, 34 (1969).
90b. P. Johns, *Spectrovision*, *24*, 6 (1970).
91. C. L. Chakrabati, *Anal. Chim. Acta*, *42*, 379 (1968).
92. H. L. Kahn and J. E. Schallis, *At. Absorption Newsletter*, *7*, 5 (1968).
93. H. L. Kahn, G. E. Peterson, and J. E. Schallis, *At. Absorption Newsletter*, *7*, 35 (1968).
93a. J. D. Kerber and F. J. Fernandez, *At. Absorption Newsletter*, *10*, 78 (1971).
93b. F. J. Fernandez and D. C. Manning, *At. Absorption Newsletter*, *10*, 86 (1971).
93c. R. B. Baird, S. Pourian, and S. M. Gabrielian, *Anal. Chem.*, *44*, 1887 (1972).
93d. H. K. Y. Lau and P. F. Lott, *Talanta*, *18*, 303 (1971).
94. S. Nakashima and K. Toei, *Talanta*, *15*, 1475 (1968).
95. Y. Shimoishi and K. Toei, *Talanta*, *17*, 165 (1970).
96. L. Schwaer and D. Suchy, *Collection Czech. Chem. Commun.*, *7*, 25 (1935).
97. G. D. Christian and E. C. Knoblock, *Anal. Chem.*, *35*, 1128 (1963).
98. G. D. Christian, E. C. Knoblock, and W. C. Purdy, *Anal. Chem.*, *37*, 425 (1965).
99. H. G. Offner and E. F. Wituck, *J. Am. Water Works Assoc.*, *60*, 947 (1968).
100. D. A. Griffin, *Anal. Chem.*, *41*, 462 (1969).
101. J. F. Norris and H. Fay, *Am. Chem. J.*, *18*, 703 (1896).
102. W. C. Coleman and C. R. McCrosky, *Ind. Eng. Chem., Anal. Ed.*, *9*, 431 (1937).
103. H. W. Hillebrand and G. E. F. Lundell, *Applied Inorganic Analysis*, Wiley, New York, 1929, p. 266.
104. F. Burriel-Marti, F. Lucena-Conde, and S. Arribas-Jimeno, *Anal. Chim. Acta*, *10*, 301 (1954).
105. P. P. Naidu and G. G. Rao, *Anal. Chem.*, *43*, 281 (1971).
106. F. Solymosi, *Chem. Anal.*, *52*, 42 (1963).
107. G. S. Deshmukh and M. G. Bapat, *Z. Anal. Chem.*, *156*, 273 (1957).
108. J. G. Gorman, E. J. Jones, and J. A. Hipple, *Anal. Chem.*, *23*, 438 (1951).
109. J. Roboz, in *Trace Analysis: Physical Methods*, G. H. Morrison, Ed., Interscience, New York, 1965, p. 435.
110. N. Sasaki and E. Watanabe, *Thirteenth Annual Conference on Mass Spectrometery and Allied Topics*, St. Louis, Mo., 1965, p. 474.
111. W. A. Walstenholme, *Nature*, *203*, 1284 (1964).
112. R. Brown, W. J. Richardson, and H. W. Somerford, *Fifteenth Annual Conference on Mass Spectrometry and Allied Topics*, Denver, Colo., 1967, p. 157.
113. C. A. Evans, Jr. and G. H. Morrison, *Anal. Chem.*, *40*, 869 (1968).
114. J. P. Yurachek, G. G. Clemena, and W. W. Harrison, *Anal. Chem.*, *41*, 1666 (1969).
115. W. W. Harrison, G. G. Clemena, and C. W. Magee, *J. Assoc. Offic. Agr. Chemists*, *54*, 929 (1971).
116. R. Alverez, P. J. Paulsen, and D. E. Kelleher, *Anal. Chem.*, *41*, 955 (1969).
117. R. C. Koch, *Activation Analysis Handbook*, Academic Press, New York, 1960.

118. W. Schulze, *Neutronenaktivierung als Analytisches Hilfsmittel*, Ferdinand Enke Verlag, Stuttgard, 1962.
119. H. J. M. Bowen and D. Gibbons, *Radioactivation Analysis*, Oxford University Press, London, 1963.
120. *Guide to Activation Analysis*, W. S. Lyons, Jr., Ed., D. Van Nostrand, Princeton, N. J., 1964.
121. A. J. Moses, *Nuclear Techniques in Analytical Chemistry*, Macmillan, New York, 1964.
122. D. Taylor, *Neutron Irradiation and Activation Analysis*, George Newnes, London, 1964.
123. L. A. Rancitelli, J. A. Cooper, and R. W. Perkins, *Natl. Bur. Std. (U.S.), Spec. Publ.*, *372*, 1969, p. 101.
124. L. P. Neithling, J. M. M. Brown, and P. J. DeWet, *J. S. African Vet. Med. Assoc.*, *39*, 93 (1968); from *Chem. Abstr.*, *70*, 131291j (1969).
125. H. J. M. Bowen and P. A. Cawse, *Analyst*, *88*, 721 (1963).
126. R. C. Dickson and R. H. Tomlinson, *Intern. J. Appl. Radiation Isotopes*, *18*, 153 (1967).
127. B. Maziere, D. Comar, and C. Kellerhohn, *Bull. Soc. Chim. France*, 3767 (1970); from *Chem. Abstr.*, *74*, 50420k (1971).
128. J. F. P. Lambert, O. Levander, L. Argrett, and R. E. Simpson, *J. Assoc. Offic. Anal. Chemists*, *52*, 915 (1969).
129. J. T. Tanner, J. F. P. Lambert, and R. E. Simpson, *J. Assoc. Offic. Anal. Chemists*, *53*, 1140 (1970).
130. D. Betteridge, *U. K. At. Energy Res. Estab. Rept. AERE-R* 4881, 1965; from *Chem. Abstr.*, *63*, 11245h (1965).
131. R. H. Filby and W. L. Yakeley, *Radiochem. Radioanal. Letters*, *2*, 307 (1969); from *Chem. Abstr.*, *72*, 87067k (1971).
132. G. S. Nixon and V. B. Myers, *Caries Res.*, *4*, 179 (1970).
133. O. Veveris, S. H. Mihelson, Z. Pelekis, L. Pelekis, and I. Taure, *Latvijas PSR Zinatnu Akad. Vestis, Fiz. Teh. Zinat. Ser.*, *2*, 25 (1969); from *Chem. Abstr.*, *71*, 78064j (1969).
134. K. Schwarz and C. N. Foltz, *J. Biol. Chem.*, *233*, 245 (1958).
135. R. A. Nadkarni and W. E. Ehmann, *Natl. Bur. Std. (U.S.) Spec. Publ. 372*, 1969, p. 190.
136. D. E. Hogue, J. F. Proctor, R. G. Warner, and J. K. Loosli, *J. Animal Sci.*, *21*, 25 (1962).
137. W. A. Haller, L. A. Rancitelli, and J. A. Cooper, *J. Agr. Food Chem.*, *16*, 1036 (1968).
138. K. P. McConnell, *Proceedings of the 1961 International Conference on Modern Trends in Activation Analysis*, p. 137.
139. V. Otvinovski and I. V. Yakolev, *Acta Chim. Acad. Sci. Hung.*, *26*, 243 (1961); from *Chem. Abstr.*, *55*, 21976b (1961).
140. K. Schwarz, J. Stesney, and C. N. Foltz, *Metab., Clin. Exptl.*, *8*, 88 (1959).
141. K. P. McConnell, H. G. Mautner, and G. W. Leddicotte, *Biochim. Biophys. Acta*, *59*, 217 (1962).
142. K. Schwarz and C. N. Foltz, *Federation Proc.*, *17*, 492 (1958).
143. B. T. Kenna, *Anal. Chem.*, *35*, 1766 (1963).
144. J. Perdijon, *Anal. Chem.*, *39*, 448 (1967).
145. D. F. C. Morris and R. A. Killick, *Talanta*, *10*, 279 (1963).
146. P. Patek and H. Sorantin, *Anal. Chem.*, *39*, 1458 (1967).
147. R. A. Nadkarni and W. E. Ehmann, *Radioanal. Chem.*, *3*, 175 (1969).
148. G. Leliaert, J. Hoste, and Z. Eeckhaut, *Anal. Chim. Acta*, *19*, 100 (1958).
149. H. P. Yule, *Anal. Chem.*, *37*, 129 (1965).

150. *Advances in Activation Analysis*, Vol. I, J. M. A. Lenihan and S. J. Thompson, Eds., Academic Press, New York, 1969.
151. T. L. Isenhour and G. H. Morrison, *Anal. Chem.*, *38*, 162 (1966).
152. I. Fujii, T. Inouye, H. Muto, K. Ondera, and A. Tani, *Analyst*, *94*, 189 (1969).
153. F. Girardi, *Talanta*, *12*, 1017 (1965).
154. S. Kodiri and L. P. Starchek, *Dokl. Akad. Nauk Tadzh. SSR*, *12* (5), 17 (1969); from *Chem. Abstr.*, *71*, 131291j (1969).
155. A. Veres, *Proc. Conf. Appl. Phys.-Chem. Methods Chem. Anal.*, Budapest, *2*, 360 (1966); from *Chem. Abstr.*, *69*, 48932g (1968).
155a. Jean-Luc Debrun, D. C. Riddle, and E. A. Schweikert, *Anal. Chem.*, *44*, 1386 (1972).
156. O. Veveris, H. Mihelsons, Z. Pelekis, L. Pelekis, and I. Taure, *Latvijas PRS Zinatnu Akad. Vestis, Fiz. Teh. Zinat. Ser.* 4, 13 (1968); from *Chem. Abstr.*, *70*, 26294u (1969).
157. K. Samsahl, *Anal. Chem.*, *39*, 1480 (1967).
158. F. J. Conrad and B. T. Kenna, *Anal. Chem.*, *39*, 1001 (1967).
159. R. Dams, J. A. Robbins, K. A. Rahn, and J. W. Winchester, *Anal. Chem.*, *42*, 861 (1970).
160. J. Leibetseder, *Trace Miner. Stud. Isotop. Dom. Anim., Proc. Panel*, 71 (1968); from *Chem. Abstr.*, *71*, 109711z (1969).
161. A. A. Arroyo and G. T. Toro, *Natl. Bur. Std. (U.S.) Spec. Publ. 312*, 1969, p. 541.
162. R. S. Maddock and W. W. Meinke, *Univ. Mich. Progr. Rept. 11*, 1962, p. 68.
163. G. W. Leddicotte and S. A. Reynolds, *Nucleonics*, *8* (3), 62 (1951).
164. *Handbook of Chemistry and Physics*, 51st ed., R. C. Weast, Ed., Chemical Rubber Co., Cleveland, 1970–71, p. B–280.
165. F. Girardi, G. Guzzi, and J. Pauly, *Anal. Chem.*, *36*, 1588 (1964).
166. E. Ricci and F. F. Dyer, *Nucleonics*, *22* (6), 45 (1962).
167. T. L. Isenhour and G. H. Morrison, *Anal. Chem.*, *36*, 1089 (1964).
168. R. E. Wainerdi, *Nucleonics*, *22* (2), 57 (1964).
169. V. P. Guinn, *J. Am. Oil Chemists' Soc.*, *45*, 767 (1968).
170. W. A. Haller, R. Filby, L. A. Rancitelli, and J. A. Cooper, *Natl. Bur. Std. (U.S.) Spec. Publ. 312*, 1969, p. 177.
171. M. Okada, *Nature*, *187*, 594 (1960).
172. O. U. Anders, *Anal. Chem.*, *33*, 1706 (1961).
173. J. Wing and M. A. Wahlgren, *Anal. Chem.*, *39*, 85 (1967).
174. F. S. Goulding, *Nucleonics*, *22* (5), 54 (1964).
175. S. M. Lombard and T. L. Isenhour, *Anal. Chem.*, *40*, 1990 (1968).
176. S. S. Nargolwalla, M. R. Crambes, and J. R. DeVoe, *Anal. Chem.*, *40*, 666 (1968).
177. Y. Kamemoto and S. Yamagishi, *Bull. Chem. Soc. Japan*, *37*, 664 (1964); from *Chem. Abstr.*, *61*, 3879c (1964).
178. S. A. Reynolds and W. T. Mullins, *Intern. J. Appl. Radiation Isotopes*, *14*, 421 (1963).
179. F. Girardi, G. Guzzi, and J. Pauly, *Anal. Chem.*, *37*, 1085 (1965).
180. H. P. Yule, *Anal. Chem.*, *38*, 818 (1966).
181. R. A. Lundgren and S. S. Nargolwalla, *Anal. Chem.*, *40*, 672 (1968).
182. R. Handley, *Anal. Chem.*, *32*, 1719 (1960).
183. L. E. Lisanti, *Ric. Sci.*, *28*, 2540 (1958); from *Chem. Abstr.*, *53*, 14236i (1959).
184. E. C. Olsen and J. W. Shell, *Anal. Chim. Acta*, *23*, 219 (1960).
185. M. Goldman and E. D. Beckman, *U. S. At. Energy Comm.* 1967 (UCD–472–210); from *Chem. Abstr.*, *69*, 8820n (1968).
186. *Official Methods of Analysis of the Association of Official Analytical Chemists*, 11th ed., W. Horowitz, Ed., Washington, D. C., 1970.

High-Performance Ion-Exchange Chromatography with Narrow-Bore Columns: Rapid Analysis of Nucleic Acid Constituents at the Subnanomole Level

CSABA HORVATH, *Yale University, New Haven, Connecticut*

I.	Introduction	80
II.	Interpretation of the Chromatogram in Ion-Exchange Chromotography	81
	1. Solute Retention and Column Efficiency	81
	2. Sensitivity of Analysis	87
	3. Quantitative Analysis	89
	A. Evaluation of the Chromatogram	89
	B. Calibration	93
	a. Calibration Curve with Standard Samples	93
	b. Determination of Mole Ratios	93
	c. Internal Standard Method	93
	d. Other Methods	94
III.	Pellicular Ion-Exchange Resins	94
IV.	Liquid Chromatograph	100
	1. Description of the Instrument	100
	A. Generation and Control of Eluent Flow	100
	B. Sample Introduction	103
	C. Ultraviolet Detector	105
	D. Components for Building an Instrument	106
	2. Preparation of Columns	109
	3. Gradient Elution	110
V.	Ion-Exchange Chromatography of Nucleic Acid Constituents	115
VI.	Rapid Analysis at the Subnanomole Level Using Pellicular Ion-Exchange Resins	117
	1. General Considerations	117
	A. Column Loading Capacity	117
	B. Effect of pH	118
	C. Effect of Temperature	120
	D. Effect of Salt Concentration	122
	E. Effect of Flow Rate on Column Efficiency	124
	2. Selection of Operating Conditions	125
	A. Optimization of Separation Efficiency	125
	B. Optimization of the Sensitivity of Analysis	127

3. Separation of the Various Classes............................ 129
   A. Purine and Pyrimidine Bases.......................... 129
   B. Nucleosides.. 130
   C. Nucleotides and Oligonucleotides...................... 131
VII. Rapid, Ultransensitive Analysis Using Conventional Resins............. 140
VIII. Sample Preparation.. 147
IX. Outlook... 148
   Acknowledgments.. 150
   List of Symbols.. 150
   References... 152

I. INTRODUCTION

Ion-exchange chromatography has been the most powerful tool for separating nucleic acid constituents since Cohn (1–3) first demonstrated the applicability and versatility of this technique in nucleic acid research. Later the development of suitable instrumentation employing ultraviolet detector systems (4–7) made it possible to perform analyses in a semiautomatic fashion and contributed to the popularization of the method, as well as to an increase in its speed and sensitivity (8).

Nevertheless, the analysis of nucleic acid constituents remained a relatively slow and cumbersome procedure in comparison with the analysis of amino acids, which has been greatly facilitated by modern amino acid analyzers. These devices, based on the fundamental work of Moore and Stein (9), employ ion-exchange chromatography and sophisticated instrumentation permitting automated quantitative analysis of the amino acid composition of protein hydrolyzates or biological fluids within a few hours (15).

The diversity of analytical problems encountered with nucleic acid derivatives, however, requires a chromatographic system with wide-ranging capabilities rather than an instrument for only a single analysis, a fact which may explain why recent developments in nucleic acid analysis have been outgrowths of progress in liquid chromatography in general. This technique is undergoing a new phase of development. Novel concepts and column materials have been introduced to improve separation efficiency, and the overall performance is being greatly enhanced by sophisticated instrumentation, especially by highly sensitive detectors (9a). It appears that the emerging "new" analytical tool, sometimes called high-performance liquid chromatography, may approach the speed, efficiency, and convenience of gas chromatography in the analysis of nonvolatile substances. With present high-performance liquid chromatographs column inlet pressures up to several thousand pounds per square inch may be routinely used. Although the high pressure is not a meritorious feature per se, it can expand the potential of liquid chromatography to achieve a separating speed and efficiency unattainable with low-pressure systems.

In the following sections, some applications of this technique to rapid analysis of nucleic acid constituents at high sensitivity are discussed. The need for highly sensitive analytical methods is dictated by the growing interest in analyzing the minute nucleic acid samples obtained from electrophoresis, zonal centrifugation, and other procedures. On the other hand, the processing of large numbers of samples calls for a fast analytical technique.

These goals can be achieved in ion-exchange chromatography by using narrow-bore columns packed with novel materials and an appropriate UV detector. The reliability of this technique, like that of any other advanced liquid chromatographic system, depends on precise flow and temperature control, which can be achieved only by using a properly constructed instrument. If such is available, then both the analytical procedure and the evaluation of the results can be carried out in a fashion similar to that used in gas chromatography. In addition, the accurately controlled conditions and the continuous effluent monitoring, as well as the employment of stable and reusable columns, make it possible in liquid chromatography to link data from the chromatogram to column parameters and operational variables in a way adopted from gas chromatography. It appears desirable, therefore, to make use of the well-established terminology of gas chromatography. The interested reader should consult the literature of gas chromatography (10,11) and of ion-exchange chromatography (12–15) for detailed information.

II. INTERPRETATION OF THE CHROMATOGRAM IN ION-EXCHANGE CHROMATOGRAPHY

1. Solute Retention and Column Efficiency

Ion-exchange chromatography includes all chromatographic processes which utilize ion-exchange materials as stationary phases and liquids as eluents. It represents the most important branch of liquid chromatography and is widely used for the separation of ionic as well as nonionic substances. Actually, "the mechanisms of the separation may be as far removed from exchange of ions as chromatography is from color" (14). In the following discussion the most important relationships governing solute retention and band spreading are briefly summarized.

Figure 1 shows the chromatogram of a single component as it appears on the recorder chart paper. The sample is injected at zero time, and the retarded component is eluted at B, that is, in time t_R, called the *retention time*. When a component is also injected which is not absorbed on the stationary

82   CSABA HORVATH

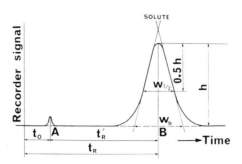

Figure 1. Chromatogram of a single solute.

phase, it appears at A, that is, in time t_0, which is then the retention time of
an unsorbed solute and is also called the *mobile-phase holdup time*. The *adjusted
retention time* of a solute, t_R', is given by t_R minus t_0. The peak height is h, and
the peak width is usually measured at half height, $w_{1/2}$, or at the base-line
intercept, w_b. For many calculations, all these values can be represented by
distances on the chart paper. When they have to be expressed as time units,
the distances must be divided by the chart speed. Sometimes it is preferable
to use volumes instead of times. At constant flow rate, F, the *retention volume*,
V_R, the *mobile-phase holdup volume*, V_M (which is equal to the volume occupied
by the mobile phase in the column), the *adjusted retention volume*, V_R', and the
peak volume can be calculated by multiplying t_R, t_0, t_R', and w_b by F,
respectively.

The peak position relative to t_0 is given by the *mass distribution ratio*,† D_m,
which is defined as

$$D_m = \frac{\text{amount of solute in stationary phase}}{\text{amount of solute in mobile phase}} \qquad [1]$$

and obtained from retention data as

$$D_m = \frac{t_R - t_0}{t_0} = \frac{t_R'}{t_0} \qquad [2]$$

The mass distribution ratio is related to the *retardation ratio*, R, as

$$D_m = \frac{1 - R}{R} \qquad [3]$$

The retardation ratio is equivalent to the R_f value used in paper and thin-
layer chromatography.

† In gas chromatography this term is denoted by k and is often called the *capacity
factor* or *distribution ratio*.

The value of the mass distribution ratio is determined by the *concentration distribution ratio*, D_c, and by the volume ratio of the stationary and mobile phases in the column as

$$D_m = \frac{D_c V_S}{V_M} \qquad [4]$$

where V_S is the volume of the stationary phase, that is, the swollen ion-exchange resin, and V_M is the volume of the mobile phase, present in the column.

The concentration distribution ratio is defined as

$$D_c = \frac{\text{total (analytical) concentration of solute in stationary phase}}{\text{total (analytical) concentration of solute in mobile phase}} \qquad [5]$$

The concentration distribution ratio is a thermodynamic equilibrium constant which is determined by the temperature, the properties of the solute, the resin, and by the mobile-phase composition. It is usually independent of the solute concentration when small quantities are analyzed because the sorbtion isotherm is linear at low concentrations. The retention of a solute is related to D_c by the following equation:

$$V_R = V_M + D_c V_S \qquad [6]$$

where V_R is the retention volume. In practice, peak retention is adjusted by varying the composition, the ionic strength, or the pH of the eluent, as well as by changing column temperature; all these parameters affect the value of D_c. It should be kept in mind, however, that changing the eluent composition or temperature may change not only D_c but V_M and V_S as well because of swelling or shrinking of the resin.

In the simplest case the eluent strength and the temperature are kept constant; thus the concentration distribution ratio is also constant for each sample component during the chromatographic run. This is the case of elution at constant eluent strength, that is, elution with a single eluent, which can conveniently be called *isocratic* elution ($\kappa\rho\alpha\tau o\sigma$ = "strength"). When V_M and V_S are known, the concentration distribution ratios can be easily calculated, using [6] from the retention volumes under these circumstances. In *gradient elution*, the eluent strength is gradually increased during the chromatographic run and also along the column, so that D_c decreases for each component during elution. As a result, the retention values, particularly for later peaks, are decreased and the time of separation is reduced, thereby achieving one of the major objectives of gradient elution. In *temperature programming* the column temperature is raised gradually during the chromatographic run with an effect similar to that of gradient elution, but this technique is seldom used in ion-exchange chromatography.

So far, the retardation of only a single component has been considered. The separation of two peaks as illustrated in Figure 2, however, depends on their *relative retention*, α, which is also called the *separation factor* in ion-exchange chromatography. It is defined as the ratio of the adjusted retention time of the later peak, t'_{R2}, to that of the earlier peak, t'_{R1}, and is equal to the ratio of the distribution ratios, that is,

$$\alpha_{2,1} = t'_{R2}/t'_{R1} = D_{c2}/D_{c1} = D_{m2}/D_{m1} \qquad [7]$$

The retention values of the components are often expressed by their relative retentions with respect to a well-defined peak as reference.

The actual separation of two components depends not only on the separation factor but also on the band broadening during elution. The performance of the column is measured by the magnitude of band spreading, which is usually expressed by the *number of theoretical plates*, N. It is actually a dimensionless measure of band spreading of Gaussian peaks and is defined by the relationship

$$N = \frac{t_R^2}{\sigma_t^2} \qquad [8]$$

that is, as the ratio of the squared retention time to the variance of the peak, σ_t^2, both expressed in time units.† The variance is the squared standard deviation, σ_t. The peak width is proportional to the standard deviation, for example,

$$w_b = 4\sigma_t \quad \text{and} \quad w_{1/2} = 2\sigma_t\sqrt{2 \ln 2} \qquad [9]$$

when the peak width at the base intercept, w_b, and the peak width at half height, $w_{1/2}$, are expressed in time units. Substituting σ_t from these expressions into [8], we obtain the following relationships, which can be used for the calculation of the plate number:

$$N = \frac{16t_R^2}{w_b^2} = \frac{5.545t_R^2}{w_{1/2}} \qquad [10]$$

where t_R and w_b are measured as distances on the chromatogram.

The efficiency of the column is also expressed by the plate height, H. This was originally defined as the height equivalent of a theoretical plate and can

---

† In order to calculate this ratio both retention and variance can be measured in units other than time. If the retention volume is used, the corresponding variance, σ_v^2, is obtained as $\sigma_t^2 \times F^2$, where F is the flow rate. The column length, L, can also be substituted for t_R, the corresponding variance, σ_l^2 (in column length units), is obtained as $\sigma_l^2 = \sigma_t^2 u_b^2$ where u_b is the band velocity. Most conveniently, however, both t_R and σ_t are taken as distances on the chromatogram.

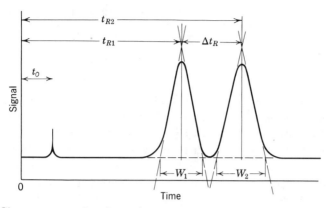

Figure 2. Chromatogram of a solute pair.
  Relative retention (separation factor):

$$\alpha = \frac{t_{R2} - t_0}{t_{R1} - t_0} = \frac{t'_{R2}}{t'_{R1}}$$

Resolution:

$$Re = \frac{\Delta t_R}{\overline{w}}, \text{ where } \overline{w} = \frac{w_1 + w_2}{2}$$

be calculated from the number of theoretical plates and from the length of the column as

$$H = \frac{L}{N} \tag{11}$$

Actually, the plate height is the ratio of the variance, σ_l^2, measured in column length units, to the column length and is given by

$$H = \frac{\sigma_l^2}{L} \tag{12}$$

Thus the plate height expresses the band spreading per unit column length which occurs as the solute is swept through the column. The major concern of the chromatographic theory is to relate H quantitatively to column parameters and operational variables.

The precise measure of the separation of two peaks is the resolution, Re, which is given by

$$Re = \frac{\Delta t_R}{\overline{w}_b} \tag{13}$$

where Δt_R is the retention time difference, and $\overline{w}_b$ is the mean peak width of the two components measured at the base intercept. The resolution necessary to separate two closely spaced peaks is a function of the column parameters discussed above and can be expressed as

$$Re = \frac{1}{4} \frac{\alpha - 1}{\alpha} \frac{D_m}{1 + D_m} \sqrt{N} \qquad [14]$$

Thus the mass distribution coefficient also affects the degree of separation in addition to the separation factor and plate number.

The foregoing considerations are valid only under isothermal and isocratic elution conditions. When gradient elution is employed, both α and D_m depend on the shape of the gradient, and the plate number and plate height, as computed by [10] and [11], fail to express the real efficiency of the column. However, the resolution remains a useful measure of the separation of two peaks under any given set of conditions. Sometimes the speed of separation is expressed by resolution per second, that is, by Re/t_R. The number of plates obtained per second, N/t_R, and the number of plates obtained in relation to the pressure drop across the column, $N/\Delta P$, are also useful measures of column performance with regard to the speed and the pressure requirement of the analysis, respectively.

The eluent flow rate through the column usually has significant influence on the value of these efficiency parameters. In many instances, however, it is preferable to use eluent velocity instead of flow rate. The chromatographic liquid velocity, u^*, is usually expressed in centimeters per second and is conveniently determined by the following relationship:

$$u^* = \frac{L}{t_0} \qquad [15]$$

provided the measurement of t_0 does not represent a technical problem, that is, an unretained tracer solute is available which can be detected at relatively low concentration. Otherwise the velocity of the mobile phase, u, can be calculated by the following expression:

$$u = \frac{F}{36A_f} = \frac{F}{9d^2\pi\epsilon} \qquad [16]$$

where F is the flow rate (ml/hr), A_f is the free cross-sectional area of the column (mm²), d is the inner diameter of the column (mm), and ϵ is the relative interstitial bed volume, given by V_M/V, where V is the total volume of the column. Column materials usually pack in a random manner, and ϵ ranges from 0.36 to 0.42. Hence the value 0.4 can often be used for ϵ to estimate u when F and d are known.

2. Sensitivity of Analysis

The advantage of small-bore columns in ion-exchange chromatography for increasing sensitivity was pointed out by Crampton et al. (16), who wrote: "Columns of smaller diameter would improve the situation from the viewpoint of sensitivity, but they present the general disadvantages of increased tendency to nonlinear flow, of inconvenience in administering the sample and the eluent, of fragility, and of loss in versatility imposed by diminishing the capacity of any given column." The feasibility of this approach for obtaining high sensitivity was soon demonstrated by Kirsten and Kirsten (17), as well as by Hamilton (18,19), who called the enhancement of the sensitivity by the reduction of the column diameter "column amplification." Hamilton also pointed out that this "amplification" is particularly advantageous because it does not increase the detector noise level, in contradistinction to optical or electrical amplifications. The advantages of a very short column having an inner diameter as small as 0.3 mm was demonstrated in ion-exchange chromatography of metal ions (20). Nevertheless this approach has not been widely adopted because of the difficulties in obtaining proper column packing with conventional ion-exchange resins in long, small-diameter tubes. In addition, the elimination of the technical problems indicated by Crampton et al. required advanced instrumentation. Recently, however, both efficient narrow-bore columns and instruments suitable for operation at low flow rates and with small samples have become available for ion-exchange chromatography. It is believed, therefore, that narrow-bore columns will find increasing use in the analysis of minute samples.

When an optical detector such as a UV photometer or refractometer is used, the minimum amount of solute which can be determined with a liquid chromatograph depends on the detector sensitivity and on the solute concentration in the column effluent. The detector sensitivity is usually expressed by the noise level and by the electric output for a given solute concentration. The linear range of the detector is given by that range of solute concentration in which the detector response is proportional to the concentration. Obviously high detector sensitivity is a prerequisite for micro- and ultramicroanalyses by liquid chromatography. In order to obtain the best results, however, column parameters and operational variables have to be optimized. In the following, their effect on the solute concentration in the effluent is discussed.

The maximum concentration of the solute in the effluent peak, C_{max}, is given by

$$C_{max} = \frac{M}{\sigma_v \sqrt{2\pi}} \qquad [17]$$

where M is the amount (moles) of solute injected, and σ_v is the standard deviation of the solute peak measured in volume units.

Using the following relationships:

$$N = \left(\frac{V_R}{\sigma_v}\right)^2 \quad \text{and} \quad V_R = (1 + D_m)V_M \qquad [18]$$

we obtain from [17] the expression

$$C_{\max} = \frac{M\sqrt{N}}{V_M(1 + D_m)\sqrt{2\pi}} \qquad [19]$$

Since

$$V_M = \frac{d^2\pi\epsilon L}{4} \qquad [20]$$

where d is the inner diameter of the column and ϵ is the packing porosity, [19] can be written as

$$C_{\max} = \frac{4M\sqrt{N}}{d^2\pi\epsilon L(1 + D_m)\sqrt{2\pi}} \qquad [21]$$

Thus, for a given injected amount of solute, the peak concentration is maximum when it is eluted at low D_m values from a small-inside-diameter column having low porosity and high efficiency. This is in agreement with the observations of Hamilton et al. (21). The most obvious means to increase the solute concentration in the effluent is, therefore, the reduction of d and ϵ and the increase of column efficiency, so that the number of plates required for the separation can be obtained with a relatively short column. Fortunately, column efficiency often increases with decreasing column diameter; therefore small-bore columns are particularly attractive. The D_m values can conveniently be decreased by changing the eluent strength. On the other hand, a decrease of the relative interstitial bed volume appears to be difficult to achieve in practice.

With the UV detector for small solute concentrations the absorbance, A, is given by the Lambert-Beer law:

$$A = \log \frac{I_0}{I} = \varepsilon C l \qquad [22]$$

where I_0 is the intensity of light passing through the eluent, I is the intensity of light passing through the solute solution, C is the solute concentration, ϵ is the molar absorptivity of the solute at the detector wavelength, and l is the length of the cell.

When the cell volume is sufficiently small relative to the peak volume, the absorbance at the peak maximum, A_{max}, is given by

$$A_{max} = \frac{\delta \varepsilon l M}{\sigma_v \sqrt{2\pi}}$$ [23]

where δ is a calibration factor specific for the particular detector and operating conditions. When δ, ε, and l are known, [23] can be used to estimate either A_{max} for known M, or M from the measured A_{max}, in the case of Gaussian or nearly Gaussian peaks, because σ_v can be calculated from the peak width and the flow rate. The molar absorptivities of a number of RNA constituents at 260 nm are listed in Table I. In addition, a great many spectrophotometric data on common nucleic acid constituents obtained at 250 nm and at different pH values have been published by Anderson et al. (5).

The increase in the apparent sensitivity of chromatographic systems featuring narrow-bore columns is due essentially to the small band volume in the effluent, as seen from [17]. This advantage can be exploited, however, only if the volume of the detector is correspondingly small and (as a rule of thumb) is not larger than one twentieth of the peak volume. A larger detector volume causes additional band spreading and distortion, which result in a decrease in the solute concentration. Thus the advantage of narrow-bore columns can be easily offset by using an unsuitable detector.

3. Quantitative Analysis

A. EVALUATION OF THE CHROMATOGRAM

Although the measurement of retention times leads to the identification of the individual sample components (qualitative analysis), their quantitative determination is based on the measurement of peak areas. The chromatogram obtained with a UV detector is the graphic representation of the absorbance of the effluent as a function of time. At constant flow rate and recorder chart speed the area under a peak, PA, that is, the integral of the A versus t curve, is proportional to the amount of solute, M, injected:

$$PA = fM$$ [24]

The proportionality factor, f, called the *response factor*, has to be evaluated in order to convert peak areas into solute quantities. The response factors vary not only from solute to solute but also with the operating conditions. Reproducible quantitative results can be expected only when the measurements are performed at the same flow rates. Relatively minor variations in eluent strength or in the slope of the gradient should not affect response factors unless the specific absorptivity of the solute also changes, for example, because of a change in the pH of the effluent.

TABLE I

Molar Absorptivities of RNA Constituents at 260 nm[a]

Symbol	Name	Molecular weight	pH	$\varepsilon_{260} \times 10^{-3}$
ADE	Adenine	135.1	1–3	13.0
ADO	Adenosine	267.2	1–2	14.3
AMP	Adenosine-5′-phosphate	329.2	2	14.5
ADP	Adenosine-5′-diphosphate	427.2	2	14.5
ATP	Adenosine-5′-triphosphate	507.2	2	14.3
CYT	Cytosine	111.1	1–3	6.0
CYD	Cytidine	243.2	1–2	6.4
CMP	Cytidine-5′-phosphate	323.2	1–2.5	6.3
CDP	Cytidine-5′-diphosphate	403.2	1–2.5	6.2
CTP	Cytidine-5′-triphosphate	483.2	2	6.1
GUA	Guanine	151.1	1–2	8.0
GUO	Guanosine	283.2	1–7	12.5
GMP	Guanosine-5′-phosphate	363.2	1–7	11.6
GDP	Guanosine-5′-diphosphate	443.2	1–7	11.8
GTP	Guanosine-5′-triphosphate	523.2	1–7	11.8
HYP	Hypoxanthine	136.1	4–7	7.9
INO	Inosine	268.2	3–6	7.1
IMP	Inosine-5′-phosphate	348.2	2	7.5
THY	Thymine	126.1	1–7	7.4
URA	Uracil	112.1	2–7	8.2
URD	Uridine	244.2	1–7	9.9
UMP	Uridine-5′-phosphate	324.2	2–7	9.9
UDP	Uridine-5′-diphosphate	404.2	2–7	9.9
UTP	Uridine-5′-triphosphate	484.2	2–7	9.9
XAN	Xanthine	152.1	2–6	8.8
XAO	Xanthosine	284.2	2	8.7

[a] Data compiled from R. M. C. Dawson et al., Eds., *Data for Biochemical Research*, 2nd ed., Oxford University Press, New York, 1969. The molar absorptivity of GUO was taken from recent measurements of Randerath and Randerath (22).

The simplest method for the evaluation of peak areas is to multiply the peak height at maximum, h, by the peak width at half height $w_{1/2}$ (cf. Figure 1). For Gaussian peaks this product is equal to 94% of the total area. A similar but less accurate method is triangulation, which consists of the multiplication of the peak height at maximum by the peak width at the base intercept, w_b. This product ix equal to 97% of the total area of a Gaussian peak. These evaluation methods are particularly useful if the peaks are symmetrical and are not very sharp or small. The precision of the peak area measurement can be enhanced and the procedure simplified by the use of integrators, such as a Disc integrator or sophisticated electronic digital integrators. Other techniques for measuring peak areas utilize planimeters or involve cutting out the peaks (preferably from a Xerox copy of the chromatogram) and weighing the paper on an analytical balance. The highest precision can be obtained by using electronic integrators (1 to 2% s.d.); the accuracy of the other techniques depends greatly on the skill of the operator. Most workers prefer to use Disc integrators or to calculate the product $h \times w_{1/2}$. The precision obtained with the latter method can be 2 to 3% s.d. when the peaks are reasonably symmetrical. Under some circumstances, the peak height, h, itself or the product of the peak height and the retention time, $h \times t_R$, is also used in quantitative analysis, but these methods are usually less accurate. If the peaks are not completely resolved, their areas can be measured only approximately. Figure 3 shows three cases of peak overlapping and the approximation of the individual peak areas. In these circumstances the error of the measurement is relatively high, particularly if the peaks are not Gaussian.

The amount of solute can be estimated in the following way from the chromatogram obtained with a UV detector which has a linear absorbance output. The detector and recorder should be properly adjusted so that the maximum absorbance, A_{max}, at h can be measured accurately, and the molar absorptivity of the solute in the effluent should be known. After rearranging [23] the number of moles of solute represented by the peak is expressed by

$$M = \frac{A_{max}\sigma_v\sqrt{2\pi}}{\delta\varepsilon l} \qquad [25]$$

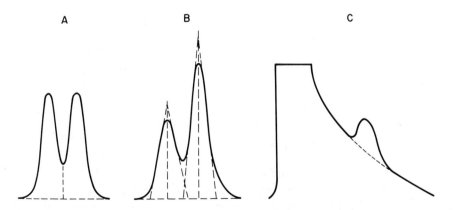

Figure 3. Estimation of peak areas in the case of unresolved peaks.

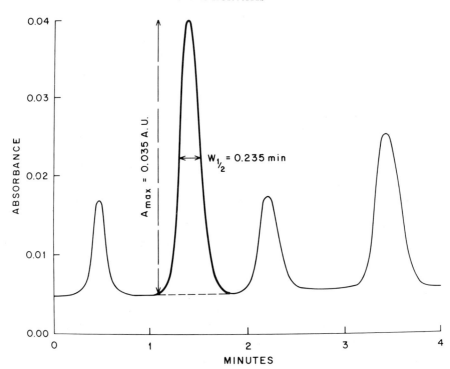

Figure 4. Estimation of the amount of a sample component. Flow rate: 30 ml/hr = 0.5 ml/min; $\delta = 1$, $l = 1$ cm, $\varepsilon_{254} = 10^4$ liters/mole/cm. From the chromatogram, $A_{max} = 0.04 - 0.005 = 0.035$ AU; $w_{1/2} = 0.235$ min, whence $\sigma_t = 0.235/2.354 \approx 0.1$ min.

As shown in Figure 4, A_{max} is obtained directly from the chromatogram. The standard deviation in volume units, σ_v, is calculated from the standard deviation in time units, σ_t, multiplying the latter by the flow rate. The value of σ_t is obtained from the chromatogram, using the relationship $\sigma_t = w_{1/2}/2.354$; for the peak in question, $\sigma_v = 0.1 \times 0.5 = 5 \times 10^{-2}$ ml $= 5 \times 10^{-5}$ liter. Taking the corresponding values from the chromatogram in Figure 4 and substituting into [25], we obtain

$$M = \frac{4 \times 10^{-2} \times 5 \times 10^{-5} \times 2.5}{1 \times 10^4 \times 1} = 5 \times 10^{-10} \text{ mole} = 0.5 \text{ nanomole}$$

In practice, the values of δ and ε are often not known accurately and the peak shape is not Gaussian. The accuracy is then rather poor, and the error may exceed 20%. However, this method may be useful to obtain a quick estimate of the amount of a known sample component.

B. CALIBRATION

Precise quantitative results can be obtained only if the relationships between the amounts of the sample components and the respective peak areas are known. That is, the response factors have to be determined experimentally for a given set of conditions.

a. Calibration Curve with Standard Samples. A number of samples containing various amounts of the components in question are analyzed under standard conditions. The peak areas are plotted against the number of moles of sample components injected. The calibration curves should be linear, and their slopes give the respective response factors. The main sources of error are variation of flow rate and the poor reproducibility of sample injection. Both the sample volume and the individual injection technique of the operator can affect the results. Therefore the volume of the injected sample and the injection technique should be unchanged in order to obtain a precision better than 3%.

b. Determination of Mole Ratios. Frequently the goal of analytical work with nucleic acid constituents is the determination of the mole ratios of the sample components. Since no absolute values are required, the procedure is relatively simple. The mole ratio of component i to a reference component r which is present in the sample is given by

$$\frac{M_i}{M_r} = \frac{f_i}{f_r} \frac{PA_i}{PA_r} \qquad [26]$$

where M_i and M_r are moles, f_i and f_r are the response factors, and PA_i and PA_r are the peak areas of components i and r, respectively. By analyzing samples of known mole ratios, the response factor ratios, f_i/f_r, are calculated from [26]. Then the mole ratios in unknown samples are obtained simply from the peak area ratios multiplied by the corresponding response factor ratios.

c. Internal Standard Method. The highest accuracy can be obtained in determining the concentrations of the individual components by the so-called internal standard method. A component (internal standard) is added to the sample which is chemically similar to the other components but was not present in the sample originally. Its concentration must be precisely known, it must be resolved completely from the other components, it must elute close to the peaks of interest, and it must give a peak similar in size. The principal advantage of this method is that the injected sample volume does not need to be known accurately. Errors due to apparatus and changing operation conditions can be eliminated and good results can be obtained even if sample is lost during preparation or is not eluted completely. Calibration curves are made by analyzing mixtures which contain various concentrations of the sample components of interest at a fixed concentration of the internal standard. Then the concentrations of the components are plotted against the ratio of the component peak area to the peak area of the internal standard.

d. Other Methods. Detailed treatment of the subject can be found in the literature of gas chromatography, for example, ref. 23. Techniques for the quantitative evaluation of chromatograms obtained in ion-exchange chromatography have been described by Uziel et al. (8) and by Jolley and Scott (24). Methods used in quantitative analysis of tissue nucleotides by high-performance liquid chromatography have been described by Brown (25).

III. PELLICULAR ION-EXCHANGE RESINS

The speed of analysis is determined by the rate of solute equilibration between the moving eluent and the stationary ion-exchange resin in the column. The faster the mass transfer is in the mobile and stationary phases, the higher can be the eluent velocity without severely increasing the band spreading. Since the speed of analysis is proportional to the velocity, fast mass transfer leads to fast separations. Under ordinary circumstances, solute diffusion in the resin particles controls the rate of solute equilibration in ion-exchange chromatography of organic substances (21).

As early as 1952 Pepper (26) discussed the possibility of confining the ion-exchange resin to the surface of a support. Shortly afterwards Weiss (27) proposed converting only the outer shell of the polymer beads into ion-exchange resin. In both cases it was assumed that the stationary-phase mass transfer resistance would be reduced by decreasing the diffusion path length in the resin phase. This concept was explored in practice with some success, using Celite coated with a cation-exchange resin for the separation of proteins (28–30). Later, on the basis of similar considerations, porous-layer-coated glass beads were introduced for use in gas chromatography (31), and Parrish described superficial ion-exchange resins for chromatography which were prepared by converting only the surface layer of polymer beads into ion exchanger (32). Recently Skafy and Lieser (33–35) studied the chromatographic properties of such superficial ion-exchange resins in greater detail. The polyethyleneimine (PEI)-coated cellulose introduced by Randerath (36) could qualify also as a superficial ion exchanger which finds wide application in thin-layer chromatography of nucleic acid constituents but is seldom used in column chromatography.

Another approach to decreasing the stationary-phase mass transfer resistance is the reduction of the particle diameter, which has the advantage

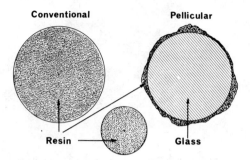

Figure 5. Cross sections of conventional and pellicular ion-exchange resin particles. The volume of the small bead is approximately equal to that of the pellicular resin coating.

that the mobile-phase mass transfer resistance is also reduced. In the practice of ion-exchange chromatography, this path was followed with remarkable success (6–8,37–39) after techniques had been developed to produce uniform particles in the 3- to 20-μ diameter range. However, columns packed with such small particles have very high flow resistance and are beset by technical difficulties arising from the gel properties of conventional ion-exchange resins† (swelling, shrinking, compressibility, etc.).

Recently, the possibility of forming thin ion-exchange resin shells on the surface of glass beads was explored, and anion- and cation-exchange resins thus prepared were used successfully for the analysis of nucleic acid constituents (40,41). Because the actual resin layer surrounds the glass sphere like a thin skin, as shown in Figure 5, the material has been termed pellicular ion-exchange resin. A photomicrograph of such particles is shown in Figure 6. Actually the uniform thin-layer configuration of the resin is an idealization and is not necessarily optimum in chromatographic practice because highly efficient column materials have also been obtained when the resin layer was not uniform.

The successful employment of such resins soon led to the introduction of a similar material, which consists of a superficially porous crust with an ion-exchange resin surrounding an impervious spherical silica core (42,43) and provides highly efficient columns for the analysis of nucleic acid constituents. This material, which can also be considered pellicular resin, is available under the trade name Zipax from Du Pont (Wilmington, Del.). Figure 7 shows a scanning electron micrograph of a Zipax anion-exchange particle. A variety

† Conventional ion-exchange resins are defined for convenience as spherical or irregularly shaped particles composed entirely of ion-exchange resin.

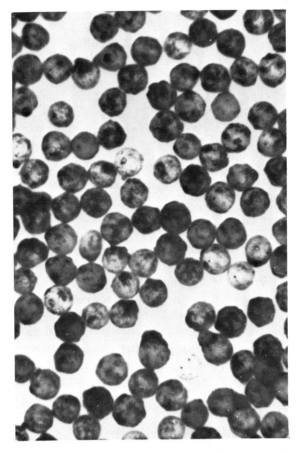

Figure 6. Photomicrograph of pellicular anion-exchange resin particles. Sieve fraction No. 270-325, particle diameter about 50 μ (Reeve Angel).

of high-efficiency pellicular resins is available under the general name Pellionex from H. Reeve Angel & Co. (Clifton, N.J.).

Pellicular resins are made by rendering the ion-exchange resin onto the surface or into a porous layer on the surface of impervious spherical particles, most conveniently glass microbeads. In one possible way of doing this, a cross-linked polymer matrix is first polymerized *in situ* to obtain a desired surface layer, and subsequently the polymer is converted into an ion-exchange resin by the introduction of appropriate functional groups (40). In Zipax resins the polymer is deposited in the cavities of the siliceous porous surface layer of the so-called controlled-surface-porosity support (Du Pont); in

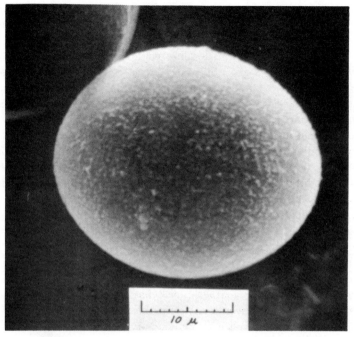

Figure 7. Scanning electron micrograph of a Zipax anion-exchange resin particle (Du Pont). From Kirkland (43).

actual pellicular resins, however, the polymer forms a coherent layer which is bonded to the surface of glass beads.

So far, strong anion- or cation-exchange resins containing quaternary ammonium or sulfonic acid groups have been used in pellicular form for most applications. Pellicular anion-exchange resins can be made by chloromethylation of cross-linked polystyrene shells and by subsequent amination with dimethylbenzylamine. One type of Pellionex strong anion-exchange resins is made by this procedure. On the other hand, the Zipax strong anion exchanger contains quaternary ammonium groups attached to cross-linked acrylic matrix. By sulfonation of cross-linked polystyrene shells, strong pellicular cation-exchange resins of the Pellionex type can be obtained. The Zipax cation exchanger is made with a fluoropolymer containing sulfonic acid groups. It appears that any kind of ion-exchange resin could be produced in pellicularized form if needed.

As only the outer part of pellicular ion-exchange particles consist of actual resin, their ion-exchange capacity is much smaller than that of conventional resins. Most of the published results were obtained with pellicular resins having ion-exchange capacities of about 10 μeq, that is, several hundred times lower than those of the conventional resins. Therefore the main use of pellicular resins will probably be in the analysis of minute quantities of sample, where no high column loading capacity is required and a resin capacity of 10 to 100 μeq/g appears to be sufficient, although pellicular resins of higher ion-exchange capacity may also find successful application. Materials other than ion-exchange resins, such as alumina and silica gel, can also be prepared in pellicular form (44), but only pellicular polyamid resins have been utilized as yet in the analysis of nucleic acid constituents (45).

Pellicular resins are characterized by the chemical composition of the resin shell, including the degree of cross linking, the ionic capacity, and so forth, by the thickness of the shell, and by the overall particle diameter, which is usually expressed by the particle size range or by the pertinent sieve fraction. The effect of all these parameters has not yet been investigated. The analytical results described here have been obtained either with pellicular resins having a mean particle diameter of 50 μ (sieve fraction No. 270–325) and an average shell thickness of about 1 μ or less, or with 20- to 37-μ Zipax resins.

In many respects, the chromatographic behavior of pellicular ion-exchange resins is unlike that of conventional resins because of their structure. The major differences as observed with Pellionex-type resins are as follows.

1. Columns packed with pellicular resins withstand large pressure gradients without undesirable changes in the packing structure; hence long columns can be operated at high liquid velocities in suitable high-pressure liquid chromatographs.

2. The packing structure of such columns does not alter even with drastic changes in temperature or eluent composition. The lifetime of the column is long, even under greatly varying elution conditions, and such columns are particularly suitable for gradient elution chromatography.

3. Since columns can be packed with pellicular resins in a dry procedure, long and narrow-bore columns (ID down to 0.5 mm) can be easily prepared. The use of small-bore columns is particularly adavntageous in the analysis of minute quantities of samples.

4. Because the resin in a pellicular ion-exchange column is in a finely dispersed form on the surface of uniform, fluid-impermeable spheres, solute equilibration takes place much faster in such columns than in conventional ion-exchange columns. As a result, at high eluent velocities pellicular ion-exchange columns have higher efficiencies than standard ion-exchange columns, and the speed of analysis can be greatly increased. In quantitative terms, the number of theoretical plates per second—a convenient measure of the speed of analysis—can be increased by one or two orders of magnitude

when using pellicular resins instead of conventional resins in ion-exchange chromatography (42). Under comparable conditions the higher column efficiency is equivalent to less peak spreading and thus to higher solute concentration in the effluent. Therefore smaller quantities can be determined with a given detector when using a more efficient column.

5. The degree of cross linking of pellicular resins can be significantly lower than that of ion-exchange resins used regularly in chromatography. Thus the analysis of large-molecular-weight solutes can be facilitated.

6. Pellicular ion-exchange resins can be purified more easily than conventional resins. Therefore column bleeding, which often hampers the analysis of small samples by gradient elution, can be greatly reduced.

7. The sample loading capacity of pellicular ion-exchange columns is lower than that of conventional ion-exchange columns having the same dimensions because of the smaller amount of resin per unit column volume. Consequently, high detector sensitivity is required.

8. The amount of stationary phase per unit column volume is smaller in pellicular ion-exchange resin columns than in conventional ion-exchange columns, but the volume of the mobile phase is practically the same. As a result, a given solute can be eluted with weaker eluents than are needed in conventional ion-exchange chromatography, and the elution of ordinarily strongly retarded solutes does not require extreme conditions. This feature of pellicular resin columns can be of great advantage in the analysis of polynucleotides, which are difficult to elute from conventional ion-exchange columns because of their high net charge and absorptivity.

The chromatographic behavior of pellicular sorbents has been discussed in detail from both the theoretical and the practical point of view (44). As it is seen from the foregoing consideration, pellicular ion-exchange resins are particularly suitable for making the efficient narrow-bore columns required for high-sensitivity analyses. Additionally, with such columns the time of analysis can be reduced. At present, a major disadvantage of pellicular resins is that their preparation is cumbersome and therefore they are expensive. The same applies, however, to narrow size fractions of ultrafine conventional ion-exchange beads. Although pellicular resins can be very stable once they are packed in the column, handling them (e.g., vigorous stirring of their suspension) may result in damage to the resin layer. The relatively low capacity of columns packed with pellicular resin also appears to be a disadvantage if large amounts have to be separated or if a sensitive detector is not available.

Since the technology of manufacturing pellicular resins is in the developmental stage, the author has encountered products which gave columns of insufficient stability, that is, the pressure resistance of the column increased continuously during operation. This phenomenon has been attributed to separation of resin fragments from the core, which then causes plugging of interparticular channels in the bed. Such difficulties occur if the resin structure cannot withstand swelling without fragmentation or if the bond between the resin layer and the core surface is inadequate.

IV. LIQUID CHROMATOGRAPH

1. Description of the Instrument

The potential of narrow-bore columns and of pellicular ion-exchange resins can be fully exploited only with an adequate liquid chromatographic system featuring accurate flow and temperature control, as well as a sensitive detector that has a small volume flow cell and can be operated at relatively high column inlet pressures. The construction of such instruments requires the successful solution of complex engineering problems. "Homemade" systems can be employed quite successfully, although the technical problems involved in building an instrument can be avoided by using a commercial product. The Varian LCS 1000 liquid chromatograph (Varian-Aerograph, Walnut Creek, Calif.) has been tested in the author's laboratory and found to be suitable for the analysis of minute quantities of nucleic acid constituents under a variety of conditions. Therefore it appears to be appropriate to follow the construction of this instrument in describing a high-pressure liquid chromatographic system.

In addition to the essential components, the LCS 1000 instrument features a number of convenient control devices, so that routine operation does not demand special skills. The instrument can be considered to consist of three fundamental parts: (a) the eluent flow generator and control system; (b) the column and sample injection port, which are enclosed in a temperature-controlled air bath; and (c) the detector and strip-chart recorder. A flow chart of the instrument is shown in Figure 8. Some features of this chromatograph are discussed in a paper of Burtis et al. (38).

A. GENERATION AND CONTROL OF ELUENT FLOW

A chromatographic run can be performed in the gradient elution mode or in the isocratic mode. In isocratic elution only one eluent is used; this is transferred from its reservoir into the deaerating and mixing chamber pneumatically as shown in Figure 8. In gradient elution, a given volume of a weak eluent (eluent I) is placed in the chamber before starting. After sample injection, the strong eluent (eluent II) is pumped from its reservoir continuously into the chamber through a flowmeter by a positive displacement pump. Thus the strength of the eluent which is pumped from the chamber

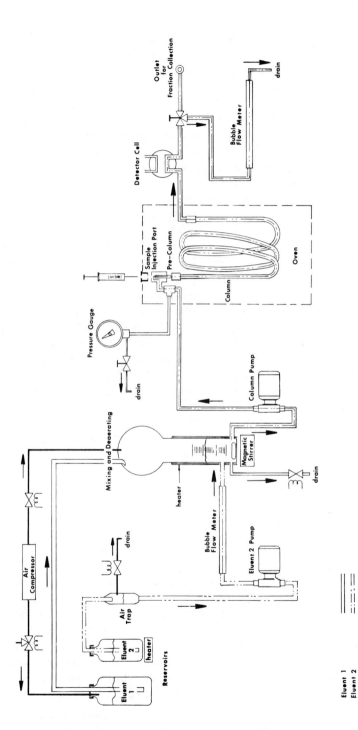

Figure 8. Flow sheet of a commercial high-pressure liquid chromatograph for nucleic acid analysis (Varian Aerograph, model LCS 1000).

101

through the column is gradually increased during the chromatographic run. The pumping of eluent II can be started at the beginning of the chromatographic run or with a delay after a desired number of the early peaks have been eluted in an isocratic fashion. The required time delay can be preset on a timer which starts the pumping of eluent II automatically.

The eluent in the deaerating and mixing chamber is heated and stirred in order to remove dissolved air and to provide uniform composition. The heater has two stages, one for quickly raising the temperature of the liquid to a predetermined limit and another for maintaining this temperature. The switching from the first stage to the second occurs automatically, and the temperature of the eluent can be monitored.

From the mixing chamber the eluent is pumped with a high-pressure pump through the sample injection port into the column. The reciprocal displacement pump of the instrument works satisfactorily at inlet pressures up to 3000 psi (200 atm) in the flow rate range from 10 to 100 ml/hr. The column inlet pressure is monitored by a pressure gauge. It is equipped with capillary bleed so that, when the eluent is changed, the liquid line leading to the column inlet as well as the gauge can be flushed out rapidly. The relatively large Bourbon tube of the pressure gauge and the long column packed with fine particles dampen very effectively the pulsating flow generated by the piston-type pump, so that neither the detector response nor the proper functioning of the flowmeter is affected. Flow pulsation appears to have no detrimental effects on the efficiency or stability of columns packed with pellicular sorbents (45).

The column effluent flows through the detector cell and the bubble flowmeter and can be diverted either to an outlet for fraction collection or to the drain. The flow rate is calculated from the time necessary for an air bubble injected into the liquid stream to pass through the calibrated volume of the glass capillary tube of the flowmeter. Hence the measurement is practically independent of the density or viscosity of the eluent. The instrument is equipped with an electrical timer and with two glass capillaries having 100- and 200-μl capacities.

One practical advantage of using narrow-bore columns is that low flow rates give high eluent velocities. The amount of eluent used, is small, therefore, in comparison with that needed in conventional systems. For instance, 1 liter of eluent can provide continuous operation for several days. The technical problems connected with high-pressure operation are also greatly reduced at low flow rates because small-outside-diameter tubing can be used throughout in the instrument and only a small amount of liquid is kept under high pressure. However, maintaining constant flow at low flow rates demands proper maintenance of the pump. The flow rate has to be checked frequently, and any malfunction of the pump valves due to entrapment of small air bubbles or dirt must be corrected immediately. Under normal conditions the flow rate variation is less than $\pm 1\%$, which is tolerable for obtaining good chromatograms. The eluent is in contact only with No. 316 stainless steel, glass, and Teflon, which are resistant to most eluents in a wide pH and temperature range. With phosphoric acid- or phosphate-containing eluents, which have been used most commonly in the separation of nucleic acid constituents, no difficulties have been observed even at pH 2.0 and 70°C.

B. SAMPLE INTRODUCTION

The sample injection port and the column are mounted in the oven, in which temperature can be maintained at a constant level between 30 and 90°C or kept at room temperature. The eluent is pumped through a small-bore tube, which is installed in the oven and serves as a heat exchanger, into the sample injection device and then through the column. The outlet end of the column is connected with a short capillary tube to the detector cell, which is outside the oven.

The introduction device is shown in Figure 9. The sample is introduced with a microsyringe, preferably at intermittent flow. After the system is ready for analysis (e.g., the column has been reconditioned, the eluent flow is stopped, and there is no pressure drop across the column), the injection port is opened and the sample is injected. The syringe needle and the sample volume are accepted either by a short precolumn, as shown in Figure 9, or by the column itself if it is inserted into the tee. In order to avert plugging of the needle, the precolumn or the column, when it is used directly, should contain a 1- or 2-cm packing of plain glass beads. If the precolumn is packed only with glass beads, it should be as short as possible in order to diminish unnecessary peak broadening. The sample can also be introduced into the eluent stream when the plug of the injection port is replaced by a cap with an elastomer septum.

Comparison of the two injection modes shows that the reproducibility of quantitative results is much better when the sample is introduced at intermittent flow into the column. Hence sample injection at continuous flow can be justified only when the time of analysis is very short and no gradient elution is used. This may not be true, however, when conventional ion-exchange resin columns are operated at high pressures because the release of pressure may have a detrimental effect on the column packing.

With narrow-bore columns (ID 1 to 1.5 mm) the sample volume is preferably less than 20 μl, but samples up to 50 μl have been used at gradient elution without deterioration of the separation. When the sample solution is very dilute or when a trace component is of interest, the solutes may be adsorbed in a preliminary step on a suitable short precolumn from a large

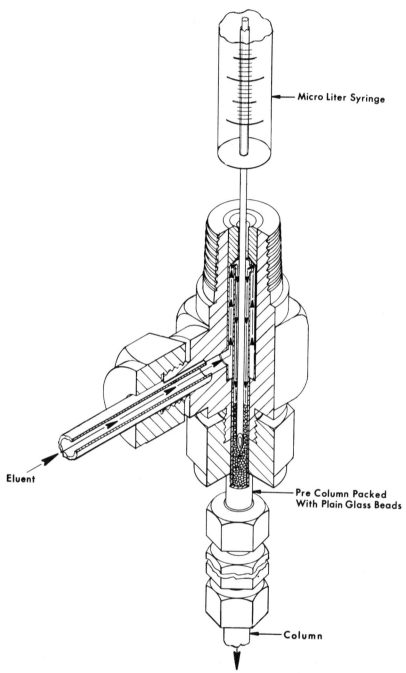

Figure 9. Sample injection port manufactured from a Swagelok TEE for injection at intermittent flow. The cap which ordinarily closes the port is removed before injection and is not shown here.

sample volume. The precolumn is then connected to the column inlet before starting elution. Optimum conditions for this comparatively tedious sampling technique have to be established for each case individually.

C. ULTRAVIOLET DETECTOR

For the analysis of nucleic acid constituents a UV detector having a low-pressure mercury lamp as the light source is often sufficient because these compounds have relatively large extinction coefficients at the single operating wavelength of 253.7 nm. When working with narrow-bore columns, however, the construction of the detector flow cell is of critical importance because the peaks are eluted in small volumes. In order to avoid significant extracolumn band broadening, a flow cell is required which has a minimum volume and minimum dead volume in the connecting tubes.

The flow cell of the UV detector is shown in Figure 10. It consists of two identical flow cells, one for the column effluent and another for the reference. Although in most applications the use of air as reference is adequate, a reference stream through a reference column may be necessary in some gradient elution techniques in order to compensate for the increasing light absorption of the effluent. The light path length and diameter of these cylindrical cells are 10 mm and 1 mm, respectively. Thus the cell volume is 8 μl, and the effective volume/light path length ratio is approximately 0.8 mm^2, which is very small in comparison with that of most commercial detectors. The noise level of the detector with stagnant liquid has been found to be about 2×10^{-5} absorbance unit and the drift less than 2×10^{-4} AU/hr. With liquid flowing through the cell, the noise level is higher, depending on the nature of the liquid and the flow conditions. The detector output is linear, and at the highest sensitivity, when the detector is used with a 10-mV recorder, full scale on the chart paper corresponds to 0.02 AU, so that a peak with a height of 0.001 AU can be easily recognized. Under carefully adjusted conditions, the detector can be used with a 1-mV recorder; thus the sensitivity can be increased by a factor of 10. The sensitivity of the detector can be adjusted in binary steps so that linear output is obtained in the range of 0 to 0.64 AU. In addition, logarithmic output can be obtained in the range of 0 to 1.3 AU for the monitoring of strongly absorbing effluent.

When the detector is operated at high sensitivity, small density and temperature fluctuations in the effluent stream increase the base-line noise, presumably because of variations in the refractive index. Large sudden changes in the refractive index of the effluent may produce ghost peaks, probably as a result of the schlieren effect. These phenomena have been observed at very high detector sensitivities and can be best avoided by using low flow rates and column temperatures close to ambient.

A miniaturized UV detector which monitors the absorbance of the effluent at 254 and 280 nm simultaneously and is suitable for working at low flow rates has been developed at Oak Ridge National Laboratory (46). At both wavelengths the detector is a double-beam instrument so that a reference stream can be used to compensate for base-line drift in gradient elution. Because of the dual-wavelength feature this detector is particularly attractive for nucleic acid analysis, and the small band spreading in the flow cells, even at very low flow rates, makes it compatible with narrow-bore columns.

D. COMPONENTS FOR BUILDING AN INSTRUMENT

A simplified version of a high-pressure liquid chromatograph can also be built in the laboratory from commercially available components. The flow chart of such a "homemade" instrument can be essentially the same as shown in Figure 8 but without the pneumatic transport for the eluent, for which a gravitational system can be substituted. For pumping the eluent through the column the Milroyal (Milton Roy Co., St. Petersburg, Fla.) pump or the Whitney Laboratory (Whitney Research Tool Co., Emeryville, Calif.) feed pump can be used. Both work satisfactorily in the low flow rate range at pressures higher than the upper limit specified by the manufacturer. The stronger eluent (eluent II) can be pumped—when gradient elution is used—by a variety of pumps which provide accurate low flow rates, for example, a syringe pump or Minipump (Milton Roy Co., St. Petersburg, Fla.). A pressure gauge with capillary bleed (ACCO, Helicoid Div., Bridgeport, Conn.) is available commercially. A mixing and deaerating chamber for such an instrument is depicted in Figure 11. The custom-made glass vessel has a capacity of about 350 ml and an outlet at the bottom connected to the suction end of the high-pressure pump. The eluent is heated by a 100-W immersion heater regulated by a variable transformer and is stirred by a magnetic stirrer (Spinfin). The narrow, lower part of the vessel is calibrated. The Teflon lid has openings for the immersion heater, thermometer, and ventilation, as well as for a Teflon tubing connected to the pump of eluent II.

The injection port can be made from a stainless steel Swagelok tee (Crawford Fitting Co., Cleveland, Ohio), as shown in Figure 9. A variety of constant-temperature baths can be used. The Lauda thermostat model No. WB-20 (Brinkman Instruments, Westbury, N.Y.) has been used in the author's laboratory. A suitable UV detector similar to that used in the Varian liquid chromatograph is available as a separate unit called the "UV Monitor" (Laboratory Data Control Co., Riviera Beach, Fla.). Flowmeters can be made of glass capillary tubings (ID 1 to 2 mm). A given volume (e.g., 100 μl) is marked. The eluent flows throgh the tube, and a gas

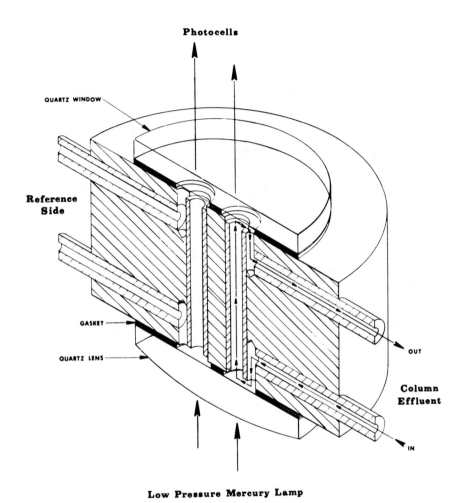

Photocells

QUARTZ WINDOW

Reference
Side

GASKET

QUARTZ LENS

OUT

Column
Effluent

IN

Low Pressure Mercury Lamp

Figure 10. Micro flow-cell of the UV-detector. Effective light path length-10 mm; ID of the cylindrical cell:1 mm. As illustrated, no reference flow is used (Laboratory Data Control).

107

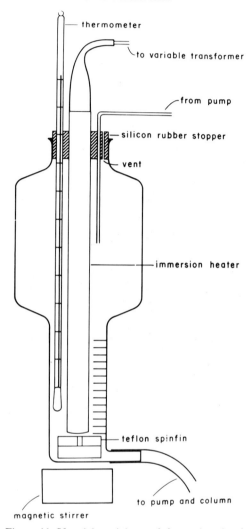

Figure 11. Vessel for mixing and deaerating the eluent.

bubble is introduced with a syringe through a septum or from a compressed gas source with a valve, for example. The time for the bubble to traverse the distance between the two calibrations is measured. From the volume of the liquid between the marks and from the displacement time the flow rate of the liquid can be calculated. As opposed to other flowmeters, this type gives the volumetric flow rate reliably even if the density or viscosity of the liquid changes during gradient elution. A commercially available flow-rate-measuring device, the "Flow Monitor" (Laboratory Data Control), operates on the same principle but with photoelectric sensors.

Stainless steel tubing (No. 316, ID 0.020 in., OD $\frac{1}{16}$ in.) can be used, preferably where the liquid is under high pressure. At low pressures Teflon tubing (Pennsylvania Fluorocarbon Co., Clifton Heights, Pa.) of various dimensions is suitable. The smallest possible inside diameter should be selected when the dead volume must be minimized. Swagelok tube fittings (Crawford Fitting Co.) can be used for making tube connections, and the column outlet is connected to the detector preferably with a $\frac{1}{16}$-in. GC union.

2. Preparation of Columns

For the analysis of minute quantities of nucleic acid constituents small-bore columns (ID less than or about 2 mm), packed with pellicular ion-exchange resins, have been used. Stainless steel tubes (No. 316) are adequate, as the analyses can be carried out with eluents which do not attack this material. Whereas the preparation of narrow-bore columns with conventional ion-exchange resins is very difficult because a wet packing procedure has to be used, pellicular ion-exchange resin columns can be packed dry in a similar way to columns in gas chromatography. Most work has been done with columns having a 1-mm inner diameter, a $\frac{1}{16}$-in. outer diameter, and lengths varying from 1 to 6 meters.

The cross-sectional area of the 1-mm-ID tubing is about 0.8 mm². On the assumption that the pellicular resin is packed in the random mode, so that the bed porosity is about 0.38, the free cross section of the column is approximately 0.3 mm². Thus the volume of the moving eluent in each 1-cm-long segment of the column is about 3 μl. Experimental data obtained with thin pellicular resins are in very good agreement with this value. As a result of the small free cross-sectional area, high eluent velocities can be obtained at relatively low flow rates. The eluent velocity, u, measured in centimeters per second can be calculated by means of [16]. If 0.3 is substituted for A_f, we find that at a flow rate of 12 ml/hr, which is frequently used in practice, the eluent velocity is roughly 1 cm /sec. This is a very high eluent velocity in comparison to the values used in conventional liquid chromatography. For instance, this liquid velocity would be achieved with a 1-cm-diameter column having the same bed porosity at a flow rate of $F = 1200$ ml/hr. The small holdup volume of narrow-bore columns and the consequently small peak volumes in the effluent make necessary the use of special instrumentation in which extracolumn dead volumes of the system (detector cell, connecting tubes, etc.) are very small as as to be in proportion to the small column holdup volumes.

The following method has been used for packing the columns. Stainless steel tubes (No. 316, ID 1 mm, OD $\frac{1}{16}$ in.) are cut into 150-cm lengths, and the necessary fittings (regular or GC Swagelok unions) are mounted on the tubes. A small $\frac{1}{16}$-in.-diameter porous Teflon disk is placed in the fitting at the outlet end of the column or into the end of the column. † The inlet end is connected to a small reservoir containing the dry resin, and the tube is packed by vibration in perpendicular position. The packing procedure is carried out most efficiently if the stainless steel tube is placed in a glass tube. When vibrated, this glass sheath transmits the vibrations very efficiently over the whole length of the metal tube. The moisture content of the resin must be controlled in order to have a free-flowing powder which gives uniform packing structure. The resin particles tend to stick, either because of electrostatic forces if they are too dry, or because of adsorbed moisture if they are too damp. The proper humidity for the resin is found by drying or humidifying the powder until it flows freely. Of course, the same applies to the screening of this material.

After the column is packed, it may be advantageous to compact the packing. This can be done by pumping an eluent, in which the resin shells do not swell appreciably, through the perpendicularly positioned column at the highest available pressure for about 1 hr. Thereafter, the pressure is released through the column outlet and the inlet end is opened. If the inlet part of the tube is void of packing, it has to be repacked or cut off to eliminate any dead volume. This column-packing technique gives highly reproducible results, as illustrated by the chromatograms obtained with five columns shown in Figure 12. Longer columns can be made by connecting separately packed shorter parts with $\frac{1}{16}$-in. Swagelok unions. Best results are obtained when the Teflon disks are removed at the connections and the bore of the fittings is also packed with the resin.

3. Gradient Elution

Although relatively simple separation problems can usually be solved satisfactorily by isocratic elution, when the separation of samples containing a large number of components is attempted preference should be given to gradient elution. With this elution mode not only can the time of analysis be reduced but the concentration of late peaks can be increased also. Thus the apparent sensitivity of the system for those solutes can be enhanced. When, for instance, nucleic acid constituents are eluted in the salt elution

---

† Thin plugs can be cut from Zitex filter membrane manufactured by Chemplast, Inc., Wayne, N. J.; thicker plugs can be cut from porous fluorocarbon sheets (grade 50–55) manufactured by Fluoro-Plastics, Inc., Philadelphia, Pa.

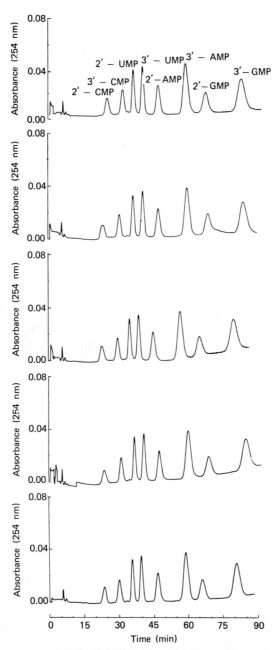

Figure 12. Chromatograms of the 2′ and 3′-ribonucleotides as obtained on five different columns under the same elution conditions. The columns were packed by the dry procedure with pellicular strong anion-exchange resin (sieve fraction No. 270-325). Columns: 1 mm ID 300 cm. Eluent: linear KH$_2$PO$_4$ gradient from 0.01M, pH 3.25, to 1.0M, pH 4.2. Inlet pressure: 320-360 psi. Flow rate: 12 ml/hr. Temperature: 70°C. Sample: 2 nmole of each component. From Burtis (70).

mode, gradient elution amounts to an increase of the salt concentration in the eluent during the chromatographic run. This is achieved in the present system by pumping a concentrated salt solution (eluent II) into the mixing and deaerating chamber, in which the dilute eluent (eluent I) is placed before the start of the run. The eluent is mixed thoroughly and pumped from the chamber through the column.

The shape of the salt concentration-time curve is calculated according to Lakshmanan and Lieberman (47) by the following equation:

$$C = C^* - C^*\left(1 + \frac{(F^* - F)t}{60V_0}\right)^{F/(F-F^*)} \qquad [27]$$

where C = momentary salt concentration (g-mole/l) in the chamber, C^* = salt concentration (g-mole/l) in eluent II, F^* = flow rate (ml/hr) of eluent II into the chamber, F = flow rate (ml/hr) of the eluent through the column, V_0 = volume (ml) of eluent I in the chamber at the start ($t = 0$), and t = elapsed time (min).

Equation [27] is not applicable if $F^* = F$, that is, if the volume of the liquid in the mixing chamber remains constant. In this case, the concentration is calculated by the following expression:

$$C = C^*(1 - e^{-Ft/60V_0}) \qquad [28]$$

From [27] it can be shown that when $F < 2F^*$ the concentration curve is concave upward, whereas when $F > 2F^*$ the curve is convex upward. When $F = 2F^*$, a linear gradient is obtained, and the concentration can be calculated simply by

$$C = C_0 + C^* \frac{Ft}{120V_0} \qquad [29]$$

where C_0 is the initial concentration of eluent I. With efficient columns, the use of linear gradients is the most convenient way to perform the separation of sample components having only slightly different chromatographic behaviors in a particular stationary phase–mobile phase system.

Figures 13 and 14 show concentration-time curves calculated by means of [27] to [29] for arbitrary conditions which are similar to those used in analytical work with this system. In both cases the salt concentration of eluent I (C_0) is zero, and the salt concentration of eluent II (C^*) is 1 g-mole/l. The flow rate of the eluent through the column, F, is taken as 15 ml/hr. The volume of Eluent I at the start, V_0, is 50 ml and 20 ml for the curves shown in Figures 13 and 14, respectively. The curves are calculated in both cases with the same set of flow rates of eluent II. It is seen that a large variety of

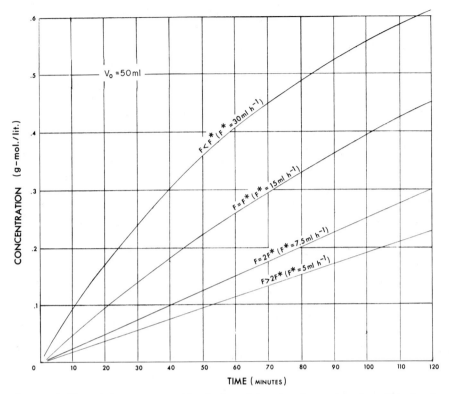

Figure 13. Plots of eluent concentration in the mixing chamber against time. Calculated for various flow rates of eluent II at fixed outlet flow rate of the eluent, $F = 15$ ml/hr; at fixed starting volume, $V_0 = 50$ ml; and at fixed initial concentrations, $C_0 = 0$, and $C^* = 1$ g-mole/l.

concentration-time curves can be obtained simply by adjustment of the flow rates and of the starting volume.

The duration of a chromatographic run is limited, of course, by the time of depletion of the eluent from the chamber. The time for removing 90% of the initial liquid volume from the chamber, t_d, is given by

$$t_d = \frac{54V_0}{F - F^*}$$ [30]

and can be used for estimating the time available for the analysis.

The concentrations calculated by [27] to [29] are those of the eluent in the mixing chamber at time t. At the inlet of the column the concentration will be different at time t because of the delay caused by the dead volume (pump,

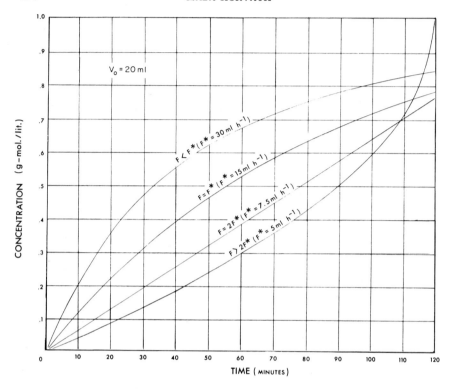

Figure 14. Plots of eluent concentration in the mixing chamber against time. Calculated for various flow rates of eluent II at fixed outlet flow rate of the eluent, $F = 15$ ml/hr; at fixed starting volume, $V_0 = 20$ ml; and at fixed initial concentrations, $C_0 = 0$, and $C^* = 1$ g-mole/l.

injection port, connecting tubing, fittings, etc.) between the outlet of the chamber and the inlet of the column. When necessary, correction can be made after calculating the dead volume or determining the delay with a tracer such as a dye. The delay time is obtained by dividing the dead volume by the flow rate. Under certain conditions the separation of early peaks is best accomplished by isocratic elution, which is then followed by gradient elution of the later peaks. In this case, the flow of eluent II has to be started at a later time.

When eluent II has a higher absorbance than eluent I or the column bleeds with increasing eluent strength, the recorder base line will drift upward during gradient elution. However, downward drift of the base line has also been observed, particularly with multicomponent eluents; this is probably caused by "salting in" of trace impurities as the salt concentration of the eluent increases. Sometimes the appearance of ghost peaks can also be attributed to impurities which accumulate on the column and are suddenly eluted at a certain eluent strength. Such difficulties can be alleviated if a matched reference column is connected parallel to the separating column so that its effluent passes through the reference side of the UV detector cell. Of course, the flow rates in the two columns should be equal. In most cases, however, the use of a single column is sufficient if the eluent does not contain UV-absorbing substances. Unfortunately, even analytical-grade KH_2PO_4 may contain polyphosphates which have rather strong UV absorption at 254 nm. The reagent can be purified by taking advantage of the fact that polyphosphates are less soluble in water than the phosphate, and making up the eluent from the mother liquor obtained from the recrystallization of analytical-grade KH_2PO_4. When gradient elution is used in ultrasensitive analysis, however, a drastic reduction of the UV-absorbing impurity content is necessary. An elaborate procedure for the purification of KH_2PO_4 has been described (48). It involves frontal chromatography of phosphate solution on a Dowex-1 anion-exchange column after a sequence of crystallization steps. After each step but one, the crystals are discarded and the mother liquor is used in the next step to gain purified material. The absorbance of $1M$ KH_2PO_4 has been reduced from 0.24 to 0.005 by this technique.

V. ION-EXCHANGE CHROMATOGRAPHY OF NUCLEIC ACID CONSTITUENTS

Ion-exchange chromatography of complex organic molecules such as nucleic acid constituents encompasses a variety of phenomena taking place in the column. Since the chromatographic process is theoretically not well understood, this technique has been developed on a largely empirical basis. The wealth of information accumulated in the past two decades, however, greatly facilitates the selection of both the proper chromatographic system and the optimum operating conditions, as well as the prediction of the chromatographic behavior of the individual sample components. The excellent reviews by Cohn (49–52) deal with the ion-exchange chromatography of nucleic acid constituents in great detail. The reviews of Boulanger and Montreuil (53) and Saukkonen (54) include ion-exchange chromatography.

Because of the ampholytic nature of these substances, both anion- and cation-exchange resins have been used successfully. The separation of small fragments is usually achieved with polystyrene-type strong ion-exchange resins, because these materials have been available with controlled properties and they possess relatively high selectivity. Cellulose ion exchangers and polyethyleneimine-cellulose (55,56), which are frequently used in thin-layer chromatography, have found only limited application in column chromatography of low-molecular-weight nucleic acid constituents. Their use in the separation of larger fragments, however, appears to be more significant.

Originally, cation- and anion-exchange columns were used with strong acidic and strong basic eluents, respectively (1–3,57). Since the introduction of the milder salt elution method (16), however, weakly acidic salt solutions have been used as eluents with both types of resins almost exclusively. Further refinement of the technique has been achieved by employing stepwise (58) or gradient (59,60) elution modes. Today, gradient elution is used most commonly when the affinities of the sample components to the resin vary over a wide range. Although the separation of bases and nucleosides has been performed on anion-exchange columns (5,6), cation-exchange chromatography has proved more efficient and convenient for this purpose (8,16,38,39,41,43,61,62). Nucleotides have been analyzed preferably on anion-exchange columns (6,38,40,63–65), but cation-exchange chromatography has also been successfully employed (57,66,67).

In the following sections the application of novel high-performance liquid chromatography to high-sensitivity analysis of nucleic acid constituents is discussed. Although the basic concepts of ion-exchange chromatography remain unchanged, this development represents a significant improvement, both in speed and sensitivity of analysis, over earlier work. It stems from the employment of novel high-efficiency resins in narrow-bore columns, from the high-pressure capability of the chromatograph, and from the use of very sensitive detectors. The results are illustrated by chromatograms, and the detailed description of operating conditions may serve as a guide to selecting the proper system. All chromatograms were obtained with sensitive UV detectors, which are most suitable for this type of work and were operated without a reference stream. The use of pellicular resins is discussed in more detail than that of conventional resins because of the unusual properties of the former types.

The analytical results presented in the following sections were obtained partly with homemade instruments and partly with commercial high-performance liquid chromatographs such as those described previously. The precision and reproducibility attained with proper instrumentation are comparable to those of gas chromatography.

VI. RAPID ANALYSIS AT THE SUBNANOMOLE LEVEL, USING PELLICULAR ION-EXCHANGE RESINS

1. General Considerations

The chemical nature of pellicular ion-exchange resins employed so far in the analysis of nucleic acid constituents has been similar to that of the conventional resins generally used. Therefore the basic chemistry of the chromatographic process has remained unchanged. As discussed in previous sections, however, pellicular resins possess some highly desirable properties, such as mechanical strength, good handling characteristics, and fast stationary-phase mass transfer rates in comparison to conventional resins of the same particle size. Since these properties are attained essentially by substituting glass for the core of the particles, a large fraction of a pellicular resin column volume is occupied by inert glass. Consequently, it contains 10 to 100 times less actual ion-exchange resin than a conventional ion-exchange column of the same volume. Although the ratio of stationary-phase volume to mobile-phase volume in the column does not differ to the same extent, the necessary eluent strength is invariably lower with pellicular resins than with conventional resins, and other chromatographic conditions have to be adjusted also because those established with conventional resins are not directly applicable.

A. COLUMN LOADING CAPACITY

The loading capacity of a column is defined as the amount of sample that causes the column efficiency to drop 10% below the value that would be obtained by extrapolating to zero sample size. Because of the relatively small amount of resin present in pellicular resin columns their loading capacity is relatively low. Figure 15 shows the effect of the sample size on the column efficiency and on the adjusted retention volume, as measured with a 1-mm-ID column packed with a pellicular cation-exchange resin. The solute (cytosine) was only slightly retarded ($D_m = 0.7$), and the loading capacity is somewhat less than 10 nmole under these rather unfavorable conditions. Other measurements indicate that for highly retarded peaks or in gradient elution the loading capacity of columns having the same diameter can be 10 to 100 times higher. The loading capacity could also be increased by increasing the resin shell thickness. However, column efficiency would then decrease, and at fixed particle size a compromise must be reached between loading capacity and column efficiency. At fixed resin layer thickness, of course, the loading capacity can be increased by decreasing the particle diameter. It is

118   CSABA HORVATH

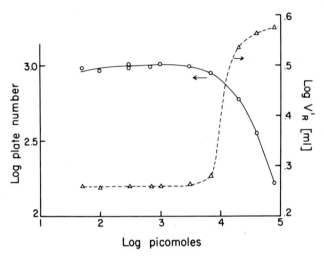

Figure 15. Column loading capacity as illustrated by plots of plate number, N, and adjusted retention volume, V'_R, against sample amount. Column: pellicular strong cation-exchange resin, sieve fraction No. 270-325; 1 mm ID, 300 cm. Eluent: 0.02M NH$_4$H$_2$-PO$_4$, pH 5.5. Inlet pressure: 1880 psi. Flow rate: 24 ml/hr. Temperature: 69°C. Sample: cytosine, D_m = 0.7. From Horvath and Lipsky (41).

seen in Figure 15 that the decrease of column efficiency with increasing loading is accompanied by an increase of the retention volume in this particular case, although in chromatography retention values usually decrease with increasing loading. This effect and the observed sharp rear boundary of the peaks at overloading indicate that the sorption isotherm is not of the usual Langmuir type but is of the anti-Langmuir type.

Usually the relatively low sample capacity of pellicular resin columns does not present a problem when small samples are analyzed. However, when trace components which are eluted in an unfavorable position close to a major component must be determined, rechromatography of the proper fraction of the effluent is advisable.

B. EFFECT OF pH

Both the retardation and the spreading of solute bands are dramatically affected by the pH of the eluent when salt elution is employed. This is exemplified by the variation of the adjusted retention volumes of nucleosides with pH as shown in Figure 16. At fixed salt concentration and temperature the equilibrium constants decrease with increasing pH at both 30° and 60°C, because ionic interactions between the solutes and the resin are reduced as a

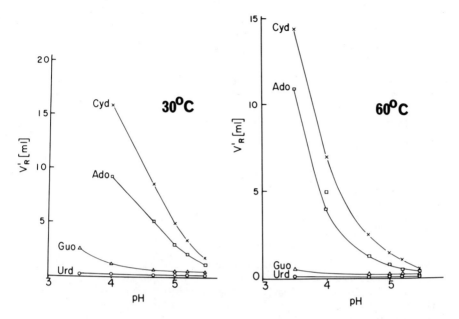

Figure 16. Effect of pH and temperature on the adjusted retention volume of nucleosides. Column: pellicular strong cation-exchange resin. Eluent:0.02M KH$_2$PO$_4$. From Horvath and Lipsky (41).

result of decreasing net charges carried by the solutes. This phenomenon is discussed by Cohn (42) in detail.

Changing the eluent pH often affects the peak shape and the broadening of the individual peaks differently because the degree of ionization appears to have an influence also on the dynamics of the solute-resin interactions. This is shown in Figure 17, which illustrates the shape and position of guanine peaks eluted at different pH values of 0.02M KH$_2$PO$_4$. The peak splitting in the pH range from 4.7 to 5.2 suggests the presence of two equilibria. Such doublets have also been observed with adenosine in a certain pH range. Although similar phenomena may also occur with conventional resins, the short elution times and the relatively high column efficiency render their observation more probable with pellicular resin columns. Whereas the appearance of double peaks when a sharp eluent gradient is employed is well established, this type of peak splitting is poorly understood.

The effect of pH on the separation has to be considered also in gradient elution. The eluent system can be designed so that only the salt concentration changes and the pH remains constant during the gradient run, or both may

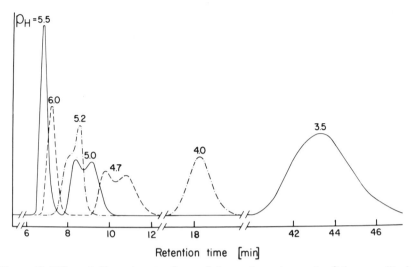

Figure 17. Effect of pH on the retention and shape of guanine peaks. Column: pellicular strong cation-exchange resin, sieve fraction No. 270-325; 1mm ID, 300.7 cm. Eluent: 0.02M KH$_2$PO$_4$. Flow rate: 14.8 ml/hr. Temperature: 60°C. From Horvath and Lipsky (41).

change. Separation efficiency can often be enhanced by employing a pH gradient simultaneously with the concentration gradient.

C. EFFECT OF TEMPERATURE

Column temperature affects the separation in three ways. First, usually sharper and more symmetrical peaks are obtained with increasing temperature because the rate of diffusion and the sorption-desorption process increase with the temperature. Second, the retention values usually decrease with increasing temperature, and hence the peaks are eluted faster. Third, the relative retention values often increase with decreasing temperature. Which effect is the most significant depends on the particular solutes and the chromatographic system, as illustrated by the chromatograms of bases in Figure 18. At 30°C the last two peaks are excessively broad and the retention times are very long, but by raising the temperature to 60°C a good separation is obtained in much less time. However, in this case the relative retention of the last two peaks (Ade-Cyt) actually increases with the temperature, whereas the relative retention of the first two peaks (Ura-Gua) decreases, following the usual pattern. Under similar conditions the relative retention of the corresponding nucleosides is affected by temperature in the same way, as indicated by the data in Figure 16.

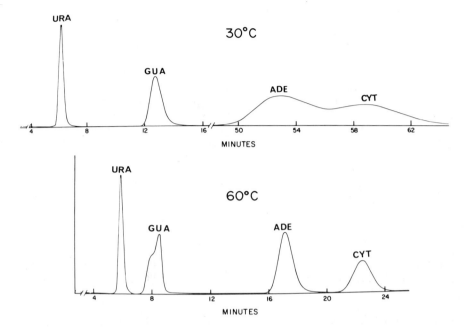

Figure 18. Effect of temperature on the separation of the bases. Column: pellicular strong cation-exchange resin, sieve fraction No. 270-325; 1 mm ID, 300.7 cm. Eluent: 0.02M KH$_2$PO$_4$, pH 5.2. Flow rate: 14.8 ml/hr. From Horvath and Lipsky (41).

Table II shows column efficiency values measured with bases at column temperatures of 39 and 69°C. About twice as many plates are obtained by raising the temperature, and the improvement is even greater in the values of plates per second and plates per atmosphere. The effect of temperature on the separation of a particular sample depends also on the pH, which, of course, changes also with temperature, and on the salt composition of the eluent and must be investigated individually.

In rapid analysis the residence time of the sample components in the column is short; and even if their decomposition is accelerated at elevated temperature, this may not be observable. When the reduction of retention time is more pronounced than the enhancement of decomposition rates as a result of raising temperature, then actually less solute decomposition occurs when the separation is performed at elevated temperature because the decrease in reaction time (residence time) overcompensates for the increase in decomposition rate. Although pellicular resin columns have frequently been employed at temperatures up to 80°C, no decomposition of simple nucleic acid constituents has yet been observed.

TABLE II

Effect of Temperature on the Efficiency of Separation[a] (41)

Com-pound	Plate number		Plates per second		Plates per atmosphere		Resolution		Resolution per second × 10³	
(°C):	39	69	39	69	39	69	39	69	39	69
Uracil	1531	1848	4.0	5.1	14.7	26.9				
							3.50	1.95	7.2	4.9
Guanine	721	1475	1.2	3.4	6.93	21.5				
							4.92	3.54	4.8	6.4
Adenine	548	1134	0.38	1.7	5.27	16.5				
							0.93	1.61	0.6	2.2
Cytosine	894	1738	0.54	2.2	8.60	24.4				

[a] Column: pellicular strong cation-exchange resin, sieve fraction No. 270–325, 1 mm ID, 300.7 cm. Eluent: 0.02M $(NH_4)H_2PO_4$, pH 5.5. Flow rate: 14.1 ml/hr.

Figure 19 shows the effect of the temperature on the separation of 2' and 3'-ribonucleotides on a pellicular cation-exchange column. The isomer pairs can be rapidly analyzed at elevated temperature, while at low temperature the individual isomers are also partially separated and their analysis becomes difficult. This example illustrates that high column temperature can be useful in achieving group separations.

The technique of raising column temperature during the chromatographic run (temperature programming) has been employed in the author's laboratory successfully in some cases. For example, the sample in Figure 18 was separated by raising column temperature from 25 to 85°C during the run. The resolution of both solute pairs was improved, and the time of separation was reduced. In liquid chromatography, however, this technique appears to be less effective than gradient elution, which can produce similar effects.

D. EFFECT OF SALT CONCENTRATION

As a consequence of the relatively small resin/liquid volume ratios in pellicular resin columns, the necessary salt concentration of the eluent is lower than in columns packed with conventional resins. This is clearly seen by comparing the eluents in Tables IV and VI. Nucleic acid constituents are usually eluted from pellicular cation-exchange resin with 0.01 to 0.1M salt solutions, although frequently dilute acid solutions can also be employed without added salt. Zipax resins usually require weak eluents with respect to

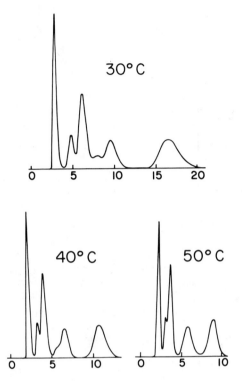

Figure 19. Effect of temperature on the separation of ribonucleoside 2′ and 3′-mono-phosphoric acids. Elution order:UMP-s, GMP-s, CMP-s, and AMP-s. Column: pel-licular strong cation-exchange resin, sieve fraction No. 270-325; 1 mm ID, 152 cm. Elu-ent: $0.01M$ $HN_4H_2PO_4$, pH 2.5. Inlet pressure: 2000 psi. Flow rates: 14.8 ml/hr. (30°C) 18.7 ml/hr (40°C), 19.3 ml/hr (50°C). Time scales: minutes. Sample: 750 pmole of each isomer pair. From Horvath (45).

salt concentration because of their lower capacity. Generally, the salt con-centration is lower by a factor of at least 10 when pellicular resins are used instead of conventional resins. This is true also in gradient elution, so that the difficulties arising from working with concentrated salt solution are minimized.

Increasing salt concentration results in shorter retention time if everything else is kept constant, but the exact relationship is usually not known. For 3′-AMP and 3′-UMP the retardation ratio, R, on pellicular anion-exchange resin has been found to depend on the concentration of KH_2PO_4 according to the following equation:

$$R = a \log C + b \qquad [31]$$

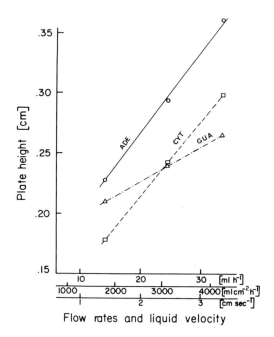

Figure 20. Effect of flow velocity on column efficiency. Column: pellicular strong cation-exchange resin, sieve fraction No. 270-325; 1 mm ID, 300 cm. Eluent: $0.02M$ KH$_2$PO$_4$, pH 5.5. From Horvath and Lipsky (41).

where C is the salt concentration, and a and b are constants. This relationship holds in the molarity range from 0.03 to 0.6 (40).

E. EFFECT OF FLOW RATE ON COLUMN EFFICIENCY

Figure 20 shows a plot of the plate height versus the flow velocity as measured with bases on a 1-mm-ID pellicular cation-exchange resin column. As expected, column efficiency decreases with increasing flow velocity.† On the other hand, the time of analysis decreases, while the necessary inlet pressure increases, with increasing flow velocity, so that in practice a compromise is sought. With small-bore (ID 1 to 3 mm) pellicular resin columns the flow velocity range from 1 to 2 cm/sec represents a practical optimum, so that most analyses have been carried out at such velocities. Figure 20 also shows the corresponding flow rates, expressed as milliliters per hour and as milliliters per square centimeter per hour, to allow com-

† The flow velocity, u(cm/sec), can be estimated from the flow rate, F (ml/hr), and column diameter, d (mm), by the formula $u = F/10d^2$ with pellicular resin columns.

parison of these flow conditions with those in other liquid chromatographic systems. The data in Table III also demonstrate that separation can be performed faster at higher flow velocities because the number of plates per second is higher, but only at the cost of increased inlet pressure as the number of plates per atmosphere of inlet pressure drops significantly.

TABLE III

Effect of Flow Rate on Column Efficiency[a] (41)

Compound ———— Flow rate (ml/hr):	Plates per second		Plates per atmosphere	
	14.1	33.4	14.1	33.4
Uracil	5.1	8.24	26.9	8.05
Guanine	3.4	5.45	21.5	6.37
Adenine	1.7	2.76	16.5	4.86
Cytosine	2.2	3.2	24.4	6.82

[a] Column: pellicular strong cation-exchange resin, sieve fraction No. 270–325; 1 mm ID, 300.7 cm. Eluent: $0.02M$ $NH_4H_2PO_4$, pH 5.5. Temperature: 69°C.

2. Selection of Operating Conditions

A. OPTIMIZATION OF SEPARATION EFFICIENCY

In ion-exchange chromatography the selection of the proper resin plays a highly important role, and this applies to pellicular resins, too. Even if the material is standardized by the manufacturer for a given separation, batch-to-batch variations in resin properties may occur and affect significantly the efficiency of other separations. It should be kept in mind that, in addition to the chemical composition and structure of the resin, the shell thickness of a pellicular resin also has a great effect on its chromatographic behavior. Therefore the specifications of these materials should include the actual resin content per gram, as well as the ion-exchange capacity and particle size for a given type of resin. The quality of column packing is another highly important factor. Improper packing can be the cause of low efficiency, severe tailing, and poor reproducibility. Since the packing is not observable in metal tubings, strict adherence to established packing and conditioning procedures is advisable. The use of metal pumps and metal columns in high-performance liquid chromatography, as well as the high-sensitivity detectors and the long column life required, imposes some limitation on the eluent composition.

A great advantage of pellicular resin columns is the fact that screening experiments can be carried out quite quickly also when gradient elution is used because of the short column-reconditioning period. An unknown sample is tested, preferably with a linear gradient, in order to estimate its composition and the necessary elution conditions. The flow velocity is kept preferably between 1 and 2 cm/sec. If isocratic elution is adequate, then the optimum pH of the eluent has to be established. Frequently, changing the pH as little as 0.1 can have a significant effect on the separation and peak shape. When the column temperature can be easily adjusted, it is advisable to explore the effect of temperature changes over a relatively broad range. High column temperatures are generally preferred because of the shorter elution times. The effect of increasing ionic strength is sometimes similar to that of increasing temperature. It can be more convenient, therefore, to fix the salt concentration and vary the temperature. The ionic composition of the eluent may have dramatic effects on the separation, but this area is less fully explored as yet. Under some circumstances the addition of organic solvents such as ethanol to the eluent may be useful to separate a critical solute pair.

Frequently, several of the parameters mentioned above have to be adjusted in order to obtain satisfactory results. When optimum conditions for a routine analysis are established, adjustment of the relative retention values by the above means is more expedient than reduction of band spreading by decreasing the flow velocity. Halving the flow velocity (and thus doubling the time of analysis) is usually less effective than changing the temperature 10° or the pH 0.1 of a unit. Thus the flow rate should be decreased only after other means to improve separation are exhausted.

For the separation of samples containing many components gradient elution gives the best results. In this case pellicular resins are particularly advantageous because the column can withstand drastic changes in the salt concentration of the eluent; hence gradient runs can be carried out very many times in reproducible fashion. In addition, the column can be rapidly reconditioned. Whereas the regeneration of columns with conventional resins may require 4 to 6 hr, pellicular resin columns are regenerated in 10 to 20 min. The optimization of gradient elution is complex, however, and requires trial and error. In most instances, the use of linear gradients is the simplest and most promising. The most important variables are the composition of the initial eluent, the slope of the gradient, and the composition of the stronger eluent, as well as the flow rate and the temperature. In the separation of nucleotides by gradient elution Brown (25) found that varying the temperature and pH has less effect on resolution than changing the flow rates and salt concentrations.

It is often useful to elute early peaks in isocratic fashion; that is, gradient elution is started with delay. In gradient elution, the slope of the gradient, the flow rate, and the length of the column together determine the elution pattern. Therefore the length of the column and the flow rate should preferably be fixed when the optimum gradient is established.

B. OPTIMIZATION OF THE SENSITIVITY OF ANALYSIS

In addition to the optimization of separation, ultrasensitive analysis requires that both the volume of the solute bands in the effluent and the detector noise be minimized.

The liquid chromatograph with narrow-bore pellicular resin columns previously described is well suited to analyze picomole quantities of nucleic acid constituents. Figure 21 shows chromatograms of adenosine phosphoric acids present in a mitochondrial reaction mixture. Under the given conditions 50 pmole of nucleotides can be determined reliably, and the lower limit of detection is about 25 pmole. As shown by the blank run, base-line drift may become significant at this sensitivity level when gradient elution is employed. Nevertheless gradient elution is particularly useful to increase the sensitivity of the technique.for later peaks. In order to obtain the highest sensitivity the detector noise has to be reduced. With UV detectors, the noise usually decreases with decreasing flow rate and temperature. On the other hand, column efficiency increases with decreasing flow velocity. Therefore the sensitivity of the analysis can usually be improved by lowering the flow rate.

The chromatogram of the four major ribonucleosides in Figure 22 was obtained with 12.5 pmole of each component at a flow rate of 3.25 ml/hr, which is much less than that recommended for the analysis of nanomole quantities, as shown in Figure 23. Both the column and the detector were kept at room temperature, and the time constant of the recorder was increased. Under these conditions the detector noise and base-line drift could be reduced so efficiently that full scale on the recorder could be expanded to give a range from 0 to 2×10^{-3} AU. The first two peaks in Figure 22 are artifacts, arising from disturbances associated with startup of the flow and sample injection. No such ghost peaks were observed when larger quantities of ribonucleosides were analyzed on the same instrument, using the same column and eluent system, as shown by the chromatogram in Figure 23. Comparison of Figures 22 and 23 demonstrates the possible trade-off between rapid analysis at lower sensitivity and high-sensitivity analysis at lower speed, using the same system. In the following sections the analysis of various classes of nucleic acid constituents, using narrow-bore pellicular resin columns, is discussed. All systems are capable of analyzing subnanamole quantities by proper adjustment of the conditions.

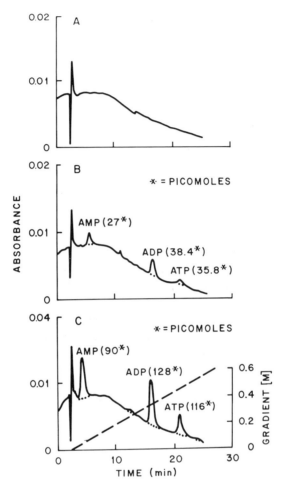

Figure 21. Chromatographs of adenine nucleotides in deproteinized mitochondrial reaction mixture. Column: pellicular strong anion-exchange resin, sieve fraction No. 270-325; 1 mm ID, 250 cm. Eluent: gradient elution with highly purified KH_2PO_4-KCl solution. Inlet pressure: 1200 psi. Flow rate: 24 ml/hr. Temperature: 70°C. (a) Blank run; (b) 3-μl sample; (c) 10-μl sample; broken line shows the eluent concentration. From Schmuckler (69).

128

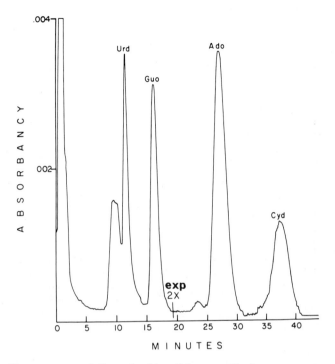

Figure 22. Chromatogram of ribonucleosides. Column: pellicular strong cation-exchange resin, sieve fraction No. 270-325; 1 mm ID, 151.7 cm. Eluent: 0.02M NH$_4$H$_2$PO$_4$,pH 5.5. Inlet pressure: 525 psi. Flow rate: 3.25 ml/hr. Temperature: 24°C. Scale was expanded by a factor of 2 in middle of chromatogram. Sample: 12.5 pmole of each component. From Horvath and Lipsky (41).

3. Separation of the Various Classes

A. PURINE AND PYRIMIDINE BASES

For the separation of bases strong cation-exchange resins and acidic eluents are most suitable. The separation of the four major base constituents of RNA on a polystyrene-sulfonic acid type of pellicular resin column is illustrated in Figure 24. The chromatogram of a more complex base mixture obtained with a fluoropolymer-sulfonic acid type of Zipax resin is shown in Figure 25. The column temperature was similar in both cases, but the Zipax column required a much weaker eluent (low pH, no salt) than the other column because the sorption capacity of the Zipax resin was lower than that of the other pellicular resin. In general the capacity of pellicular resins varies not only with their chemical composition but also with the particle

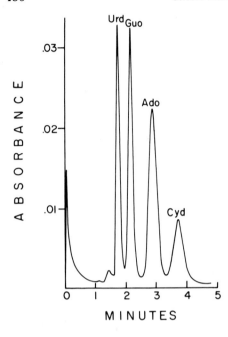

Figure 23. Chromatogram of ribonucleosides. Column: pellicular strong
cation-exchange resin, sieve fraction
No. 270-325; 1 mm ID, 151.7 cm.
Eluent: 0.02M $NH_4H_2PO_4$, pH 5.6.
Inlet pressure: 1920 psi. Flow rate:
25.5 ml/hr. Temperature: 39°C. Sample:
300 pmole of each component. From
Horvath and Lipsky (41).

size and the thickness of the resin shell; therefore the optimum eluent composition may be different for different products.

B. NUCLEOSIDES

Conditions for the analysis of nucleosides are similar to those used for the
separation of the bases, as shown in Figures 22 and 23. Uziel et al. (8) have
pointed out the advantages of the determination of the nucleic acid base
composition of the nucleoside level; both the isomeric peaks arising from
alkaline hydrolysis of nucleic acid material and the drastic treatment required to obtain the bases can be avoided when enzymic degradation
leading to nucleosides is employed.

Although strong cation-exchange resins represent the best choice for this
separation, it must be kept in mind that uridine and guanosine are not
ionized in the usual pH range of the eluent. Since the sorption affinity of
uridine to the resin is low, it is eluted at the front and under some circumstances can be difficult to separate from sample components which are not,
or only very slightly, retarded. Here the low sorption capacity of pellicular
resins is clearly a disadvantage. Since nonionic interactions are rather
pronounced with polystyrene-type resins, in order to obtain adequate
retardation of uridine such resins should be given preference.

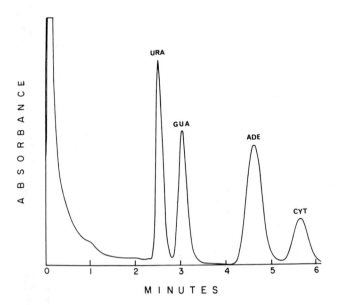

Figure 24. Chromatogram of bases. Column: pellicular strong cation-exchange resin, sieve fraction No. 270-325; 1 mm ID, 300.7 cm. Eluent: 0.02M NH$_4$H$_2$PO$_4$, pH 5.5. Inlet pressure: 2480 psi. Flow rate: 33.4 ml/hr. Temperature: 68°C. Sample: 800 pmole of each component. From Horvath and Lipsky (41).

C. NUCLEOTIDES AND OLIGONUCLEOTIDES

The separation of nucleotides on conventional ion-exchange resin columns is achieved by gradient elution, using a wide eluent strength range with regard to salt concentration (52). This method is associated with long reconditioning periods after each chromatographic run and with a deterioration of column performance, caused by swelling and shrinking of the resin.

On the other hand, nucleotides can be separated on pellicular resin columns, either in isocratic fashion or by employing gradient elution, without these setbacks. It is natural, therefore, that most of the work with pellicular resins has been centered around the analysis of nucleotides.

Both cation- and anion-exchange resins have been used with isocratic elution. Figure 19 shows the separation of 2′ and 3′ isomeric pairs of the four major ribonucleoside monophosphoric acids. When the analysis is carried out at high temperatures, the four component pairs can be easily determined. The separation of the four 5′-ribonucleoside monophosphoric acids on an anion-exchange resin column is shown in Figure 26. The Zipax column packing employed for this separation has an acrylic resin matrix.

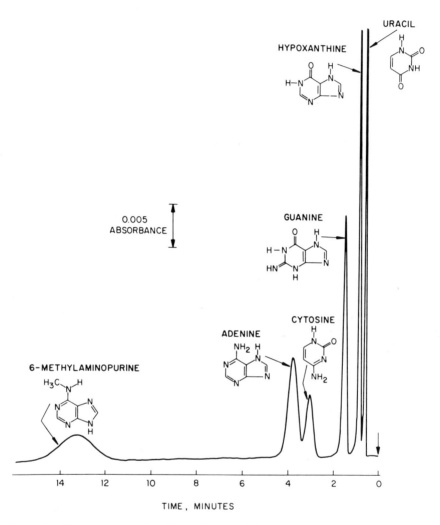

Figure 25. Chromatogram of nucleic acid bases. Column: Zipax strong cation-exchange resin, 20-37 μ; 2.1 mm ID, 100 cm. Eluent: 0.01N HNO₃. Inlet pressure: 735 psi. Flow rate: 120 ml/hr. Temperature: 63°C. Sample: 3 nmole of each component. From Kirkland (43).

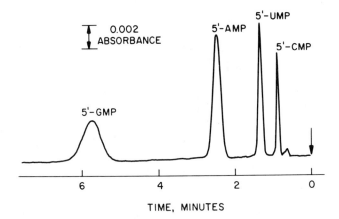

Figure 26. Chromatogram of 5'-ribonucleotides. Column: Zipax strong anion-exchange resin, 20-37 μ; 2.1 mm ID, 100 cm. Eluent: $0.006M$ H_3PO_4-$0.002M$ KH_2PO_4, pH 3.75. Inlet pressure: 900 psi. Flow rate: 114.6 ml/hr. Temperature: 60°C. Sample: about 1 nmole of each component. From Kirkland (43).

Consequently, nonionic interactions play a lesser role in solute retention with this resin than with polyaromatic ion exchangers. This explains why isocratic elution suffices with the Zipax column whereas gradient elution is necessary to separate these components on a polystyrene-type pellicular anion-exchange column of similar ion-exchange capacity (40). The 5'-deoxynucleotides have been rapidly separated on pellicular strong anion-exchange resin columns, using a concave phosphate gradient at high temperature, as shown by Figure 27.

Mixtures of ribonucleoside 2' and 3' monophosphoric acids can be completely separated on a pellicular anion-exchange resin by gradient elution, as shown in Figure 28. Kennedy and Lee (68) have used similar conditions to analyze ribonucleotides and deoxyribonucleotides at the nanomole level. Partial separation of the isomeric pairs has also been obtained in 18 min on Zipax anion-exchange resin columns with isocratic elution, using dilute acidic phosphate solutions (43).

Table IV shows chromatographic data typical for the analysis of mixtures of 2' and 3'-ribonucleoside monophosphoric acids on both conventional and pellicular strong anion-exchanger columns at inlet pressures above 1000 psi. As seen, retention times are much shorter when a pellicular resin column is employed. The greatest advantage of such columns is, however, their reusability, since conventional resin columns usually have to be replaced after each gradient run. In the author's laboratory pellicular resin columns were used for this analysis without deterioration over a year.

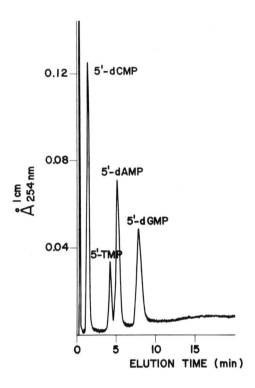

Figure 27. Chromatogram of 5'-deoxynucleotides. Column: pellicular strong anion-exchange resin, sieve fraction No. 270-325; 1 mm ID, 300 cm. Eluent: concave phosphate gradient. Inlet pressure: 2400-2800 psi. Flow rate: 57 ml/hr. Temperature: 80°C. Sample: 4-6 nmole of each component. From Burtis et al. (38).

The separation of the mono-, di-, and triphosphates of nucleosides on conventional ion-exchange columns is particularly time consuming and cumbersome (5,52). These substances can be rapidly analyzed, however, with narrow-bore pellicular anion-exchange resin columns, as shown in Figure 29. The separation improves, but the time of analysis increases, when phosphate solutions are used instead of formate buffer as eluent. The stronger eluent may be a $1M$ KH_2PO_4 solution or KH_2PO_4 in concentrated KCl solution in order to enhance ionic strength (69). Using a similar chromatographic system, Brown (25) has investigated in great detail the optimum conditions for obtaining nucleotide profiles of various cell extracts. Complete recovery of nucleotides could be obtained at column temperatures as high as 75°C with pellicular anion-exchange resin columns.

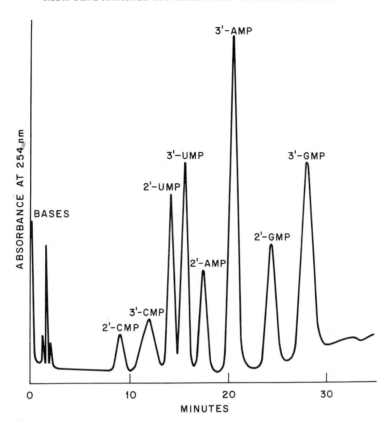

Figure 28. Chromatogram of ribonucleoside 2'- and 3'-monophosphoric acids. Column: pellicular strong basic anion-exchange resin, sieve fraction No. 270-325; 1 mm ID, 330 cm. Eluent: linear gradient from $0.01M$ KH_2PO_4, pH 3.35, to $1.0M$ KH_2PO_4, pH 4.2. Inlet pressure: 2900 psi. Flow rate: 48 ml/hr. Temperature: 75°C. Sample: 1 nmole of each isomer pair. From Horvath (45).

Adenosine mono-, di-, and triphosphates have been analyzed on strong anion-exchange resin columns using gradient elution as shown in Figure 21. In order to determine ^{32}P-labeled compounds, Schmuckler (69) placed a continuous-flow radioactivity monitor downstream from the UV detector. Figure 30 shows the chromatogram obtained with this tandem detector system. The radioactivity peaks are much broader than the absorbance peaks because of the large cell volume of the radioactivity detector necessitated by the low specific activity of ^{32}P used in the mitochondrial reaction system. Therefore the first radioactivity peak representing inorganic phosphate cannot be well distinguished from the first UV peak, AMP. The two

TABLE IV

Analysis of the 2', and 3'-Ribonucleoside Monophosphoric Acids on Conventional
and Pellicular Strong Anion-Exchange Columns

Resin	Eluent	Column		Inlet pressure (psi)	Flow rate (ml/hr)	Temp. (°C)	Sample amount per component (nmole)	Retention time (min)								Ref.
		ID (cm)	Length (cm)					2'-CMP	3'-CMP	2'-UMP	3'-UMP	2'-AMP	3'-AMP	2'-GMP	3'-GMP	
Conventional Dowex 1–X8 7–28 μ	Sodium acetate gradient from 0.05M to 1.0M (pH 4.1)	0.32	91.4	—	60	45	500	32.4	38.0	—	85.9	105.6	146.5	174.6	222.5	6
Pellicular 44–53 μ	KH₂PO₄ gradient from 0.01M (pH 3.35) to 1.0M (pH4.2)	0.1	300	2900	48	75	1	8.9	11.9	14.1	15.6	17.4	20.7	24.3	27.9	45

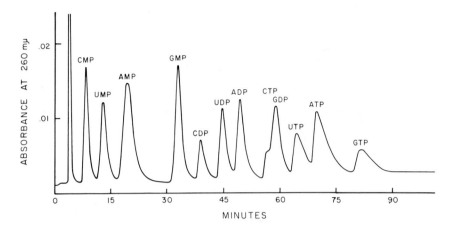

Figure 29. Chromatogram of ribonucleoside mono-, di-, and triphosphoric acids. Column: pellicular strong anion-exchange resin, sieve fraction No. 270-325, 1.0 mm ID, 193 cm. Eluent: linear gradient of ammonium formate buffer, pH 4.35 at 25°C, from 0.04M to 1.5M. Inlet pressure: 750 psi. Flow rate: 12 ml/hr. Temperature: 71°C. Sample: 1.5-3.5 nmole of each component. From Horvath et al. (40).

peaks can be better separated, however, by a preliminary run with dilute buffer before the start of gradient elution.

Brooker (71) has worked out a rapid method for the quantitative determination of 10 to 100 pmole of cyclic AMP formed by incubation of adenyl cylase with ATP. A pellicular strong anion-exchange resin column (1 mm ID, 300 cm) is used at 80°C in the isocratic elution mode with dilute HCl solution at pH 2.20 and a flow rate of 12 ml/hr. This highly sensitive and rapid technique is also useful for the assay of adenyl cylase activity in brain tissue.

Brown (25) measured the retention times of a large number of nucleotides; her data are given in Table V. Pellicular strong anion-exchange columns and linear elution gradients were employed. Retention times can vary significantly with the column material, with column conditioning, with the makeup of the eluent, or with the flow rate and temperature. However, the data in Table V give useful information on the relative retention behavior of various nucleotides under similar conditions.

Pellicular resins would be particularly suitable for the separation of large nucleic acid fragments, as suggested by Cohn (72), but this field has not yet been explored. Nevertheless the chromatogram of cytidine-containing dinucleotides in Figure 31 demonstrates that rapid separation of oligonucleotides can be achieved with such columns. The elution sequence is determined

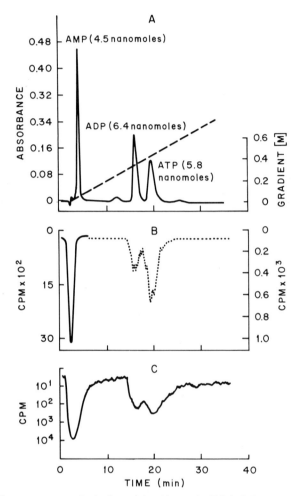

Figure 30. Chromatograms of adenine nucleotides and ^{32}P-labeled compounds in deproteinized mitochondrial reaction mixture obtained with tandem UV-radioactivity detector. Column: pellicular strong anion-exchange resin, sieve fraction No. 270-325; 1 mm ID, 250 cm. Eluent: gradient elution with highly purified KH_2PO_4-KCl solution. Inlet pressure: 1400 psi. Flow rate: 24 ml/hr. Temperature: 30°C. (a) Ultraviolet chromatogram: broken line shows salt concentration in the eluent; (b) linear radiogram; (c) logarithmic radiogram. From Schmuckler (69).

TABLE V

Retention Times of Nucleotides[a,b] (25)

Monophosphates	Time (min)	Diphosphates	Time (min)	Triphosphates	Time (min)
CMP	10	NAD	8.5	CTP	46
UMP	12	UDPG	24	UTP	49.5
TMP	14	CDP	29	TTP	54
IMP	18	UDP	31	ATP	60
AMP	19	TDP	33	GTP	65
GMP	23	ADP	38		
c–AMP	24	NADH	39		
6MMPRMP	27	GDP	43		
TGMP	32	UDPGA	44		
XMP	34				

[a] Chromatographic conditions: Varian LCS–1000 liquid chromatograph. Column: pellicular strong anion-exchange resin, sieve fraction No. 270–325, 1 mm ID, 300 cm. Linear gradient elution; eluent I: $0.015M$ KH_2PO_4, starting volume (V_0) = 50 ml; eluent II: $0.25M$ KH_2PO_4 in $2.2M$ KCl. Flow rate: 12 ml/hr. Temperature: 75°C.

[b] Abbrevations are explained in Table I or in "List of Symbols" at the end of this chapter.

by the 5′ terminal residues, since the 3′ terminals are cytidine in all four components. The elution pattern of dinucleotides from a pancreatic ribonuclease digest of RNA has been also investigated, using exponential elution gradient of acidic phosphate solutions and a pellicular strong anion-exchange column (68).

In addition to the standard methods, "enzymic peak shifts" can be employed for the identification of nucleotide peaks by an original method of Brown (25), which utilizes the specificity of certain enzyme reactions. For example, the identity of ATP and ADP peaks on the chromatogram can be verified by making a subsequent chromatographic run under identical conditions after the sample has been incubated with hexokinase in the presence of glucose and magnesium ions. According to the following reaction:

$$\text{ATP} + \text{glucose} \xrightarrow[\text{Mg}^{2+}]{\text{hexokinase}} \text{ADP} + \text{glucose-6-phosphate}$$

the ATP peak disappears and the ADP peak increases proportionally on the second chromatogram, while the peaks of guanine nucleotides and AMP remain the same. Thus the former two components can be identified. This type of enzymic peak shift is demonstrated by the chromatograms shown in Figure 32. The technique is also useful in "unmasking" a chromatogram when one nucleotide is present in relatively large quantity and its peak masks small peaks having similar retention time. In this case the specific removal of

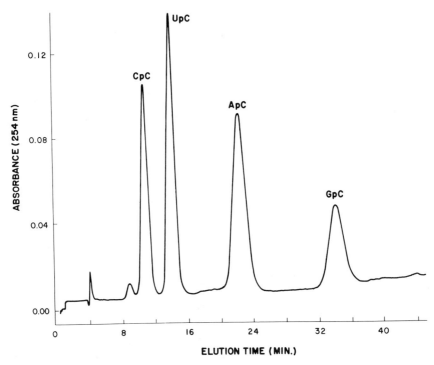

Figure 31. Chromatogram of cytidine containing dinucleotides. Column: pellicular strong anion-exchange resin, sieve fraction No. 270-325; 1 mm ID, 300 cm. Eluent: linear KH_2-PO_4 gradient from $0.01M$, pH 3.25, to $1.0M$, pH 4.2. Inlet pressure: 1300-1400 psi. Flow rate: 18 ml/hr. Temperature: 80°C. Sample: 2-4 nmole of each component. From Burtis et al. (38).

the excess component facilitates the analysis of the trace component or components. The specific removal of one of two unresolved components in a similar fashion makes quantitative analysis possible without changing the chromatographic condition in order to improve resolution. A number of possible enzyme reactions for peak shifts in nucleotide analysis, as well as technical details, are given in the paper of Brown (25).

VII. RAPID, ULTRASENSITIVE ANALYSIS USING CONVENTIONAL RESINS

The first rapid and sensitive analytical method with conventional resins (8) was introduced for the determination of nucleosides at the nanomole level. As shown in Figure 33, the four major ribonucleosides are analyzed by

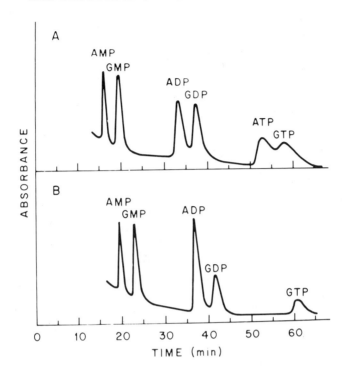

Figure 32. Enzymic shifts in trichloroacetic extract of adenine and guanine nucleotides. (A) Chromatogram of original sample; (B) Chromatogram of sample incubated with hexokinase and glucose. Column: pellicular strong anion-exchange resin, sieve fraction No. 270-325; 1 mm ID, 300 cm. Eluent: linear gradient from $0.015M$ KH_2PO_4 with $0.25M$ KH_2PO_4 in $2.2M$ KCl. Flow rate: 12 ml/hr. Temperature: 75°C. Sample: 10 μl of cell extract. From Brown (25).

isocractic elution within an hour, using a column packed with 11- to 15-μ cation-exchange resin at relatively low inlet pressures.

Recently, the availability of very fine spherical resins and high-pressure liquid chromatographs with sensitive detectors has given great impetus to further developments in the use of conventional resins for nucleic acid analysis. The introduction of the dynamic column packing method (73) makes it possible to obtain uniform and stable columns with resin particles as small as 3 to 5 μ in diameter. In agreement with the theory, the smaller the resin particles the faster is the separation if the inlet pressure can be increased, as shown by the results of Burtis et al. (38). Figure 34 shows a typical chromatogram of nucleosides as obtained with a narrow-bore column packed with 3 to 7 μ cation-exchange resin and operated at an inlet pressure of 4800 psi.

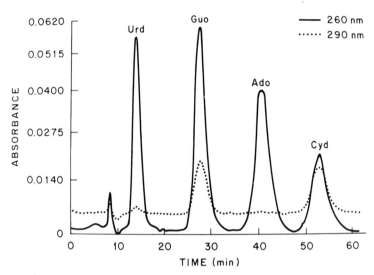

Figure 33. Chromatogram of ribonucleosides obtained with dual-wavelength UV detector. Column: Bio-Rad Aminex A-6 cation-exchange resin; 6 mm ID, 19 cm. Eluent: 0.4M ammonium formate, pH 4.65. Flow rate: 24 ml/hr. Temperature: 50°C. Sample: 4–6 nmole of each component. From Uziel et al. (8).

With the same system but a shorter column and slightly higher inlet pressure the four major ribonucleotides can be analyzed within a minute at the subnanomole level.

This technique has been applied to rapid determination of RNA base composition at the micro- and nanogram levels (74) with high precision. The results suggest that the method can be extended to the analysis of picogram quantities of the major nucleosides. The conditions employed in the analysis of the four major ribonucleosides, using conventional resins and pellicular resins, are given in Table VI. The data indicate that rapid analyses can be obtained with both systems, although conventional resins may offer some advantages in this case because uridine is only slightly retarded on pellicular resin. Hence its quantitative determination is complicated by unretained components or, at high detector sensitivity, by transient effects. The chromatogram of substituted adenosines shown in Figure 35 was obtained at lower inlet pressure with a similar system because the column efficiency had to be increased by reducing the flow rate in order to resolve the critical solute pair.

So far the application of high-performance liquid chromatography with conventional ion-exchange resins to the analysis of nucleic acid constituents is most promising when isocratic elution can be used. Nevertheless, the high

TABLE VI

Separation of the Four Major Ribonucleosides on Conventional and Pellicular Strong Cation-Exchange Columns

Resin	Eluent	Column ID, (cm)	Column Length (cm)	Inlet pressure (psi)	Flow rate (ml/hr)	Temp. (°C)	Retention times (min) URD	GUO	ADO	CYD	Ref.
Conventional Aminex A–6	0.4M ammonium formate, pH 4.65	0.6	19	—	24	50	14.2	27.3	40.2	53.2	8
Conventional Durrum DC–8X 3–7μ	0.4M ammonium formate, pH 4.75	0.24	25	4800	40	55	0.41	1.68	2.49	3.65	74
Pellicular 44–53μ	0.02M NH$_4$H$_2$PO$_4$, pH 5.6	0.1	151.7	1920	25.5	39	1.76	2.19	2.90	3.76	41

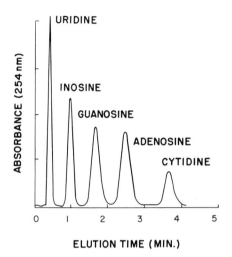

Figure 34. Chromatogram of ribonucle-osides. Column: Durrum DC-X (8X, 3-7 μ) conventional cation-exchange resin; 2.4 mm ID, 25 cm. Eluent: 0.4M ammonium formate, pH 4.75. Inlet pressure: 4800 psi. Flow rate: 40 ml/hr. Temperature: 55°C. Sample: 300-400 pmole. From Burtis et al. (70).

column efficiency makes it possible to carry out conveniently by isocratic elution some separations which would require gradient elution if traditional columns were used. The degree of resin cross linking plays an important role in determining column stability at high-pressure gradients when convention-al resins are employed. Under the operating conditions shown in the figure captions the lower limit of cross linking of polystyrene-type resins is around 8%, as expressed by the divinylbenzene content. This, of course, imposes a limit on the extension of this technique to the analysis of large nucleic acid fragments. The results demonstrate that pellicular and very fine conventional resins are about equally efficient in the rapid analysis of relatively small nucleic acid constituents when isocratic elution can be used. On the other hand, when gradient elution is necessary, pellicular resins have a distinct advantage because of the stability of the column and the rapidity of the reconditioning procedure.

Comparison of chromatograms shows that in the analysis of a given sample much higher inlet pressures are needed to obtain the same speed of analysis when conventional resin is used instead of pellicular resin, both having the same chemical composition. A comparison of the column efficiency values listed in Table VII with those in Tables II and III also supports this obser-vation. The plates per second values, which measure the speed of analysis, are very similar with pellicular resins (Tables II and III) and with con-ventional resins (Table VII). On the other hand, the plate per atmosphere values are significantly lower with conventional resins. Since these values bear a reciprocal relationship to the inlet pressure necesssary to perform the analysis at the given speed, they indicate that the inlet pressure requirement is lower for pellicular resin columns to perform the separation at the same

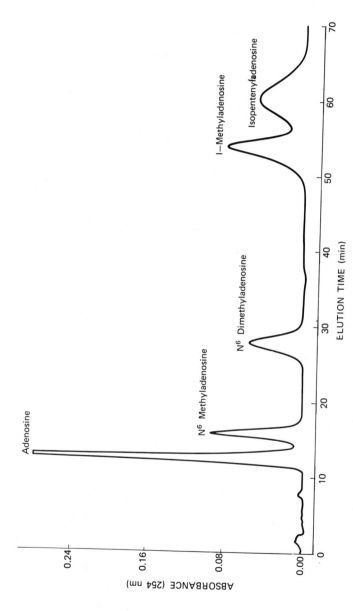

Figure 35. Chromatogram of nucleosides with substituted adenine bases. Column: Bio-Rad A-7 cation-exchange resin, particle size 7-10 μ; 2.4 mm ID, 15 cm. Eluent: 0.4M ammonium formate, pH 4.5. Inlet pressure: 2175 psi. Flow rate: 11.4 ml/hr. From Burtis (70).

TABLE VII

Efficiency Values Obtained with Three Conventional Cation-Exchange Resin Columns Used for the Separation of Nucleosides with 0.4M Ammonium Formate[a]

| Resin | Particle size (μ) | Column | | Temp. (°C) | Flow rate (ml/hr) | Flow velocity (cm/sec) | Pressure (atm) | Plate number | Plate height (mm) | Plates per second | Plates per atmosphere |
		ID (mm)	Length (cm)								
A–7[b]	7–10	2.4	25	55	35	0.54	300	806	0.31	2.6	2.69
DC–X[c]	3–7	2.4	25	55	40	0.61	326	1042	0.24	4.6	3.20
DC–X[c]	3–7	2.4	15	85	60	0.92	306	455	0.33	9.2	1.49
DC–X[c]	3–7	2.4	15	85	55	0.84	306	484	0.31	5.8	1.58
VC–10[d]	7–14	2.4	25	85	85	1.30	334	338	0.74	4.6	1.01

[a] From data of Burtis et al. (38).
[b] Aminex resin of Bio-Rad Laboratories, Richmond, Calif.
[c] Durrum Chemical Corporation, Palo Alto, Calif.
[d] Sondell Scientific Instruments, Palo Alto, Calif.

speed. In turn, we can infer that at a fixed inlet pressure pellicular resin columns can facilitate faster analysis than columns with conventional resins.

It appears that each type of resin will adopt its proper role in this type of analytical work. Obviously the relatively high capacity of conventional resins is of great advantage in the separation of large amounts, in trace analysis, or in the analysis of sample components which are not retarded sufficiently on pellicular resins or do not give strong detector signals.

VIII. SAMPLE PREPARATION

The preparation of the sample is an integral part of the analytical procedure and may greatly affect the final results. It is particularly critical when samples of biological origin are analyzed on a chromatographic system designed for rapid analysis of minute quantities. Therefore considerable attention has to be paid to proper sample preparation in order to exploit the full potential of the previously discussed techniques for nucleic acid analysis.

In order to avoid excessive band spreading, the volume of the injected sample should not exceed the volume of $\sqrt{N}$ plates, where N is the number of plates in the column. This rule of thumb is applicable to isocratic elution. Fortunately, larger sample volumes can be tolerated in gradient elution without deterioration of separation efficiency. In practice, the sample volume is between 5 and 50 μl with narrow-bore columns. Therefore the sample components have to be present at such a concentration that they can be detected but no column overloading occurs when they are injected in such a volume. The sample solution always contains other substances besides the components of interest. Some of them can interfere with the chromatographic analysis, particularly if their relative concentration is high. Most commonly they appear together at the beginning of the chromatogram as a large tailing peak and preclude the separation or quantitative evaluation of less well retained sample components. Therefore one of the major problems of sample preparation is to obtain the components to be analyzed at a sufficiently high concentration in a solution which contains the interfering substances at a low concentration.

Usually a number of preliminary steps are performed until the sample components are obtained in the desired form. Different types of samples or the same types but of different origins demand individual consideration based on the literature data (see, e.g., refs. 75–77) and also on the skill and ingenuity of the biochemist. Techniques used in paper and thin-layer chromatography are often directly applicable, or those used in conventional column chromatography can be scaled down with appropriate modifications.

The preparation of minute sample amounts means working either with very small volumes or with dilute solutions. In the first case, the syringe may be a convenient reaction vessel. A chemical reaction can also be carried out in a small tube which is connected to the inlet of the chromatographic column. After proper incubation the sample is swept into the column by the eluent. Under these circumstances, the reaction can be carried out in a few microliters of solution. Suitable immobilized enzymes could be also used in this arrangement. If necessary, a small column packed with an appropriate filter material can be connected between the reaction tube and the chromatographic column in order to prevent undesired usbstances from entering the column. Of course, the filter must not have affinity to the sample components of interest. Such a technique mitigates any difficulties arising from filtration and the transfer of minute volumes and can also be automated.

When the sample components are present in very dilute solution, they can be trapped on a precolumn concentrator. For example, nucleotides can be concentrated at high pH when the solution is pumped slowly through a short, small-bore tube packed with an anion-exchange resin. After a sufficient volume has passed through the precolumn—the effluent can be monitored with a UV detector—it is connected upstream from the separating column, and the chromatographic run is performed in the usual manner. Similar "on-line" techniques can be devised for a number of applications with small-bore columns.

IX. OUTLOOK

High-performance liquid chromatography has already become a major analytical tool in nucleic aid research, although its potential is far from being fully exploited. Further progress is expected as a result of broadening the field of application of the presently available techniques, as well as from introducing novel column materials which extend the applicability of the method to the separation of high-molecular-weight species. In addition, the coupling of a liquid chromatograph with other analytical instruments could revolutionize nucleic acid analysis.

The quantitative determination of tissue nucleotides has been shown to benefit enormously from the speed, efficiency, and reproducibility of the method, on both the qualitative and the quantitative level. The employment of auxiliary techniques such as "enzymic peak shifts" and of multiwavelength detectors, as well as the adaptation of the chromatographic system to work with radioactively labeled sample components, greatly facilitates the rapid mapping of nucleotides and metabolites in tissue extracts.

The versatility, speed, and excellent sensitivity of high-performance liquid chromatography permit rapid determination of the base compositions of polynucleotides by various methods requiring only minute samples. This technique is expected to become an invaluable tool for the study of the so-called minor bases and of the nucleic acid degradation process itself.

As sequence analysis of RNA and DNA remains one of the major objectives in nucleic acid reaearch, there is a great need for a rapid separation technique which can cope with the formidable task of separating a large number of oligo and polynucleotide fragments. Here, the application of high-performance liquid chromatography is still in its infancy, and new developments are awaited. We can envisage a system consisting of a multiplicity of columns in which the sample components are first separated into groups, which are then further separated into the individual components on different columns. Backflushing, column-switching techniques have been successfully used in gas chromatography, and there is no reason why a liquid chromatograph incorporating these features could not be built. Of course, development of novel column materials suitable for efficient separation of high-molecular-weight fragments is imperative. Here again the pellicular structure appears to be promising.

At present, the sensitivity of high-performance liquid chromatography facilitates analysis at the picomole level. This technique competes very favorably with methods which utilize radioactive derivatives, such as the ultrasensitive tritium method of Randerath (22,78). This elaborate technique is, however, applicable to ribonucleotides only, whereas high-performance liquid chromatography can be used in base or sequence analysis of both deoxynucleotides and ribonucleotides. Research on DNA can profit greatly from the fact that the high sensitivity is achieved without the employment of radioactive isotopes. It appears that the sensitivity of analysis can be enhanced further by miniaturizing the column and the detector and by increasing system stability so that the nucleic acid constituent can be accurately and conveniently determined at the subpicomole level.

It is probable that pellicular column materials will play a major role in the further development of high-performance liquid chromatography. A variety of sorbents which do not have suitable mechanical and handling properties (e.g., inorganic gels and low cross-linked polymers having proper functional groups) may give efficient column materials in pellicular form. The greatest improvement could be achieved by such developments in the separation of large-molecular-weight sample components. Liquid-liquid chromatography with pellicular-type supports may also play an important role in the separation of tRNAs and large fragments, since reverse-phase chromatography has been found to be a powerful technique in this field (79).

On the other hand, the use of very fine column materials could lead to further revolutionary improvements in separation efficiency. To explore the potential of such columns, however, liquid chromatographs will be needed which can be operated reliably at much higher column inlet pressures (above 3,000 psi) than those of the present-day instruments and which are equipped with detectors having sufficiently small sensing volumes. Here a particularly challenging task will be the preparation of suitable sorbents which not only have favorable equilibrium and dynamic properties to achieve high separation efficiency but also give columns which withstand high pressure gradients and remain stable under various elution conditions. The enormous gain in separation efficiency which could be derived from such a system is so attractive that we can expect this approach to be explored in the not too distant future.

The pleasant expectations are clouded only by the realization that the effort required for the preparation of the sample to be analyzed is augmented by increasing the efficiency and sensitivity of the technique. Major innovations concerning sample preparation on a microscale will be required to take full advantage of the increasing analytical potential. In a similar vein, the more the speed and the efficiency of the chromatographic technique increase, the greater will be the need for immediate readout of results, as well as for rapid identification of unknown components. This need is already leading to the development of sophisticated hardware, and it also may promote the introduction of techniques which will enable us to couple an appropriate analytical tool, such as a nuclear magnetic resonance or mass spectrometer, directly to the liquid chromatograph.

Acknowledgments

The author would like to thank S. R. Lipsky, as well as E. Randerath and K. Randerath, for helpful discussions. The author's research work was been supported by the following grants from the U.S. Public Health Service: HE-03558, RR00356, and GM16681.

List of Symbols

a	constant in [31]
A	absorbance
A_f	free cross section of column
$A_{\max}$	absorbance at peak maximum
b	constant in [31]
C	concentration
C^*	salt concentration in eluent II
$C_{\max}$	solute concentration at peak maximum
C_0	initial concentration of eluent I
d	column inner diameter
D_c	concentration distribution ratio
D_m	mass distribution ratio
f	response factor
F	flow rate

F^*	flow rate of eluent II into mixing chamber
H	plate height
l	light path length in detector cell
L	column length
M	amount of solute injected
N	plate number
PA	peak area
R	retardation ratio
Re	resolution
t	time
t_d	time for 90% depletion of V_0 when linear gradient is used
t_0	retention time of an unsorbed solute
t_R	retention time
t_R'	adjusted retention time
u	mobile-phase velocity
u^*	chromatographic mobile-phase velocity
u_b	band velocity
V	total column volume
V_M	volume of mobile phase in column
V_0	starting volume in gradient mixing chamber
V_R	retention volume
V_R'	adjusted retention volume
V_S	volume of stationary phase in column
w_b	peak width at base intercept
$w_{1/2}$	peak width at half height
α	relative retention, separation factor
δ	detector calibration factor
σ_l	standard deviation of peak in column length units
σ_t	standard deviation of peak in time units
σ_v	standard deviation of peak in volume units
ε	molar absorptivity
ϵ	relative interstitial bed volume

The following abbreviations, in addition to those listed in Table I, are used in the text and tables:

c-AMP	cyclic 3',5'-adenosine monophosphate
c-GMP	cyclic 2',3'-guanosine monophosphate
TMP, TDP, TTP	5'-phosphate of deoxythymidine
6MMPRMP	5'-monophosphate of 6-methylmercaptopurineriboside
TGMP	5'-monophosphate of 6-thioguanosine
XMP	5'-monophosphate of xanthosine
IDP	inosine-5'-diphosphate
UDPG	uridine diphosphoglucose
UDPGA	uridine-5'-diphosphoglucoronic acid
NAD, NADH	the oxidized and reduced forms of nicotinamide-adenine dinucleotide

The abbreviations and symbols for nucleic acids, polynucelotides, and their constituents in this text follow the recommendations (1970) of the IUPAC-IUB Commission on Biochemical Nomenclature (80).

References

1. W. E. Cohn, *Science*, *109*, 377 (1949).
2. W. E. Cohn, *J. Am. Chem. Soc.*, *72*, 1471 (1950).
3. W. E. Cohn and C. E. Carter, *J. Am. Chem. Soc.*, *72*, 4273 (1950).
4. G. Agren, *Årsskrift*, Uppsala University, Vol. 5, 1958.
5. N. G. Anderson, J. G. Green, M. L. Barber, and F. C. Ladd, Sr., *Anal. Biochem.*, *6*, 153 (1963).
6. J. G. Green, C. E. Nunley, and N. G. Anderson, *Natl. Cancer Inst. Monogr.*, *21*, p. 431.
7. C. D. Scott, J. E. Attril, and N. G. Anderson, *Proc. Soc. Exptl. Biol. Med.*, *125*, 181 (1967).
8. M. Uziel, C. K. Koh, and W. E. Cohn, *Anal. Biochem.*, *25*, 77 (1968).
9. S. Moore and W. H. Stein, *J. Biol. Chem.*, *192*, 663 (1957).
9a. J. J. Kirkland, Ed., *Modern Practice of Liquid Chromatography*, Wiley-Interscience, New York, 1971.
10. L. S. Ettre and A. Zlatkis, Eds., *Practice of Gas Chromatography*, Interscience, New York, 1967.
11. H. Purnell, *Gas Chromatography*, Wiley, New York, 1962.
12. O. Samuelson, *Ion Exchange Separations in Analytical Chemistry*, Wiley, New York, 1963.
13. H. F. Walton, in *Chromatography*, E. Heftmann, Ed., Reinhold, New York, 1967, pp. 287–342.
14. F. Helfferich, in *Advances in Chromatography*, Vol. I, J. C. Giddings and R. A. Keller, Eds., Dekker, New York, 1965, pp. 3–60.
15. P. B. Hamilton, in *Advances in Chromatography*, Vol. II, J. C. Giddings and R. A. Keller, Eds., Dekker, New York, 1966, pp. 3–62.
16. C. F. Crampton, F. R. Frankel, A. M. Benson, and A. Wade, *Anal. Biochem.*, *1*, 249 (1960).
17. E. Kirsten and R. Kirsten, *Biochem. Biophys. Res. Commun.*, *7*, 76 (1962).
18. P. B. Hamilton, *Ann. N. Y. Acad. Sci.*, *102*, 55 (1962).
19. P. B. Hamilton, *Nature*, *205*, 284 (1965).
20. M. Mohnke, R. Schmunk, and H. Schutze, *Z. Anal. Chem.*, *219*, 137 (1966).
21. P. B. Hamilton, D. C. Bogue, and R. A. Anderson, *Anal. Chem.*, *31*, 1504 (1959).
22. K. Randerath and E. Randerath, *Anal. Biochem.*, *28*, 110 (1969).
23. L. S. Ettre, in *The Practice of Gas Chromatography*, L. S. Ettre and A. Zlatkis, Eds., Interscience, New York, 1967, pp. 373–406.
24. R. L. Jolley and C. D. Scott, *J. Chromatog.*, *47*, 272 (1970).
25. P. R. Brown, *J. Chromatog.*, *52*, 257 (1970).
26. K. W. Pepper, *Chemistry Research*, Her Majesty's Stationery Office, London, England, 1952.
27. D. E. Weiss, *Australian J. Appl. Sci.*, *4*, 510 (1953).
28. J. Feitelson and S. M. Partridge, *Biochem. J.*, *64*, 607 (1956).
29. N. K. Boardman, *J. Chromatog.*, *2*, 388 (1959).
30. N. K. Boardman, *J. Chromatog.*, *2*, 398 (1959).
31. C. Horvath, Dissertation, University of Frankfurt, 1963.
32. J. R. Parrish, *Nature*, *207*, 402 (1965).
33. M. Skafi and K. H. Lieser, *Z. Anal. Chem.*, *249*, 182 (1970).
34. M. Skafi and K. H. Lieser, *Z. Anal. Chem.*, *250*, 306 (1970).
35. M. Skafi and K. H. Lieser, *Z. Anal. Chem.*, *251*, 177 (1970).

36. K. Randerath, *Thin Layer Chromatography*, 2nd ed., Academic Press, New York, 1966, pp. 229–243.
37. P. Hamilton, *Anal. Chem.*, *35*, 2055 (1963).
38. C. A. Burtis, N. M. Munk, and F. R. MacDonald, *Clin. Chem.*, *16*, 667 (1970).
39. C. D. Scott, *Clin. Chem.*, *14*, 521 (1968).
40. C. Horvath, B. Preiss, and S. R. Lipsky, *Anal. Chem.*, *39*, 1422 (1967).
41. C. Horvath and S. R. Lipsky, *Anal. Chem.*, *41*, 1227 (1969).
42. J. J. Kirkland, *J. Chromatog. Sci.*, *7*, 361 (1969).
43. J. J. Kirkland, *J. Chromatog. Sci.*, *8*, 72 (1970).
44. C. Horvath and S. R. Lipsky, *J. Chromatog. Sci.*, *7*, 109 (1969).
45. C. Horvath, unpublished results.
46. L. H. Thacker, C. D. Scott, and W. Pitt, Jr., *J. Chromatog.*, *51*, 175 (1970).
47. T. K. Lakshmanan and S. Lieberman, *Arch. Biochem. Biophys.*, *53*, 258 (1954).
48. H. W. Shmuckler, *J. Chromatog. Sci.*, *8*, 581 (1970).
49. W. E. Cohn, in *The Nucleic Acids*, Vol. I, E. Chargaff and J. N. Davidson, Eds., Academic Press, New York, 1955, p. 211.
50. W. E. Cohn, in *Ion Exchangers in Organic and Biochemistry*, C. Calmon and T. R. E. Kressman, Eds., Interscience, New York, 1957.
51. W. E. Cohn, in *Methods in Enzymology*, Vol. II, S. P. Colowick and N. O. Kaplan, Eds., Academic Press, New York.
52. W. E. Cohn, in *Chromatography*, E. Heftmann, Ed., Reinhold, New York, 1967, pp. 627–660.
53. P. Boulanger and J. Montreuil, in *Chromatographie*, Vol. 2, E. Lederer, Ed.,, Masson, Paris, 1960.
54. J. J. Saukkonen, *Chromatog. Rev.*, *6*, 53 (1964).
55. K. Randerath, *Angew. Chem.*, *74*, 780 (1962); Intern. Ed., *1*, 553 (1962).
56. D. D. Christianson, J. W. Paulis, and J. S. Walls, *Anal. Biochem.*, *22*, 35 (1968).
57. S. Katz and D. G. Comb, *J. Biol. Chem.*, *238*, 3065 (1963).
58. R. Bergkvist and A. Deutsch, *Acta Chem. Scand.*, *8*, 1877 (1954).
59. R. B. Hurlbert, H. Schmitz, A. F. Brumm, and V. R. Potter, *J. Biol. Chem.*, *209*, 23 (1954).
60. H. Schmitz, R. B. Hurlbert, and V. R. Potter, *J. Biol. Chem.*, *209*, 41 (1954).
61. E. W. Busch, *J. Chromatog.*, *37*, 518 (1968).
62. B. E. Bonnelycke, K. Dus, and S. L. Miller, *Anal. Biochem.*, *27*, 262 (1969).
63. J. Lerner, A. I. Schepartz, *J. Chromatog.*, *35*, 37 (1968).
64. I. C. Caldwell, *J. Chromatog.*, *44*, 331 (1969).
65. P. Virkola, *J. Chromatog.*, *51*, 195 (1970).
66. F. Maley and G. J. Maley, *J. Biol. Chem.*, *235*, 2968 (1960).
67. F. R. Blattner and H. P. Erickson, *Anal. Biochem.*, *18*, 220 (1967).
68. W. P. Kennedy and J. C. Lee, *J. Chromatog.*, *51*, 203 (1970).
69. H. W. Schmuckler, *J. Chromatog. Sci.*, *8*, 653 (1970).
70. C. Burtis, private communication.
71. G. Brooker, *Anal. Chem.*, *42*, 1108 (1970).
72. W. E. Cohn, private communication.
73. C. D. Scott and N. E. Lee, *J. Chromatog.*, *42*, 263 (1969).
74. C. A. Burtis, *J. Chromatog.*, *51*, 183 (1970).
75. G. L. Cantoni and D. R. Davies, Eds., *Procedures in Nucleic Acid Research*, Harper & Row, New York, 1966.
76. R. N. Nazar, H. G. Lawford, and J. T. F. Wong, *Anal. Biochem.*, *35*, 305 (1970).

77. D. Mandel, in *Progress in Nucleic Acid Research and Molecular Biology*, Vol. VIII, J. Davidson and N. Cohen, Eds., Academic Press, New York, 1694, pp. 304–306.
78. K. Randerath and E. Randerath, in *Procedures in Nucleic Acid Research*, Vol. 2, G. L. Cantoni and D. R. Davis, Eds., Harper & Row, New York, in press.
79. J. F. Weiss and A. D. Kelmers, *Biochemistry*, *6*, 2507 (1967).
80. IUPAC-IUB Commission on Biochemical Nomenclature, "Abbreviations and Symbols for Nucleic Acids, Polynucleotides and Their Constituents," Recommendations 1970, *European J. Biochem.*, *15*, 203 (1970), and elsewhere.

ADDENDA

Further details regarding the subject of this chapter can also be found in the following references:

P. R. Brown, *High Pressure Liquid Chromatography: Biochemical and Biomedical Applications*, Academic Press, New York, 1973.

C. Horvath, in *Ion Exchange and Solvent Extraction*, J. Marinsky and Y. Marcus, Eds., Vol. 5, Dekker, New York, 1973, pp. 207–260.

Newer Developments in Enzymic Determination of D-Glucose and Its Anomers

JUN OKUDA AND ICHITOMO MIWA, *Faculty of Pharmaceutical Science,*
Meijo University, Nagoya, Japan

I. Introduction... 156
II. Determination of D-Glucose and Its Anomers with β-D-Glucose Oxidase.... 157
    1. Properties of β-D-Glucose Oxidase and Mutarotase................ 157
        A. β-D-Glucose Oxidase..................................... 157
        B. Mutarotase... 158
    2. Determination of D-Glucose and Its Anomers Using the Oxygen
       Electrode.. 159
        A. D-Glucose.. 159
        B. D-Glucose Anomers..................................... 165
        C. Effect of Mutarotase on Determination of D-Glucose and Its
          Anomers... 167
    3. Colorimetric Determination of D-Glucose and Its Anomers.......... 169
        A. D-Glucose.. 169
        B. D-Glucose Anomers..................................... 172
    4. Fluorometric Determination of D-Glucose........................ 173
    5. Electrochemical Determination of D-Glucose..................... 175
III. Determination of D-Glucose with Hexokinase.......................... 177
    1. Properties of Hexokinase and D-Glucose-6-Phosphate Dehydrogenase 177
        A. Hexokinase... 177
        B. D-Glucose-6-Phosphate Dehydrogenase.................... 178
    2. Spectrophotometric Determination of D-Glucose.................. 178
    3. Fluorometric Determination of D-Glucose........................ 180
    4. Colorimetric Determination of D-Glucose........................ 182
    5. Radioisotopic Determination of D-Glucose....................... 183
IV. Determination of D-Glucose with Acylphosphate: D-Glucose-6-Phosphotrans-
    ferase... 184
V. Conclusion.. 185
    References... 186

I. INTRODUCTION*

In water, D-glucose equilibrates among three structures (1,2): α-anomer (36.5%), β-anomer (63.5%), and a so-called aldehyde form (0.003%), as shown in [1]:

$$\alpha\text{-Anomer} \rightleftharpoons \text{aldehyde form} \rightleftharpoons \beta\text{-anomer} \qquad [1]$$
$$\text{(36.5\%)} \qquad \text{(0.003\%)} \qquad \text{(63.5\%)}$$

In many biological systems, the interconversion among these three forms is catalyzed by mutarotase (aldose 1-epimerase, EC 5.1.3.3) (3,4), phosphate ions (5), and other factors.

Many kinds of chemical determinations of D-glucose, depending generally on its reducing power, have been reported. These methods require protein-free filtrates, however, and are not specific for D-glucose in the presence of certain other reducing substances.

On the other hand, enzymic methods have a number of theoretical and practical advantages, and therefore it is natural that much attention has been devoted to the subject of enzymic determination of D-glucose. Enzymic methods with β-D-glucose oxidase (β-D-glucose : oxygen oxidoreductase, EC 1.1.3.4) or hexokinase (ATP: D-hexose 6-phosphotransferase, EC 2.7.1.1) are found to be more specific and sensitive than the chemical methods, and are now accepted generally by biochemists for the determination of D-glucose. The enzymic methods have been reviewed by Free (6), Slein (7), Kitamura (8), and Martinek (9).

In the last decade, the manometric technique used in the early stage of the development of the β-D-glucose oxidase method has been replaced by the oxygen electrode technique, and rapid colorimetric determinations of D-glucose in serum or plasma without deproteinization have been investigated. Fluorometric and electrochemical methods have been also developed with β-D-glucose oxidase.

It was recently found that the addition of mutarotase to the reaction mixture in the β-D-glucose oxidase method causes a rapid and complete oxidation of total D-glucose (both β-D-glucose and α-D-glucose). Hence the use of mutarotase makes the β-D-glucose oxidase method more accurate and rapid.

D-Glucose can be also determined with the system of hexokinase and D-glucose-6-P dehydrogenase (D-glucose-6-P: NADP oxidoreductase, EC 1.1.1.49) by measuring the absorbance or the fluorescence of NADPH formed in the course of the reaction. The enzymic cycling method using the

---

* The following abbreviations are used in this chapter: ATP, adenosine triphosphate; ADP, adenosine diphosphate; D-glucose-6-P, D-glucose-6-phosphate; NADP, nicotina-mide-adenine dinucleotide phosphate; NADPH, reduced form of nicotinamide-adenine dinucleotide phosphate; FAD, flavin-adenine dinucleotide.

same system and the hexokinase–ATP-γ-^{32}P method are the most sensitive of the techniques hitherto reported.

In regard to the determination of D-glucose anomers, physical methods using polarimetry (2), NMR spectrometry, and gas chromatography (10), as well as a chemical method based on the difference in the speeds of oxidation of the two anomers by bromine (2), are available. However, these physical and chemical methods require a rather large amount of sample and a long time for the estimation, and are not specific for D-glucose. Recently, simple and specific microdeterminations of D-glucose anomers with only β-D-glucose oxidase or with a β-D-glucose oxidase–mutarotase system were reported.

This chapter reviews mainly the two enzymic methods mentioned above and also refers to special problems which have been encountered with these methods. The experimental procedures are described in detail, followed by discussions and criticisms of the methods. Although automatic D-glucose determinations with β-D-glucose oxidase or hexokinase have been introduced especially in the clinical field, this chapter does not deal with these applications.

II. DETERMINATION OF D-GLUCOSE AND ITS ANOMERS WITH β-D-GLUCOSE OXIDASE

Enzymic determinations of D-glucose with β-D-glucose oxidase are classified into four groups: oxygen electrode method, colorimetric method, fluorometric method, and electrochemical method.

The oxygen electrode method is now used in many laboratories by reason of its technical simplicity. The colorimetric method has been especially developed in the clinical field because of its extensive use in blood analysis. In this chapter, some physicochemical properties of β-D-glucose oxidase and mutarotase used in the determination of D-glucose and its anomers are first discussed, followed by descriptions of the four methods mentioned above.

1. Properties of β-D-Glucose Oxidase and Mutarotase

A. β-D-GLUCOSE OXIDASE

β-D-Glucose oxidase is a flavoprotein having two molecules of FAD per molecule of enzyme. The enzyme was extracted from *Aspergillus niger*, *Penicillium amagasakiense*, *Penicillium notatum*, and honey. Kusai et al. (11,12) reported the purification and crystallization of the enzyme from *Penicillium amagasakiense*. The same investigators found that the pure enzyme has a molecular weight of 154,000.

In the absence of catalase, the reaction catalyzed by β-D-glucose oxidase proceeds according to [2]:

$$\beta\text{-D-Glucose} + O_2 + H_2O \xrightarrow{\beta\text{-D-glucose oxidase}} \text{D-gluconic acid} + H_2O_2 \quad [2]$$

Hence D-glucose may be determined by estimating either the amount of consumed oxygen or the amount of newly formed H_2O_2.

Keilin and Hartree (13) and Kusai (11) recognized that the oxidation of sugars other than D-glucose by β-D-glucose oxidase is negligible. No oxidation of D-fructose, D-galactose, maltose, sucrose, lactose, or melibiose was observed. However, a slight oxidation of D-mannose, D-xylose, 6-methylglucose, and 4,6-dimethylglucose with the enzyme from *Penicillium notatum* (13) was reported, although the rates of oxidation were only 1 to 2% of that of D-glucose.

When catalase is present, the H_2O_2 formed in [2] is decomposed to 1 mole of water and $\frac{1}{2}$ mole of oxygen. This oxygen may serve for the oxidation of another molecule of β-D-glucose. Consequently, the overall reaction in the presence of catalase is represented by [3]:

$$\beta\text{-D-Glucose} + \tfrac{1}{2}O_2 \xrightarrow[\text{catalase}]{\beta\text{-D-glucose oxidase}} \text{D-gluconic acid} \quad [3]$$

The consumption of oxygen during the enzymic oxidation of D-glucose in the absence of catalase, or under conditions in which the H_2O_2 formed is decomposed without liberation of oxygen, is, therefore, twice as great as that in the presence of catalase.

As shown in [2], pure β-D-glucose oxidase oxidizes only β-D-glucose, not α-D-glucose. However, if the reaction mixture contains enough mutarotase, α-D-glucose is converted instantaneously into β-D-glucose, which is then oxidized to D-gluconic acid (see Sections II.2.B and II.2.C).

The optimum pH of the enzyme from *Penicillium amagasakiense* is 5.6 to 5.8, and the enzyme is stable in the pH range of 4 to 6. Kusai (11) also reported that 1 mole of this enzyme oxidizes 34,000 moles of D-glucose per minute. This means that the Q_{O_2} of the enzyme (in the pure state) at 30°C is 148,360. The enzyme is strongly inhibited by SH chelators.

B. MUTAROTASE

The enzyme mutarotase catalyzes the interconversion [1] of the anomeric forms of D-glucose and other configurationally related aldoses. Since Keilin and Hartree (3), found mutarotase in *Penicillium notatum*, it has been known that the enzyme is widely distributed in animals, plants, and microorganisms. Keston (4) and Bailey et al. (14) recognized that in the rat the kidney contains more mutarotase than any other organ. This fact was confirmed by

Miwa (15). Lapedes and Chase (16) and Bailey et al. (17) reported the purification of the enzyme from the kidney. The optimum pH of mutarotase is 7.0, and the enzyme is stable for several weeks if kept in a freezer. Mutarotase is strongly inhibited by phlorizin. Miwa (15) also reported a simple and rapid method for the determination of mutarotase with β-D-glucose oxidase, using the oxygen electrode. The activity of the enzyme is determined from the rate of the oxygen consumption due to the oxidation of β-D-glucose formed from α-D-glucose in the presence of both mutarotase and β-D-glucose oxidase.

As mutarotase is not commercially available, a simple method for preparation of the enzyme from hog kidney is described.

Preparation of Hog Kidney Mutarotase (16–18). One hundred grams of hog kidney cortex is homogenized for 2 min with a mixture of 100 ml of 0.1M Tris-HCl buffer (pH 7.2) and 40 ml of chloroform at 53°C. After centrifugation for 10 min at 12,000g, powdered ammonium sulfate is gradually added to the supernatant, and the fraction precipitating between 48 and 80% saturation is collected by centrifugation, dissolved in 0.01M Tris-HCl buffer (pH 7.2), and dialyzed at 4°C for 10 hr against several changes of the same buffer. The dialyzed solution is condensed to about 5 ml, using the ultrafiltration technique, and the condensed extract is applied to a column (2 × 25 cm) of Sephadex G-75 (Pharmacia, Uppsala, Sweden), previously equilibrated with 0.01M Tris-HCl buffer (pH 7.2). After the turbid, brownish fraction containing large quantities of protein but no mutarotase activity is eluted with the same buffer, the clear, pale yellow fraction containing high mutarotase activity (about 6000 total units) is eluted and the fraction is used as the mutarotase solution for the determination of D-glucose. This solution should be kept frozen or lyophilized to dryness.

2. Determination of D-Glucose and Its Anomers Using the Oxygen Electrode

A. D-GLUCOSE

Since a proportionate amount of oxygen is consumed from a solution during oxidation of D-glucose with β-D-glucose oxidase, measurement of the oxygen consumed serves as a direct determination of D-glucose. Keilin and Hartree (19) first reported the analysis of blood D-glucose based on this principle, using a Barcroft differential manometer.

Recently polarographic oxygen electrode methods have been utilized for this purpose (since Clark-type oxygen electrodes have become commercially available). These have the advantages of simplicity in handling and of not requiring prior treatment, such as deproteinization, of the sample.

The Clark-type oxygen electrode measures the diffusion flow of oxygen through a plastic membrane and gives a signal proportional to the partial pressure of oxygen which would be in equilibrium with the concentration of oxygen present in the solution. At fixed conditions (i.e., temperature, ionic strength, etc.), this reading is proportional to the concentration of dissolved oxygen. As the oxidation of D-glucose proceeds under the catalytic influence of β-D-glucose oxidase, the oxygen level can be observed to decrease gradually. The relation between oxygen consumed and amount of D-glucose in a sample can be studied by determining total oxygen consumption, provided that a closed system is employed so that oxygen cannot enter the solution from the atmosphere. With a slight modification of the polarographic oxygen-measuring instrument, however, it is possible to determine the relation by measuring the rate of oxygen consumption, instead of the oxygen level itself.

The oxygen electrode method is insensitive to the factors (reducing substances, colored materials, turbidity, etc.) which have limited the usefulness of the colorimetric β-D-glucose oxidase system. Moreover, it is rapid and technically simple.

Kadish and Hall (20) first reported the polarographic determination of D-glucose with a Clark-type oxygen electrode (Beckman Instruments, Fullerton, Calif.) and β-D-glucose oxidase. In this method, the D-glucose concentration was determined by using a β-D-glucose oxidase procedure in which the amount of oxygen depletion from a solution saturated with atmospheric oxygen is observed in a direct oxygen-consumption mode. Blood D-glucose was continuously monitored by this technique.

Kajihara and Hagihara (21) described a polarographic determination for serum D-glucose, using a rotating platinum electrode constructed by Hagihara (22). The electrode was equipped with a glass cell designed to reduce the access of atmospheric oxygen to a negligible level. The D-glucose in 50 μl of serum was determined with an error below 0.5% in 5 min by measuring the amount of oxygen consumed in the presence of β-D-glucose oxidase and catalase.

A Clark-type oxygen electrode (Yellow Springs Instrument Co., Yellow Springs, Ohio) was used to determine D-glucose in whole blood by Makino and Konno (23). They converted the native hemoglobin in the red blood cells to cyanmethemoglobin by incubating whole blood in potassium ferricyanide solution and then adding potassium cyanide solution in order to eliminate the influence of hemoglobin on the oxygen-consumption measurement. Since catalase contained in β-D-glucose oxidase preparations and blood samples is also inhibited by adding potassium cyanide, D-glucose in whole blood is accurately determined in 10 min using this treatment.

Makin and Warren (24) also determined D-glucose in whole blood, using a modified Clark-type oxygen electrode (Rank Bros., Bottisham, Cambridge), by measuring the amount of total oxygen consumed. They stated that it was easier and quicker to add sodium nitrite to blood samples before analysis so as to convert the active hemoglobin to inactive methemoglobin, and no steps to prevent the breakdown of H_2O_2 were needed. The method is capable of giving an estimation of the D-glucose concentration in 50 μl of venous or capillary blood or plasma within 2 min. Makin and Warren stated that the analysis of D-glucose in urine has also given reasonable results (30 mg/24 hr) but has been found to be complicated by the fact that substances which undergo spontaneous oxidation, and thus take up oxygen in the absence of β-D-glucose oxidase, may be present in urine.

A determination for D-glucose, suitable in principle for continuous use *in vivo*, has been developed by Updike and Hicks (25). They employed a so-called enzyme electrode which was made by immobilizing the enzyme β-D-glucose oxidase in a layer of acrylamide gel, 25 to 50 μ thick, on the surface of a polarographic oxygen electrode. The output current of the enzyme electrode was measured after allowing sufficient time for the diffusion process to reach the steady state. This interval varied from about 30 sec to 3 min for 98% of the steady-state response, depending primarily on the thickness of the plastic and gel membranes.

Okuda et al. (26,27) described a polarographic microdetermination of D-glucose (0.2 to 600 μg), using a Clark-type oxygen electrode (Beckman Instruments) and β-D-glucose oxidase, by recording the total oxygen consumption. In this method, a bias voltage supply unit (Tōa Electronics, Suwa, Shinjuku, Tokyo) was employed only in the determination of 0.2 to 3 μg of D-glucose to increase the sensitivity, and a trace amount of sodium azide was used to inhibit catalase. The results obtained on serum D-glucose by this method, using an internal standard, were practically consistent with those obtained by the *o*-toluidine-borate method. This method is technically simple, and the detailed experimental procedure is described below.

D-Glucose levels in serum, plasma, and urine were determined with β-D-glucose oxidase and a polarographic oxygen analyzer (Beckman Instruments) modified to record the rate of oxygen consumption by Kadish et al. (28). In this method, peak values in the recording give the maximum rate of consumption and are directly proportional to the D-glucose concentration; results are obtainable within 20 sec after sample (100 μl) addition with a standard deviation of less than 1.5%. Kadish et al. used an iodide-ammonium molybdate-ethanol mixture to prevent the catalase-catalyzed decomposition of H_2O_2.

Jemmali and Rodriguez-Kabana (29) also reported a polarographic technique for determining D-glucose (5 to 800 µg), using a Clark-type oxygen electrode (Yellow Springs Instrument Co.) and β-D-glucose oxidase, by estimating the initial rate of oxygen consumption, that is, the tangent ($-\Delta O_2/\Delta t$) of the angle formed by the recorder trace with the horizontal at the origin.

Recently Okuda and Okuda (30) described a successive microdetermination of oxygen and D-glucose in 15 µl of arterial or venous blood using oxygen electrode and β-D-glucose oxidase in 3 to 4 min. The procedure consists of the liberation of oxygen of blood in 1 ml of the oxygen-depleted potassium ferricyanide aqueous solution, and of oxygen consumption due to D-glucose of blood with additional β-D-glucose oxidase.

Experimental Procedure of Okuda et al. (27)

REAGENTS. 1. β-D-*Glucose oxidase*, 8 mg/ml. Dissolve the almost pure enzyme from *Penicillium amagasakiense* (Nagase & Co., Amagasaki, Osaka) in distilled water. The crude enzyme preparation, which is commercially available, may also be used for D-glucose determination.

2. *Equilibrium D-glucose solution.* Prepare an aqueous solution of equilibrium D-glucose by dissolving α-D-glucose in distilled water and keeping it in a cold room for more than 24 hr before use in order to complete the mutarotation. If time is a factor, it is recommended that aqueous α- or β-D-glucose solution be heated for 10 min at 90°C to complete the mutarotation.

3. *Sodium azide*, 0.1%. The solution is used as an inhibitor for catalase which exists in the test material and β-D-glucose oxidase preparation (26).

4. *Sodium acetate buffer*, pH 5.6, 0.2M.

APPARATUS

Oxygen electrode. The Polarographic (Clark-type) oxygen electrode is available from Beckman Instruments. The Galvani-type oxygen electrode (Kyusui Kagaku Kenkyusho, Sasazuka, Shibuya, Tokyo) can also be used for determination of D-glucose.

Recorder. The recorder (2 mV–50 V full scale, Tōa Electronics) is sold under the trade name Electronic Polyrecorder model EPR-2TC.

Bias voltage supply unit. This unit is used to supply bias voltage to increase the sensitivity (Tōa Electronics, EPR-Prebox model PB-30A).

Reaction vial. A 3-ml glass vial (2.0 cm in height, 1.4 cm ID) is used when the oxygen electrode (model 777, diameter of electrode 1.2 cm, Beckman

Instruments) is employed. The diffusion of atmospheric oxygen is not observed during the experiment when this vial is used. A plastic holder or a proper cushion is interposed between the vial and magnetic stirrer to avoid the temperature influence of the magnetic stirrer during the long experiment. In the microdetermination of D-glucose, below 3 μg, a water-circulating system for controlling temperature is necessary to keep the temperature of the reaction mixture in the vial strictly constant.

Microsyringes. Volumes of 10, 25, and 50 μl.

Magnetic stirrer.

Procedure.

1. D-Glucose (3 to 600 μg).

In a 3-ml glass vial with a plastic holder, stir 10 μl of 3.3 mg/ml equilibrium D-glucose solution (33 μg), 1.0 ml of 0.2M sodium acetate buffer (pH 5.6), and 10 μl of 0.1% sodium azide continuously with a stainless steel stirring bar. Immerse the oxygen electrode in the reaction mixture up to 0.3 cm from the tip of the electrode. After the meter reading of the recorder (at 50 mV full scale) is adjusted to approximately the 90% position, add 10 μl of 8 mg/ml β-D-glucose oxidase. Rapid oxygen consumption can be observed, as shown in Figure 1a, and the consumption will reach maximum within about 1 min. Then add 10 μl of 0.25% D-glucose (25 μg) as an internal standard. The second oxygen consumption occurs as shown in Figure 1b. From the values (a and b) of the oxygen consumption due to the oxidation of D-glucose in the sample and the internal standard, calculate the amount of D-glucose in the sample:

$$\text{D-Glucose } (\mu g) \text{ in sample} = 25 \times \frac{a}{b} \qquad [4]$$

The method with the internal standard described here is more accurate than the method with an external standard. The former method is recommended for the determination of D-glucose in crude mixtures, and the latter is generally used for the determination of pure D-glucose.

For the measurement of 70 to 600 μg of D-glucose, the procedure is the same as described above, but the volume of the reaction mixture is increased with the increase in the amount of D-glucose in the sample; hence a bigger cylindrical glass vial should be used. It should be noted that the oxygen dissolved (8.1 μg at 25°C) in 1.0 ml of 0.2M sodium acetate buffer (pH5.6) is theoretically insufficient for oxidation of over 70 μg of D-glucose (oxygen in pure water at 25°C, 8.4 μg/ml) (31,32). The relation between various amounts of D-glucose (3 to 600 μg) and oxygen consumption gives a straight line. It was also found that the H_2O_2 formed did not interfere in the determination.

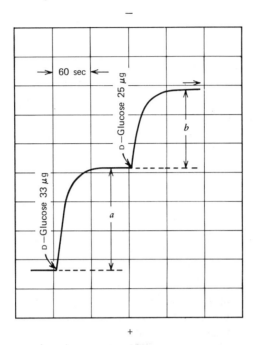

Figure 1. Internal standard method for measurement of D-glucose with β-D-glucose oxidase and the oxygen electrode.

2. D-Glucose (0.2 to 3 μg)

Put a mixture of 1.0 ml of 0.2M sodium acetate buffer (pH 5.6), 10 μl of 8 mg/ml β-D-glucose oxidase, and 10 μl of 0.1% sodium azide into a 3-ml vial under controlled temperature at 25°C, using a water-circulation system. After the oxygen electrode is immersed in the reaction mixture, adjust the meter reading of the recorder to about the 90% position under continuous stirring of a reaction mixture, and then add 10 μl of 0.1 mg/ml equilibrium D-glucose solution (1 μg) with a microsyringe. Record the oxygen consumption as shown in Figure 2a. When the same amount of D-glucose is successively added, a similar oxygen consumption is observed (Figure 2b). A third addition of D-glucose results in another, similar oxygen consumption (Figure 2c). In this microdetermination, the bias voltage of EPR-Prebox (PB-30A) is set so that the recorder can be used at 2 mV full scale.

3. D-Glucose in Blood

In this case, dilute 40 μl of peripheral blood in 0.5 ml of distilled water to hemolyze it completely. Mix the diluted sample with 0.8 ml of 0.2M sodium acetate buffer (pH 5.6), and then heat at 90°C for 30 sec to decompose the

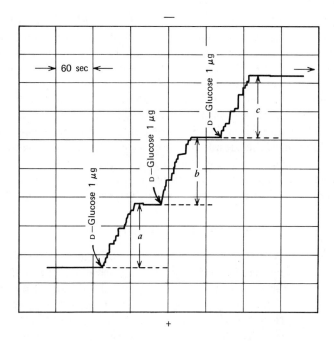

Figure 2. Microdetermination of D-glucose with β-D-glucose oxidase and the oxygen electrode.

hemoglobin, which interferes with the oxygen-consumption measurement. An alternative is to first mix the same volume of peripheral blood with 0.2 ml of $0.7N$ NaOH to denature the hemoglobin, and add 0.2 ml of $0.7N$ HCl and 0.9 ml of $0.2M$ sodium acetate buffer (pH 5.6) to neutralize the reaction mixture. Then either the heated sample, sufficiently aerated after cooling, or the neutralized sample is employed for the determination. The denatured hemoglobin does not interfere with the estimation. A single estimation is complete within 5 min.

B. D-GLUCOSE ANOMERS

Keston and Brandt (33) first reported colorimetric determination of D-glucose anomers, using a β-D-glucose oxidase-peroxidase-chromogen system. Hill (34) developed an automated procedure employing a similar system and measured deviations from equilibrium of anomers in the blood of dogs. Keston's method, which will be described in detail later, requires more than 30 min for determination of each sample.

Okuda and Miwa (35) recently devised a rapid polarographic micro-determination of D-glucose anomers, using an oxygen electrode and β-D-glucose oxidase. The method requires less than 10 min for each sample. They reported the ratio of β-anomer to total D-glucose as 63.5%, which is very close to the values in the literature (2,10), and this value is used in the method for determination of D-glucose anomers. In this determination, an aliquot of the sample is used for measurement of oxygen consumption due to β-D-glucose, employing β-D-glucose oxidase and the oxygen electrode. Another aliquot of the sample is heated to equilibrate the D-glucose in the sample, and the oxygen consumption is also measured. From the values of the two estimations of oxygen consumption, the amount of D-glucose anomers in the sample is obtained as described in the experimental procedure.

Experimental Procedure of Okuda and Miwa (35)

REAGENTS. Commercial samples of α-D-glucose and β-D-glucose are used. However, if pure α- and β-D-glucose are required, the commercial material is further purified with aqueous ethanol (4). Equilibrium D-glucose solution, 0.2M sodium acetate buffer (pH 5.6), 0.1% sodium azide solution, and 8 mg/ml β-D-glucose oxidase solution are the same as described in Section II.2.A.

APPARATUS. The oxygen electrode, recorder, magnetic stirrer, micro-syringes, and reaction vial are also the same as described in Section II.2.A.

Procedure. Place 1 ml of 0.2M sodium acetate buffer (pH 5.6), 10 μl of 5 mg/ml unequilibrated D-glucose solution, and 10 μl of 0.1% sodium azide in a 3-ml vial at 20°C. Record the oxygen consumption (a) due to β-D-glucose in the sample in the presence of β-D-glucose oxidase. Then record the oxygen consumption (b) due to β-D-glucose in the equilibrated sample, which was heated for 30 sec at 90°C, cooled, and aerated sufficiently at 20°C. The ratio, y, of β-D-glucose to total D-glucose in the sample is calculated by [5] with the values a and b:

$$y = \frac{a}{b} \times 0.635 \qquad [5]$$

If the amount of total D-glucose, S, is calculated from the value b by using a calibration curve for D-glucose, the amounts of α- and β-D-glucose can be expressed as $S \cdot (1 - y)$ and $S \cdot y$, respectively.

A similar method can be applied for the determination of D-glucose anomers in blood before mutarotational equilibrium.

C. EFFECT OF MUTAROTASE ON DETERMINATION OF D-GLUCOSE AND ITS ANOMERS

As described in Section II.2.A, the amount of D-glucose in the sample is usually obtained by estimation of the oxygen consumption due to only β-anomer, not α-anomer.

Okuda and Miwa (36) reported a clear effect of the mutarotase on the determination of D-glucose and its anomers with β-D-glucose oxidase. It was found that the mutarotase added previously to the reaction mixture acts on the α-anomer in the sample to convert it rapidly into β-D-glucose and that the total D-glucose, not only the β-anomer but also the α-anomer, is finally oxidized to D-gluconic acid in the presence of β-D-glucose oxidase and mutarotase:

α-D-Glucose

$$\Big\updownarrow \text{mutarotase}$$

$$\beta\text{-D-Glucose} + O_2 + H_2O \xrightarrow[\text{oxidase}]{\beta\text{-D-glucose}} \text{D-gluconic acid} + H_2O_2 \quad [6]$$

Thus the sensitivity increases by 57% compared with the result in the absence of mutarotase. In this method, there is no need to consider the influence of such factors as pH, phosphate ions, and temperature, which affect the rate of D-glucose mutarotation.

Mutarotase is also employed for the determination of D-glucose anomers. An aliquot of the sample is used for estimation of the oxygen consumption of β-D-glucose catalyzed by β-D-glucose oxidase. After the oxygen consumption is complete, mutarotase is added to the reaction mixture and the second stage of oxygen consumption, due to α-D-glucose remaining in the reaction mixture, is observed. The ratio of D-glucose anomers in the sample is calculated directly, using the two values of oxygen consumption. This method for the estimation of D-glucose anomers requires only 3 min for each sample. Recently Okuda and Okuda (30) applied the method to the determination of D-glucose anomers in blood.

Experimental Procedure of Okuda and Miwa (36)

REAGENTS. β-D-Glucose oxidase solution, 8 mg/ml (free from mutarotase), equilibrium D-glucose solution, and 0.1% sodium azide solution are the same as described in Section II.2.A, except for 0.05M sodium acetate buffer (pH 5.6). Mutarotase solution, 100 units/ml (15), was prepared as described in Section II.1.B.

APPARATUS. The oxygen electrode, recorder, magnetic stirrer, microsyringes, and reaction vial are the same as described in Section II.2.A.

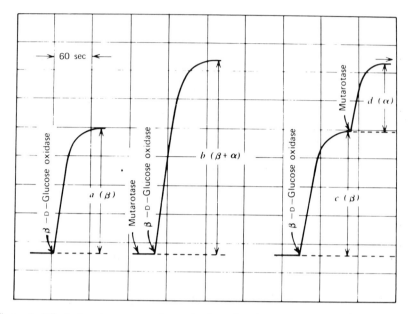

Figure 3. Effect of mutarotase on determination of D-glucose and its anomers with β-D-glucose oxidase and the oxygen electrode.

Procedure. Place in a 3-ml vial 1.0 ml of $0.05M$ sodium acetate buffer, 10 μl of 4.0 mg/ml equilibrium D-glucose solution (40 μg), and 10 μl of 0.1% sodium azide (26), which is an inhibitor for catalase. Immerse the oxygen electrode in the reaction mixture up to about 0.3 cm from the tip. After the meter reading is adjusted to about the 90% position on the recorder (at 50 mV full scale), under continuous stirring with a stainless steel stirring bar, add 50 μl of β-D-glucose oxidase solution to give an oxygen consumption (Figure 3a) corresponding to the β-D-glucose in the sample. In this case, if 5 μl of mutarotase solution is added before the addition of β-D-glucose oxidase, a further oxygen consumption (Figure 3b) corresponding to the total D-glucose (α-anomer + β-anomer) can be recorded in the presence of the same amount (40 μg) of equilibrium D-glucose. The difference $(b - a)$ in oxygen consumption in the two experiments is due to the rapid conversion of α-D-glucose to β-D-glucose with mutarotase and to the successive oxidation, by β-D-glucose oxidase, of the β-D-glucose formed.

If the same amount of mutarotase is added to the reaction mixture after the oxygen consumption (Figure 3c) corresponding to β-D-glucose in the same amount (40 μg) of equilibrium D-glucose, D-glucose anomers would be determined. Since the oxygen consumption (Figure 3d) after the addition of mutarotase is due to α-D-glucose in the sample, each ratio of two anomers to total D-glucose can be calculated from the two oxygen consumptions in the stepwise oxidation. The value of 35.5% obtained for α-D-glucose in equilibrium D-glucose by the present method was in good agreement with the results obtained by the polarimetric (36.2%) (2) and polarographic (36.5%) (35) methods.

3. Colorimetric Determination of D-Glucose and Its Anomers

A. D-GLUCOSE

Keston (37) and Teller (38) first introduced a coupled β-D-glucose oxidase–peroxidase system for colorimetric determination of D-glucose in biologic fluids. In this system, peroxidase catalyzes the oxidation of a chromogenic oxygen acceptor such as o-tolidine or o-dianisidine by H_2O_2. The following reactions are involved:

$$\beta\text{-D-Glucose} + O_2 + H_2O \xrightarrow{\ \beta\text{-D-glucose oxidase}\ } \text{D-gluconic acid} + H_2O_2 \quad [2]$$

$$H_2O_2 + \text{chromogen} \xrightarrow{\ \text{peroxidase}\ } \text{dye} + H_2O \quad [7]$$

After the reports of Keston and Teller, many kinds of oxygen acceptors and catalysts were studied by numerous workers.

These colorimetric methods for H_2O_2 with peroxidase or other catalysts used in the determination of D-glucose can be classified as follows:

1. Peroxidase–o-tolidine system (39–43)

$$H_2O_2 + o\text{-tolidine} \xrightarrow{\ \text{peroxidase}\ } \text{oxidized } o\text{-tolidine}$$

2. Peroxidase–o-dianisidine system (44–59,87)

$$H_2O_2 + o\text{-dianisidine} \xrightarrow{\ \text{peroxidase}\ } \text{oxidized } o\text{-dianisidine}$$

3. Peroxidase—$K_4Fe(CN)_6$ system (60)

$$H_2O_2 + K_4Fe(CN)_6 \xrightarrow{\ \text{peroxidase}\ } K_3Fe(CN)_6$$

4. Molybdenic acid—iodide system (42,61–63)

(1) $$H_2O_2 + I^- \xrightarrow{\ \text{molybdenic acid}\ } I_3^-$$

(2) $$H_2O_2 + I^- + o\text{-tolidine} \xrightarrow{\ \text{molybdenic acid}\ } \text{oxidized } o\text{-tolidine}$$

(3) $$H_2O_2 + I^- + \text{starch} \xrightarrow{\ \text{molybdenic acid}\ } I_2\text{—starch}$$

5. Catalase—chromotropic acid system (64)

$$H_2O_2 + CH_3OH \xrightarrow{\text{catalase}} HCHO$$

HCHO + [chromotropic acid structure with OH, OH, HO$_3$S, SO$_3$H groups] $\xrightarrow{H_2SO_4}$ stable colored complex (blue–violet)

6. Catalase—acetyl acetone—NH_4^+ system (65,66)

$$H_2O_2 + CH_3OH \xrightarrow{\text{catalase}} HCHO$$

$$CH_3COCH_2COCH_3 + HCHO + NH_4^+ \rightarrow \text{3,5-diacetyl-1,4-dihydrolutidine}$$

7. Catalase—MBTH system (67)

$$H_2O_2 + CH_3OH \xrightarrow{\text{catalase}} HCHO$$

HCHO + 3-methyl-2-benzothiazolone hydrazone → stable, colored complex
(MBTH)

8. Cu^{2+}·histamine—indigocarmine system (68)

H_2O_2 + [indigocarmine structure] $\xrightarrow{Cu^{2+}\cdot\text{histamine}}$ Oxidized substance (colorless)

9. Cu^{2+}·phenanthroline—o-dianisidine system (58)

$$H_2O_2 + \text{o-dianisidine} \xrightarrow{Cu^{2+}\cdot\text{o-phenanthroline}} \text{oxidized o-dianisidine}$$

Other chromogens—2,6-dichlorophenol indophenol (69,70), Ti^{4+}—xylenol orange (71), 4-aminophenazone (72), benzothiazoline derivatives (73), guajac (34,74), adrenalin (75), and phenolphthalin (76)—have also been used as chromogenic oxygen acceptors.

Of the systems listed, the peroxidase–o-dianisidine is most familiar in the clinical field. The oxidized product of o-dianisidine is more stable than that of o-tolidine (77) and is colloidal; hence it is easily precipitated, and means of stabilization have been suggested. McComb and Yushok (78) stated that addition of a strong mineral acid, such as sulfuric acid, changes the color of the reaction from amber to a deep pink that is stable for several hours. This caused a shift in the absorption maximum from 400–420 nm to 520–540 nm with an increase in sensitivity.

The stabilization of the colored product by addition of ghatti gum (49) or detergent (BRIJ-35) (79), or by extraction with isobutyl alcohol (49), was recognized.

Greenberg and Glick (80) reported a microdetermination of D-glucose using the β-D-glucose oxidase–peroxidase system. They measured 0.1 μg of D-glucose with only 60 μl of total reaction mixture.

With the increase in sensitivity of these colorimetric determinations, the volume of the sample can be decreased to 10 to 20 μl in the case of serum or plasma; hence deproteinization has become unnecessary as described in "Experimental Procedure." However, as many interfering substances are known to be present in serum, plasma, and other biological fluids, treatment (e.g., deproteinization) for eliminating the interference before the determination of D-glucose is recommended when accuracy is required. Deproteinization techniques, other than that of Somogyi (81), fail to remove uric acid (82), and peroxidase may develop in acid filtrates to yield positive errors (48). Concentrations of ascorbic acid greater than 5 mg/100 ml will decrease the D-glucose color (46). Bilirubin and hemolysis also interfere (83,84). Each 1 mg uric acid per 100 ml causes a decrease in color from D-glucose equivalent to 1 mg D-glucose per 100 ml (78,85). Glutathione, cysteine, thymol, and catechols are other reducing agents wihch inhibit the β-D-glucose oxidase-catalyzed reaction (53,83). Interferences from bilirubin, hemolysis, and uric acid are removed by use of the Somogyi zinc filtrate. The British (86) have been concerned about the alleged carcinogenicity of the aromatic amine chromogens, o-tolidine and o-dianisidine (which has not been proved to everyone's satisfaction), and such chromogens are being replaced by other oxygen acceptors (69,70,75,76).

Miwa et al. (87) reported the mutarotase effect on the colorimetric determination of D-glucose, using the β-D-glucose oxidase–peroxidase system. They found that in the presence of mutarotase the color-developing reaction was complete within 5 min, and also both sensitivity and accuracy increased somewhat in comparison with the results obtained without mutarotase, because of the complete oxidation of all D-glucose (not only β-anomer but also α-anomer).

Experimental Procedure of Miwa et al. (87)

REAGENTS. 1. *Enzyme-chromogen mixture.* One thousand units (12) (25 units/test) of β-D-glucose oxidase, 200 units (5 units/test) of hog kidney mutarotase partially purified using Sephadex G-75 as described in Section II.1.B, and 120 units (3 units/test) of horseradish peroxidase, Type I (Sigma

Chemical Co., St. Louis, Missouri), are dissolved in 100 ml of 0.05M sodium phosphate buffer (pH 7.0). The enzyme solution can be stored at $-10°C$ for at least 1 month. One part of 0.66% solution of o-dianisidine in methanol and 100 parts of the enzyme solution are mixed as needed. This enzyme-chromogen mixture keeps for several hours.

2. D-*Glucose standard.* Twenty milligrams of D-glucose is dissolved in saturated benzoic acid solution, and the volume is brought to 100 ml with the same solution.

Procedure. Add 0.1 ml each of D-glucose sample (2 to 50 μg) and D-glucose standard to 2.5 ml of enzyme-chromogen mixture in separate test tubes. Allow each tube to stand for 5 to 10 min at room temperature. Add 2.5 ml of 25% sulfuric acid, mix well to stop the reaction, and cool in tap water. Transfer the contents of each test tube to cuvettes, and measure the absorbance at 535 nm against a blank prepared by adding 2.5 ml of 25% sulfuric acid and 2.5 ml of enzyme-chromogen mixture to 0.1 ml of sample. The color formed is stable for several hours. The reaction rate obtained by use of 0.1 ml (20 μg) of D-glucose standard is shown in Figure 4.

Analyses of samples containing considerable amounts of interfering substances or proteins require preparation of a Somogyi filtrate, using a mixture containing 5% zinc sulfate and 0.3N barium hydroxide (81).

If mutarotase is absent in the reaction mixture, the reaction is slower and the color intensity rather low, as shown in Figure 4. In the absence of mutarotase, it is recommended that the reaction mixture be incubated for 20 min.

B. D-GLUCOSE ANOMERS

Keston and Brandt (33) first performed the enzymic analysis of D-glucose anomers using β-D-glucose oxidase. In this method, an aliquot of a sample solution containing about 50 μg of D-glucose is cooled in an ice bath, added to a cold β-D-glucose oxidase solution (containing a small amount of catalase) and shaken in the ice-water bath for 5 min to destroy the β-D-glucose in the sample. Then a peroxidase-chromogen (o-dianisidine) solution is rapidly added to determine residual α-D-glucose in the reaction mixture. For the untreated sample, the aliquot is added to a tube containing the complete analytical D-glucose reagent (a mixture of β-D-glucose oxidase solution and the peroxidase-chromogen mixture) to determine the total D-glucose. Incubation at 37°C for 20 min with shaking is used for color development. In this way the amounts of the anomers of D-glucose in the sample are determined. Since the catalase contained in the β-D-glucose oxidase solution and the peroxidase are competing for the H_2O_2 produced during the phase of

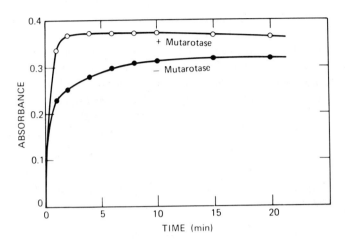

Figure 4. Effect of mutarotase on colorimetric determination of D-glucose with the β-D-glucose oxidase–peroxidase–o-dianisidine system.

the procedure when the chromogen is present, and the partition of H_2O_2 between catalase and peroxidase is slightly influenced by the H_2O_2 concentration, the final color of the solution would not be expected to be exactly the same for β-D-glucose, equilibrium D-glucose, and α-D-glucose initially present. A certain factor, therefore, must be used for the accurate determination of D-glucose anomers to correct the analytical values dependent on the final absorbance produced by α- and equilibrium D-glucose.

Hill (34) reported an automated method for measuring deviations from equilibrium of D-glucose anomers in blood. The method is dependent on the specificity of β-D-glucose oxidase for β-D-glucose and the catalytic effect of hydroxyl ions on mutarotation, and allows for continuous monitoring of the blood stream or a reaction mixture. In brief, the sample to be analyzed is dialyzed and the D-glucose which passes through the membrane is divided into two portions. Each portion is then analyzed for β-D-glucose, one with and one without equilibration of the anomers with alkali. This facilitates the finding of short periods of deviation from equilibrium such as might be expected to occur in rapidly changing systems.

4. Fluorometric Determination of D-Glucose

It is a well-known fact that fluorometry is more sensitive than colorimetry. Guilbault et al. (88) developed a fluorometric determination of D-glucose using the β-D-glucose oxidase–peroxidase system. The method is based on the peroxidatic conversion of the nonfluorescent homovanillic acid to the highly fluorescent 2,2'-dihydroxy-3,3'-dimethoxybiphenyl-5,5'-diacetic acid.

Later, Guilbault et al. (89) screened 25 indicator substrates for this determination. Of these compounds, p-hydroxyphenylacetic acid (A) was judged to be best because it is completely stable to autoxidation and has the advantage over homovanillic acid of lower cost and a higher fluorescent coefficient (fluorescence/molarity). It is converted on oxidation to the highly fluorescent compound 2,2'-dihydroxybiphenyl-5,5'-diacetic acid (B), which has an excitation of 317 nm and a fluorescence of 414 nm.

The reactions are as follows:

$$\beta\text{-D-Glucose} + O_2 + H_2O \xrightarrow[\text{oxidase}]{\beta\text{-D-glucose}} \text{D-gluconic acid} + H_2O_2 \quad [2]$$

The fluorometric technique developed by Guilbault and his coworkers can be applied to the determination of many oxidative enzymes and their substrates.

Experimental Procedure of Guilbault et al. (89)

REAGENTS. The following solutions are prepared by dissolving the substance in triple-distilled water: peroxidase (Calbiochem. Co., LaJolla, Calif., B grade, RZ = 0.3), 1 mg/ml; β-D-glucose oxidase (Calbiochem. Co., 20 units/mg), 2 mg/ml; p-hydroxyphenylacetic acid (Columbia Organic and Chemical Co., Columbia, South Carolina) 3.7 mg/ml. Tris-HCl buffer, pH 8.5, 0.1M; Tris (Sigma Chemical Co.), is also dissolved in triple-distilled water, and the pH is adjusted with concentrated HCl.

APPARATUS. Fluorescent measurements were made with an Aminco fluoromicrophotometer filter instrument. The lamp used was a Turner 110-855 general-purpose UV lamp. A Kodak Wratten 47B secondary filter, a circulating water bath to control the temperature at 30°C, and a Beckman linear recorder for automatic readout were used.

Calibration. The sensitivity of the instrument is set at the highest level, using the Turner 110–855 lamp. The 0.01 μg/ml solution of quinine sulfate in 0.1N H$_2$SO$_4$ gave a fluorescent coefficient (fluorescence/molarity) of 1.12 × 10^8 under the conditions used.

Procedure. Add 2.0 ml of $0.1M$ Tris-HCl buffer (pH 8.5) to 0.1 ml of D-glucose solution (1 to 100 μg total), 0.1 ml of p-hydroxyphenylacetic acid, 0.1 ml of peroxidase, and 0.1 ml of β-D-glucose oxidase. Record the maximum reaction rate (excitation: 317 nm, fluorescence: 414 nm), and make calibration plots of $\Delta F_{max/min}$ versus D-glucose concentration. In the case of low D-glucose concentrations (1 to 10 μg total) a 10- to 15-min induction period is required before a change in fluorescence is observed. At much higher concentrations (10 to 100 μg total), analysis can be performed within 3 to 5 min, based on the initial reaction rate.

5. Electrochemical Determination of D-Glucose

Malmstadt and Pardue (90) reported the quantitative analysis of D-glucose by an automatic potentiometric reaction rate method. As H_2O_2 forms in the β-D-glucose oxidase reaction, it reacts immediately with iodide in the presence of a molybdenum catalyst to form iodine, and the change of iodine concentration causes a change in the voltage of a concentration cell, which is detected by platinum electrodes and followed automatically by combining a commercial comparator and an auxiliary relay system to provide results within 1 min from the start of the reaction. D-Glucose, 10 to 1000 μg in a total volume of 2 ml, is determined with a relative error of about 1% throughout the range.

Pardue (91) reported an automatic amperometric determination of D-glucose in 20 μl of serum, plasma, and whole blood with measurement times ranging from 10 to 100 sec. This method utilizes the same reaction sequence that was employed by Malmstadt and Pardue (90), but the rate of increase in iodine concentration is measured amperometrically, using a polarized rotating platinum electrode.

Pardue (92) also developed automated potentiometric methods, using the same reaction sequence for quantitative analysis of D-glucose, based on the measurement of slopes of linear reaction rate curves (93) or nonlinear rate curves (92). D-Glucose, 5 to 500 μg, was determined with a relative standard deviation within 1.5%.

The electrochemical method developed by Guilbault et al. (94) for determining D-glucose is also based on the air oxidation of this substance by β-D-glucose oxidase. A small and constant current of 40 μA is applied across two platinum thimble electrodes, and the change in potential of the anode with time caused by the H_2O_2 formed during the reaction is recorded. Because the substrate is electrochemically inactive in the system used, diphenylamine sulfonic acid is used to establish a selected, well-poised starting potential. By this procedure, 4.15 to 45 mg of D-glucose in 25 ml of the reaction mixture was determined with measurement times ranging from 3 to 5 min and a relative standard deviation of 1.5%.

Blaedel and Olson (95) determined D-glucose by using a differential amperometric procedure, based on continuous measurement of the rate of the β-D-glucose oxidase reaction in a flowing system. The H_2O_2 produced oxidizes ferrocyanide to ferricyanide, which is measured with a tubular platinum electrode. The equipment used may be calibrated with a standard D-glucose solution and gives direct readout of the D-glucose concentration of the sample. The recorded response is linear with D-glucose concentration up to 100 μg/ml. When flow rates of sample solutions ranging between 1 and 2 ml/min are employed, sensitivity can be adjusted to give full-scale response for 1, 10, or 100 μg/ml of D-glucose, and the relative standard deviation is <1% of full scale in all three cases. Samples can be run at the rate of 20 per hour. The use of continuous flow is especially well suited for determinations that utilize reaction rates, automation is convenient, and direct readout of the answer is obtainable.

Williams et al. (96) reported another type of electrochemical enzymic analysis of blood D-glucose. The β-D-glucose oxidase reaction used in this method replaces oxygen, the natural enzyme acceptor, with benzoquinone without expelling dissolved oxygen. D-Glucose is measured by monitoring the electro-oxidation of hydroquinone with the electrochemical sensor (platinum electrode). The reactions are as follows:

D-Glucose + benzoquinone

$$+ H_2O \xrightarrow[\text{oxidase}]{\beta\text{-D-glucose}} \text{D-gluconic acid} + \text{hydroquinone} \qquad [9]$$

$$\text{Hydroquinone} \xrightarrow{\text{Pt}} \text{benzoquinone} + 2H^+ + 2e \qquad [10]$$

($E = 0.4$ V versus standard calomel electrode).

The advantage of this method is that the oxidant concentration is readily controlled, and the D-glucose measurement has a wide linear concentration range which would not be possible when monitoring oxygen depletion. This method needs to be greatly improved in sensitivity, however, for general use in D-glucose determination, since it is applicable at present only to samples containing at least 1 mg of D-glucose.

In general, the electrochemical methods for determining D-glucose have definite advantages in that the measurement time is very short, wide concentration ranges are measurable by scale switching without sample dilution, and the test sample need not be optically clear. They also have some disadvantages in that the sensitivity may not be good, and expertise in electrochemical technique is required to operate the apparatus. If a more easily obtainable and simpler apparatus is developed in the future, the electrochemical determination of D-glucose will be more widely used, mainly because of its rapidity.

III. DETERMINATION OF D-GLUCOSE WITH HEXOKINASE

Hexokinase is known to catalyze the phosphorylation of D-glucose to D-glucose-6-P in the presence of ATP. If D-glucose-6-P is quantitatively determined, therefore, this reaction may serve for the determination of D-glucose.

$$\text{D-Glucose} + \text{ATP} \xrightarrow[\text{Mg}^{2+}]{\text{hexokinase}} \text{D-glucose-6-P} + \text{ADP} \qquad [11]$$

In the presence of NADP, D-glucose-6-P is specifically oxidized by D-glucose-6-P dehydrogenase to form 6-phospho-D-gluconic acid and NADPH. Therefore, by using a coupled hexokinase–D-glucose-6-P dehydrogenase system, the amount of D-glucose is linked to that of NADPH, which is easily measurable by spectrophotometry (at 340 or 366 nm) or by fluorometry.

$$\text{D-Glucose-6-P} + \text{NADP} \xrightarrow[\text{dehydrogenase}]{\text{D-glucose-6-P}} \text{6-phospho-D-gluconic acid}$$
$$+ \text{NADPH} \qquad [12]$$

The enzymic cycling method using the same system may be utilized for a highly sensitive fluorometric determination of D-glucose.

If ATP-γ-^{32}P is used in the hexokinase reaction, D-glucose-6-P is labeled with ^{32}P. D-Glucose can also be determined, therefore, with only hexokinase by measuring the radioactivity of D-glucose-6-^{32}P.

1. Properties of Hexokinase and D-Glucose-6-Phosphate Dehydrogenase

A. HEXOKINASE

Recently, studies on hexokinase isoenzymes have been actively carried forward by many investigators.

Enzyme preparations of various degrees of purity are commercially available; these are usually obtained from yeast and are composed of a mixture of isoenzymes. Although the preparation need not be crystalline, it must be free of the following enzymes: 6-phospho-D-gluconic acid dehydrogenase; phosphoglucomutase (with D-glucose-1-phosphate); enzymes which oxidize or destroy NADPH; phosphohexose isomerase, which converts D-fructose-6-phosphate to D-glucose-6-P; and glycosidases, which liberate D-glucose by hydrolyzing glucosides. Hexokinase requires Mg^{2+} for full activation, and the optimum pH of the enzyme obtained from baker's yeast is between 7 and 9. Since the substrate specificity of hexokinase is rather low, it acts on various hexoses (e.g., D-glucose, D-fructose, D-mannose, 2-deoxy-D-glucose and D-glucosamine).

B. D-GLUCOSE-6-PHOSPHATE DEHYDROGENASE

Many commercial preparations of this enzyme are available. The crystalline preparation from yeast has an optimum pH between 7 and 9, and its activity is enhanced by Mg^{2+}. This enzyme requires NADP as a coenzyme and acts only on D-glucose-6-P to form 6-phospho-D-gluconic acid and NADPH.

The enzyme preparation used for the D-glucose assay must be free from the same interfering enzymes as listed in Section III.1.A. Since phosphate ions inhibit D-glucose-6-P dehydrogenase, phosphate buffers should be avoided or diluted out (97).

2. Spectrophotometric Determination of D-Glucose

Whereas hexokinase catalyzes the phosphorylation of D-glucose and other hexoses as described above, D-glucose-6-P dehydrogenase is specific for D-glucose-6-P among the hexose-6-phosphates produced by the hexokinase reaction. Therefore it is possible by combining these two enzymes to determine D-glucose with high specificity.

Slein et al. (98) showed the possibility of applying these enzyme reactions to the quantitative determination of D-glucose. Cori and Larner (99) first determined D-glucose with hexokinase and D-glucose-6-P dehydrogenase by measuring the stoichiometric formation of NADPH spectrophotometrically at 340 nm.

Although the colorimetric β-D-glucose oxidase method for the determination of D-glucose in blood has been well established for some time, its application for the determination of urinary D-glucose has been limited. This is due to the presence of inhibitory substances in urine which interfere with the enzyme reaction. Removal of such substances by adsorption (47,50,100), using activated charcoal, Lloyd's reagent, or ion-exchange resin, has improved the method but has not yet led to satisfactory results. Since, by using hexokinase and D-glucose-6-P dehydrogenase, this source of error is essentially eliminated, the spectrophotometric method employing this enzyme system has been frequently applied for the determination of urinary D-glucose. The application of this method to urine was first suggested by Barthelmai and Czok (101), but only for the estimation of high D-glucose levels (0.3 to 3.0 g/100 ml), and these workers compared it with the β-D-glucose oxidase–ferrocyanide method for the determination of D-glucose in blood. Keller (102) studied this method with urine from dogs, tested a number of sugars and other compounds for interference, and demonstrated the application of the method to a diuresis study of a normal human being. With this method, Renschler et al. (103) and Scherstén and Fritz (104) found that the upper normal limit for D-glucose concentration in human urine was 30 and 20 mg/100 ml, respectively. This spectrophotometric enzymic method was extensively evaluated in regard to its suitability for urine analysis by Peterson (105) and Peterson and Young (106). They reported that it is highly specific and interference-free without purification of the samples, and the only possible urinary interference that could be found was a diminution of the D-glucose measured in the presence of extremely abnormal D-fructose concentrations. Furthermore this method could be used for screening purposes without determining the background ultraviolet absorbance of the sample.

An application of the method for the separate determinations of D-glucose and D-fructose in blood was reported by Schmidt (107). The D-fructose-6-phosphate formed with hexokinase reaction is converted to D-glucose-6-P with phosphohexose isomerase and is measured in a separate reaction.

Mager and Farese (108) and Mager (109) compared the hexokinase–D-glucose-6-P dehydrogenase method with the β-D-glucose oxidase and Somogyi-Nelson procedures for the analysis of deproteinized plasma and whole blood. They reported that a close relationship was found between the data obtained by these three methods.

A blood D-glucose determination using this enzyme system was also reported by Stork and Schmidt (110). In their method, a hemolyzate (0.5 ml) prepared by adding 5 μl of capillary blood to 0.5 ml of water served as the sample, and the absorbance was measured at 366 nm. This method needs no deproteinization.

Finch et al. (111) reported an enzymic method for the assay of D-glucose and other sugars in acid hydrolyzates of glycoproteins. With this method D-glucose is determined spectrophotometrically with hexokinase and D-glucose-6-P dehydrogenase as follows.

Experimental Procedure of Finch et al. (111)

REAGENTS. Hexokinase, D-glucose-6-P dehydrogenase, ATP, and NADP are commercial samples (C. F. Boehringer, Mannheim, Germany).

Procedure. Prepare solutions containing the following components (μmoles): ATP, 1.0; Tris-HCl (pH 8.0), 25.0; MgCl$_2$, 1.0; D-glucose, 0.01 to 0.2; and 0.28 unit of hexokinase, in a final volume of 0.30 ml. Incubate the solutions at 37°C for 15 min and stop the reaction by boiling at 100°C for 2 min. Add to these solutions the following (μmoles): NADP, 0.5; MgCl$_2$, 20.0; and Tris-HCl (pH 7.8), 100, in a final volume of 2.0 ml. Determine the absorbance of each solution at 340 nm, add 0.53 unit of D-glucose-6-P dehydrogenase in a volume of 0.075 ml, and follow the reaction at

340 nm at 22°C until complete (usually within 20 min). The absorbance then is taken as the final reading.

Calculation. The change in absorbance at 340 nm is calculated for every incubation by subtracting initial from final readings for both the enzyme blank and the test samples. The change in absorbance (enzyme blank) for the incubation in the absence of D-glucose is then subtracted from changes obtained in the presence of various concentrations of D-glucose, and these values are used to construct standard curves.

3. Fluorometric Determination of D-Glucose

Scherstén and Tibbling (112) described an enzymic fluorometric method for the determination of urinary D-glucose. In this method, the nonspecific fluorescence of urine was almost completely removed by treating the urine with a mixed ion-exchange resin (Amberlite IR-120 plus Amberlite IRA-400), and the fluorescence of NADPH produced by the enzyme reactions was measured (excitation: 340 nm, fluorescence: 458 nm). It was also pointed out that even 0.04 μg of D-glucose was distinguishable from blank, and the cost of the reagents per analysis was reduced 40-fold by adapting the spectrophotometric method to fluorometry.

Zwiebel et al. (113) reported a very sensitive method for determining D-glucose, utilizing the principle of the enzymic cycling method described by Lowry et al. (114). In this method, the NADPH produced by the hexo-kinase–D-glucose-6-P dehydrogenase reaction is cycled:

$$\text{NADPH} + \alpha\text{-ketoglutarate} + \text{NH}_4^+ \xrightarrow[\text{dehydrogenase}]{\text{glutamic}} \text{NADP} + \text{glutamate} \quad [13]$$

$$\text{D-Glucose-6-P} + \text{NADP} \xrightarrow[\text{dehydrogenase}]{\text{D-glucose-6-P}} \text{6-phospho-D-gluconic acid} + \text{NADPH} \quad [12]$$

Under the proper conditions (pH 8.0, 38°C), each molecule of NADPH catalyzes the formation of 5000 to 10,000 molecules of 6-phospho-D-gluconic acid within 30 min. The 6-phospho-D-gluconic acid is then measured with 6-phospho-D-gluconate dehydrogenase and extra NADP:

$$\text{NADP} + \text{6-phospho-D-gluconic acid} \xrightarrow[\text{dehydrogenase}]{\text{6-phospho-D-gluconate}} \text{NADPH}$$

$$+ \text{D-ribulose-5-phosphate} + \text{CO}_2 \quad [14]$$

The NADPH produced is determined fluorometrically (115). This method is extremely sensitive (1 × 10^{-12} g) and therefore especially advantageous for histochemical studies, but it is less suitable for routine use because of its complexity.

Guilbault et al. (116) showed that an increased sensitivity can be achieved by coupling the hexokinase–D-glucose-6-P dehydrogenase reactions with the resazurin (nonfluorescent)-resorufin (fluorescent) indicator reaction developed by Guilbault and Kramer (117):

$$\text{NADPH} + \quad \xrightarrow[\text{methosulfate}]{\text{phenazine}} \quad + \text{ NADP} \tag{15}$$

Resazurin   Resorufin

The production of the highly fluorescent resorufin (excitation: 560 nm, fluorescence: 580 nm) was monitored with a spectrophotofluorometer connected with a linear recorder, and its rate of formation, $\Delta F/\text{min}$, was directly proportional to the concentration of D-glucose. From 0.03 to 150 μg D-glucose was assayed with a precision of 1% and an accuracy of about 1.2%. The experimental procedure is as follows.

Experimental Procedure of Guilbault et al. (116)

REAGENTS. All solutions are prepared using triple-distilled water: peroxidase, 1 mg/ml horseradish peroxidase (type I, Sigma Chemical Co.); β-D-glucose oxidase, 5 mg/ml fungal β-D-glucose oxidase (type II, Sigma) in 0.05M sodium acetate buffer (pH 5.0); D-glucose-6-P dehydrogenase, 0.093 mg/ml enzyme (type X, yeast, Sigma); hexokinase, 0.1 mg/ml pure enzyme (type V, yeast, Sigma). McIlvaine buffer, 0.1M, pH 7.0, is used. Stock solutions of ATP (Sigma, 0.03M), phenazine methosulfate (Sigma, 0.001M), MgCl$_2$ (0.01M), and NADP (Sigma, 0.005M) are prepared. A $2 \times 10^{-4}M$ solution of resazurin (Eastman Organics, Rochester, New York) is prepared in methyl Cellosolve.

APPARATUS. Fluorescence measurements (excitation: 560 nm, fluorescence: 580 nm) made with an Aminco-Bowman spectrophotofluorometer.

Procedure. To 2.0 ml of 0.1M McIlvaine buffer add 0.1 ml each of D-glucose, MgCl$_2$, ATP, and hexokinase solutions, and incubate the resulting mixture for 10 min. To this mixture then add 0.1 ml each of NADP, D-glucose-6-P dehydrogenase, resazurin, and phenazine methosulfate solutions. Measure the rate of reaction, $\Delta F/\text{min}$, and determine the concentration of test sample (D-glucose) from a calibration plot of $\Delta F/\text{min}$ versus D-glucose concentration.

4. Colorimetric Determination of D-Glucose

The colorimetric serum D-glucose assay utilizing hexokinase-D-glucose-6-P dehydrogenase reactions coupled with the concomitant reduction of a tetrazolium salt was developed by Carroll et al. (118). The following reactions summarize the assay:

$$\text{D-Glucose} + \text{ATP} \xrightarrow[\text{Mg}^{2+}]{\text{hexokinase}} \text{D-glucose-6-P} + \text{ADP} \qquad [11]$$

$$\text{D-Glucose-6-P} + \text{NADP} \xrightarrow[\text{dehydrogenase}]{\text{D-glucose-6-P}} \text{6-phospho-D-gluconic acid}$$
$$+ \text{NADPH} \qquad [12]$$

$$\text{NADPH} + \begin{matrix}\text{phenazine}\\ \text{methosulfate}\\ \text{(PMS)}\end{matrix} \rightarrow \text{NADP} + \begin{matrix}\text{reduced}\\ \text{phenazine methosulfate}\\ \text{(PMSH)}\end{matrix} \qquad [16]$$

$$\text{PMSH} + \text{iodonitro tetrazolium chloride} \rightarrow \text{PMS} + \text{formazan} \qquad [17]$$
$$\text{(520 nm)}$$

The major advantages of this procedure are as follows: (a) there is no need to use an ultraviolet spectrophotometer or a spectrophotofluorometer; (b) only catalytic quantities of NADP, an expensive ingredient, are required because [16] regenerates NADP; (c) individual serum blanks need not be run; and (d) deproteinization is not required.

Experimental Procedure of Carroll et al. (118)

REAGENTS. 1. *Buffered-enzyme reagent.* The following constituents are contained in 100 ml of $0.1M$ Tris-HCl buffer (pH 7.8): 50.8 mg $\text{MgCl}_2 \cdot 6\text{H}_2\text{O}$, 18 mg NADP, 50 mg ATP disodium salt, 10.4 mg ethylenediamine tetraacetate tetrasodium salt, 44,500 IU D-glucose-6-P dehydrogenase, 280,000 IU hexokinase, 150 mg protein as dialyzed plasma, and 20 mg methyl *p*-hydroxybenzoate as a preservative. Additional protein is required to stabilize the enzymes and to solubilize the formazan. This reagent is stable for several days when refrigerated. However, the enzymes cannot be frozen and thawed repeatedly without significant inactivation. A stable reagent is obtained by lyophilizing this reagent.

2. *Color developer.* Prepare by dissolving 200 mg iodonitro-tetrazolium chloride in approximately 80 ml of distilled water, to which 50 mg phenazine methosulfate and 20 mg methyl *p*-hydroxybenzoate are added; the final solution is diluted to 100 ml with distilled water. This reagent must be protected from light.

Procedure. Add 20 μl of serum to 1.0 ml of the buffered-enzyme reagent and warm to 37°C in a water bath. Add 0.2 ml of the color developer, mix, and incubate for 10 min at 37°C (incubation periods beyond 10 min result in higher blanks). After the incubation, add 5 ml of $0.1N$ HCl and mix. Read the absorbance of the samples versus a reagent blank at 520 nm. The color is stable for at least 30 min. (This procedure can be performed by making a prior one-tenth sample dilution and using 0.1 ml of diluted sample with half quantities of reagents.)

5. Radioisotopic Determination of D-Glucose

Chick and Like (119) used tracer methodology in the determination of D-glucose with hexokinase and ATP-γ-^{32}P. In this micromethod D-glucose was first converted into D-glucose-6-^{32}P by reaction with ATP-γ-^{32}P in the presence of hexokinase. Then the radioactivity of D-glucose-6-^{32}P separated from the ^{32}P-labeled precursor and contaminants was counted to estimate the D-glucose.

Experimental Procedure of Chick and Like (119)

REAGENTS. 1. *Hexokinase.* Dissolve crystalline yeast hexokinase (Sigma Chemical Co.), specific activity 300 μM units/mg protein to be 6 μg/ml in $1.0M$ Tris-HCl buffer (pH 8.0) containing 25 mM MgCl$_2$ and 0.5% crystalline bovine serum albumin; 20 μl is used in the final reaction mixture.

2. D-*Glucose standards.* Prepare dilutions of D-glucose, 2.5×10^{-4} to $5 \times 10^{-7}M$ prepared from the same stock of $0.1M$; 20 μl is used in the final reaction mixture (5×10^{-9} to 1×10^{-11} mole.

3. *ATP-γ-^{32}P.* Dry ATP-γ-^{32}P (International Chemical & Nuclear Corp., Irvin, California), specific radioactivity of 187 mCi/mmole, under vacuum and reconstitute with double-distilled water to 0.2 μCi/ml.

Procedure. Carry out the enzymic conversion of D-glucose to D-glucose-6-^{32}P in a total volume of 100 μl in 6×50 mm tubes, incubated for 10 min at 30°C. The final reaction mixture contains albumin, 0.1%; hexokinase, 1.2 μg/ml; MgCl$_2$, 5 mM; Tris-HCl, 0.2 M; ATP, $5 \times 10^{-5}M$ (ATP-γ-^{32}P, 0.02 to 0.4 μCi); and D-glucose, 5×10^{-9} to 1×10^{-11} mole. Start the reaction by adding the ATP, and stop it by plunging the tubes into powdered Dry Ice. Add 10 μl of $10^{-3}M$ carrier D-glucose-6-P, and inactivate the hexokinase by heating the tubes in a boiling-water bath for 5 min. Separate ATP-γ-^{32}P and ^{32}P-labeled contaminants from D-glucose-6-^{32}P by a two-step procedure. First, sediment 1 mg of acid-washed Norit A (Amend Drug and

Chemical Co., Irvington, New Jersey) by centrifugation for 30 min at 2400g. Transfer 50 μl of supernatant fluid to 6 $\times$ 50 mm tubes, and convert traces of remaining ^{32}P in ATP, pyrophosphate, and polyphosphate to labeled orthophosphate (^{32}P$_i$) by the addition of 10 μl of 10N HCl, followed by heating in a boiling-water bath for 30 min. The D-glucose-6-^{32}P, however, is not hydrolyzed by this procedure. Precipitate the ^{32}P$_i$ by the addition of 10 μl of $10^{-2}M$ carrier P$_i$ as KH$_2$PO$_4$, followed by 20 μl of an orthophosphate precipitation mixture consisting of 2 vols 10% ammonium molybdate and 1 vol 0.2M triethylamine hydrochloride. After sedimenting the yellow precipitate for 30 min at 2400g, plate 50 μl of the supernatant fluid containing the D-glucose-6-^{32}P on aluminum planchets and count in a 2 π gas-flow counter fitted with a thin end window.

Of the D-glucose determinations reported in this chapter, this enzymic labeling method is one of the most sensitive. Glycogen can also be determined by this method after its conversion to D-glucose by using an acid mixture consisting of 0.5N HCl and 5% trichloroacetic acid. Although this method can determine D-glucose to 1 $\times$ 10^{-11} mole, it may not be practicable to apply it to the determination of less than 1 $\times$ 10^{-10} mole, since the count per minute for 1 $\times$ 10^{-11} mole is only 50% of the no-D-glucose blank. The specificity of this method for D-glucose is not as good as can be obtained with other enzymic methods, since hexokinase phosphorylates other hexoses as well, as described previously. Furthermore, the procedure for D-glucose-6-^{32}P isolation does not include separation of the acid-stable phosphorylated hexoses such as D-mannose-6-phosphate-^{32}P.

IV. DETERMINATION OF D-GLUCOSE WITH ACYLPHOSPHATE: D-GLUCOSE-6-PHOSPHOTRANSFERASE

Bergmeyer and Moellering (120) reported the highly specific determination of D-glucose with acylphosphate: D-glucose-6-phosphotransferase and D-glucose-6-P dehydrogenase. The following reactions summarize this assay:

$$\text{D-Glucose} + \text{RCOOPO}_3\text{H}_2 \xrightarrow[\text{6-phosphotransferase}]{\text{acylphosphate: D-glucose}} \text{D-glucose-6-P}$$
$$\text{(acyl phosphate)}$$

$$+ \text{RCOOH} \quad [18]$$

$$\text{D-Glucose-6-P} + \text{NADP} \xrightarrow[\text{dehydrogenase}]{\text{D-glucose-6-P}} \text{6-phospho-D-gluconic acid}$$

$$+ \text{NADPH} \quad [12]$$

As a phosphoryl donor, acetyl phosphate, benzoyl phosphate, nicotinyl phosphate, or carbamyl phosphate was used, and the NADPH formed by the second reaction was spectrophotometrically measured at 340 or 366 nm to estimate the D-glucose.

Since the enzyme acylphosphate: D-glucose-6-phosphotransferase, obtainable from *Aerobacter aerogenes PRL-R3* (121), is absolutely specific for D-glucose, this method is scarcely influenced by the presence of other hexoses (D-fructose, D-mannose, etc.) even if the two enzymes used are contaminated by hexosephosphate isomerase. This method is superior in specificity to any other enzymic method for D-glucose, but the fact that acylphosphate: D-glucose-6-phosphotransferase is not commercially available poses a problem.

V. CONCLUSION

The enzymic methods for determining D-glucose are classified into three principal groups: β-D-glucose oxidase method, hexokinase method, and acylphosphate: D-glucose-6-phosphotransferase method.

In the β-D-glucose oxidase category are included the oxygen electrode, colorimetric, fluorometric, and electrochemical methods. The oxygen electrode method is recommended for the analysis of a small number of samples because of its rapidity (about 1 to 3 min for each sample) and simplicity (e.g., no deproteinization and a simple reaction mixture). On the other hand, the colorimetric method is suitable when a large number of estimations are required, as in clinical laboratories. This method, however, has the disadvantage that interfering substances in biological fluids affect coupling with the peroxidase reaction. The fluorometric method is applicable for no more than 0.5 μg of D-glucose, so that its sensitivity is almost the same as those of the colorimetric and oxygen electrode methods. It would be desirable, therefore, to exploit the high sensitivity of fluorometry by finding better indicator substrates and by eliminating fluorescence quenching substances and interfering fluorescence-producing substances which are likely to be present in the samples. The electrochemical method has the advantages that measurement time is very short (about 1 min) and the test sample need not be optically clear; however, the sensitivity is not great, and fairly expert knowledge of electrochemical technique is required to operate the apparatus.

One of the shortcomings of the β-D-glucose oxidase method frequently comes from the gradual oxidation, due to the spontaneous conversion of α- to β-anomer that occurs before the oxidation with β-D-glucose oxidase. This disadvantage can be eliminated, however, by adding mutarotase to the reaction mixture to convert the α- to β-anomer rapidly.

The hexokinase method may be based on spectrophotometric, fluorometric, colorimetric, or radioisotopic procedures. The first three using the hexokinase–D-glucose-6-P dehydrogenase system are highly specific for D-glucose and are scarcely influenced by extraneous constituents in blood and urine, but they require expensive ingredients—hexokinase, D-glucose-6-P dehydrogenase, and NADP. Since the fluorometric method using resazurin has a fairly high sensitivity (0.05 μg of D-glucose), it is useful for the microanalysis of a large number of samples. The fluorometric method using enzymic cycling is the most sensitive (1×10^{-12} g of D-glucose) and is suitable for histochemical studies on an ultramicroscale, notwithstanding its complexity. The radioisotopic method using the hexokinase-ATP-γ-^{32}P system is also extremely sensitive (1×10^{-9} g of D-glucose), but it is not applicable to samples containing hexoses which can be phosphorylated by the hexokinase reaction.

The acylphosphate: D-glucose-6-phosphotransferase method, which is the most specific of those reported hitherto, has the disadvantage that the enzyme is not available commercially.

The enzymic methods for determining D-glucose anomers with β-D-glucose oxidase are the oxygen electrode and colorimetric methods. The oxygen electrode method without mutarotase is rather rapid (within 10 min), adaptable to turbid samples, and free from the interferences which may affect the colorimetric method. The method using mutarotase requires no more than 3 min, and the sample needed is half that of the colorimetric method, which takes about 30 min and has limitations due to the peroxidase reaction. However, this method is the only colorimetric one for the determination of D-glucose anomers.

References

1. W. Pigman and H. S. Isbell, in *Advances in Carbohydrate Chemistry*, Vol. 23, M. L. Wolfrom and R. S. Tipson, Eds., Academic Press, New York, 1968, p. 4.
2. H. S. Isbell, in "Polarimetry, Saccharimetry and the Sugars," F. J. Bates, Ed., *Natl. Bur. Std. (U.S.) Circ. C-440*, 1942, p. 455.
3. D. Keilin and E. F. Hartree, *Biochem. J.*, 50, 341 (1952).
4. A. S. Keston, *Science*, 120, 355 (1954).
5. J. M. Bailey, P. H. Fishman, and P. G. Pentchev, *Biochemistry*, 9, 1189 (1970).
6. A. H. Free, in *Advances in Clinical Chemistry*, Vol. 6, H. Sobotka and C. P. Stewart, Eds., Academic Press, New York, 1963, p. 67.
7. M. W. Slein, in *Methods of Enzymatic Analysis*, H. U. Bergmeyer, Ed., Academic Press, New York, 1963, p. 117.
8. M. Kitamura, *Japan. J. Clin. Pathol., Suppl.*, 15, 35 (1968).
9. R. G. Martinek, *J. Am. Med. Technol.*, 31, 530 (1969).
10. R. Bentley, *Ann. Rev. Biochem.*, 41, 812 (1972).
11. K. Kusai, *Ann. Rept. Sci. Works, Fac. Sci. Osaka Univ. 8*, 1960, p. 43.
12. K. Kusai, I. Sekuzu, K. Okunuki, B. Hagihara, M. Nakai, and S. Yamauchi, *Biochim. Biophys. Acta*, 40, 555 (1960).
13. D. Keilin and E. F. Hartree, *Biochem. J.*, 42, 221 (1948).
14. J. M. Bailey, P. G. Pentchev, and J. Woo, *Biochim. Biophys. Acta*, 94, 124 (1965).
15. I. Miwa, *Anal. Biochem.*, 45, 441 (1972).
16. S. L. Lapedes and A. M. Chase, *Biochem. Biophys. Res. Commun.*, 31, 967 (1968).
17. J. M. Bailey, P. H. Fishman, and P. G. Pentchev, *J. Biol. Chem.*, 244, 781 (1969).
18. Lu-Ku Li, *Arch. Biochem. Biophys.*, 110, 156 (1965).

19. D. Keilin and E. F. Hartree, *Biochem. J.*, *42*, 230 (1948).
20. A. H. Kadish and D. A. Hall, *Clin. Chem.*, *11*, 869 (1965).
21. T. Kajihara and B. Hagihara, *Japan. J. Clin. Pathol.*, *14*, 322 (1966).
22. B. Hagihara, *Biochim. Biophys. Acta*, *46*, 134 (1961).
23. Y. Makino and K. Konno, *Japan. J. Clin. Pathol.*, *15*, 391 (1967).
24. H. L. J. Makin and P. J. Warren, *Clin. Chim. Acta*, *29*, 493 (1970).
25. S. J. Updike and G. P. Hicks, *Nature*, *214*, 986 (1967).
26. J. Okuda and G. Okuda, *Clin. Chim. Acta*, *23*, 365 (1969).
27. J. Okuda, G. Okuda, and I. Miwa, *Chem. Pharm. Bull.*, *18*, 1945 (1970).
28. A. H. Kadish, R. L. Litle, and J. C. Sternberg, *Clin. Chem.*, *14*, 116 (1968).
29. M. Jemmali and R. Rodriguez-Kabana, *Anal. Biochem.*, *37*, 253 (1970).
30. J. Okuda and G. Okuda, *Biochem. Med.*, *7*, 257 (1973).
31. *Standard Methods for the Examination of Water and Waste-Water*, American Public Health Association, New York, 1965, p. 409.
32. J. Okuda, T. Inoue, and I. Miwa, *Analyst*, *96*, 858 (1971).
33. A. S. Keston and R. Brandt, *Anal. Biochem.*, *6*, 461 (1963).
34. J. B. Hill, *J. Appl. Physiol.*, *20*, 749 (1965).
35. J. Okuda and I. Miwa, *Anal. Biochem.*, *39*, 387 (1971).
36. J. Okuda and I. Miwa, *Anal. Biochem.*, *43*, 312 (1971).
37. A. S. Keston, *Abstracts of Papers, 129th Meeting, Am. Chem. Soc.*, p. 31C (1956).
38. J. D. Teller, *Abstracts of Papers, 130th Meeting, Am. Chem. Soc.*, p. 69C (1956).
39. J. E. Middleton and W. J. Griffiths, *Brit. Med. J.*, *5048*, 1525 (1957).
40. L. L. Salomon and J. E. Johnson, *Anal. Chem.*, *31*, 453 (1959).
41. W. J. Blaedel and G. P. Hicks, *Anal. Chem.*, *34*, 388 (1962).
42. R. H. Thompson, *Clin. Chim. Acta*, *13*, 133 (1966).
43. J. E. Middleton, *Clin. Chim. Acta*, *22*, 433 (1968).
44. A. St. G. Huggett and D. A. Nixon, *Lancet*, *273*, 368 (1957).
45. A. Saifer and S. Gerstenfeld, *J. Lab. Clin. Med.*, *51*, 448 (1958).
46. L. P. Cawley, F. E. Spear, and R. Kendall, *Am. J. Clin. Pathol.*, *32*, 195 (1959).
47. G. R. Kingsley and G. Getchell, *Clin. Chem.*, *6*, 466 (1960).
48. F. W. Fales, J. A. Russell, and J. N. Fain, *Clin. Chem.*, *7*, 389 (1961).
49. G. Guidotti, J. P. Colombo, and P. P. Foa, *Anal. Chem.*, *34*, 388 (1962).
50. J. E. Logan and D. E. Haight, *Clin. Chem.*, *11*, 367 (1965).
51. M. Mager and G. Farese, *Am. J. Clin. Pathol.*, *35*, 104 (1965).
52. Y. Sakae and T. Kanno, *Japan. J. Clin. Pathol.*, *16*, 473 (1968).
53. D. M. Kilburn and P. M. Taylor, *Anal. Biochem.*, *27*, 555 (1969).
54. J. B. Lloyd and W. J. Whelan, *Anal. Biochem.*, *30*, 467 (1969).
55. G. Asrow, *Anal. Biochem.*, *28*, 130 (1969).
56. J. W. Rosevear, K. J. Pfaff, F. J. Service, G. D. Molnar, and E. Ackermant, *Clin. Chem.*, *15*, 680 (1969).
57. P. H. Lenz and A. J. Passannante, *Clin. Chem.*, *16*, 427 (1970).
58. K. Konobu, K. Okamura, S. Murao, and T. Komaki, *Abstracts of Papers, 91st Meeting, Pharm. Soc. Japan*, p. 502 (1971).
59. J. F. Goodwin, *Clin. Chem.*, *16*, 85 (1970).
60. J. W. Hall and D. M. Tucker, *Anal. Biochem.*, *26*, 12 (1968).
61. H. V. Malmstadt and S. I. Hadjiioannou, *Anal. Chem.*, *34*, 452 (1962).
62. A. G. Ware and E. P. Marbach, *Clin. Chem.*, *14*, 548 (1968).
63. S. E. Aw, *Clin. Chim. Acta*, *26*, 235 (1969).
64. F. W. Sunderman, Jr. and F. W. Sunderman, *Am. J. Clin. Pathol.*, *36*, 75 (1961).
65. S. Ikawa and T. Obara, *Japan. J. Clin. Pathol.*, *13*, 197 (1965).

66. K. Motegi and K. Shōji, *Japan. J. Clin. Pathol.*, *16*, 589 (1968).
67. A. Tsuji, T. Kinoshita, A. Sakai, and M. Hoshino, *Abstracts of Papers, 89th Meeting, Pharm. Soc. Japan*, p. 403 (1969).
68. R. Asato, T. Komano, T. Okuyama, and K. Satake, *J. Chem. Soc. Japan*, *88*, 560 (1967).
69. E. Kawerau, *Z. Klin. Chim.*, *4*, 224 (1966).
70. A. Clark and B. G. Timms, *Clin. Chim. Acta*, *20*, 352 (1968).
71. A. R. Tammes and C. D. Nordschow, *Am. J. Clin. Pathol.*, *49*, 613 (1968).
72. P. Trinder, *Ann. Clin. Biochem.*, *6*, 24 (1969).
73. K. Kahle, L. Weiss, M. Klarwein, and O. Wieland, *Fresenius' Z. Anal. Chem.*, *252*, 288 (1970).
74. J. M. C. Gutteridge and E. B. Wright, *J. Med. Lab. Technol.*, *25*, 385 (1968).
75. P. Trinder, *J. Clin. Pathol.*, *22*, 158 (1969).
76. Mosk. Inst. Biol. Med. Khim., *Mikrobiol. Zh.*, *31*, 84 (1969).
77. J. B. Hill and G. Kessler, *J. Lab. Clin. Med.*, *57*, 970 (1961).
78. R. B. McComb and W. D. Yushok, *J. Franklin Inst.*, *263*, 417 (1958).
79. D. A. Van der Heiden, *Pharm. Weekblad.*, *103*, 1133 (1968).
80. L. J. Greenberg and D. Glick, *Biochemistry*, *1*, 452 (1962).
81. M. Somogyi, *J. Biol. Chem.*, *160*, 69 (1945).
82. R. J. Henry, in *Clinical Chemistry*, Hoeber, New York, 1964, p. 625.
83. F. W. Fales, in *Standard Methods of Clinical Chemistry*, Vol. 4, D. Seligson, Ed., Academic Press, New York, 1963, p. 101.
84. R. G. Martinek, *J. Am. Med. Technol.*, *29*, 257 (1967).
85. E. Raabo and T. C. Terkildsew, *Scand. J. Clin. Lab. Invest.*, *12*, 402 (1960).
86. Chester Beaty Research Institute, *Precautions for Laboratory Workers Who Handle Carcinogenic Aromatic Amines*, Institute of Cancer Research, London, 1966.
87. I. Miwa, J. Okuda, K. Maeda, and G. Okuda, *Clin. Chim. Acta*, *37*, 538 (1972).
88. G. G. Guilbault, P. J. Brignac, Jr., and M. Zimmer, *Anal. Chem.*, *40*, 190 (1968).
89. G. G. Guilbault, P. J. Brignac, Jr., and M. Juneau, *Anal. Chem.*, *40*, 1256 (1968).
90. H. V. Malmstadt and H. L. Pardue, *Anal. Chem.*, *33*, 1040 (1961).
91. H. L. Pardue, *Anal. Chem.*, *35*, 1240 (1963).
92. H. L. Pardue, *Anal. Chem.*, *36*, 1110 (1964).
93. H. L. Pardue, *Anal. Chem.*, *36*, 633 (1964).
94. G. G. Guilbault, B. C. Tyson, Jr., D. N. Kramer, and P. L. Cannon, Jr., *Anal. Chem.*, *35*, 582 (1963).
95. W. J. Blaedel and C. Olson, *Anal. Chem.*, *36*, 343 (1964).
96. D. L. Williams, A. R. Doing, Jr., and A. Korosi, *Anal. Chem.*, *42*, 118 (1970).
97. L. Glaser and D. H. Brown, *J. Biol. Chem.*, *216*, 67 (1955).
98. M. W. Slein, G. T. Cori, and C. F. Cori, *J. Biol. Chem.*, *186*, 763 (1950).
99. G. T. Cori and J. Larner, *J. Biol. Chem.*, *188*, 17 (1951).
100. E. F. Beach and J. J. Turner, *Clin. Chem.*, *4*, 462 (1958).
101. W. Barthelmai and R. Czok, *Klin. Wochschr.*, *40*, 585 (1962).
102. D. M. Keller, *Clin. Chem.*, *11*, 471 (1965).
103. H. E. Renschler, H. Weicher, and H. V. Baeyer, *Deut. Med. Wochschr.*, *90*, 2349 (1965).
104. B. Scherstén and H. Fritz, *J. Am. Med. Assoc.*, *20*, 949 (1967).
105. J. I. Peterson, *Clin. Chem.*, *14*, 513 (1968).
106. J. I. Peterson and D. S. Young, *Anal. Biochem.*, *23*, 301 (1968).
107. F. H. Schmidt, *Klin. Wochschr.*, *39*, 1244 (1961).
108. M. Mager and G. Farese, *Tech. Bull. Reg. Med. Tech.*, *35*, 1965, p. 104.

109. M. Mager, *Scand. J. Clin. Lab. Invest.*, *18*, 58 (1966).
110. H. Stork and F. H. Schmidt, *Klin. Wochschr.*, *46*, 789 (1968).
111. P. R. Finch, R. Yuen, H. Schachter, and M. A. Moscarello, *Anal. Biochem.*, *31*, 291 (1969).
112. B. Scherstén and G. Tibbling, *Clin. Chim. Acta, 18*, 383 (1967).
113. R. Zwiebel, B. Höhmann, P. Frohnert and K. Baumann, *Pflügers Arch.*, *307*, 127 (1969).
114. O. H. Lowry, J. V. Passonneau, D. W. Schulz, and M. K. Rock, *J. Biol. Chem.*, *236*, 2746 (1961).
115. O. H. Lowry, N. R. Roberts, and J. I. Kapphahn, *J. Biol. Chem.*, *224*, 1047 (1957).
116. G. G. Guilbault, M. H. Sadar, and K. Peres, *Anal. Biochem.*, *31*, 91 (1969).
117. G. G. Guilbault and D. N. Kramer, *Anal. Chem.*, *37*, 1219 (1965).
118. J. J. Carroll, N. Smith, and A. L. Babson, *Biochem. Med.*, *4*, 171 (1970).
119. W. L. Chick and A. A. Like, *Anal. Biochem.*, *32*, 340 (1969).
120. H. U. Bergmeyer and H. Moellering, *Clin. Chim. Acta, 14*, 74 (1966).
121. M. Y. Kamel and R. L. Anderson, *J. Biol. Chem.*, *239*, 3608 (1964).

Radiometric Methods of Enzyme Assay

K. G. OLDHAM, *The Radiochemical Centre, Amersham, Buckinghamshire, England*

I. Introduction.. 193
II. The General Principles of Enzyme Assay.............................. 194
 1. The Definition of a Unit of Enzyme Activity..................... 194
 2. The Measurement of Enzyme Activity........................... 195
  A. The Initial Velocity..................................... 195
  B. The Choice of Method.................................. 197
 3. The Classification of Enzymes................................. 198
III. Techniques... 198
 1. Solvent Extraction Methods................................... 199
 2. Methods Involving the Precipitation of Macromolecules............ 200
 3. Ion-Exchange Methods....................................... 202
  A. Ion-Exchange Resin Methods............................ 203
  B. Ion-Exchange Paper Methods............................ 204
   a. Ion-Exchange Paper Chromatographic Methods....... 205
   b. Ion-Exchange Paper Disk Methods.................. 205
 4. Methods Depending on the Release or Uptake of Activity in a Volatile Form... 207
  A. Release of Tritium as Tritiated Water.................... 207
  B. Release or Uptake of Carbon-14 Dioxide.................. 208
  C. Release or Uptake of Other Forms of Volatile Activity....... 210
 5. Paper and Thin-Layer Chromatographic Methods................. 211
 6. Electrophoretic Methods...................................... 213
 7. Methods Involving Adsorption on Charcoal or Alumina........... 213
  A. Methods Involving Adsorption on Charcoal............... 213
  B. Methods Involving Adsorption on Alumina............... 214
 8. Coupled Enzyme Assay Methods.............................. 215
  A. Radioisotopic Derivative Assays......................... 215
  B. Generation of Labeled Substrates *in situ*.................. 216
  C. Methods Involving Coupling to Give Improved Separation ... 216
 9. Methods Involving Precipitation of Derivatives.................. 217
 10. Dialysis and Gel Filtration Methods........................... 218
 11. Methods Involving Reverse Isotope Dilution Assay............... 218
IV. Radiometric Enzyme Assays..................................... 219
 1. The Assay of Enzymes of Class 1: the Oxidoreductases............ 219
  A. Subclass 1, Oxidoreductases Acting on the CH—OH Group of Donors.. 219

B. Sublcass 2, Oxidoreductases Acting on the Aldehyde or Keto Group of Donors.. 220
C. Subclass 3, Oxidoreductases, Acting on the CH—CH Group of Donors.. 220
D. Subclass 4, Oxidoreductases, Acting on the CH—NH₂ Group of Donors.. 220
E. Subclass 5, Oxidoreductases, Acting on the C—NH Group of Donors.. 221
F. Subclass 13, Oxygenases............................... 221
G. Subclass 14, Hydroxylases............................. 222
2. The Assay of Enzymes of Class 2: the Transferases................. 224
A. Subclass 1, Enzymes Transferring One-Carbon Groups....... 224
B. Subclass 2, Enzymes Transferring Aldehydic or Ketonic Residues.. 226
C. Subclass 3, Enzymes Transferring Acyl Groups............. 226
D. Subclass 4, Enzymes Transferring Glycosyl Groups.......... 227
E. Subclass 5, Enzymes Transferring Alkyl or Related Groups... 232
F. Subclass 6, Enzymes Transferring Nitrogenous Groups....... 232
G. Subclass 7, Enzymes Transferring Phosphorus-Containing Groups.. 233
H. Subclass 8, Enzymes Transferring Sulfur-Containing Groups.. 243
3. The Assay of Enzymes of Class 3: the Hydrolases................. 244
A. Subclass 1, Hydrolases Acting on Ester Bonds.............. 244
B. Subclass 2, Hydrolases Acting on Glycosyl Compounds....... 247
C. Subclass 4, Hydrolases Acting on Peptide Bonds............ 248
D. Subclass 5, Hydrolases Acting on C—N Bonds Other Than Peptide Bonds.. 248
E. Subclass 6, Hydrolases Acting on Acid Anhydride Bonds...... 249
4. The Assay of Enzymes of Class 4: the Lyases.................... 249
A. Subclass 1, C—C Lyases............................... 250
B. Subclass 2, C—O Lyases............................... 252
C. Subclass 3, C—N Lyases............................... 252
5. The Assay of Enzymes of Class 5: the Isomerases................. 252
A. Subclass 1, Racemases and Isomerases................... 252
B. Subclass 3, Intramolecular Oxidoreductases............... 253
6. The Assay of Enzymes of Class 6: the Ligases................... 253
A. Subclass 1, Ligases Forming C—O Bonds................. 253
B. Subclass 3, Ligases Forming C—S Bonds................. 254
C. Subclass 3, Ligases Forming C—N Bonds................. 254
D. Subclass 4, Ligases Forming C—C Bonds................. 255
V. Some Important Considerations in Radiometric Enzyme Assays.......... 255
1. Factors Affecting the Sensitivity of Enzyme Assays................. 256
A. High Blank Readings Due to Inadequate Separation Procedures 256
B. High Blank Readings Due to the Presence of Impurities in the Substrate... 256
C. High Blank Readings Caused by Concomitant Nonenzymatic Reactions... 257
2. Selection of Optimum Conditions for Assay..................... 258
A. Enzyme Assays at High Substrate Concentration............ 258
B. Enzyme Assays at Low Substrate Concentration............. 259

	C.	Selecting the Optimum Volume for the Assay Reaction Mixture	259
	D.	Selecting the Optimum Substrate Concentration	260
3.		How High Should the Specific Activity of the Substrate Be?	260
4.		How Accurately Should the Specific Activity of the Substrate Be Known?	262
5.		How Pure Should the Substrate Be?	262
6.		Isotope Effects	263
7.		Problems Associated with the Handling of Small Masses and Very Dilute Solutions	265
8.		Special Effects with Tritium-Labeled Substrates	266
	A.	False Rates Due to the Labilization of Tritium	267
	B.	Intramolecular Migration of Tritium, the "NIH Shift"	268
	C.	Errors Resulting from the Use of Substrates in Which the Position of Labeling Is Not Exactly as Specified	268
	D.	Low Results Due to Self-Absorption Losses during the Liquid Scintillation Counting of Tritium-Labeled Compounds	269
9.		Miscellaneous Problems Arising during Assays by Sampling Methods	270
10.		Miscellaneous Problems Associated with the Supply, Cost, and Counting of Labeled Substrates	271
VI.		Summary: The Advantages of Radiometric Methods of Enzyme Assay	272
		Acknowledgments	274
		References	274

I. INTRODUCTION

During the past decade the inherent sensitivity, specificity, and simplicity of radiometric enzyme assays have resulted in a rapid increase in their development and application in many fields of biochemistry. This has led, in turn, to a proliferation of methods which are documented only in scientific journals. Because the technical details are to be found only in the "methods" sections of most published works, and no reference to the use of these techniques is made in either the title or the abstract, it is difficult for research workers to find the necessary information about the various methods without searching through a vast amount of published work. As a result, many workers have continued to use complicated and tedious assay methods when quicker and simpler ones are available. The lack of suitably collated practical information has also inhibited the acceptance of methods developed in research laboratories as clinically useful routine diagnostic procedures.

It is the purpose of this chapter to present a precise, up-to-date account of the principles, experimental techniques, and applications of radiometric enzyme assays and to review critically their present use and potential. The bibliography includes three recent review articles on this subject (1–3), but the very wide application of these methods means that only a fraction of the relevant published work can be cited. No attempt has been made to include full working directions for each of the several hundred assays already described in the literature. Instead the various techniques are covered in detail, and their practical applications are described more generally. A selection has been made of references which will enable the reader to obtain any additional detailed technical information he may require for a particular assay.

II. THE GENERAL PRINCIPLES OF ENZYME ASSAY

1. The Definition of a Unit of Enzyme Activity

The presence of an enzyme is recognized by the chemical reaction it catalyzes, and in most cases it is neither possible nor even desirable to measure the amount of enzyme present in absolute terms. Instead one defines a unit of an enzyme as that amount which produces a certain rate of reaction under a specified set of conditions.

Unfortunately in the past there has been no standard method of relating the unit employed to the reaction rate. Consequently a large number of arbitrary units have been defined for various enzymes, and often several different ones for the same enzyme. Frequently these units have not been made consistent or comparable with one another, and in many cases the assay conditions have not been properly defined. The resulting chaos led the Enzyme Commission of the International Union of Biochemistry (4) to issue in 1964 the following recommendation: "One unit of any enzyme is that amount which will catalyse the transformation of 1 micromole of the substrate per minute under standard conditions."

Two special cases require separate mention.

1. When the substrate is a molecule in which more than one bond is attacked, "1 microequivalent of the group concerned" should be substituted for "1 micromole of the substrate." In other words, the number of bonds broken is taken as a measure of the reaction.

2. In a bimolecular reaction, $A + B = C + D$, either 1 micromole of A or 1 micromole of B can be taken as the basis; but in the special case of a reaction between two identical molecules, when $B = A$, the basis should obviously be 2 micromoles of A.

Thus, in all cases, one cycle of the reaction is taken as the measure of the rate.

As far as practicable, the assay conditions should be optimal, especially with respect to the pH, substrate concentration, and any activators which may be necessary.

Whenever possible, assays should be based on measurements of initial rates of reaction, and not on the amount of substrate transformed in a fixed period of time, unless it is known that the velocity remains constant throughout the period.

In order to facilitate measurements of the initial velocity, it was recommended that the substrate concentration should be high enough to saturate the enzyme, so that the kinetics approach zero order. This may not always be desirable (see Section V.2). Sometimes, too, it may not be possible for various reasons, such as the limited availability or solubility of the substrate, the low affinity of the substrate for the enzyme, or, in special cases, the degradation of the enzyme (e.g., catalase) by its substrate. In these situations it is recommended that the Michaelis constant, K_m, should be determined, if possible, so that the velocity at enzyme saturation can be calculated from the equation

$$V = v\left(1 + \frac{K_m}{s}\right)$$

where v is the observed velocity at substrate concentration s. The assay temperature should also be stated and, where practicable, should be 30°C. This temperature is high enough to give reasonable enzymatic activity and low enough to avoid thermal denaturation with most enzymes.

The definition of an International Unit of enzyme activity is a great step forward, but at present few enzyme activities have been reported in such units. However, many of the arbitrary units used at present can be converted into International Units, and conversion tables for these have been published (5).

Ultimately it is to be hoped that all enzyme activities will be reported in terms of International Units. This would help to reduce the present confusion by making observations by different workers more easily comparable. Such uniformity is particularly important in clinical biochemistry, where the assignment of "normal" and "abnormal" levels of enzyme activity is often made after consideration of results obtained in several laboratories.

In this respect it is unfortunate that the International Committee on Enzyme Nomenclature changed its recommended temperature to 30°C after recommending 25°C in the first edition of its report (6). This has caused much confusion and has resulted in at least one textbook (7) stating erroneously that the recommended temperature is 25°C. Because enzyme rates change by approximately 10% for each degree centigrade change in temperature, there could be an error of approximately 50% in the quoted figure if authors report enzyme activities in International Units without specifying the temperature used.

2. The Measurement of Enzyme Activity

A. THE INITIAL VELOCITY

The measurement of the reaction velocity is a most important part of the technique of enzyme investigation. Progress curves of most enzyme reactions are of the general form shown in Figure 1, in which the velocity decreases

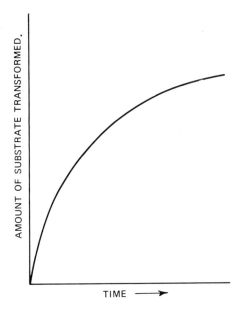

Figure 1. General form of progress curves for most enzyme reactions.

with time, and it is rarely possible to allocate an enzyme reaction to one of the classical kinetic orders of uncatalyzed reactions. Various factors contribute to this fall-off in the reaction rate. For instance, the products of reaction may inhibit the enzyme, the degree of saturation of the enzyme with substrate will fall as the substrate concentration declines during reaction, the reverse reaction may become more important as the product concentration increases, and the enzyme or coenzyme may become inactivated under the conditions of assay. Some or all of these factors may operate at the same time, and it is therefore usual in enzyme assays to measure the initial velocity. It is only at the initial point, when the various interfering factors just mentioned have not had time to operate, that the conditions are accurately known. However, most reaction curves are straight lines up to about 20% reaction, and, as long as measurements are made in this linear region, the rate of substrate utilization is a true measure of enzyme activity.

The initial rate of an enzymatic reaction should be proportional to the enzyme concentration, and experience has shown that this is usually the case. Whenever a new enzyme assay system is set up, this linear relationship between rate and enzyme concentration, and the range of conditions under which it applies, should be checked. The various causes of nonlinearity have been discussed in detail by Dixon and Webb (8).

B. THE CHOICE OF METHOD

Methods of following enzyme reactions are of two types: continuous and sampling. In both methods either the product of the reaction or the residual substrate may be measured. If the reaction is simple, either product formation or substrate utilization may be measured. Product formation methods are normally used because they are inherently more sensitive than substrate utilization methods. However, for multistage reactions in which intermediates accumulate, only substrate disappearance will measure the true rate. This condition may also apply when the reaction products are further metabolized by other enzymes (9).

Continuous assay techniques are thought to be used for following the majority of enzyme reactions (8). Most of these methods rely on coupling the enzyme reaction being assayed to another which involves an NAD or NADP oxidoreductase system. This results in a large change in the optical density during reaction. Such methods have often been preferred on account of their supposed ease, simplicity, and sensitivity. Although this may be true with purified enzymes, it is very often far from the case with crude extracts. For a coupled reaction to give an accurate measure of the primary enzyme reaction, it is essential that the primary product be utilized immediately by the coupling enzyme or enzymes. This requires a large excess of coupling enzyme, and it has been shown (10) that assays involving one coupling enzyme need at least a 100-fold excess of this enzyme to provide 4% accuracy. Because many coupling enzymes may contain small amounts of other enzymes as impurities, the use of large excesses of coupling enzymes means that all coupled enzyme assay systems are likely to contain unknown amounts of contaminating enzymes. In such situations interfering reactions can easily lead to gross errors. These and other problems arising in the use of coupled enzyme assays are discussed in refs. 10–12.

In sampling methods the reaction is stopped after a suitable time interval, and then either the amount of product formed or the amount of substrate remaining is measured. It is to the latter group of sampling methods that radiometric techniques of enzyme assay belong. In the past the chief objection to sampling methods of enzyme assay has involved the time needed for each assay. The improved techniques now used in radiometric enzyme assays have overcome these objections; rapid and simple sampling and separation procedures, which can easily be performed by technical assistants, are now available for most enzyme assays. As a result radiometric methods offer many advantages over conventional methods for nearly all enzyme assays other than those carried out during the study of the kinetics of reaction of highly purified enzymes, where continuous assays may be essential.

The advantages of radiometric methods are discussed, and these techniques are compared with other methods of enzyme assay, in more detail in Section VI.

3. The Classification of Enzymes

The International Union of Biochemistry Commission on Enzyme Nomenclature and Classification recommended (4) that enzymes should be classified, according to the chemical reaction which they catalyze, into six major classes:

Class 1. Oxidoreductases
Class 2. Transferases
Class 3. Hydrolases
Class 4. Lyases
Class 5. Isomerases
Class 6. Ligases

Lyases are enzymes which remove groups nonhydrolytically from their substrates, leaving double bonds, or which, conversely, add groups to double bonds. Ligases (also known as synthetases) are enzymes which catalyze the joining together of two molecules, coupled with the breakdown of a pyrophosphate bond in ATP or a similar nucleoside triphosphate.

This chapter will consider examples from each enzyme class, describe the types of radiometric assay which have been used in their study, and comment on the advantages and disadvantages of the radiometric methods in comparison with the conventional methods. The various techniques available will be reviewed critically with particular reference to possible sources of error both in the assay and in the interpretation of results.

III. TECHNIQUES

Apart from a few examples in which enzyme reactions have been followed by forming radioactive derivatives of their products (see Section III.8.A), radiometric enzyme assays are based on the conversion of a radioactive substrate to a radioactive product. The rate of the enzyme-catalyzed reaction is determined by measuring the radioactivity of either the product or the residual substrate after separation from the reaction mixture. The amount of product formed during the reaction period is then calculated using the known specific activity of the substrate. Thus the two major requirements of a radiometric enzyme assay are that the specific activity of the substrate be known, and that a good method of quantitatively separating substrate and product exists. In radiometric enzyme assays substrates are normally used at low specific activities which can easily be determined accurately (see Section V.4); hence the success of any radiometric enzyme assay depends on the quantitative separation of labeled substrate and products. Various separation techniques have been used, and eleven of these are described in Sections III.1 to III.11.

Unless otherwise stated, all the radioactive procedures discussed in Section III employ liquid scintillation counting for the measurement of radioactivity.

1. Solvent Extraction Methods

A simple and rapid method of enzyme assay is possible when substrate and product can be separated by solvent extraction. In some reactions, when substrate and product are not directly separable in this way, separation can be achieved after forming a derivative of one component. For example, inorganic phosphate can be converted to the silico-molybdate complex, which can be extracted in isobutanol (13), acetylcholine forms a ketone-soluble complex with tetraphenylboron (14), and keto acids can be converted into 2,4-dinitrophenylhydrazones, which are soluble in ethyl acetate (15,16).

Sometimes, when speed is more important than precision, solvent extraction methods can be used to follow enzyme reactions by separating compounds which are not usually regarded as separable in this way. A rapid assay for adenosine kinase, in which the substrate, adenosine, is separated from the product, ATP, by extraction with n-butanol, has recently been described (17).

For maximum sensitivity the conflicting demands of quantitative recovery and complete separation of substrate and product must be met (see Section V.1). Unfortunately this is rarely achieved.

In many assays a complete separation of substrate and product is not possible, but usually less than 1% of the substrate is extracted with the product. The resulting reduction in sensitivity is normally more than compensated for by the speed and simplicity of the method. If a higher sensitivity is required, this can often be achieved by using a different extraction procedure or a different substrate which results in a more efficient isolation of product. This technique was successfully adopted by McCaman (18,19), who achieved a 1000-fold increase in sensitivity in an assay of catechol O-methyltransferase by using dihydroxybenzoic acid as substrate instead of the usual adrenaline, and by substituting ethyl acetate for toluene–isoamyl alcohol as the extractant.

Many extraction procedures give less than 100% recovery. Reproducibility is the most important requirement, however; and provided that recovery is about 70% and is reproducible, the method can be used without any serious loss of sensitivity.

Radiometric methods are often used to study the effects of drugs, antimetabolites, inhibitors, and activators on enzyme activity, and it is obviously desirable to use an assay procedure which is unaffected by the presence of such additives. When solvent extraction procedures are employed in such studies, it is advisable to check that the presence of the additives does not affect either the separation or the recovery of products.

Many organic solvents can cause serious quenching in scintillation counting. If this problem occurs, it can easily be overcome by removing the solvent by evaporation before counting, by using microtechniques so that the volume of quenching solvent added is minimized, or by selecting a different, nonquenching solvent.

The usual procedure is to carry out the extraction in a centrifuge tube, mixing with a vortex mixer, and separating the layers by centrifugation. This provides the activity in a clear, sharply defined organic phase in a form very suitable for counting. It is an advantage to use a solvent lighter than water to avoid the risk of contamination with the aqueous layer when removing samples for counting.

The process has been speeded up by extracting the products in a solvent, such as toluene, containing scintillant, so that the extract can be counted directly (20,21). Because this permits larger samples to be taken for counting, it can also make the assay more sensitive and accurate. Some workers have further accelerated these assays by immersing the centrifuge tubes in a freezing bath after separating the phases. This freezes the lower, aqueous phase and allows the upper, organic layer to be poured directly into a vial for scintillation counting, thus obviating the need to use (and clean) a large number of pipettes (20,21).

A recent innovation which should prove useful in solvent extraction procedures is the availability of silicone-treated papers (Whatman IPS papers, H. Reeve Angel & Co., London), described by the manufacturers as "disposable separating funnels." These are permeable to organic solvents but not to aqueous solutions.

2. Methods Involving the Precipitation of Macromolecules

Many enzymes involved in the synthesis of macromolecules, such as the nucleic acid polymerases (22–31) and the polypeptide (32–37), polysaccharide (38,39), and amino-acyl tRNA (40–42) synthesizing enzymes, have been assayed by using labeled substrates and measuring the activity incorporated into a form insoluble in acid or organic solvents.

Often samples have been processed individually, each sample being washed several times before being transferred, as a suspension, to a planchet or filter before activity measurement. Some workers have solubilized the precipitates with enzymes, hyamine, or formic acid before counting. These procedures are tedious, however, when a large number of samples require analysis; they are also open to the risk of loss of active product during the transfer of insoluble material.

A greatly improved method, which allows large numbers of samples to be processed simultaneously, has been described in detail by Bollum (25,43). Samples are applied to disks of filter paper, and reaction is stopped by dropping the disks into a bath of the appropriate cold solvent. The macromolecular product is precipitated evenly throughout the filter paper, and impurities can be removed easily and efficiently by washing by decantation. After the disks have been dried in air, they are counted by liquid scintillation methods. Tests have shown that the precipitates are so firmly embedded in the fibers that samples can be left in the stopping bath for up to 4 days without transferring any activity into the solution. When tritium-labeled substrates are used, the efficiency of scintillation counting of the precipitated products can be increased by using the more transparent glass-fiber filter disks (44,45).

When protein or polypeptide synthesis is assayed, the precipitates must be heated in 5% trichloroacetic acid for 10 min at 90°C to solubilize any labeled amino-acyl tRNA which may have been formed (from the donor amino acids) by contaminating enzymes, or may be present (as unreacted substrate) when labeled amino-acyl tRNA is used as the amino acid donor (43).

Recent publications (46,47) have drawn attention to the fact that radiation decomposition products can be spuriously precipitated in many macromolecular synthesis assays. Although only small amounts of these products are likely to be present in most labeled substrates, their effect will be magnified if only a small proportion of the labeled substrate is converted to product in the assay. As a result, an interfering impurity which might represent less than 0.1% of the total activity can give rise to a very high blank reading and markedly reduce the sensitivity of the assay. A simple method of removing such troublesome trace impurities has been described (46,47).

Precipitation methods may also be used for the rapid assay of macromolecule-degrading enzymes (43). For the highest sensitivity and accuracy, the best method is to precipitate the unreacted substrate and remove it by centrifugation before measuring the activity of the product in the supernatant. This can be tedious, however, when a large number of samples need analysis, and it is easier and quicker to use the filter disk technique and to measure the activity of the residual substrate.

Precipitation methods are not completely satisfactory for the assay of endonucleases whose products include large oligonucleotides, which are also precipitated with the acid-insoluble substrates (43,48). For an accurate assay of endonuclease activity, the acid-stopped samples should be filtered through cellulose nitrate membrane filters whose pore diameter is such that the macromolecules, but not the oligonucleotides, are retained (48). On the other hand, the filter disk method, although less accurate, may be used for a rapid, rough survey of endonuclease activity (e.g., during enzyme purification). Alternatively, it should be possible to employ gel filtration methods to separate the oligonucleotides from the substrate. Although such methods have not yet been used for endonuclease assays, they are capable of giving a sharper separation than is possible with membrane filters and might be applied if a more precise assay is required.

Filter paper disk techniques are also used extensively for the determination of tRNA by charging the acceptor tRNA with the appropriate labeled amino acid (43,49,50). During column chromatographic separation of tRNAs a considerable analytical load is generated, and there has been a great demand for a very rapid method of assaying large numbers of samples. Two modifications of the filter disk technique have been applied successfully for the assay of large numbers of samples containing tRNA and could prove equally useful in the assay of many macromolecule-synthesising enzymes.

In the first method, Cherayil et al. (51,52) dried aliquots of column eluates on paper disks which were then washed free of urea and salts before adding a solution containing labeled amino acid, activating enzyme, and other cofactors, and allowing reaction to proceed in the moist filter paper disk for 20 to 30 min. The disks were then dropped into trichloracetic acid to stop the reaction, and processed in the usual way. Other workers (53), using a slight modification of this technique, claimed that it gave reproducible results which were 70% of the value obtained when the reaction mixture was incubated in free solution.

In the second modification, Goldstein et al. (54) concentrated on the automation of the filter disk procedure and developed a technique which is automated to the stage at which an aliquot of the reaction mixture is placed on the filter paper disk. This automated procedure allowed the analysis of 40 samples per hour with a precision comparable to that of the manual method.

3. Ion-Exchange Methods

Ion-exchange techniques are used extensively in preparative organic chemistry, and it is not surprising that modifications of the preparative techniques have proved extremely useful for the separation of substrates and products in radiometric enzyme assays. Both ion-exchange resin and ion-exchange paper methods are widely used.

A. ION-EXCHANGE RESIN METHODS

In the simplest methods very small columns are used; one component is retained, while the other passes straight through. If necessary, the retained material may be removed by a suitable eluent. In reactions where several products are formed, as, for example, with certain kinases, which produce a mixture of mono-, di-, and triphosphates, the various products may be eluted separately, if necessary, by using the appropriate eluents in turn (55).

In reactions where it is desired to count the activity not retained by the ion-exchange resin it is better, instead of using a column, to add the sample to an appropriate amount of resin in a centrifuge tube, shake to equilibrate, and then count the activity in the supernatant after removing the resin by centrifugation (56).

When tritium-labeled substrates are used, advantage may be taken of the quenching, by the resin, of the soft β-emissions of adsorbed compounds. The ion-exchange separation is carried out inside the scintillation vial, and quenching by the resin reduces the counts due to the adsorbed substrate to a negligible level, so that, on addition of the scintillant, only the activity of the unadsorbed product is counted. This technique has been used by three groups of workers (57–59) to assay cyclic nucleotide phosphodiesterases by means of a method which requires the determination of the activity of tritium-labeled nucleosides in the presence of tritium-labeled cyclic nucleotides. One group (58) reported that this technique gives a blank reading of about 5 to 7% of the total activity added to the vial. It is probable that this blank reading arises because quenching by the resin does not completely eliminate the counts due to the bound substrate. Although the blank reading is high and must reduce the sensitivity and accuracy of the assay, it is acceptable in many cases where speed and simplicity are more important than sensitivity and absolute precision. Obviously this simple and rapid technique has great potential application in the assay of a wide variety of enzymes.

Other recent improvements in ion-exchange separations are of use in radiometric enzyme assays. Ribonucleosides and deoxyribonucleosides are difficult to separate by conventional ion-exchange chromatography, but ribonucleosides, unlike deoxyribonucleosides, can form complexes with borate ions. Steeper and Steuart (60) have taken advantage of this property and used borate anion-exchange chromatography to effect a very simple and speedy separation of cytidine and deoxycytidine in an assay of CDP reductase. Goldstein (61) has described the technique of ligand-exchange chromatography and has used it to effect a rapid separation of nucleotides, nucleosides, and nucleic acid bases on a Chelex-100 (Cu^{2+}) resin. This technique could prove applicable in the assay of nucleoside phosphorylases and kinases.

Ion-exchange resin separations are simple, quick, and efficient, and yield the product in aqueous solution, a convenient form for counting by liquid scintillation. Bray's scintillant (62) is probably the most widely used liquid scintillant for counting relatively large volumes of aqueous solutions. It is quite expensive, however, and a much cheaper scintillant, containing the surface-active agent Triton X-100 (Rohm and Hass Co., Philadelphia, Penn.), introduced by Patterson and Greene (63) for the emulsion counting of aqueous solutions, has been shown to be satisfactory for counting up to 3 ml of aqueous solution, containing ^{14}C or tritium, in 10 ml of scintillant (64,65). The use of this scintillant will increase the sensitivity of many assays by allowing a larger proportion of the eluted active solution to be counted.

B. ION-EXCHANGE PAPER METHODS

The availability of ion-exchange papers in the last decade has led to a marked increase in the use of ion-exchange separation techniques in radiometric assays of a wide range of enzymes. These include several kinases (66–75), nucleotidyltransferases (76,77), aminotransferases (78), phosphoribosyltransferase (79), amidohydrolases (80,81), adenyl cyclase (82–85), and methionine decarboxylase and adenosyltransferase (86).

Two types of ion-exchange paper are available: the substituted celluloses, and papers loaded with ion-exchange resins. DEAE-cellulose was the first ion-exchange paper to be used in radiometric enzyme assays, and, not surprisingly, the majority of methods so far described employ this paper (66–71,73–77,79,80,82,83). ECTEOLA-cellulose (81,84), PEI-cellulose (85), CM-cellulose (86), and papers loaded with Amberlite cation-exchange resins (72,78) have also been successfully used. Their application will undoubtedly increase as their usefulness becomes more widely known.

When ion-exchange paper techniques are used, the active material is firmly bound to the support medium. This leads inevitably to self-absorption losses during the counting of soft β-emitters such as tritium and ^{14}C. The problem is, of course, much more serious for tritium than for ^{14}C. Nevertheless tritium-labeled substrates have been successfully used (67,67b,74). When tritium-labeled compounds bound to ion-exchange papers are counted, the efficiency of counting is very low, and it is advisable to elute the active material and to count it in homogeneous solution (67b). Whenever possible, ^{14}C-labeled substrates should be used, but even then about 25% of the counts can be lost as a result of self-absorption by the paper (72,79,87). If the scintillation solvent elutes the active material from the paper, the counts will steadily rise as more and more of the sample is eluted and counted in homogeneous solution at higher efficiency than prevails for the material remaining bound to the paper (72,87,88). This makes it extremely difficult to obtain reproducible results. It is essential, therefore, to use a scintillant which will elute either all or none of the active material. Because elution techniques are tedious, and it is often difficult to obtain quantitative elution, it is customary to count samples in scintillants which do not elute the active material and to accept the small, but reproducible, loss of counts by self-absorption. An added advantage of using such scintillants is that, because the active material remains firmly bound to the paper and does not contaminate the scintillant, the paper may be removed after counting and the scintillant and vial reused (77,87). This can often be done 20 or 30 times before the background count of the scintillant rises significantly, and can result in a great reduction in the cost of assays.

When active materials are counted on paper, unless both product and residual substrate are counted, it is necessary to correct for self-absorption losses by counting an aliquot of the substrate (72) dried on paper.

Some workers (70,78) using ion-exchange paper assay methods have commented on high blanks caused by impurities in the labeled substrates, which were retained by the paper and counted with the product. A simple and rapid way of reducing such blank readings is to remove the troublesome impurities by filtering the substrate, before use, through a disk of the ion-exchange paper used in the assay. This will retain the impurities and should reduce the blank reading to an insignificant figure, thus resulting in an appreciable increase in sensitivity (46).

Ion-exchange paper techniques involving both conventional paper chromatographic and filter disk methods have been extensively used.

a. Ion-Exchange Paper Chromatographic Methods. Conventional ion-exchange paper chromatographic methods give very rapid (10 to 120 min) and very effective separation of a wide range of compounds. They are particularly useful, for example, for the separation of nucleotides (89) and have become widely applied in the difficult assay of adenyl cyclase (82–85). Having carried out a separation, it is necessary to locate substrate and product. Autoradiography is usually too slow at the activity levels used in radiometric enzyme assays, and other methods are employed to locate substrate and product. Many compounds are easily visualized under UV light, and this technique has been used in many assays of enzymes which catalyze nucleoside and nucleotide conversions. Such methods are otherwise applied mostly for the assay of enzymes whose products are firmly located at, or very near, the origin, and whose substrates move with the solvent front. For these separations it is not necessary to visualize substrate and product, which can be measured by counting appropriately sized areas cut out from the origin and adjacent to the solvent front.

b. Ion-Exchange Paper Disk Methods. A greatly improved method of ion-exchange separation was first used in 1963 by Sherman (66) and Furlong (66b) and has since been developed and tested in detail by Newsholme and his coworkers (68). In this ion-exchange paper disk method, samples of kinase reaction mixtures were filtered through DEAE-cellulose disks, which retained phosphorylated products. Residual substrate was removed by washing with water before measuring the activity retained by the disk.

This method has also been used for the assay of nucleoside kinases. Because nucleosides are partially retained by DEAE-cellulose, however, they must be removed by washing with dilute ammonium formate, which elutes nucleosides but not nucleotides (67,67b,69,74,75).

Nucleotide kinases (66b) and nucleotidyltransferase (76) have also been assayed by a modification of this technique. Samples were removed, enzyme action was stopped, and a phosphatase was added to dephosphorylate the monophosphate substrate but not the phosphorylated products. The phosphatase-treated samples were then filtered through DEAE-cellulose paper disks, which retained the phosphorylated products of the reactions for counting.

Although the disk technique has so far been used almost exclusively for DEAE-cellulose, it has great potential and could be employed for many enzyme assays where an ion-exchange paper satisfactory for the separation of product and substrate exists. For example, CM-cellulose has recently been used in assays of methionine adenosyltransferase and methionine decarboxylase (86).

A recent publication (90) has drawn attention to the fact that, as Newsholme and his coworkers pointed out (68), other anions can compete with labeled products for the exchange sites on the disks and that, as a result, the product may not be retained completely at high salt concentrations. Roberts and Tovey (90) found that, in radiometric assays of glucose-1-phosphate adenylyltransferase, the DEAE-cellulose disk assay consistently gave lower results than three other assays. Since most kinases and nucleotidyltransferases are assayed in the presence of various concentrations of magnesium salts, these authors investigated the effect of magnesium chloride concentration on the retention of uridine diphosphoglucose-^{14}C by DEAE-cellulose disks. They found that retention was satisfactory only at concentrations below 1.25mM. Higher concentrations of magnesium chloride caused retention to fall to about 40% at 0.25M. Magnesium chloride concentrations higher than 0.25M had no further effect on the retention of activity.

An alternative explanation of these results is, however, possible (91). Low retention of activity might be due, not to the competition of chloride ions for the exchange sites, but rather to the formation, at high magnesium concenrations, of a magnesium–UDP-glucose complex, which is retained by the DEAE-cellulose disk less effectively than is the free UDP-glucose. If all the

UDP-glucose is present as a magnesium complex at magnesium concentrations greater than $0.25M$, this would be consistent with the finding that increasing the magnesium concentration above this figure did not result in a further decrease in the retention of activity. Dilution of a solution of such a complex would cause the complex to dissociate and would result in an increase in the proportion of UDP-glucose present in the free, uncomplexed form. This, in turn, would result in a greater retention of activity and is, therefore, again consistent with the finding by Roberts and Tovey that dilution of samples, before application to the disks, caused an increased retention of activity.

In all probability, magnesium complex formation and anion competition can each play a part in reducing retention, with complex formation perhaps being more important in the presence of moderately high concentrations of magnesium salts. The cautionary article by Roberts and Tovey is, therefore, most timely. Workers using ion-exchange paper disk techniques should be aware of the potential hazards and should investigate the effects of salts on the retention of products by the disks (67b). The magnesium concentrations at which serious loss of retention occurs are, however, much higher than those normally used in enzyme assays, and the most important factor in most assays is likely to be the effect, on the retention of activity, of other salts, such as ATP, buffer, and ammonium sulfate (present in many reaction mixtures because the phosphatases used in the assays are supplied in concentrated ammonium sulfate solution).

Whatever the causes of reduced retention of product are, the problem might be reduced by using, instead of DEAE-cellulose, disks made from paper loaded with an anion-exchange resin which would have a much higher exchange capacity than DEAE-cellulose. The results of Roberts and Tovey show that better retention is obtained in DEAE-cellulose disk assays if the sample is diluted before application to the disk. It is better, for example, to dilute a sample fivefold and apply a 50-μl aliquot than to apply a 10-μl aliquot of the undiluted sample. If very high magnesium concentrations are necessary in assays, it may be possible to reduce any formation of a magnesium-nucleotide-sugar complex by adding an appropriate amount of EDTA before processing the sample, provided that the amount added is sufficient to complex with the free magnesium ions but insufficient to compete significantly for the exchange sites on the DEAE-cellulose disks.

4. Methods Depending on the Release or Uptake of Activity in a Volatile Form

A. RELEASE OF TRITIUM AS TRITIATED WATER

The release of tritium from specifically labeled substrates has been used as the basis of the rapid and simple assay of a large number of enzymes. These include the aromatic amino acid hydroxylases, arylhydroxylases, folate reductase, monoamine oxidase, thymidylate and hydroxymethyl deoxy-cytidylate synthetases, aminotransferases, and the cobamide-dependent ribonucleotide reductases.

Tritiated water can easily be recovered for counting, either by simple distillation if the substrate is not volatile, or by using ion-exchange columns, alumina, or charcoal to remove labeled compounds.

Three pitfalls exist when using this method, and they can result in apparently lower rates of reaction being observed with tritium-labeled substrates than with other labeled, or inactive, substrates. As a result the unwary may erroneously infer isotope effects. These pitfalls, which are discussed in greater detail in Section V.8 and in ref. 92, are as follows.

1. Low rates of tritium release will result if the position of labeling is not exactly as specified.

2. Errors may also be caused by the labilization of the tritium label.

3. Intramolecular migration of tritium—the so-called NIH shift—which causes tritium to be retained in the product instead of being released as tritiated water, has been shown to occur in many hydroxylations. Although this can cause apparently low rates to be observed, it is easy to detect and correct for. Low rates have been reported in tritium-release assays for other enzymes, and these may also be due to similar intramolecular shifts.

B. RELEASE OR UPTAKE OF CARBON-14 DIOXIDE

Decarboxylating enzymes can be easily and accurately assayed by measuring $^{14}CO_2$ evolution from appropriately labeled substrates. The carbamoyl-transferases, such as ornithine carbamoyltransferase (93), which form carbamoyl phosphate when the reaction is carried out in phosphate buffer, can easily be assayed by carrying out the reaction, with ^{14}C-carbamoyl-labeled substrate, in arsenate buffer. The product, carbamoyl arsenate, is unstable and decomposes spontaneously, liberating $^{14}CO_2$. Other reactions which do not themselves produce $^{14}CO_2$ can be assayed by coupling them enzymatically to reactions which release $^{14}CO_2$. Hexokinase (94,95) has been assayed by using glucose-1-^{14}C as the substrate, and coupling the reaction with glucose-6-phosphate dehydrogenase and 6-phosphogluconate dehydro-genase (decarboxylating). Arginosuccinate synthetase (96) and arginase (97) have been assayed by using carbamoyl-^{14}C-labeled substrates and de-carboxylating the product, urea-^{14}C, with urease. Xanthine oxidase (98) has been assayed by using xanthine-6-^{14}C as substrate and decarboxylating the product with uricase.

Released $^{14}CO_2$ may be counted after being trapped in solution, or on filter paper, using alkali (99–102), or hyamine or a similar base (102–108).

Unsatisfactory results have sometimes followed from insufficient acidification of the reactants to ensure complete release of $^{14}CO_2$, and from insufficient time being allowed for its complete diffusion from solution. Some workers (108), who trapped $^{14}CO_2$ on filter paper soaked in hyamine reported that hyamine bicarbonate was slowly eluted by the scintillant and that it was necessary to allow the papers to soak for 8 hr in the scintillant in the counting vials to ensure complete elution and reproducible results. Choice of a scintillant in which hyamine bicarbonate is insoluble would eliminate this problem.

There has been much interest in developments which will permit radiometric methods of enzyme assay to be used for the continuous monitoring of reactions. In one very rapid method, which permitted almost continuous sampling (107), $^{14}CO_2$ was trapped in 2-phenylethylamine adsorbed onto diatomaceous earth. This mixture was packed into small tubes, each containing 0.5 g, and these tubes were connected to the reaction vessel in such a way that the outflowing gas stream could be diverted to pass through each tube in turn. The adsorbent was then transferred to a vial containing scintillant solution, and the activity measured after it had been extracted from the adsorbent into this solution. The counting efficiency was 65%, and the recovery of activity was 97 to 99%. Sample times as low as 10 sec were achieved. More recently, several authors (109–111) have described methods of continuously recording $^{14}CO_2$ evolution, using commercially available ionization-chamber air monitors. Glutamic decarboxylase has already been assayed in this way (109).

Many enzyme assays, radiometric and otherwise, suffer from the presence of concomitant nonenzymatic reactions, which can reduce the sensitivity of the assays by causing high blank readings (see Section V.1 and ref. 46). Decarboxylase assays are well known to be subject to this interference, and the evidence available suggests that the use of phosphate buffers is a major cause of high blanks in decarboxylase assays (46). In one assay of histidine decarboxylase the rate of nonenzymatic decarboxylation was eight times higher in phosphate buffer than in Tris [2-amino-2-(hydroxymethyl)-propane-1,3-diol] buffer (46,112). The use of phosphate buffers should be avoided, therefore, in decarboxylase assays.

Many enzymes have been assayed by using sodium bicarbonate-^{14}C and measuring the fixed activity after removal of the unused substrate. Malate dehydrogenase (113), pyruvate decarboxylase (114), pyruvate carboxylase (115), and carbamylphosphate synthetase (116–118) have been assayed in this way. The residual substrate is often removed by acidification and drying, but when the product is unstable it must be stabilized, either chemically or enzymatically, before the acid-volatile substrate is removed. Unstable oxo acids may be stabilized by forming the 2,4-dinitrophenylhydrazones (115) or, for example, by converting oxaloacetate to malate with malic dehydrogenase (119,120). Alternatively, oxaloacetate may be stabilized by conversion to aspartate with aspartate aminotransferase (121). Carbamyl phosphate has been stabilized by conversion to citrulline with ornithine carbamoyltransferase (116–118).

A recent report (122) has indicated that pyruvate carboxylase, which was assayed by measuring the incorporation of $^{14}CO_2$ into tricarboxylic acid cycle intermediates, yielded products which had lower specific activities than the original bicarbonate. The author suggested that carbon dioxide derived from the decarboxylation of pyruvate was preferentially incorporated into the products. This suggests that in many tissues pyruvate carboxylase activity may be significantly higher than previously reported in the literature. Many oxo acids are unstable in solution and may decompose, chemically or enzymatically, during enzyme assays and so liberate unlabeled carbon dioxide, which would lower the specific activity of the bicarbonate-^{14}C used in the assay. Such an effect may easily be detected by varying the ratio of unlabeled oxo acid to bicarbonate-^{14}C.

Some workers have been troubled by relatively high blank readings (46). These are probably due to the use of unnecessarily high concentrations of bicarbonate-^{14}C. This results in the incorporation of only a minute fraction of the total activity into the product during the assay, thus magnifying the effect of any trace of nonvolatile activity. These nonvolatile trace impurities are probably the products of radiation decomposition on storage because the method of preparation of sodium bicarbonate-^{14}C makes it unlikely that nonvolatile active impurities will be present in the freshly prepared material. These assays require only low-specific-activity bicarbonate-^{14}C substrate, and storage of bulk stocks of substrate at low specific activity should, therefore, reduce the rate of formation of troublesome nonvolatile trace impurities (46).

Many workers determine the activity of solutions of bicarbonate-^{14}C by liquid scintillation counting. Experience in our laboratories has shown, however, that appreciable (and variable) losses of activity can occur unless a trace of alkali is added to the scintillation fluid in the counting vial before the sodium bicarbonate solution is added. Our results show that the addition of 200 μl of ethanolamine-methyl cellosolve (1:2), for example, consistently enables reproducible results to be obtained. The activities of solutions measured in the absence of alkali varied between 70 and 90% of the true value.

C. RELEASE OR UPTAKE OF OTHER FORMS OF VOLATILE ACTIVITY

Acetylcholine hydrolase (123) and an acetylthymidine acylhydrolase (124) have been assayed by a method which is generally applicable to enzymes that have a volatile substrate or produce a volatile product. Acetyl-1-^{14}C, labeled esters were used as substrates, and samples were removed, acidified- and dried to volatilize the free acetic acid. The activities of the residual substrates were then measured.

An enzyme having a substrate or producing a product which, although not volatile, can easily be decomposed to form a volatile compound can also be assayed in this way. For example, acetyltransferases, which transfer acetyl groups from acetyl-CoA to form an acid-stable product, may be assayed by using acetyl-labeled acetyl-CoA and measuring the fixed activity after hydrolyzing the residual substrate and volatilizing the free acetic acid (125). Aspartate carbamoyltransferase has also been assayed in a similar way, using the acid-labile substrate carbamoyl phosphate (126).

5. Paper and Thin-Layer Chromatographic Methods

In the past, enzymologists regarded these methods as a last resort to be used only when no other technique was available. Their objections were based on the time taken to separate and to estimate residual substrate and products by chemical or spectrophotometric methods (8). These objections have been overcome, and many enzymes are now assayed by paper and thin-layer chromatographic methods.

The development and use of faster-running chromatographic solvents and papers have considerably reduced the time required for separation, and adequate separation can usually be achieved in a few hours or less. The measurement of radioactive substrates and products is both easy and accurate. Although spots can be located by autoradiography, in general this is not done as it requires too much activity per assay and is time consuming. The usual, as well as the easiest, method is to use inactive markers which can be visualized by optical or chemical methods. Alternatively, it is unnecessary to locate spots if the activity is measured by strip scanning.

One advantage of these chromatographic methods is that, by occasionally using autoradiography to locate the products, they allow a check to be kept on the stoichiometry of the reaction. This ensures that any degradation of substrate or product, during the assay, by other enzymes present in the extract can easily be detected.

In some methods the separation of products and substrates has been vastly simplified by selecting a solvent system in which the active material to be measured (product or residual substrate) is immobile. A spot of the reaction mixture can be applied to a filter paper disk and dried. The disk is then swirled in a flask, containing the appropriate solvent, for sufficient time to remove the faster-running compounds. The papers are then removed, dried, and counted (127). In this way a large number of samples may be processed simultaneously, rapidly, and easily. This technique is very similar to Bollum's filter paper disk method, used in the analysis of precipitable macromolecules (see Section III.2).

When tritium- or ^{14}C-labeled materials are counted on paper, allowance must be made for self-absorption losses (see Section III.3.B). If liquid scintillation is used, samples should be counted in a scintillant which either does not elute the active material or else elutes it completely.

During the paper and thin-layer chromatographic separations of tritium-labeled compounds in strongly acidic or alkaline solvents, care should be taken to detect, and allow for, any loss of tritium which may occur by exchange with the solvent (see Section V.8). Although this problem is not likely to occur often, a significant (10%) loss of tritium from tyrosine was reported (128) when phenylalanine and tyrosine were separated by paper chromatography in the strongly alkaline solvent system 2-propanol-ammonia-water (8:1:1).

Many nucleotide sugars are readily degraded in alkaline media to yield cyclic sugar phosphates. A recent publication (129) has drawn attention to the fact that the solvent system ethanol-$1M$ ammonium acetate, pH 7.5, commonly used for the paper chromatographic separation of sugar phosphates and nucleotide sugars, releases sufficient ammonia to cause extensive degradation of many nucleotide sugars during the equilibration of papers in chromatography tanks before elution. If undetected, this degradation could result in the underestimation of nucleotide sugar synthesis, or in a false inference of enzymatic synthesis of cyclic sugar phosphate. It is easy, however, to avoid this pitfall by equilibrating papers in an atmosphere of ethanol-water, instead of in an atmosphere of the solvent system ethanol-ammonium acetate.

Because widely different compounds can have identical mobilities, it is essential to ensure that the product being measured is the true product of the enzyme reaction under study, and not a degradation product of the substrate. Bär and Hechter (130) have drawn attention to the fact that hypoxanthine, which can be formed by the enzymatic degradation of ATP, has the same mobility in some paper chromatographic solvents as cyclic AMP, the product of the enzyme adenyl cyclase. As a result, some workers who had thought that they were following cyclic AMP production by adenyl cyclase were in fact following the production of hypoxanthine. Even if cyclic AMP had been produced in the reaction, the formation of hypoxanthine would have resulted in a gross overestimation of the adenyl cyclase activity. This type of problem can easily be overcome by choosing a different solvent system or, for the adenyl cyclase assay, by using ATP labeled with ^{32}P instead of ^{14}C, so that any hypoxanthine formed will be unlabeled and so will not interfere in the assay.

6. Electrophoretic Methods

The advent of high-voltage electrophoresis has considerably reduced the time required for the electrophoretic separation of substrates and products, and electrophoretic methods of separation are now used in many radiometric enzyme assays. Because uncharged molecules remain on the origin, electrophoretic methods are particularly useful for the assay of enzymes which form neutral products from charged substrates, and many glycosyl-transferases have been assayed in this way (131–134).

Sometimes neutral sugars are produced by the degradation of nucleotide sugar substrates during assays. If this happens the uncharged degradation products will remain on the origin of the electrophoretogram and will be counted with the true product of the reaction. This can cause high blanks which reduce the sensitiviry of the method. If the glycosyl acceptor is a macramolecule, low-molecular-weight sugars, formed by degradation of the substrate, can easily be removed by washing the piece of paper cut from the origin of the electrophoretogram in 50% ethanol before counting (131). Alternatively, the paper strip can be developed in a chromatographic solvent after electrophoresis, to separate the true product from degradation products of the substrate (134).

As with other assays, when the activity of materials adsorbed on paper is measured, allowance must be made for self-absorption losses. If liquid scintillation is used, samples should be counted in a scintillant which does not elute the active material (see Section III.3.B).

7. Methods Involving Adsorption on Charcoal or Alumina

The adsorption of either product or substrate by an insoluble adsorbent is the basis of a large number of very simple and rapid radiometric enzyme assays. Charcoal and alumina have been used extensively in such assays.

A. METHODS INVOLVING ADSORPTION ON CHARCOAL

The ability of charcoal selectively to adsorb nucleotides and nucleosides has been used in the assay of nucleotidases (135,136), cyclic phosphodiesterase (136), nucleotide kinases (137,138), nucleotidyltransferases (26,139), and nucleoside phosphorylase (140). Amino-acid-activating enzymes, such as the amino-acyl tRNA synthetases, have also been assayed by measuring the amino-acid-dependent release of pyrophosphate from ATP-γ-^{32}P or the exchange of pyrophosphate-^{32}P with ATP (40,141,142).

As a rule, ^{32}P-labeled substrates are preferred because the hard β-emission makes it possible to measure the activity of material adsorbed on the charcoal without self-absorption losses. The activity remaining in solution as free ortho- or pyrophosphate may also be measured. Errors caused by partial adsorption of inorganic phosphates by the charcoal can be reduced by adding carrier phosphates before adsorption. Care must be taken not to add so much phosphate that complete adsorption of nucleoside phosphate is prevented.

When ^{14}C- or tritium-labeled substrates are used, self-absorption makes direct counting of adsorbed compounds impossible for the latter substrates, and gives very low efficiency with the former. It is customary with these substrates either to count the activity not adsorbed by the charcoal, or to elute the activity from the charcoal with alcoholic ammonia. It is rarely possible to obtain complete elution, and the amount eluted is a function of the total weight of nucleotide present (143). This problem can be overcome by adding carrier nucleotide, determining the percentage eluted in separate experiments, and then applying a correction factor to determine the total activity adsorbed. Alternatively, carrier nucleotide may be added, and the specific activity of the eluted nucleotide determined by UV spectroscopy and activity measurements. A simple calculation will then give the total activity adsorbed.

Nucleotide kinases have been assayed by hydrolyzing the unreacted nucleoside monophosphate substrate enzymatically before adsorbing the di- and triphosphate products on charcoal (137,138). In general, however, the charcoal adsorption methods of assaying these enzymes have been superseded by the DEAE-cellulose disk methods.

Uridine phosphorylase (140) has been assayed by using as substrate ^{14}C-labeled uridine, adsorbing uracil and uridine on charcoal, and measuring the activity of the ribose remaining in solution. Because nucleosides and bases are not as strongly adsorbed as nucleotides, it is necessary to add relatively large amounts of charcoal to adsorb the uracil and uridine completely. Failure to do so will result in high blank readings. However, large excesses of charcoal can result in adsorption of the product, ribose. It is necessary, therefore, to select the amount of charcoal which will give maximum adsorption of uracil and uridine and minimum adsorption of ribose. This amount will probably vary from batch to batch of charcoal and will require routine checking.

B. METHODS INVOLVING ADSORPTION ON ALUMINA

Alumina has been used to separate substrate and product in assays of tyrosine hydroxylase and asparagine synthetase. In both assays separation is carried out on small columns, and separation is not quantitative. An eluent has to be chosen to give maximum elution of product, with minimum contamination by the substrate, which would result in high blank readings (46).

Carrier unlabeled product is added, before column separation, to improve recovery of the labeled product. In separations where the residual substrate is eluted before the product, it may be advantageous to add carrier substrate as well, to ensure good removal of substrate before the product is eluted.

Tyrosine hydroxylase has been assayed using ^{14}C-labeled tyrosine as substrate (144–146). Residual tyrosine was removed by washing the column with water before eluting the product, dihydroxyphenylalanine, with dilute acetic acid. A correction factor was used to correct the results for the low recovery (about 65%) of product (144). This method has now been superseded by the simpler tritium-release assay (see Section III.4.A).

Asparagine synthetase has been assayed using ^{14}C-labeled aspartate as substrate (147,148). Aspartate is retained by alumina, and the product, asparagine, may be eluted with water (147) or dilute acetic acid (148). Water is reported to give a low recovery of asparagine (148) free from aspartate (147), whereas acetic acid yields a higher recovery of asparagine contaminated ($<2\%$) with aspartate (148). This method is probably less useful than ion-exchange or electrophoretic methods of assay.

8. Coupled Enzyme Assay Methods

Coupled radiometric enzyme assays have proved useful in three ways: to form labeled derivatives of unlabeled reaction products, to generate substrates *in situ*, and to give improved separations in assays. These three uses will be described separately.

A. RADIOISOTOPIC DERIVATIVE ASSAYS

Some enzymes have been assayed by carrying out the reaction with unlabeled substrates, and determining the amount of reaction by making a radioactive derivative of the product. This technique may be used to advantage in three cases:

1. Suitable labeled substrates are unavailable.
2. The primary products of the enzyme reaction are unstable.
3. The substrate specificity of enzymes is to be determined. Only a few assays of this type have so far been described.

Palmityl-CoA synthetase has been assayed by using tritium-labeled carnitine and an excess of carnitine palmityltransferase to convert the product of the first reaction, palmityl-CoA, into labeled carnityl palmitate, which was isolated by solvent extraction (149).

Bilirubin glucuronyltransferase has been assayed by diazotizing the inactive product with ^{35}S-labeled sulfanilic acid and isolating the labeled product by paper chromatography (150).

Many hydroxylation reactions result in the formation of unstable dihydroxy compounds. By coupling the hydroxylation reactions with catechol-*O*-methyltransferase, and using methyl-labeled *S*-adenosylmethionine as the methyl donor, the unstable dihydroxy compounds can be converted into stable *O*-methyl derivatives which can easily be isolated by solvent extraction (151–154). This technique is also particularly useful for measuring the substrate specificity of single-substrate enzymes. For example, the substrate specificity of a number of hydroxylases can be studied by using one coupling enzyme, catechol-*O*-methyltransferase, and one labeled substrate, *S*-adenosylmethionine.

B. GENERATION OF LABELED SUBSTRATES *in situ*

When labeled substrates are unstable, unavailable commercially, or expensive, they may be prepared *in situ* by using the appropriate synthesizing enzymes and readily available, cheap, labeled precursors. If the pH optima of the synthetase and of the enzyme being assayed are too far apart for the substrate to be prepared *in situ*, it may be synthesized immediately before use; the pH is then adjusted to the assay pH, and the other components necessary for the enzyme assay are added.

Acetyl-CoA, the expensive (both labeled and unlabeled) substrate of many acetyltransferases, can readily be synthesized from labeled sodium acetate and the enzyme acetyl-CoA synthetase (155,156).

Carbamoyl phosphate, the substrate of many carbamoyltransferases, can be synthesized from ^{14}C-labeled sodium bicarbonate and the enzyme carbamoylphosphate synthetase.

The highly unstable oxaloacetate may easily be synthesized *in situ* from labeled aspartate and the enzyme aspartate aminotransferase.

C. METHODS INVOLVING COUPLING TO GIVE IMPROVED SEPARATION

Coupled enzyme assays can result in improved separation in two ways. They can stabilize the product so that losses during separation are reduced, or they can convert a primary product into a secondary product which is easier to separate from the residual substrate. Usually, in order to avoid many of the objections to coupled assays mentioned in Section II.2.B, the primary enzyme reaction is stopped before adding the coupling enzyme to generate the secondary product.

Many examples were discussed in Section III.4.B. For example, enzymatic coupling has been used to stabilize oxo acids and carbamyl phosphate, and hexokinase, xanthine oxidase, arginosuccinate synthetase, and arginase have been assayed by simple $^{14}CO_2$-release techniques, utilizing suitably position-labeled substrates and coupling with decarboxylases. Other assays in which coupled enzyme reactions have been used to improve separation include the following.

Racemases and epimerases form products which are very difficult to separate from residual substrates. Enzymatic modification of either product or residual substrate is an obvious way to facilitate separation. UDPgalactose-4-epimerase has been assayed by oxidizing the product, UDP-glucose, to UDP-glucuronic acid with UDPglucose dehydrogenase. Then UDP-galactose and UDP-glucuronic acid were separated by thin-layer chromatography on PEI-cellulose (157).

The enzyme tryptophan hydroxylase produces hydroxytryptophan, which is difficult to separate from tryptophan other than by tedious paper chromatographic methods. The radiometric assay was greatly improved by coupling with the enzyme 5-hydroxytryptophan decarboxylase. This converted the product to 5-hydroxytryptamine, which was easily isolated by ion exchange (158).

ATP citrate lyase, which converts citrate and CoA into oxaloacetate and acetyl-CoA, has been assayed by coupling the reaction with toluidine and acetyltransferase to convert the relatively unstable product, acetyl-CoA, into p-acetyltoluidine, which can easily be extracted with organic solvents (159).

Tryptophan oxygenase, which converts tryptophan into formylkynurenine, has been assayed by using tryptophan-(ring-2-^{14}C) and coupling the reaction with the enzyme formylkynurenine formylase, which releases labeled formate, easily separated by ion exchange (160).

9. Methods Involving Precipitation of Derivatives

These methods are usually time consuming and are not often used. The difficulty of ensuring complete precipitation and recovery of products means that frequently the accuracy and sensitivity are low unless carrier is added, and specific activity determinations are carried out on the isolated derivative so that correction may be made for any losses of material during processing.

Kinases have been assayed by isolating the phosphorylated products by precipitation as the heavy metal salts from alcoholic solution (161–163). Workers using this method have employed unnecessarily tedious methods of purification involving repeated dissolution and reprecipitation and hence a great risk of loss of activity during the repeated washing of small precipitates. This method for kinase activity has been superseded in speed, simplicity, accuracy, and sensitivity by the DEAE-cellulose paper disk method (see Section III.3.B.ii), but is nevertheless still being used by some workers.

Oxo acids have been precipitated as their 2,4-dinitrophenylhydrazones after adding a known weight of inactive oxo acid as a carrier (164). The precipitated derivative was filtered off, dried, weighed, and dissolved before counting. A simple calculation then gave the total activity of the oxo acid.

Acetylcholine forms insoluble derivatives with several reagents. Choline acetyltransferase has been assayed by precipitating the product, acetylcholine, as the reineckate salt (165) or as the tetraphenyl borate (155). The reineckate precipitation method was vastly improved by using solvent extraction methods which made it sufficiently sensitive for the determination of choline acetyltransferase in individual cells (166).

Another example in which precipitation methods have been used is in the assay of adenyl cyclase (167). This assay normally results in extremely low conversion of substrate, and the precipitation technique is used to remove traces of residual substrate from the product. In this assay the product, cyclic AMP, is separated from the substrate, ATP, by ion exchange but contains about 0.1% of the substrate. This can be removed by precipitation with barium hydroxide and zinc sulfate. It is claimed that this modification makes the method so sensitive that 1 part of cyclic AMP in 50,000 parts of ATP can be detected.

10. Dialysis and Gel Filtration Methods

Enzymes catalyzing the formation of mucopolysaccharides have been assayed by measuring the incorporation of activity into a nondialyzable form (168). This required 48 hr of dialysis to achieve complete removal of low-molecular-weight materials. Other workers have speeded up this separation by using gel filtration on Sephadex G-200 (169). Although this technique is much better than dialysis, it is probable that the filter paper disk precipitation method (see Section III.2) would prove even better.

11. Methods Involving Reverse Isotope Dilution Assay

Such methods are tedious and insensitive and have been used only as a last resort.

The assay of an enzyme system synthesizing inositol from ^{14}C-labeled glucose-6-phosphate has been carried out in this way (170). After the reaction was stopped, a large excess of carrier inositol was added, and a sample of pure inositol was isolated and was purified by crystallization. The specific activity of this inositol was then calculated and used to obtain the total inositol activity.

Histidine decarboxylase has also been assayed in this way, the product, histamine, being isolated as the benzene sulfonyl derivative and purified by crystallization before determination of its specific activity (171).

IV. RADIOMETRIC ENZYME ASSAYS

1. The Assay of Enzymes of Class 1: the Oxidoreductases

This group of enzymes is well suited for conventional spectroscopic assay because most of them utilize the NAD, NADP, FMN, or FAD coenzymes, whose oxidation and reduction are accompanied by large spectral shifts.

Often, however, because conventional methods have proved insufficiently sensitive and too susceptible to interference from contaminating enzymes and endogenous materials, radiometric methods have been used in their place. Also, some reactions have been studied only with crude extracts, and their cofactor and hydrogen donor or acceptor requirements are not yet known. In these situations radiometric assay methods appear to be the only possibility.

Illustrative examples of assays of enzymes in this class are listed below.

A. SUBCLASS 1, OXIDOREDUCTASES ACTING ON THE CH—OH GROUP OF DONORS

The decarboxylating malate dehydrogenases (EC 1.1.1.38–40) have been assayed, in the reverse direction, by measuring the incorporation of $^{14}CO_2$ into malate (113,172,173).

CDP glucose reductase is very difficult to assay spectrophotometrically because the reaction is followed by three further enzyme reactions, involving oxidoreductases, which all contribute to the spectrophotometric change observed. The reaction is easy to follow radiometrically, however, products and substrate being readily separated by paper chromatography (174).

UDP glucose reductase, which converts UDP glucose into UDP rhamnose, occurs in very small amounts in the presence of large amounts of other enzymes which degrade UDP glucose. It cannot be assayed spectrophotometrically, but has been assayed using glucose-labeled UDP glucose (175). The nucleotide sugars were first separated from the products of other UDP glucose-degrading enzymes by electrophoresis, and then hydrolyzed to labeled glucose and rhamnose. These sugars were then separated by paper chromatography.

Hydroxymethylglutaryl-CoA reductase (EC 1.1.1.34) is difficult to assay because the substrate, mevalonate, reacts with many contaminating enzymes, and the product, hydroxymethylgutaric acid, can undergo nonenzymatic degradation during assay. It has been assayed in the reverse direction by a radiometric method using a specifically labeled substrate so chosen that nonenzymatic degradation did not yield a labeled product. The product was extracted in ether and purified by thin-layer chromatography before counting (176).

The clinically significant enzyme glucose-6-phosphate dehydrogenase (EC 1.1.1.49) has been assayed by measuring the $^{14}CO_2$ released when 1-^{14}C-glucose-6-phosphate was used as the substrate, and the product was decarboxylated by an excess of the decarboxylating enzyme, phosphogluconate dehydrogenase (177).

Ribonucleotide reductases, which convert ribonucleotides to the corresponding deoxyribonucleotides, have been assayed by two radiometric methods. In one method, labeled cytidine diphosphate was enzymatically reduced to deoxycytidine diphosphate. Product and residual substrate were then hydrolyzed chemically to their monophosphates, which were separated by borate anion-exchange chromatography (60). In the second assay, cytidine triphosphate was used as the substrate, and the product and residual substrate were again chemically hydrolyzed to their monophosphates before separation by thin-layer chromatography (178).

B. SUBCLASS 2, OXIDOREDUCTASES ACTING ON THE ALDEHYDE OR KETO GROUP OF DONORS

Inosinic acid dehydrogenase (EC 1.2.1.14) has been assayed by using labeled inosinic acid as the substrate and separating the product, xanthosine monophosphate, by paper chromatography (179).

Xanthine oxidase (EC 1.2.3.2) has been assayed by two radiometric methods (98,180). In one assay, 8-^{14}C-xanthine served as the substrate, and the product, uric acid, was isolated by ion-exchange column chromatography (180). In the other assay, 6-^{14}C-xanthine was used to produce uric acid, which, on decarboxylation with uricase, liberated $^{14}CO_2$.

C. SUBCLASS 3, OXIDOREDUCTASES ACTING ON THE CH—CH GROUP OF DONORS

Succinate dehydrogenase (EC 1.3.99.1) has been assayed by a tritium-release technique. Tritiated water was separated from residual product on a small DEAE-cellulose column (181).

The enzymatic β-oxidation of oleic acid has been assayed using ^{14}C-labeled oleic acid as substrate (182). The $^{14}CO_2$ released was trapped on alkaline filter paper and measured by liquid scintillation counting.

D. SUBCLASS 4, OXIDOREDUCTASES ACTING ON THE CH—NH₂ GROUP OF DONORS

Drugs which inhibit the enzyme monoamine oxidase (EC 1.4.3.4) have a widespread therapeutic use, and consequently there has been a demand for a simple and rapid assay method for this enzyme. Many radiometric assays have been described. Most are based on solvent extraction methods (20,21,

146,183–188), but ion-exchange (187,189,190), tritium-release (191), and thin-layer chromatographic (192) procedures have also been used.

The solvent extraction methods have been speeded up by using an extracting solvent containing scintillant, and decanting the organic layer after freezing the aqueous layer (20,21,193, and Section III.1). An elegant microscale solvent extraction assay has been described by McCaman et al. (184).

The initial products of the monoamine oxidase reaction are aldehydes, which are usually partially oxidized to their acids by contaminating enzymes in the reaction mixture. Some workers (187) have reported that they obtained better recovery of products in their assays (ion exchange and solvent extraction) by adding excess aldehyde oxidase to convert all the product to acid (187).

Diamine oxidase (EC 1.4.3.6) has also been assayed by solvent extraction methods. Kobayashi et al. (193) reported that the usual method (addition of sodium bicarbonate) did not stop the reaction completely, with the result that reaction continued and spurious high readings were obtained if extraction was delayed. They modified the assay by using a mixture of aminoguanidine and sodium bicarbonate. This mixture stopped the reaction completely and did not interfere with the extraction procedure.

E. SUBCLASS 5, OXIDOREDUCTASES ACTING ON THE C—NH GROUP OF DONORS

Folate reductase (tetrahydrofolate dehydrogenase, EC 1.5.1.3) is difficult to assay by conventional spectrophotometric methods (194), which require high substrate concentrations to give a significant optical density change and which are insensitive because of light absorption by tetrahydrofolate and dihydrofolate at 340 mμ. A radiometric assay, in which product and residual substrate are separated by paper chromatography, and which uses high-specific-activity tritiated folic acid, is extremely sensitive and can serve accurately for very low levels of enzyme activity (195,196). Another assay, based on the tritium-release technique, has also been described (197,198) and is claimed to be quicker and simpler than the paper chromatographic method.

F. SUBCLASS 13, OXIDOREDUCTASES ACTING ON SINGLE DONORS WITH INCORPORATION OF OXYGEN (OXYGENASES)

Tryptophan oxygenase (EC 1.13.1.12) has been assayed by using tryptophan-(ring-2-^{14}C) as the substrate and coupling the reaction with the enzyme formylkynurenine formylase. This released labeled formate, which was easily isolated by ion exchange (160).

G. SUBCLASS 14, OXIDOREDUCTASES ACTING ON PAIRED DONORS WITH INCOR-
PORATION OF OXYGEN INTO ONE DONOR (HYDROXYLASES)

The hydroxylases form an important subclass of enzymes. They are involved in catecholamine and melanin biosynthesis and in many detoxication
reactions; drug metabolism, for example, often involves aryl hydroxylation.

Most hydroxylases occur only at very low levels in tissues, and the low
conversions obtained in assays are difficult to measure by conventional
methods. Also, 2-mercaptoethanol and other agents which are used to
protect the unstable enzymes and cofactors, such as tetrahydropteridine,
interfere in most spectrophotometric assays.

Simple, sensitive, and accurate radiometric assays, using ^{14}C- or tritium-
labeled substrates, have been described for most hydroxylases. Although
ion-exchange, solvent extraction, alumina adsorption, paper and thin-layer
chromatographic methods have been used, the very simple and quick
tritium-release technique (see Section III.4.A) has been preferred by most
workers.

This method has been somewhat complicated, however, by an intramolecular migration of the tritium atom displaced by the hydroxyl group,
which results in varying amounts of tritium being retained in the product
instead of being released into solution. These intramolecular migrations have
been studied in detail by Udenfriend and his coworkers (199,200) and are
now referred to as the "NIH shift." Their studies showed that the shift also
occurs with deuteriated and chlorinated substrates, and that it is a general
consequence of all enzyme-catalyzed aromatic hydroxylations (201).

The use of this simple assay method is not precluded by these migrations.
If the amount of retention is small, a simple correction can be applied after
determining the amount of activity retained in the product. If a large proportion of the activity is retained, the problem can be overcome by displacing
the retained tritium chemically (202), or by using a different substrate
which results in less retention of tritium (203,204).

Many hydroxylase assays, radiometric and otherwise, are complicated by
concomitant chemical reactions (46) which often proceed at greater rates
than the enzymatic reactions. The existence of the enzyme tyrosine hydroxylase, for example, was difficult to prove because the customary use of
boiled enzyme extracts in controls released some unknown impurities which
catalyzed the nonenzymatic reaction to such an extent that the enzymatic
reaction was completely masked. Nagatsu and his coworkers (144,205)
showed that, unlike the enzymatic reaction, the chemical reaction was not
stereospecific, and that boiled tissue extracts would convert both
D- and L-tyrosine to dihydroxyphenylalanine. They were able to overcome
the problem, therefore, by using a control containing the untreated enzyme

and D-tyrosine as the substrate. This technique—using the unnatural isomer as substrate in controls—may be of general application in the assay of enzymes which are accompanied by significant interfering chemical reactions (46).

A somewhat similar problem arose in the assay of dopamine-β-hydroxylase (206). Boiled tissue extracts catalyzed the nonenzymatic production of 2,4,5-trihydroxyphenethylamine from dopamine. The former compound is isomeric with noradrenaline, the enzymatic product, and the two compounds were not separated by the analytical procedure used in the assay. In this case the difficulty could not be overcome by substituting a different isomer as substrate. It was resolved quite simply, however, by using a different analytical separation procedure in which the chemical product was not included with the enzymatic product. Goldstein and his coworkers (207) have, for example, successfully used a tritium-release assay.

Another problem which can arise in hydroxylase assays is that decomposition of substrates on storage often produces the same products as the enzymatic reaction. If these decomposition products are not removed from the substrates before use, they will result in high blanks and consequent low sensitivity. Simple methods of removing such impurities are described in ref. 46.

The following are some examples of radiometric assays of hydroxylases.

3,4-Benzpyrene hydroxylase catalyzes a complex reaction leading to the formation of several products. Little is known about the various steps, but by taking advantage of the fact that the products, unlike the substrate, are insoluble in hexane, this enzyme has easily been assayed by using tritium-labeled 3,4-benzpyrene and measuring the activity of the residual substrate after extraction with hexane (208).

The chemical similarities of substrates and products make it difficult to assay steroid hydroxylases by conventional methods. A rapid (3-hr) paper chromatographic separation technique has been used in the radiometric assay of steroid 9α-hydroxylase, employing ^{14}C-labeled androst-4-ene-3,17-dione as the substrate (209).

Aryl 4-hydroxylase (EC 1.14.1.1) has been assayed, using various tritium-labeled substrates, by tritium-release methods (203,204,210). Retention of tritium, by the NIH shift, occurs with all substrates but is minimal with benzene sulfon-(4-^{3}H)anilide (203,204).

Cinnamic acid 4-hydroxylase has been assayed by using a ^{14}C-labeled substrate, and isolating the product, coumaric acid, by paper chromatography (211). A tritium-release assay, in which 88% of the tritium was retained by an NIH shift, has also been reported (212).

Tryptophan hydroxylase has been assayed by a tritium-release method in

which an NIH shift resulted in only 5 to 20% of the tritium being released
(213). This markedly reduces the sensitivity, and other methods (158,214,
215), in which a ^{14}C-labeled substrate is used, seem preferable. In one assay
(158,214), the product, 5-hydroxytryptophan, is decarboxylated to 5-hy-
droxytryptamine, which is isolated by solvent extraction. In another, substrate and product are separated by thin-layer chromatography on silica
gel (215).

Phenylalanine 4-hydroxylase (EC 1.14.3.1) has been assayed with either
a ^{14}C- or a tritium-labeled substrate. In the ^{14}C assay (216), substrate and
product were separated by overnight paper chromatography. In the much
quicker and simpler tritium-release assay (202), 92% of the tritium was
retained in the product, but this was readily displaced by treating the
product with N-iodosuccinimide before isolating the tritiated water.

Tyrosine hydroxylase has also been assayed with ^{14}C- and tritium-labeled
substrates. In the former assay (145,146), substrate and product were
separated by adsorption on alumina. This method has now been replaced by
a tritium-release assay, in which <10% of the tritium is retained in the
product (205,217,218).

Collagen proline hydroxylase has also been assayed by a tritium-release
assay, using a synthetic polypetide substrate labeled with 3,4-^{3}H-L-proline
(219).

Many enzymes which form unstable dihydroxy products have been
assayed by coupling the hydroxylation reaction with the enzyme catechol-O-
methyltransferase and methyl-labled S-adenosylmethionine. This results in
the formation of stable O-methyl derivatives, which are easily isolated by
solvent extraction (152–154).

2. The Assay of Enzymes of Class 2: the Transferases

Class 2 is the largest class of enzymes, and radiometric methods have been
extensively employed in their assay. Ion-exchange methods involving
column separations have often been used; it is probable that many of these
assays could be accomplished more easily and quickly by means of the
ion-exchange paper method (see Section III.3.B).

A. SUBCLASS 1, ENZYMES TRANSFERRING ONE-CARBON GROUPS

Many methyltransferases have been assayed using S-adenosylmethionine,
labeled in the methyl group with ^{14}C or tritium, as the methyl donor.

Protein (220,221), hemi-cellulose (222), RNA (223–225), and DNA
(226–228) methyltransferases have been assayed by macromolecular precipitation methods, using ^{14}C-labeled S-adenosylmethionine as the substrate.

In such assays, false results will be obtained if the macromolecular acceptor is contaminated with other macromolecules capable of being methylated (47). In order to avoid this danger, some workers have added the enzyme RNAase to their DNA methyltransferase assay reaction mixture to solubilize any methyl RNA formed by contaminating RNA and RNA methyltransferase (228).

Acetylserotonin methyltransferase (EC 2.1.1.4) (229), catechol O-methyltransferase (EC 2.1.1.6) (188,230) histamine methyltransferase (EC 2.1.1.8) (231), phenylethanolamine-O-methyltransferase (232), and phenylethanolamine-N-methyltransferase (233) have all been assayed by solvent extraction methods.

Phenylethanolamine-N-methyltransferase has also been assayed by a method in which the product, adrenaline, was isolated by a combined ion-exchange and alumina adsorption technique (234). This gave lower blanks and freed the product from labeled contaminants, which were not separated in the usual solvent extraction assay.

Homocysteine methyltransferase (EC 2.1.1.10) has been assayed by an ion-exchange column separation method (235).

Assays involving other methyl donors have also been carried out radiometrically. N^5-Methyltetrahydrofolate: L-homocysteine S-methyltransferase (236) and betaine: L-homocysteine S-methyltransferase (237) have been assayed using methyl-labeled donors. The product, methionine, was isolated by ion exchange.

Some hydroxymethyltransferases have been assayed radiometrically.

Serine hydroxymethyltransferase (EC 2.1.2.1), which transfers an hydroxymethyl group from serine to tetrahydrofolate, has been assayed by using uniformly labeled serine-^{14}C and separating the products by paper chromatography (238). In another assay 3-^{14}C-serine served as the substrate and labeled formaldehyde was isolated from the product by solvent extraction as the dimedon derivative (239).

The syntheses of thymidylate from deoxyuridylate (75,240) and of hydroxymethyl deoxycytidylate from deoxycytidylate (241) have been followed by using labeled substrates and separating the products by ion exchange or paper chromatography. Simpler and quicker methods involving the measurement of tritium release into solution from substrates labeled with tritium in the 5-position have also been described (72, 198,242–245).

Aspartate carbamoyltransferase (EC 2.1.3.2) has been assayed by using ^{14}C-labeled carbamoyl phosphate and measuring the activity of the acid-stable product after acidification and heating to eliminate the residual substrate as $^{14}CO_2$ (126). Another assay used ^{14}C-labeled aspartate as the substrate, products being separated by ion exchange (246). This method, although slightly more tedious than the $^{14}CO_2$-release assay, has the advantage of using a commercially available and stable substrate.

The clinically significant enzyme ornithine carbamoyltransferase (EC 2.1.3.3), has been assayed by using carbamoyl ^{14}C-labeled citrulline and carrying out the reaction in arsenate, instead of phosphate, buffer. The product, carbamoyl arsenate, decomposed instantaneously, releasing $^{14}CO_2$, which was trapped and counted (93).

B. SUBCLASS 2, ENZYMES TRANSFERRING ALDEHYDIC OR KETONIC RESIDUES

Transketolase (EC 2.2.1.1) is significant clinically. Its activity is decreased in thiamine deficiency, resulting in pentose accumulation. Other enzymes present in the erythrocytes cause the decarboxylation of this pentose. A clinical assay that has been described (247) uses 2-^{14}C-glucose, which is converted to 1-^{14}C-ribose. This, in turn, releases $^{14}CO_2$ on decarboxylation.

C. SUBCLASS 3, ENZYMES TRANSFERRING ACYL GROUPS

All of the enzymes in this group use acyl-CoA as the acyl donor. Acyl-labeled donors have often been synthesized in situ (see Section III.8.B) because they were either unavailable commercially or expensive. In most assays acyl-labeled acyl-CoAs have been used as substrates; occasionally, labeled acceptors have been employed with inactive acyl-CoA donors; and in one assay the reaction was followed in the reverse direction.

Arylamine acetyltransferase (EC 2.3.1.5) has been assayed with p-toluidine as the acetyl acceptor. The product, 1-^{14}C-acetyltoluidine, was isolated by solvent extraction (159).

Choline acetyltransferase (EC 2.3.1.6) is the most important acetyltransferase, and many radiometric assays have been described. These replace the original tedious pharmacological assay, which was very sensitive to changes in the medium and could not be used in studies on the effects of drugs and salts on the enzyme.

In the original radiometric assay advantage was taken of the fact that acetyl-CoA is acid labile and decomposes to produce volatile acetic acid. The activity of the residual nonvolatile product, acetylcholine, was then measured (125).

Other workers (165,166) have used a method in which labeled acetylcholine is precipitated as the reineckate salt, which is isolated by solvent extraction and counted. This technique gives much lower blanks than the original simple method and is, therefore, much more sensitive. An elegant microscale (1-μl) version of this assay allowed the determination of the choline acetyltransferase activity of individual cells (166).

Fonnum (14,155) has used what is essentially a combination of these two techniques to develop an assay which gives low blanks and is reported to be easier to manipulate than earlier methods. Residual acetyl-CoA is decomposed with hydroxylamine, and the acetylcholine purified by precipitation with sodium tetraphenyl borate.

Sensitive ion-exchange column methods, claimed to be very sensitive and less tedious than the precipitation assays, have also been described (248,249).

Galactoside acetyltransferase (EC 2.3.1.18) is an enzyme which catalzyes the acetylation of thiogalactosides (and other simple galactosides). It is important on account of its involvement with the three structural genes of the lactose operon of *Escherichia coli* and has been assayed by an ion-exchange column technique (156).

Enzymes which acylate macromolecules can easily be assayed by macromolecular precipitation methods. An assay for histone acetyltransferase, by this technique, has been described (250).

Two acetyltransferases have been assayed by using an inactive acetyl-CoA donor and labeled acceptors. In the first case, homoserine acetyltransferase was assayed, with tritium-labeled homoserine as the acceptor, by an ion-exchange column technique (251). In the second, chloramphenicol acetyltransferase was assayed by a combined solvent extraction and thin-layer chromatographic technique, using ^{14}C-labeled chloramphenicol as the acceptor (252).

Carnitine palmityltransferase has been assayed in the reverse direction by using carnitylpalmitine, tritium-labeled in the carnitine moiety, as the substrate, and isolating one of the products, carnitine, by solvent extraction (253).

D. SUBCLASS 4, ENZYMES TRANSFERRING GLYCOSYL GROUPS

The hexosyltransferases, sub-subclass 1, form a very large group of enzymes in which nucleotide sugars are the most common glycosyl donors. They are readily (and almost exclusively) assayed by radiometric methods. Most of the enzymes are highly specific, specificity being conferred both by the nucleotide sugar donor and by the glycosyl acceptor. Most of the assays so far described employ UDP-glucose as the glycosyl donor, but a few others, utilizing UDP-glucuronic acid, UDP-galactose, UDP-*N*-acetylglucosamine, UDP-*N*-acetylgalactosamine, ADP-glucose, GDP-glucose, GDP-mannose, and GDP-fucose, have also been described. Whether this represents the true pattern in nature, or is merely a reflection on the commercial availability of labeled nucleotide sugars, is open to question. There is no doubt that the increasing number of labeled nucleotide sugars available commercially is a good indication of the large and increasing interest in this field.

The enzymes which catalyze glycosyl transfer from nucleotide sugars to macromolecular acceptors are readily assayed by using glycosyl-labeled nucleotide sugars, and measuring the activity incorporated into macromolecular products precipitated by acid or alcohol (see Section III.2). This method has the advantage that any labeled degradation products of the substrate, produced enzymatically or chemically during the assay, will not be included with the true product.

In some assays, particularly when the product (e.g., teichoic acid) is not completely insoluble in the precipitating solvent, separation has been achieved by paper chromatography using solvent systems in which the product remains on the origin while the substrate and any degradation products migrate (127,254,255, and Section III.5).

Enzymes which catalyze the transfer of glycosyl groups from nucleotide sugars to mono- and disaccharide acceptors are easily assayed by using labeled nucleotide sugars and isolating the neutral products by high-voltage electrophoresis (see Section III.6). Any monosaccharide produced by the degradation of the nucleotide sugar will remain on the origin and be included with the polysaccharide product, but can easily be detected, and corrected for, by paper chromatography.

Ion-exchange methods (see Section III.3.A,B) readily separate the neutral products from the charged substrates and have been used in many assays.

Enzymes which catalyze glycosyl transfer from a macromolecular polysaccharide to a monosaccharide acceptor may easily be assayed by following the reverse reaction.

Many of the enzymes in this sub-subclass have been assayed radiometrically; brief details follow.

Glycogen phosphorylase (EC 2.4.1.1) was assayed by using ^{14}C-labeled glucose-1-phosphate as the substrate and isolating the product by precipitation with alcohol (256).

Levansucrase (EC 2.4.1.10), which catalyzes the synthesis of sucrose by fructosyl transfer to glucose, has been assayed in the reverse direction by measuring the incorporation of activity from ^{14}C-labeled sucrose into a macromolecular product, which was isolated by paper chromatography (257).

Glycogen synthetase (glycogen-UDP glucosyltransferase, EC 2.4.1.11) is important in glycogen storage disease. A large number of radiometric assays have been described. In all of these, UDP-glucose, ^{14}C-labeled in the glucose moiety, has been used as substrate, and products have been isolated by precipitation techniques (38,258–264). Thomas and his coworkers (264) have described a very simple and rapid filter paper assay which allowed many samples to be processed simultaneously. The product was bound so firmly to the filter paper that the scintillation vials could be reused for several months before rejection when the background count rose to twice the normal value. In many assays glycogen is precipitated in ethanol, dissolved in water, and counted by liquid scintillation. DeWulf and Hers (38) have drawn attention to the fact that glycogen is insoluble in many scintillants and becomes precipitated in the counting vials. They found it necessary to allow samples to stand for 30 min to obtain complete precipitation and reproducible results. Alternatively, glycogen should be counted in a scintillant in which it is not precipitated (64,88), counted on paper (264), or hydrolyzed before counting.

Glucan synthetase (β-glucan–UDP glucosyltransferase, EC 2.4.1.12) has been assayed by precipitating labeled polysaccharides with hot alkali; the reaction mixture was then heated to remove the alkali-soluble polysaccharide before filtering the insolubleβ-1,4-glucosyl product on a glass fiber filter disk and measuring its activity by liquid scintillation counting (265).

Sucrose synthetase (sucrose-UDP glucosyltransferase, EC 2.4.1.13) and sucrosephosphate synthetase (sucrosephosphate-UDP glucosyltransferase, EC 2.4.1.14) have been assayed by measuring the incorporation of labeled fructose or fructose-6-phosphate into sucrose or sucrose-6-phosphate (266). Substrates and products were separated by paper chromatography.

Trehalosephosphate synthetase (trehalosephosphate-UDP glucosyltransferase, EC 2.4.1.15), which transfers glucose from UDP-glucose (267–269) and also from GDP-glucose (268–269), has been assayed by dephosphorylating the product with alkaline phosphatase and then separating the neutral products from the residual nucleotide sugar substrate by ion exchange (267–269) or electrophoresis (269). The trehalose activity was measured after the trehalose had been separated from other labeled sugars by paper chromatography. Perhaps the simplest technique would be to combine electrophoresis, after enzymatic dephosphorylation of the trehalose phosphate, with paper chromatography. Both separations could be carried out consecutively on a single strip of cellulose paper.

Chitin synthetase (chitin-UDP acetylglucosaminyltransferase, EC 2.4.1.16) has been assayed by measuring the incorporation of activity from UDP-N-acetylglucosamine. The product, which remained on the origin, was isolated by paper chromatography (255).

Two methods have been described for the assay of UDP glucuronyltransferase (EC 2.4.1.17). In one method, glucuronyl transfer to bilirubin was carried out with nonradioactive substrates, and the product was diazotized with ^{35}S-labeled sulfanilic acid. The labeled derivative was then isolated by paper chromatography (150). In the second method, glucuronyl transfer to ^{14}C-labeled estradiol was assayed by a solvent extraction technique (270).

The biosynthesis of lactose is an important metabolic reaction catalyzed by the enzyme lactose synthetase (UDPgalactose-glucose galactosyltrans-ferase, EC 2.4.1.22). High-voltage electrophoretic (132,271), paper chroma-tographic (132), and ion-exchange column chromatographic (132,272–274) methods have been used in assays.

Fucosyl-lactose synthetase (GDPfucose-lactose fucosyltransferase) has been assayed by measuring fucosyl transfer from ^{14}C-labeled GDP-fucose to lactose (275). Neutral sugars were isolated by ion exchange, and fucosyl-lactose was separated from contaminants by paper chromatography.

Fucosyl transfer is also important in the determination of blood group characteristics. It has been assayed by using GDP-fucose as the donor, and isolating neutral products by high-voltage electrophoresis, followed by paper chromatography to remove contaminating labeled sugars (134).

The synthesis of fucosyl-labeled glycoproteins has also been assayed by using labeled GDP-fucose as the donor, and isolating the product by the macromolecular precipitation technique (276).

Cellulose synthetase (GDPglucose-β-glucan glucosyltransferase, EC 2.4.1.29) has been assayed by means of a macromolecular precipitation technique (277).

Mannose transfer from GDP-mannose to a phospholipid acceptor has been assayed by a solvent extraction method (278), and to a polysaccharide acceptor by a paper chromatographic method (131,279).

Synthesis of a galactan has been assayed by measuring the transfer of activity from UDP-galactose to a polysaccharide acceptor. The product was isolated by high-voltage electrophoresis (280).

Acetylglucosaminyl transfers to lipopolysaccharide (281) and to steroid (282) acceptors have been assayed. In the first assay, the incorporation of activity from labeled UDP-N-acetylglucosamine into an acid-precipitable product was determined. In the second, tritium-labeled estradiol was used as the acceptor, and the product was removed by solvent extraction before measuring the activity of the residual substrate.

Acetylgalactosaminyl transfer, from UDP-N-acetylgalactosamine, is important in glycoprotein synthesis and has been assayed by an high-voltage electrophoretic method, using a ^{14}C-labeled donor (183,283). Acetyl-galactosaminyl transfer to lactosyl-fucose is an important factor in the determination of human blood groups and has been assayed by a combined high-voltage electrophoretic and paper chromatographic method (284). Paper chromatography was necessary to remove N-acetylgalactosamine (formed by degradation of the substrate during the assay), which remained near the origin, with the true product, during electrophoresis.

The pentosyltransferases, sub-subclass 2, are a large and important group of enzymes which in the past have been assayed largely by spectrophotometric methods. For some reactions, however, the difference in the UV spectra of base and nucleoside is insufficient for these methods to be effective. For this reason and because of the demand for higher sensitivity, a large number of radiometric assays have been reported in recent years. DEAE disk and high-voltage electrophoretic methods are simpler and quicker than the paper chromatographic methods which have frequently been used.

Examples of radiometric assays of the pentosyltransferases are briefly reviewed below.

A pyrimidine nucleoside phosphorylase (EC 2.4.2.2) has been assayed in each direction by using ^{14}C- or tritium-labeled uridine or uracil, and separating substrate and product by high-voltage electrophoresis in borate buffer (285).

Uridine phosphorylase (EC 2.4.2.3) has been assayed by paper chromatographic (286), borate complex ion-exchange chromatographic (287), and charcoal adsorption (140) methods.

Thymidine phosphorylase (EC 2.4.2.4) has been assayed by measuring thymidine synthesis from ^{14}C-labeled thymine. Substrate and product were separated by paper chromatography (288).

Nucleoside deoxyribosyltransferase (EC 2.4.2.6), which catalyzes the transfer of deoxyribose from a deoxyribonucleoside donor to a purine or pyrimidine acceptor, has been assayed by using 2-^{14}C-thymine as the acceptor and inactive thymidine as the deoxyribosyl donor (289). The assay was carried out in the presence of a high concentration of phosphate to inhibit thymidine phosphorylase activity, which could also catalyze the formation of labeled thymidine. Products were separated by paper chromatography.

The purine phosphoribosyltransferases (EC 2.4.2.7,8), which catalyze the synthesis of phosphoribosyl pyrophosphate from purine mononucleotides and pyrophopshate, have been assayed in the reverse direction by following the formation of labeled mononucleotides from labeled adenine or hypoxanthine. Separation was effected by paper chromatographic (290–292), high-voltage electrophoretic (293), or DEAE disk (79,290,294) methods.

Orotate phosphoribosyltransferase (EC 2.4.2.10), which catalyzes the formation of orotidine-5'-monophosphate (OMP), has been assayed by measuring the $^{14}CO_2$ released by decarboxylating the product, OMP, synthesized from carboxyl-^{14}C-orotic acid (295).

Nicotinate phosphoribosyltransferase (EC 2.4.2.11) has been assayed, using ^{14}C-labeled nicotinic acid, by a paper chromatographic method (296).

Amidophosphoribosyltransferase (EC 2.4.2.14) has been assayed by measuring the conversion of labeled glutamine to glutamate, and separating substrate and product by paper chromatography (297,298).

Quinolinic acid phosphoribosyltransferase has been assayed by using appropriately labeled quinolinic acid, and measuring the $^{14}CO_2$ release from the unstable product, quinolinic acid ribonucleotide (299).

E. SUBCLASS 5, ENZYMES TRANSFERRING ALKYL OR RELATED GROUPS

This small subgroup contains few important enzymes.

Prenyltransferase (EC 2.5.1.1) has been assayed by using labeled isopentenyl pyrophosphate and inactive dimethylallyl pyrophosphate as substrates, and isolating the product, geranyl pyrophosphate, by solvent extraction (300).

Methionine adenosyltransferase (EC 2.5.1.6), which catalyzes the synthesis of the important methyl donor S-adenosylmethionine, has been assayed by ion-exchange column (301) and cation-exchange paper disk (86) methods. The recently published CM-cellulose disk assay (86) has many advantages. It is very simple and rapid. Interference from contaminating decarboxylases is avoided by using carboxyl-labeled methionine, so that decarboxylation yields inactive products. Reaction is stopped by adding an aliquot of reaction mixture to a disk pretreated with inactive methionine, which reduces the specific activity of the labeled methionine and so stops the synthesis of labeled product. At the same time this dilution reduces blanks caused by the irreversible adsorption of substrate by the disk (see Section V.7).

F. SUBCLASS 6, ENZYMES TRANSFERRING NITROGENOUS GROUPS

Aminotransferases are readily assayed by conventional spectrophotometric methods, but these have many disadvantages. Radiometric assays are much more sensitive and far less susceptible to interference by endogenous oxo acids. They are also easily adapted for the routine assay of large numbers of samples.

A rapid and sensitive survey of tissues for the presence of several aminotransferases may be carried out by incubating tissue samples with 1-^{14}C-2-oxoglutarate, glutamic decarboxylase, and the appropriate amino acid. If the aminotransferase is present, 1-^{14}C-glutamate is formed and is decarboxylated, releasing $^{14}CO_2$.

The clinically important glutamic-oxaloacetic transaminase (aspartate aminotransferase, EC 2.6.1.1) has been assayed by several methods. In one assay, 4-^{14}C-oxaloacetate was used as the substrate, and reaction stopped by heating in the presence of a cupric salt. This decarboxylated the residual substrate, allowing the activity of the stable product to be measured (113). This aminotransferase (and others) may also be assayed by using labeled 2-oxoglutarate as substrate, and isolating the product by paper chromatography (302) or cation-exchange paper chromatography (78).

Several very sensitive assays for tyrosine (EC 2.6.1.5) and other aromatic aminotransferases have been described. In one assay high-specific-activity ^{14}C-labeled amino acids were used as substrates, and products were separated by ion-exchange column chromatography (303). Cation-exchange paper methods may also be used (78). Another assay, in which the product, p-hydroxyphenylpyruvic acid, was separated from the substrate, tritium-labeled tyrosine, by solvent extraction, has been described (304). Litwack and Squires (305) have developed a tritium-release assay using 2-^{3}H-tyrosine as substrate.

Leucine aminotransferase (EC 2.6.1.6) has been assayed by a method in which the labeled oxo acid product was isolated, as its 2,4-dinitrophenyl-hydrazone, by solvent extraction (16).

G. SUBCLASS 7, ENZYMES TRANSFERRING PHOSPHORUS-CONTAINING GROUPS

This is a very large group of enzymes and includes three important sub-subclasses.

a. **Sub-subclass 1, Phosphotransferases with an Alcohol Group as Acceptor.** The enzymes which transfer a phosphoryl group to an alcohol acceptor form one of the largest and most important groups. More than 50 separate enzymes, with varying degrees of specificity, have been described. Almost exclusively, ATP is the phosphoryl donor. Many quick, simple, sensitive, and accurate radiometric assays have been described. Those based on ion-exchange paper chromatographic and paper disk methods seem most generally applicable and useful. It has been claimed, for example, that the DEAE disk method for glucokinase is 2 orders of magnitude more sensitive than the fluorimetric assay, and that this fluorometric assay is itself some 2 to 3 orders of magnitude more sensitive than the conventional spectrophotometric assay.

Usually, ^{14}C- or tritium-labeled substrates have been used in assays, but γ-^{32}P-labeled ATP is suitable for all enzymes for which ATP is the phosphoryl donor. However, because most enzyme extracts contain contaminating enzymes which degrade ATP, methods using γ-^{32}P-ATP are less useful than those involving labeled acceptors.

An added advantage of radiometric assays is that, if a large excess of ATP-degrading enzymes is present, ATP-regenerating systems can be created by adding creatine phosphate and creatine kinase, or phosphoenol pyruvate and pyruvic kinase. This cannot be done, of course, for the conventional spectrophotometric assays based on reactions coupled to the formation of ADP.

A brief review of the published radiometric assays of this group of enzymes follows.

Glucokinase (EC 2.7.1.2) has been assayed by the DEAE-paper disk method, which is generally applicable to all hexokinases (68). DEAE-paper chromatographic methods may equally well be used (66,87). Separation takes less than 2 hr, and blanks are often lower than with the paper disk method (87).

Galactokinase (EC 2.7.1.6) has been assayed, with a ^{14}C-labeled substrate, by ion-exchange paper chromatographic (306) and paper disk (307) methods.

Adenosine kinase (EC 2.7.1.20) has been assayed by rapid (4 hr) paper chromatographic and solvent extraction methods (17).

Thymidine kinase (EC 2.7.1.21) has been assayed by DEAE-paper chromatographic (67,308), DEAE-paper disk (67), and the slower paper chromatographic (309) and ion-exchange column (310) methods. Many enzyme extracts contain thymidylate kinase, which converts part of the thymidine monophosphate product into the di- and triphosphates. In order to calculate the true thymidine kinase rate, the total activity of all phosphorylated products must be known. An advantage of the DEAE-paper methods is that, unlike paper chromatographic and ion-exchange column methods, they retain the mono-, di-, and triphosphates, which are counted together.

The similar enzyme deoxycytidine kinase has been assayed by an ion-exchange paper (DEAE and Amberlite) disk technique, using tritium-labeled deoxycytidine as the substrate (67b,311). Because of the difficulty of counting tritium-labeled products on paper, a simple and effective method of eluting the products in the scintillation vial so that they could be counted in homogeneous solution was employed. This gave a sixfold increase in sensitivity and also made the technique more reproducible. The factors affecting the use of ion-exchange paper techniques with tritium-labeled substrates are discussed in detail (67b).

Glycerol kinase (EC 2.7.1.30) has been assayed by the DEAE disk method (68,312). This enzyme has an extremely low Michaelis constant, and only this radiometric assay was sufficiently sensitive to allow the accurate determination of reaction rates at very low substrate concentrations.

Protein kinase (EC 2.7.1.37) has been assayed by a macromolecular precipitation method, using γ-^{32}P-ATP as the phosphoryl donor (313).

Uridine kinase (EC 2.7.1.48) has been assayed by a paper chromatographic method (314) and by a simpler ion-exchange paper chromatographic method (71,315).

Two other enzymes in this group, for which ATP is not the phosphoryl donor, have been assayed using ^{32}P-labeled donors.

Phosphoroamidate: D-hexose 1-phosphotransferase has been assayed using ^{32}P-phosphoroamidate as the donor and glucose as the acceptor (316).

The reaction was stopped in acid. This hydrolyzed residual substrate to inorganic orthophosphate, which was removed as the molybdate complex.

Another enzyme, which transfers a phosphoryl group from inorganic pyrophosphate to sugars and polyols, has been assayed by two methods. In one, inorganic ^{32}P-pyrophosphate was the donor and glucose the acceptor. The reaction was stopped in acid which did not affect the product, glucose-6-phosphate, but hydrolyzed residual pyrophosphate to inorganic phosphate, which was removed as the molybdate complex (317). In the other, glycerol was the acceptor, and the product, glycerol-1-phosphate, was isolated by paper chromatography (318).

b. Sub-subclass 2, Phosphotransferases with a Carboxyl Group as Acceptor. Only one enzyme in this small group, carbamoylphosphate synthase (EC 2.7.2.5), has been assayed radiometrically. Many methods have been described. All of them depend on the incorporation of $^{14}CO_2$ into an unstable product, carbamoyl phosphate, which has been stabilized by enzymatic conversion to citrulline (116–118,319,320) or by chemical conversion to urea (321). The nonvolatile material has been shown, however, to contain an unknown labeled product, and in some assays this impurity has been removed by thin-layer chromatography (118,319) or ion-exchange column chromatography (320,321).

c. Sub-subclass 3, Phosphotransferases with a Nitrogenous Group as Acceptor. Only one enzyme in this small group has been assayed radiometrically. The clinically important enzyme creatine kinase (EC 2.7.3.2) has been assayed by using 1-^{14}C-creatine as the substrate, and isolating the product by a DEAE disk method (70). The authors observed very high blanks (4% of the total activity) in the assay and suggested that these were due to an impurity in the substrate. If this explanation was valid, the blanks could have been reduced markedly, and the sensitivity increased, by removing the impurity by filtering the substrate through a DEAE-cellulose disk, or by employing DEAE-cellulose paper chromatography, before use (46).

d. Sub-subclass 4, Phosphotransferases with a Phospho Group as Acceptor. The nucleotide kinases are the most important enzymes in this group. Either ATP or some other nucleoside triphosphate acts as the phosphoryl donor. Many radiometric assays have been described. The presence of ATP-degrading enzymes in most extracts has precluded the use of labeled ATP as a donor, and ^{14}C-, tritium-, and ^{32}P-labeled acceptors have been employed almost exclusively.

When the acceptors are diphosphates, the triphosphate products have been separated by electrophoresis (322) or thin-layer chromatography (323).

When monophosphate acceptors are used, the diphosphate products are often further phosphorylated to triphosphates by other kinases present in the extract. In such assays, the combined activity of the di- and triphosphates represents the amount of substrate converted by the monophosphate kinase being assayed. In these assays it is preferable, therefore, to use analytical techniques which do not separate di- and triphosphates.

Radiometric assays of the mononucleotide kinases have been carried out in two ways. In one, the residual labeled nucleoside monophosphate substrates are dephosphorylated enzymatically with a phosphomonoesterase after the kinase reaction has been stopped. The di- and triphosphate products are then isolated by charcoal adsorption (137,138) when ^{32}P-labeled substrates are used, and by DEAE-cellulose paper disk methods when ^{14}C- or tritium-labeled substrates are used (66b). In the second type of assay, substrate and products have been separated by paper chromatographic (324), electrophoretic (325,326), ion exchange column (310,327), and ion-exchange paper chromatographic (328,329) methods.

Methods which depend on the conversion of labeled monophosphates to phosphomonoesterase-resistant products appear to afford the simplest and quickest assays. Because ^{14}C- and tritium-labeled substrates are now more readily available commercially and have longer shelf-lives, methods using them, rather than the short-lived ^{32}P-labeled substrates, are preferred. Ion-exchange paper chromatographic and paper disk methods may be employed. Although self-absorption problems have led to objections to the use of tritium-labeled substrates in ion-exchange paper methods, a recent report discusses in detail a very effective method of elution from DEAE papers, which overcomes these objections by allowing the products to be counted in homogeneous solution (67b).

Radiometric assays of enzymes of this sub-subclass are briefly reviewed below.

Myokinase (adenylate kinase, EC 2.7.4.3) has been assayed by using ^{14}C-labeled AMP or ADP as the substrate, and separating substrate and products by electrophoresis (325).

Deoxycytidylate kinase (EC 2.7.4.5) has been assayed by using a ^{32}P-labeled substrate, dephosphorylating the residual substrate, and adsorbing the products on charcoal (137,138). In another assay a ^{14}C-labeled substrate was employed, and the products were isolated by ion-exchange paper chromatography (328).

Nucleosidediphosphate kinase (EC 2.7.4.6) has been assayed, with ^{14}C-labeled UDP as the substrate, by an electrophoretic method (322).

Guanylate kinase (EC 2.7.4.8) has been assayed by using ^{14}C-labeled GMP or ATP as substrate, and isolating the product by ion-exchange paper chromatography on AE-81 formate paper (329).

Thymidinemonophosphate kinase (EC 2.7.4.9) has been assayed by paper chromatographic (324), electrophoretic (326), and ion-exchange column (310,323,327) methods, as well as by the enzymatic dephosphorylation method using charcoal adsorption (137,138) and DEAE disk (66b) separation.

Thymidinediphosphate kinase has been assayed by a thin-layer chromatographic method, with tritium-labeled thymidine diphosphate as the substrate (323).

e. Sub-subclass 6, Pyrophosphotransferases. Only one enzyme in this very small group has been assayed radiometrically. Thiamine pyrophosphokinase (EC 2.7.6.2) has been assayed by using ^{14}C-labeled thiamine as substrate, and isolating the product, thiamine pyrophosphate, by paper chromatography (330). A quicker and simpler separation could probably be achieved by a DEAE-paper technique.

f. Sub-subclass 7, Nucleotidyltransferases. This very large and important group includes the enzymes which catalyze the synthesis of DNA, RNA, polynucleotides, and nucleotide sugars. They have been assayed almost exclusively by radiometric methods.

The macromolecule-synthesizing enzymes have been assayed by using ^{14}C-, tritium-, or ^{32}P-labeled nucleotide donors, and measuring the activity incorporated into an acid-precipitable product (see Section III.2).

In some crude extracts the presence of appreciable amounts of polynucleotide hydrolyases can cause degradation of the newly synthesized polynucleotides; an effect which reduces the accuracy of many assays. This interference has been overcome by using terminal ^{32}P-labeled nucleotides, adsorbing unused nucleoside phosphates onto charcoal, and measuring the activity of the free inorganic phosphate (26). Alternatively, when a diphosphate donor was used, the orthophosphate liberated was extracted into isobutanol as the phosphomolybdate complex (139).

Sometimes, when ADP is the donor, the presence of myokinase in crude extracts can interfere by converting the ADP to AMP and ATP. This difficulty can be avoided by following the reaction in the reverse direction and measuring the phosphorolysis of the polynucleotide by labeled orthophosphate (139). The phosphorylated product can be isolated by charcoal adsorption, or, alternatively, residual phosphate can be removed by solvent extraction as the phosphomolybdate complex. Because these methods isolate AMP, ADP, and ATP together, the presence of myokinase does not cause erroneously low results.

Nucleotide sugar synthesis has been assayed by many radiometric methods, using ^{14}C- or tritium-labeled glycosyl donors, or by following the reaction in the reverse direction with labeled phosphate or pyrophosphate. The choice of labeled substrate has often been dictated by commercial availability. In the past, when labeled glycosyl phosphates were not readily available, most assays involved the incorporation of labeled inorganic phosphates, but now that a large number of labeled glycosyl phosphates are on the market they are most commonly used in assays.

The formation of nucleotide sugars has been assayed by many methods. Products have been isolated by electrophoresis, by charcoal adsorption, or by enzymatically dephosphorylating residual glycosyl phosphate with a phosphomonoesterase and separating the nucleotide sugar prouct from the free sugar. This separation can be carried out by adsorption of the nucleotide sugar onto charcoal or by ion exchange. Ion-exchange paper methods appear to be the simplest and most rapid.

A brief review of the methods used for this group of enzymes follows.

Sulfate adenylyltransferase (EC 2.7.7.4), which catalyses the reversible formation of adenosine-5'-phosphosulfate, has been assayed by using ^{35}S-sulfate as the substrate, and isolating the product by electrophoresis (331). Interference from the reverse reaction was prevented by adding pyrophosphatase to degrade the second product, inorganic pyrophosphate.

The enzyme ATP:glutamine synthetase adenylyltransferase, which adenylates and hence inactivates glutamine synthetase, has been assayed by using ^{14}C-labeled ATP as the adenylyl donor, and isolating the product by precipitation (332).

Many assays of RNA polymerase (RNA nucleotidyltransferase, EC 2.7.7.6) have been described. Most use macromolecule precipitation techniques (27,28,43,334). In one method, however, in which the acceptor is a polynucleotide instead of the more usual RNA, the assay has been accomplished by isolating the smaller than usual polynucleotide product by paper chromatography (333) or by ion-exchange paper chromatography (30).

Usually, ^{14}C- and tritium-labeled nucleotides, which have longer shelf-lives, are preferred to ^{32}P-labeled donors. Although tritium-labeled substrates are cheaper than ^{14}C-labeled substrates, self-absorption problems often make the macromolecular products more difficult to count than the corresponding ^{14}C-labeled products (see Section V.9). Most self-absorption problems can be overcome by solubilizing the products and counting in homogeneous solution, but this can be tedious when large numbers of samples need processing. For most investigations filter disk or membrane filter methods are preferable.

However, many people use precipitation methods of counting tritium-labeled precipitates and are willing to accept the resulting low counting efficiency. Care must be exercised when making allowance for self-absorption losses in these assays. Often workers have attempted to correct for self-absorption losses by drying a sample of the tritium-labeled precursor on a filter disk, and comparing the count rate with that obtained for a similar sample counted in homogeneous solution. This can grossly overestimate self-absorption losses, particularly when filtration methods are used, because the small precursors are adsorbed throughout the disk and self-absorption losses are caused by the support medium. When macromolecular products are filtered, or precipitated on the surface of fibers, self-absorption losses are largely caused by the thin layer of labeled product on the surface, and not by the more dense support medium.

A recent publication (333) has shown that ethanol is a potent inhibitor of some RNA polymerases; 2% ethanol in the reaction mixture was sufficient to cause 80% inhibition. Many labeled nucleotides are supplied in 50% ethanol to reduce radiation decomposition on storage. The potent inhibition by ethanol makes it essential to remove this substance before use unless the concentration in the assay mixture is reduced, by dilution, to less than 0.1%.

Minute traces of radiation decomposition products can be incorporated into macromolecular precipitates and thus cause relatively large blank readings in polymerase assays (46,47), particularly when only a small proportion of the nucleotide substrate is incorporated into the product. A recent publication (46) has shown that these minute trace impurities can often be removed simply by filtering the substrate through a nitrocellulose filter before use. The impurities "stick" to the macromolecular filter and are retained. For example, a sample of high-specific-activity ^{14}C-labeled ATP, used in an RNA polymerase assay, gave a blank reading of 6900 cpm because of the presence of radiation decomposition products, which were coprecipitated with RNA. The blank reading fell to 200 to 300 cpm when the ATP was filtered through a membrane filter before use in the assay (46).

DNA polymerase (DNA nucleotidyltransferase, EC 2.7.7.7) has been assayed by macromolecular precipitation techniques similar to those used for RNA polymerase, with ^{14}C- (22,23,335,336), tritium- (335,337,338), and ^{32}P- (22,23,25,335–337,339) labeled substrates. Self-absorption, blank, and alcohol inhibition problems similar to those encountered in RNA polymerase assays may arise.

Polynucleotide phosphorylase (EC 2.7.7.8), which catalyzes nucleotidyl transfer from nucleoside diphosphates to polyribonucleotide acceptors, may be assayed by techniques similar to those used for RNA polymerase (340), or by following the reaction in the reverse direction and measuring the incorporation of activity from labeled orthophosphate (139,140).

Another adenylyltransferase, poly(adenosinediphosphate ribose) synthase, has been assayed by measuring adenylyl incorporation, from ^{14}C-adenine-labeled nicotinamide adenine dinucleotide, into an acid-insuluble product (341,342).

Two ribonucleases (EC 2.7.7.16,17) have been assayed by using tritium-labeled polyuridylic acid as substrate and measuring the residual substrate by an alcohol precipitation method (343), and also by using ^{14}C-benzyl esters of 3'-nucleotides and extracting the product, labeled benzyl alcohol, in toluene (344). An advantage of the latter method is that it allows easy differentiation between base-specific RNAases.

Polyadenylate nucleotidyltransferase (EC 2.7.7.19) has been assayed by measuring the incorporation of activity from ^{14}C-ATP into an acid-precipitable product (31). Crude extracts contained ATP-degrading enzymes, and a creatine phosphate-creatine kinase ATP-regenerating system was used.

A tRNA nucleotidyltransferase (EC 2.7.7.25) has been assayed by the precipitation technique, using various ^{14}C-labeled nucleoside triphosphates as the nucleotidyl donors (345).

Many nucleotide sugar-synthesizing enzymes have been assayed radiometrically. The best and most generally applicable method is to use a glycosyl-labeled acceptor, and to separate the products either by ion-exchange paper chromatography or by dephosphorylating the residual substrate and isolating the nucleotide sugar by ion-exchange paper chromatographic or paper disk methods. The assays will be improved by adding pyrophosphatase to hydrolyze inorganic phosphate and so convert the normally reversible reaction into a unidirectional one. If appreciable amounts of ATP-degrading enzymes are present in crude enzyme extracts, an ATP-regenerating system should be added.

UDPglucose pyrophosphorylase (EC 2.7.7.9) has been assayed by using ^{14}C-glucose-1-phosphate and separating the products by paper chromatography (346), by dephosphorylating the residual substrate and isolating the labeled UDP-glucose by precipitation as the mercury salt (347), and in the reverse direction by using a charcoal adsorption method to measure the incorporation of activity from labeled pyrophosphate (348).

UDPglucose: galactose-1-phosphate uridyltransferase is a clinically significant enzyme, deficient in congenital galactosemia. It has been assayed, with labeled galactose-1-phosphate as substrate, by two methods. In one, an ion-exchange paper chromatographic separation technique was used (77). An advantage of this method is that the presence of UDPglucose epimerase in the enzyme extracts does not interfere because all nucleotide sugars are retained on the origin. In the second, more tedious, assay, the labeled nucleotide sugar was isolated by charcoal adsorption, followed by elution with ammoniacal ethanol, and lyophilization before counting (349).

GDPmannose pyrophosphorylase (EC 2.7.7.13) has been assayed by a DEAE-cellulose paper disk method after dephosphorylating the residual substrate with a phosphomonoesterase (76). Because the substrate, ^{14}C-mannose-1-phosphate, was not available commercially, it was synthesized in situ from labeled mannose.

GDPglucose pyrophosphorylase has also been assayed, using ^{14}C-glucose-1-phosphate as the substrate. The product was isolated both by paper chromatographic and ion-exchange paper chromatographic methods (350). The rapid (1 to 3 hr) ion-exchange paper chromatographic method is preferable to the slower paper chromatographic method.

CDPglucose pyrophosphorylase has also been assayed by an ion-exchange technique after enzymatic dephosphorylation of the substrate (351).

ADPglucose pyrophosphorylase has been assayed by following the reverse reaction and measuring the incorporation of activity from labeled pyrophosphate into ATP, which was isolated by charcoal adsorption, and by using ^{14}C-glucose-1-phosphate, adding pyrophosphatase to drive the reaction in one direction, dephosphorylating the residual substrate, and isolating the product by filtration through a DEAE-cellulose disk (352).

Adenyl cyclase is an enzyme which has not yet been classified but which might be regarded as an internal nucleotidyltransferase. It converts ATP to adenosine-3',5'-cyclicmonophosphate (cyclic AMP), which is a critical regulatory intermediary in the action of many mammalian hormones and biogenic amines. The concentration of cyclic AMP in many tissues is determined by the balance between its formation from ATP by adenyl cyclase, its degradation by a specific phosphodiesterase, and the degradation of ATP by ATPase and other enzymes. Precautions should be taken, therefore, to minimize the effect of these interfering enzymes in all assays performed with crude extracts. In addition to suffering from interference by these degrading enzymes, adenyl cyclase assays are complicated by the fact that the enzyme is unstable and occurs at such a very low level that it is rarely possible to obtain more than 0.5% conversion of substrate to product. Often, too, there is no linear relationship between product formation and time. Testifying to the difficulty of this assay is the very large number of different methods which have been described without any one becoming universally accepted.

Before any method can be successfully applied, degradation of both ATP and cyclic AMP must be prevented. There are three ways of reducing ATP degradation.

1. By adding an ATP-regenerating system, such as creatine phosphate and creatine kinase, or phosphoenol pyruvate and pyruvate kinase. However, caution may be needed in the use of the latter regenerating system because it produces pyruvate, which has been reported to activate some microbial adenyl cyclases (353).

2. By adding fluoride to inhibit ATPase. However, fluoride can activate adenyl cyclase by a mechanism independent of its inhibition of ATPase (354), and this may cause complications in the assay.

3. By using relatively high ATP concentrations, much greater than K_m (1mM, ref. 353), so that degradation of a large proportion of the ATP will

not have a marked effect on the rate of reaction of the adenyl cyclase, which will remain saturated with substrate during the assay. The chief disadvantage of this approach is that, unless a very effective method of isolating cyclic AMP, free from impurities, is available, the use of high ATP concentrations will decrease the sensitivity by reducing the proportion of substrate converted to product during the assay.

The use of a creatine phosphate-creatine kinase ATP-regenerating system appears to be the best method of minimizing ATP degradation. Whatever method is used, routine checks should be made to confirm that a suitable ATP concentration is maintained during assays, and that low cyclic AMP production does not result from excessive ATP degradation.

Cyclic AMP degradation by phosphodiesterase during assays can be reduced in two ways.

1. By adding inhibitors such as the methyl xanthines, caffeine, and theophylline. Recent work (355) has shown that caffeine is much less effective than theophylline, and that $0.04M$ theophylline (an almost saturated solution) was needed to reduce cyclic AMP degradation to an acceptable level.

2. By adding inactive cyclic AMP to the reaction mixture in order to inhibit degradation of the labeled cyclic AMP. This method is effective only when the concentration of inactive cyclic AMP in the reaction mixture is greater than its K_m for the diesterase, so that the diesterase is completely saturated with inactive substrate. Values of K_m ranging from $1.3 \times 10^{-4}M$ to $10^{-6}M$ have been reported (356).

Very high levels of phosphodiesterase activity have been reported in some tissues, and in some assays it may be necessary, to use both theophylline and inactive cyclic AMP to reduce this activity to acceptable levels. Only by reducing cyclic AMP degradation will it be possible to obtain proportionality between the amount of enzyme extract added and the adenyl cyclase activity, and a linear relationship between enzyme activity and time (355). Because the level of phosphodiesterase activity can vary markedly from source to source, periodic checks should be made to confirm that degradation of cyclic AMP during assays is minimal.

When the effects of the degrading enzymes have been neutralized, the success of any adenyl cyclase assay will depend on the technique used to isolate the product, cyclic AMP. Many radiometric methods have been employed.

Perhaps the most common is the one devised by Krishna (357,358), in which α-^{32}P-ATP is used as the substrate, and ^{3}H-cyclic AMP is added, when reaction is stopped, to measure the recovery of product. Cyclic AMP is isolated by ion-exchange column chromatography, and minute traces of residual substrate are removed by precipitation with $ZnSO_4$-$Ba(OH)_2$. This separation method is claimed to be sufficiently sensitive to allow the detection of 1 part of cyclic AMP in 50,000 parts of ATP (167). However, some workers (359) have reported that reproducible results are not obtained with this method when assaying very low levels of activity. This criticism has been refuted by the original workers (357).

Paper chromatographic methods of separation have been used, but complex separation procedures are often necessary to isolate pure cyclic AMP (355,359). Also, it has been shown (359) that hypoxanthine, produced by the degradation of ^{14}C-adenine-labeled ATP, can be mistaken for cyclic AMP in some paper chromatographic solvents.

Some excellent ion-exchange paper and thin-layer chromatographic methods have been described (83,359–361). Care should be taken in these and other assays, however, to characterize the product as cyclic AMP, possibly by enzymatic degradation to AMP.

The similar enzyme guanyl cyclase has been assayed (362,363) by using ^{14}C-GTP as the substrate, adding ^{3}H-cyclic GMP to correct for the recovery of product, and then isolating cyclic GMP by thin-layer chromatography. The product was eluted from the chromatogram and divided into two parts. One was treated with phosphodiesterase and alkaline phosphatase to convert cyclic GMP to guanosine. The other was treated only with alkaline phosphatase, which does not degrade cyclic GMP. Anion-exchange resin, which does not adsorb guanosine, was added to each sample, and the difference in counts between the samples was taken as a measure of cyclic GMP formation. Another assay has been described by Hardman and Sutherland (364), who, after terminating the reaction, treated the reaction mixture with snake venom and alkaline phosphatase to degrade all phosphates other than cyclic, GMP, which was then isolated by ion exchange.

H. SUBCLASS 8, ENZYMES TRANSFERRING SULFUR-CONTAINING GROUPS

All of the sulfate-transferring enzymes utilize 3′-phosphoadenosine-5′-phosphosulfate as the sulfate donor. Several have been assayed radiometrically, using a ^{35}S-labeled donor.

Aryl sulfotransferase (EC 2.8.2.1) has been assayed by two methods, using tyrosine methyl ester or phenol as the acceptor (365). In one assay the residual substrate was removed by precipitation as the barium salt; in the other, substrate and products were separated by electrophoresis or paper chromatography.

Steroid sulfotransferase (EC 2.8.2.2,4) has been assayed by a method in which the residual substrate was removed by precipitation as the barium salt (366).

Arylamine sulfotransferase (EC 2.8.2.3) has been assayed by using tyramine as the acceptor, and separating substrate and product by barium salt precipitation or electrophoresis (365).

Chondroitin sulfotransferase (EC 2.8.2.5) has been assayed by a paper chromatographic method (367).

Sulfate transfer to N-desulfoheparin has been followed by separating the product from residual substrate by dialysis for 48 hr (168). This method has since been speeded up by isolating the high-molecular-weight product by gel filtration (169). However, since both procedures involve the measurement of activity incorporated into macromolecules, the best method of assay should be the filter paper disk method (see Section III.2).

3. The Assay of Enzymes of Class 3: the Hydrolases

A. SUBCLASS 1, HYDROLASES ACTING ON ESTER BONDS

This very large subclass of enzymes includes several important subsubclasses. In the past, these enzymes have been assayed chiefly by estimating the acid formed, by pH titration or by measuring CO_2 release from bicarbonate buffers, or by colorimetric estimation of products. A large number of radiometric assays have been described; in general, these are more rapid, simple, accurate, and sensitive than the conventional assays.

a. Sub-subclass 1, Carboxylic Acid Hydrolases. Lipase (EC 3.1.1.3), for which no simple conventional assay exists, has been assayed by solvent extraction methods using triglycerides labeled, in the fatty acid moiety, with ^{14}C (368,369), tritium (370), and ^{131}I (371,372). Residual substrates and products were separated by solvent extraction.

Phospholipase A (EC 3.1.1.4) has been assayed by thin-layer chromatographic (373–377) and solvent extraction (378) methods, using ^{14}C- (373,375,376,378), tritium- (373,375,376), and ^{32}P- (374,377) labeled substrates.

Phospholipase B (EC 3.1.1.5) has also been assayed by a thin-layer chromatographic method, using a ^{32}P-labeled substrate (377).

A microscale assay of an acetylesterase (EC 3.1.1.6), which deacetylated the 3',5'-diesters of fluorodeoxyuridine and thymidine, has been carried out by using ^{14}C-acetyl-labeled 3',5'-diacetylfluorodeoxyuridine as the substrate, and measuring the activity of the residual, nonvolatile substrate after volatilization of the volatile product, acetic acid (124).

The most important enzyme in this group is acetylcholinesterase (EC 3.1.1.7), for which several radiometric assays have been described. In the original assay (123), 1-^{14}C-acetyl-labeled acetylcholine was used as the substrate. Samples were removed, acidified, and dried to volatilize the free acetic acid. The activity of the residual substrate was then measured. This method was sufficiently sensitive to be used with a 1-μl blood sample, and is 1000 times as sensitive as the conventional manometric assay. Objections that the sensitivity was unnecessarily reduced by measuring the residual substrate instead of the product were overcome in an assay in which the reaction was stopped by adding an ion-exchange resin in alcohol. Residual substrate was retained, and the product (acetic acid) was left in the ethanolic extract, which was removed for scintillation counting (56). Another method, in which the product was extracted in toluene-isoamyl alcohol, has also been described (379).

In vitro assays are often used to measure the effect of drugs on the *in vivo* activity of acetylcholinesterase and other enzymes. Attention has been drawn to the fact that, because many drugs are reversible inhibitors of enzymes, enzyme activities measured *in vitro* will depend both on the dilution of the sample and on the substrate concentration used in the assay (380,381). The *in vitro* results will inevitably underestimate the true *in vivo* activity, and the closest approximation will be obtained by using substrate concentrations as low as possible, with minimum dilution of the sample.

b. Sub-subclass 3, Phosphoric Monoester Hydrolases. The phosphoric monoester hydrolases are easy to assay by the colorimetric estimation of either product. Radiometric methods have been used in a few assays where high sensitivity is needed, as well as for assaying some of the more specific phosphatases. For example, when two phosphomonoesterases, one (such as glucose-6-phosphatase) of narrow specificity and the other (such as alkaline phosphatase) of broad specificity, occur together, the specific phosphatase may be assayed by using an appropriate labeled substrate and inhibiting the nonspecific phosphatase with a large excess of phenyl phosphate (382).

Because phosphomonoester hydrolysis often results in the formation of an uncharged product from a charged substrate, ion-exchange paper methods, using ^{14}C- or tritium-labeled substrates, provide a very simple and generally applicable technique of assay. Ives and his coworkers (67b) have published the details of such a method, which they used with tritium-labeled nucleotide substrates, but which could serve equally well with ^{14}C-labeled substrates. Assay samples were removed and spotted onto DEAE-cellulose disks, which were then placed in scintillation vials containing 1 ml of water to elute the nucleoside products. After a suitable inteval, the disks, containing bound nucleotide substrate, were removed and transferred to second vials containing 1 ml of 0.1M HCl-0.2M KCl. This eluted the nucleotide. Triton-based scintillant was then added to the vials, and the activities of residual substrate and product were measured by liquid scintillation. The enzyme activity was then calculated from the ratio of the two counts.

An interesting histochemical assay of alkaline phosphatase (EC 3.1.3.1) activity in cells has been described (383,384). Tissue slices were incubated with 2-glycerophosphate and then with a solution of ^{45}Ca-calcium nitrate. The ^{45}Ca was retained as the insoluble calcium phosphate proportionately to the degree of hydrolysis of the glycerophosphate. The tissue was washed to remove free calcium ions and then autoradiographed to measure the amount of ^{45}Ca-calcium phosphate retained.

Phosphoserine hydrolase (EC 3.1.3.3) has been assayed by using a ^{14}C-labeled substrate and isolating the product by ion-exchange column chromatography (385).

5'-Nucleotidase (EC 3.1.3.5) has been assayed with a ^{32}P-labeled substrate by a charcoal adsorption method (386), with a ^{14}C-labeled substrate by ion-exchange column (387) and ion-exchange paper (71) methods, and with tritium-labeled substrates by an ion-exchange paper disk method (67b).

3'-Nucleotidase (EC 3.1.3.6) has been assayed by using ^{32}P-labeled substrates and separating residual substrates and products by charcoal adsorption (388) or solvent extraction (389) methods.

c. Sub-subclass 4, Phosphoric Diester Hydrolases. Several phosphodiesterases (EC 3.1.4.1) exist. They have a wide specificity, with respect to both substrate and position of bond fission, which varies from source to source. The current interest in metabolic regulation by cyclic AMP has resulted in the publication of a large number of assays for cyclic AMP phosphodiesterase, the enzyme responsible for the degradation of cyclic AMP in tissues. Most assays use tritium-labeled cyclic AMP as the substrate, probably because it is more readily available than other labeled forms.

Three methods of separating substrate and product have been described. These include paper chromatography (390–392); precipitation of the product AMP with $Ba(OH)_2$-$ZnSO_4$, followed by determination of residual substrate activity (393); and dephosphorylation of the product with 5'-nucleotidase, followed by ion-exchange separation of adenosine from residual cyclic AMP (394,395). In one assay the ion-exchange separation took place in the scintillation vial. Tritium-labeled cyclic AMP was absorbed by the anion-exchange resin beads, and its counts were lost by self-absorption, so that only the counts due to adenosine were recorded (394).

The phosphoric diester hydrolases include the important DNA- and RNA-degrading enzymes. Several simple and accurate radiometric assays, using ^{14}C-, tritium-, and ^{32}P-labeled substrates, have been described (23,48, 138,337,338,386). Such methods require measurement of the release of activity in an acid-soluble form. Alternatively, when speed and simplicity are more important than absolute accuracy, the activity of the residual acid-insoluble substrate can be measured by the filter paper disk method (25,43).

d. Sub-subclass 6, Sulfuric Ester Hydrolases. Most of the sulfatases of low specificity are assayed by the colorimetric estimation of phenols liberated by the hydrolysis of phenolic substrates. These methods are not suitable for the assay of some of the more specific enzymes, however, and radiometric assays have therefore been made.

A steroid sulfatase (EC 3.1.6.2) has been assayed by using tritium-labeled dehydroisoandrosterone sulfate as the substrate, and isolating the product by solvent extraction (21).

Choline sulfatase (EC 3.1.6.6) has been assayed by using ^{35}S-labeled choline-O-sulfate as the substrate, and separating product and substrate by paper chromatography (398).

3'-Phosphoadenosine-5'-phosphosulfate labeled with ^{35}S has been used as substrate in the assay of a sulfatase. The product was isolated by electrophoresis (168).

B. SUBCLASS 2, HYDROLASES ACTING ON GLYCOSYL COMPOUNDS

This large class of enzymes is well served by conventional techniques of assay. Radiometric methods are used to advantage when high sensitivity is required, as well as for assaying the degradation of macromolecules and substrates which give complex products.

Glycocerebroside glycohydrolases have been assayed by using ^{14}C- (399) or tritium- (400) glycosyl-labeled substrates and measuring the activity of the glycosyl product after removing the residual substrate with an organic solvent.

A glycogen hydrolase (EC 3.2.1.1–3) has been assayed by using ^{14}C-glycogen and following the formation of ethanol-soluble products (401).

A modification of this method has served for the clinical assay of amylo-1,6-glucosidase (dextrin-1,6-glucosidase, EC 3.2.1.33), which is important in the diagnosis of glycogen storage disease. The reaction is partially reversible and was assayed by measuring the incorporation of ^{14}C-glucose into glycogen, which was precipitated in aqueous ethanol (402).

The hydrolysis of tri(N-acetyl)chitotriose by lysozyme (EC 3.2.1.17) has been assayed using a tritium-labeled substrate. The di- and monosaccharide products were separated by paper chromatography (403).

A microassay of α-glucosidase (EC 3.2.1.20), suitable for use with very small biopsy samples, has been described (404). In this method, ^{14}C-labeled sucrose and maltose were used as substrates, and the product, glucose, was isolated by paper chromatography.

β-Galactosidase (EC 3.2.1.23) has been assayed by using ^{14}C-labeled o-nitrophenyl β-D-galactoside as the substrate, and isolating the labeled nitrophenol product by solvent extraction (405).

In an assay of the enzyme nucleosidase (EC 3.2.2.1), which catalyzes the hydrolysis of adenosine to adenine and ribose, 8-^{14}C-adenosine was used as the substrate. The labeled product was isolated by paper chromatography (291).

Nicotinamide adenine dinucleotide nucleosidase (EC 3.2.2.5) has been assayed by using NAD(7-^{14}C-nicotinamide) as the substrate and isolating the product, nicotinamide, by paper chromatography (341).

C. SUBCLASS 4, HYDROLASES ACTING ON PEPTIDE BONDS

Despite the fact that these enzymes are not easily assayed by conventional methods, very few radiometric assays have been described, chiefly because of the difficulty of obtaining suitably labeled substrates.

A microassay of lecine aminopeptidase (EC 3.4.1.1), using ^{14}C-leucinamide as the substrate, has been described (406). The product was isolated by paper chromatography.

A similar assay has been described for glycyl-leucine dipeptidase (EC 3.4.3.2), using a ^{14}C-glycine-labeled substrate (407).

Presumably both of these separations could be speeded up by using electrophoresis or thin-layer chromatography. Alternatively, it might be possible to decarboxylate the labeled amino acid product, enzymatically or chemically, and to trap and count the liberated $^{14}CO_2$.

Some proteolytic enzymes have been assayed by measuring the rate of production of acid-soluble products from labeled polypeptides. Only iodine-labeled polypeptides are freely available commercially, and in every assay, with one exception, radioiodinated polypeptide substrates have been used. The sensitivity of these assays may be reduced by free iodine produced on storage, but this iodine can easily be removed by ion exchange or gel filtration.

Pepsin (EC 3.4.4.1) has been assayed with iodine-labeled albumin and casein (408), papain (EC 3.4.4.10) with iodine-labeled albumin (409), other unspecified polypeptidases with iodine-labeled albumin (371) and ^{14}C-algal protein (410), and a pancreatic insulinase with iodine-labeled insulin (411).

D. SUBCLASS 5, HYDROLASES ACTING ON C—N BONDS OTHER THAN PEPTIDE BONDS

Asparaginase (EC 3.5.1.1) is much used in cancer therapy. Four radiometric assays have been described; all use ^{14}C-labeled asparagine as the substrate. The product, aspartic acid, was separated by high-voltage electrophoresis (412), or by ion-exchange paper chromatography using DEAE (80), ECTEOLA (81), or Amberlite SB2 (413) ion-exchange papers. In one assay (80), location of the product was facilitated by means of a picric acid marker which had a mobility only slightly greater than that of aspartic acid.

A similar DEAE-paper assay has been described for glutaminase (EC 3.5.1.2).

Arginase (EC 3.5.3.1) is a clinically significant enzyme. It has been assayed by using guanido-labeled arginine and measuring the $^{14}CO_2$ released when the product, urea, is decarboxylated with urease (97). An advantage of this method is that, unlike other assays, it is not subject to interference by the large concentrations of endogenous urea found in most enzyme extracts.

Three nucleoside deaminases (EC 3.5.4.–)—deoxyadenosine deaminase (414,415), deoxycytidine deaminase (415), and cytosine arabinoside deaminase (416)—have been assayed by paper chromatographic (414,416) and electrophoretic (415) methods, using ^{14}C- (414,415) and tritium- (415, 416) labeled substrates.

Deoxycytidylate deaminase (EC 3.5.4.12), an enzyme important in the synthesis of DNA, has been extensively studied. A large number of radiometric assays have been described. Separation techniques have included ion-exchange column (75,417–419) and ion-exchange paper (420) chromatography and electrophoresis (415). Substrates have been labeled with ^{14}C (75,415,417,420) and ^{32}P (418,419).

E. SUBCLASS 6, HYDROLASES ACTING ON ACID ANHYDRIDE BONDS

A few enzymes in this group have been assayed radiometrically.

Inorganic pyrophosphatase (EC 3.6.1.1) and ATPase (EC 3.6.1.8) have been assayed by using ^{32}P-labeled substrates and isolating the product, orthophosphate, by solvent extraction as the molybdate complex (421).

The enzyme GTPase has been assayed by using γ-^{32}P-GTP as the substrate, and either removing residual substrate by charcoal adsorption (422) or isolating labeled inorganic phosphate by solvent extraction as the molybdate complex (421).

Deoxynucleoside triphosphate nucleotidohydrolases have been assayed by using tritium-labeled thymidine triphosphate and isolating the product by DEAE-paper chromatography (424), and also by using α-^{32}P-labeled deoxyuridine triphosphate and isolating the product by paper chromatography (328). Ion-exchange paper methods are quicker and simpler than conventional paper chromatographic techniques.

4. The Assay of Enzymes of Class 4: the Lyases

Apart from their use in the assay of decarboxylases, radiometric methods have been applied only infrequently to this class of enzymes.

A. SUBCLASS 1, THE CARBON—CARBON LYASES

a. Sub-subclass 1, Carboxy-lyases. This group contains a very large number of enzymes, usually of high substrate specificity. At least 50 separate enzymes have been described, and about 25% of them have been assayed radiometrically. The release of $^{14}CO_2$ from carboxyl-labeled substrates is the obvious and most often used technique. When substrates labeled other than in the carboxyl group are used, solvent extraction, ion-exchange, derivative precipitation, and paper chromatographic separation methods may be employed. Some enzymes have also been assayed in the reverse direction by measuring $^{14}CO_2$ incorporation.

Many decarboxylases are pyridoxal phosphate dependent, and their assays are often complicated by concomitant chemical decarboxylation of the substrate. This chemical decarboxylation is markedly stimulated by phosphate buffers, which should, therefore, be avoided whenever possible (46). Similar degradation, particularly of histidine, can occur on storage, and the sensitivity of assays can be increased by removing these impurities before use (46,425). Some workers (112) who studied the ascorbic acid-catalyzed chemical decarboxylation of histidine observed different results when the reaction was followed by $^{14}CO_2$ release and by histamine formation methods. They suggested that the usefulness of the $^{14}CO_2$-release assay is limited because $^{14}CO_2$ release could apparently take place without concurrent formation of histamine. However, this result is probably an artifact caused by carrying out the histamine formation assay in a nitrogen atmosphere and the $^{14}CO_2$-release assay in an oxygen-containing atmosphere (46).

The chemical decarboxylation of some substrates may be stimulated by unknown factors released from boiled enzyme extracts. The use of unnatural D-isomers or potent enzyme inhibitors instead of the usual boiled enzyme extract is recommenced in controls when very low levels of activity are to be measured (46).

Paper chromatographic methods have been used in the assay of nucleotide sugar decarboxylation. A recent paper (129) has drawn attention to the fact that one solvent system, ethanol-ammonium acetate, pH 7.5, commonly used for the separation of nucleotide sugars, can cause excessive degradation of some of these compounds. Care should be exercised, therefore, in the use of this solvent system. The authors recommend a simple way of preventing this degradation.

The following is a selection of references to methods of assay depending on $^{14}CO_2$ release from carboxyl-labeled substrates.

Malonyl-CoA decarboxylase (EC 4.1.1.9), ref. 427.
Glutamate decarboxylase (EC 4.1.1.15), refs. 108, 109, 428–430.
Ornithine decarboxylase (EC 4.1.1.17), refs. 431, 432.

Histidine decarboxylase (EC 4.1.1.22), refs. 112, 426, 433–435.
Orotidine-5'-phosphate decarboxylase (EC 4.1.1.23), refs. 295, 436.
DOPA decarboxylase (EC 4.1.1.26), ref. 183.
Hydroxytryptophan decarboxylase (EC 4.1.1.28), ref. 184.
S-Adenosylmethionine decarboxylase (EC 4.1.1.–), ref. 437.

A selection of references to assays which have been carried out by following the reverse direction and measuring the incorporation of $^{14}CO_2$ is given below. Often the products are unstable oxo acids, which require stabilization either by formation of the 2,4-dinitrophenylhydrazones or by enzymatic conversion to malate or aspartate.

Pyruvate decarboxylase (EC 4.1.1.1), ref. 438.
Phosphopyruvate carboxylase (EC 4.1.1.31), refs. 113, 119.
Phosphopyruvate carboxylase (EC 4.1.1.32), ref. 439.
Ribulosediphosphate carboxylase (EC 4.1.1.39), refs. 113, 440.

Solvent extraction methods have been used for the following.

Histidine decarboxylase (EC 4.1.1.22), ref. 171.

DOPA decarboxylase (EC 4.1.1.26), refs. 441, 442.

Paper chromatographic methods have been used for the following.

Orotidine-5'-phosphate decarboxylase (EC 4.1.1.23), ref. 443.
DOPA decarboxylase (EC 4.1.1.26), ref. 192.
UDPglucuronate decarboxylase (EC 4.1.1.35), ref. 444.

Ion-exchange column techniques have been used for the assays of the following.

Histidine decarboxylase (EC 4.1.1.22), ref. 425.
Hydroxytryptophan decarboxylase (EC 4.1.1.28), ref. 445.

A very simple and rapid ion-exchange paper disk method has been used for the assay of methionine decarboxylase (EC 4.1.1.–), ref. 86.

Derivative precipitation methods have been used in the assay of histidine decarboxylase (EC 4.1.1.22), refs. 171, 446.

b. Sub-subclass 3, Keto-Acid Lyases. Three enzymes in this small group have been assayed radiometrically.

ATP citrate lyase (EC 4.1.3.8) has been assayed in the forward direction by converting the labeled acetyl-CoA product into acetyltoluidine, which was extracted in ether (159), and in the reverse direction by measuring the incorporation of $^{14}CO_2$ into citric acid (172).

Two lyases (EC 4.1.3.9,10) which catalyze the condensation of glyoxalate with acetyl-CoA and other acyl-CoA derivatives have been assayed with a ^{14}C-labeled glyoxalate substrate. In one assay, reaction was stopped by adding 2,4-dinitrophenylhydrazine and extracting the resulting hydrazone in ethyl acetate (447). In the other assay, the product was isolated by thin-layer chromatography on silica gel (448).

B. SUBCLASS 2, CARBON—OXYGEN LYASES

Only two enzymes in this large group have been assayed radiometrically.

Cystathionine synthase (L-serine dehydratase, EC 4.2.1.13), which forms cystathionine from serine and homocysteine, is clinically important because of its involvement in some inherited enzyme-deficiency diseases. It has been assayed by using ^{14}C-serine as the substrate, and isolating the product by ion-exchange column (449,450) or paper chromatographic (451) methods.

Cystathionase (homoserine dehydratase, EC 4.2.1.15), which degrades cystathionine to 2-oxobutyrate, ammonia, and cysteine, is also clinically important. It has been assayed by an ion-exchange column method with ^{14}C-cystathionine as substrate (450).

C. SUBCLASS 3, CARBON—NITROGEN LYASES

Two enzymes in this small subclass have been assayed radiometrically.

Phenylalanine ammonia-lyase (EC 4.3.1.5), which catalyzes the non-hydrolytic deamination of phenylalanine, has been assayed by a method in which the product, ^{14}C-cinnamic acid, was isolated by solvent extraction (452).

Adenylosuccinate lyase (EC 4.3.2.2) has been assayed by using ^{14}C-labeled adenylosuccinate and isolating the labeled product, fumaric acid, by solvent extraction (453).

5. The Assay of Enzymes of Class 5: the Isomerases

Although radiometric methods have been applied only infrequently to the assay of this class of enzymes, they have been found to be much more sensitive and simple than the conventional techniques.

A. SUBCLASS 1, RACEMASES AND ISOMERASES

Radiometric assays for three enzymes acting on carbohydrates and their derivatives have been described. UDPglucuronate-4-epimerase (EC 5.1.3.6) (454), N-acylglucosamine-2-epimerase (455), and N-acylglucosamine-6-phosphate-2-epimerase (455), have been assayed using ^{14}C-labeled substrates. Because these substrates and products are chemically very similar, their separation is difficult and is best achieved by high-voltage electro-phoresis, usually in borate buffers. In order to achieve satisfactory separation in the assay of N-acylglucosamine-6-phosphate-2-epimerase, it was necessary to dephosphorylate the substrate and product enzymatically before separat-ing the acylamino sugars by electrophoresis.

B. SUBCLASS 3, INTRAMOLECULAR OXIDOREDUCTASES

Isopentenylpyrophosphate isomerase (EC 5.3.3.2) catalyzes the conversion of isopentenyl pyrophosphate into dimethylallyl pyrophosphate. It has been assayed, using ^{14}C-labeled isopentenyl pyrophosphate, by solvent extraction methods (300,456,457).

6. The Assay of Enzymes of Class 6: the Ligases

A. SUBCLASS 1, LIGASES FORMING C—O BONDS

This subclass includes the important aminoacyl-tRNA synthetases, which catalyze the ATP-dependent formation of aminoacyl-tRNA from amino acids and specific acceptor transfer ribonucleic acids. These enzymes are highly specific, each enzyme being specific for only one natural amino acid. They have been assayed almost exclusively by radiometric methods.

The best method is to measure the incorporation of ^{14}C- or ^{3}H-labeled amino acids into an acid-insoluble form (40–43,141, and Section III.2).

Recently (46,47) attention has been drawn to the fact that minute traces of radiation decomposition products can be coprecipitated with macro-molecules. In macromolecule synthesis assays this will result in high blanks which reduce the sensitivity. These trace impurities can often be removed by simply filtering the labeled amino acid through a nitrocellulose membrane filter before use (46,47). The impurities which would otherwise have caused high blanks "stick" to the macromolecular membrane filter and are removed.

It is often thought that the controls used in aminoacyl-tRNA synthesis assays correct for the spurious incorporation of impurities. Unfortunately, this is not always so (47). If the blank reading is very small in comparison with the true incorporation, any failure of the control to compensate com-pletely for spurious incorporation will be unimportant. However, if the incorporation is small in comparison with the blank reading, errors in the controls can cause very serious overestimations of synthesis. Ways of over-coming or avoiding this problem have been discussed in detail elsewhere (47).

An alternative, slower but cheaper method of assaying these enzymes is to measure the amino-acid-dependent exchange of ^{32}P-pyrophosphate with ATP, which they catalyze (40,141,458). The ATP has been separated from residual pyrophosphate by charcoal adsorption (40,141) or by rapid paper

(458) or thin-layer (459) chromatography. The reaction could also be followed by measuring the amino-acid-dependent release of pyrophosphate from terminal-labeled ATP.

These ATP-exchange assay methods are subject to interference, however, from contaminating enzymes which react with ATP or pyrophosphate and hence are less satisfactory than the macromolecule precipitation assays.

B. SUBCLASS 2, LIGASES FORMING C—S BONDS

Only one enzyme in this small group, acyl-CoA synthetase (EC 6.2.1.3), has been assayed radiometrically. In one assay (460), ^{14}C-palmitate was used as substrate, and the product, palmitoyl-CoA, was precipitated with trichloracetic acid and isolated by a filter paper disk technique. In a second assay (149), a coupled enzyme method was used, the inactive product, palmitoyl-CoA, being coupled enzymatically to tritium-labeled carnitine. The labeled product, palmityl carnatine, was isolated by solvent extraction.

C. SUBCLASS 3, LIGASES FORMING C—N BONDS

Protein and polypeptide synthesis has been assayed by measuring the incorporation of activity from ^{14}C-, tritium-, and ^{35}S-labeled amino acids (32–35), or from labeled aminoacyl-tRNA (36,37), into a form insoluble in hot trichloracetic acid (see Section III.2). During processing, the precipitates must be heated in 5% trichloracetic acid to dissolve any labeled aminoacyl-tRNA which might either be present as unused substrate or be formed from labeled amino acids by contaminating enzymes.

Polyphenylalanine synthetase (EC 6.3.2.–) catalyzes the polyU-directed synthesis of polyphenylalanine. It has been assayed using ^{14}C-labeled phenylalanyl-tRNA as substrate (36,464–466). When synthetic polyU is used to direct the synthesis of polyphenylalanine, the rate of reaction increases with increasing molecular size of the polyU. A recent publication (464) has reported a wide variation in the molecular weights of the polyU available from various commercial sources. As a result, the incorporation of activity in assays depends on the source of the polyU. In order to get reproducible results, it was found necessary to fractionate polyU on Sephadex G-200 before use.

Proline polypeptides are soluble in the trichloracetic acid precipitant normally used in polypeptide synthesis assays. Proline polypeptide synthesis was assayed, therefore, by a method in which the product was isolated by paper chromatography, instead of by the customary precipitation techniques (466).

A peptide synthetase, D-alanylalanine synthetase (EC 6.3.2.4), has been assayed by using ^{14}C-labeled D-alanine as substrate, and separating the products by electrophoresis or paper chromatography (462).

UDP-N-acetylmuramoyl-L-alanyl-D-glutamyl-L-lysine : D-alanyl-D-alanine ligase (ADP) (EC 6.3.2.10) has been assayed by following the incorporation of ^{14}C-D-alanyl-D-alanine (463). The product was isolated both by charcoal adsorption and by paper chromatographic methods.

Three other C—N ligases have been assayed radiometrically.

CTP synthetase (EC 6.3.4.2), which catalyzes the amination of UTP, has been assayed with ^{14}C-UTP as substrate. Residual substrate and product were enzymatically dephosphorylated, and the resulting nucleosides were then separated by ion-exchange column chromatography (467).

Adenylosuccinate synthetase (EC 6.3.4.4) has been assayed by measuring the incorporation of ^{14}C-aspartate. The product was isolated by ion-exchange column chromatography (453).

Arginosuccinate synthetase (EC 6.3.4.5), which catalyzes the formation of arginosuccinate from citrulline and aspartate, has been assayed by a coupled enzyme method using carbamoyl-labeled citrulline as the substrate (96). The product was enzymatically degraded to arginine with arginosuccinase; then the arginine was degraded to urea with arginase, and finally the urea was decarboxylated with urease, liberating $^{14}CO_2$.

D. SUBCLASS 4, LIGASES FORMING C—C BONDS

Three enzymes in this small subclass have been assayed, almost exclusively by measuring the fixation of $^{14}CO_2$.

Pyruvate carboxylase (EC 6.4.1.1) has been assayed by following the incorporation of $^{14}CO_2$ (115,119,120,135,368) or 2-^{14}C-pyruvate (469) into oxaloacetic acid. Because oxaloacetic acid is unstable, it is usually stabilized by conversion to the 2,4-dinitrophenylhydrazone (115) or to malate (119,120) or aspartate (469). The simplest method appears to be that of Keech and Barritt (115), who stopped the reaction with acidified 2,4-dinitrophenyl-hydrazine, removed residual $^{14}CO_2$ by gassing, and then spotted aliquots onto paper disks which were dried before being counted.

Acetyl-CoA carboxylase (EC 6.4.1.2) (470) and propionyl-CoA carboxylase (EC 6.4.1.3) (471) have also been assayed by $^{14}CO_2$-fixation methods.

V. SOME IMPORTANT CONSIDERATIONS IN RADIOMETRIC ENZYME ASSAYS

Despite the many advantages of radiometric methods of enzyme assay, several problems are associated with their use. These have caused trouble in the past because of the failure of users to recognize and to allow for them.

1. Factors Affecting the Sensitivity of Enzyme Assays

Although radiometric enzyme assays are widely used because of their high sensitivity, inevitably there has been an increasing demand in special cases for even greater sensitivity. This demand will arise if only very small amounts of biological material are available, as with tissue biopsy samples, and also if very low levels of enzyme activity are present in biological samples which are available in relatively large amounts.

McCaman (18) has defined the "limit sensitivity" of an analytical method as being the amount of material which will give rise to a signal (e.g., counts per minute) which is twice that obtained for the blank. An obvious way of increasing the sensitivity (i.e., decreasing the limit sensitivity) is to reduce the blank reading.

In radiometric assays high sensitivity will be achieved by employing the most sensitive detection method, liquid scintillation counting, and by using scintillation vials and scintillants which give minimal background counts. Contributions to the blank reading will also be made by analytical procedures which do not result in the complete separation of substrate and product, by impurities in the substrate, and by concomitant nonenzymatic reactions. These problems and methods of reducing their effect are discussed in detail in ref. 46 and are summarized in the following sections.

A. HIGH BLANK READINGS DUE TO INADEQUATE SEPARATION PROCEDURES

High blank readings will result if the analytical procedure fails to give complete separation of product and substrate. Such blank readings can, of course, be reduced only by improving the separation procedure and this will probably result in less than 100% recovery of the product. If necessary, this can be corrected for by the double-label technique in which an aliquot of product, labeled with a nuclide different from that used to label the substrate, is added to the reaction mixture when reaction is stopped. A sample of pure product is then isolated, and the activity due to each of the two nuclides is determined. The amount of activity due to the second nuclide may then be used to correct for losses during isolation and purification of the product.

B. HIGH BLANK READINGS DUE TO THE PRESENCE OF IMPURITIES IN THE SUBSTRATE

The sensitivity of enzyme assays can be reduced by impurities in substrates. Often when low levels of enzyme activity are being assayed, and high sensitivity is required, only a very small proportion of the substrate is converted to product during the assay. As a result, the amount of impurity necessary to cause a "high" blank is usually a quite small (0.25% or less) proportion of the total substrate activity and hence is well within the manufacturer's specifications. Often it can be measured only by carrying out the enzyme assay of interest, and frequently only the user can say whether or not a labeled substrate meets the requirements for a specific assay.

An impurity which causes a high blank might be a product of radiation or chemical decomposition on storage, or might have been present in the freshly prepared material. All radiometric assays require the separation of substrate and product. A high blank, due to the presence of an impurity in the substrate, means that the impurity is included with the product in the separation procedure. An obvious and simple way of reducing the blank reading in such circumstances is to subject the substrate to the separation procedure to be used in the enzyme assay. In this way, the impurities causing the high blank will be removed, and the pretreated substrate may then be used in the enzyme assay. This technique has proved satisfactory in many enzyme assays (46,47,472).

C. HIGH BLANK READINGS CAUSED BY CONCOMITANT NONENZYMATIC REACTIONS

Many enzyme assays, radiometric or otherwise, suffer from the presence of concomitant nonenzymatic reactions which can cause high blanks. These blank readings, unlike those due to the presence of impurities, do not remain constant throughout an assay, but increase with time. Unless correction is made for these nonenzymatic reactions, their presence can cause the enzymatic activity of the extract to be overestimated.

In most assays compensation is made by carrying out control experiments in the presence either of boiled enzyme extracts or of the undenatured enzyme and a potent irreversible inhibitor (105). The latter method, however, will not be satisfactory if the added inhibitor inhibits both the enzymatic and the nonenzymatic reactions. The use of boiled enzyme extracts can sometimes lead to erroneous conclusions when these extracts release unknown factors which catalyze the nonenzymatic reactions. In some assays it has been found that controls have given higher readings than the actual enzyme assay samples (46). Caution should be exercised, therefore, when boiled enzyme extracts are used in controls.

If any enzyme assay is complicated by the presence of a concomitant nonenzymatic reaction, the need to perform many control experiments is wasteful of both time and material. By selecting conditions where the effect of nonenzymatic reactions is minimal, it is often possible to simplify the assay and at the same time to markedly increase its sensitivity. No detailed study of methods of reducing such interference has yet been made, but the evidence available suggests that the nature of the buffer, the presence of trace metal ions, the presence of oxygen, and the pH can be important. Phosphate buffers, in particular, seem to be major contributors to these problems, and their use should therefore be avoided whenever possible.

2. Selection of Optimum Conditions for Assay

Because many factors must be considered when setting up an enzyme assay, it is difficult to define unequivocably the optimum conditions for all enzyme assays. The most important factor affecting the choice of conditions is the affinity of the substrate for the enzyme. A second factor is the choice of the optimum volume for the assay reaction mixture. These factors are discussed in detail in ref. 46 and are summarized in the following sections.

A. ENZYME ASSAYS AT HIGH SUBSTRATE CONCENTRATION

Many enzymes are assayed at high substrate concentration. This simplifies the assay procedure by giving zero-order kinetics over a wide range and reduces the possibility of interference from other enzymes or inhibitors which might be present in crude extracts and which could compete for or with the substrate. The rate of formation of products under these conditions (substrate concentration $\gg$ the Michaelis constant, K_m) is a maximum when expressed as micromoles of product formed per unit time; but, when expressed as the proportion of substrate utilized per unit time, the rate decreases as the substrate concentration increases. In this zero-order region, increasing the substrate concentration does not increase the rate of the enzymatic reaction, and carrying out an assay in this region has five important consequences.

1. If the substrate contains an impurity which separates with the product, carrying out the assay at very high substrate concentration will increase the blank reading without raising the rate of formation of products.

2. If the separation technique used in the assay fails to give 100% separation of substrate and product, carrying out the assay at very high substrate concentration will increase the blank without increasing the rate of formation of products.

3. If a concomitant nonenzymatic reaction occurs, its rate is likely to be proportional to the substrate concentration. Carrying out the assay at very high substrate concentration will raise the blank reading by increasing the rate of the nonenzymatic reaction without increasing the rate of the enzymatic reaction.

4. When very low levels of enzymatic activity are assayed, only a small fraction of the substrate will be utilized during the assay. This is wasteful when rare or expensive substrates are used.

5. At high concentrations substrate inhibition may occur.

B. ENZYME ASSAYS AT LOW SUBSTRATE CONCENTRATION

Enzymes are sometimes assayed at very low substrate concentration. Under these conditions (substrate concentration $\ll K_m$), the rate, expressed as the proportion of substrate converted to product per unit time, reaches a maximum value. This can be important economically when rare or expensive substrates are used. There are, however, some disadvantages.

1. The rate of reaction, expressed as micromoles of product formed per unit time, is markedly reduced at low substrate concentrations where the enzyme is not saturated with substrate. This reduces the sensitivity of the assay.

2. The kinetics are more complex, and accurate measurement of the initial rate of the reaction is more difficult, than at high substrate concentration.

3. Interference from endogenous substrate and inhibitors in the enzyme extract is more pronounced than at high substrate concentration.

4. Interference from contaminating enzymes, which can compete for the substrate, is more pronounced at low substrate concentration.

5. The presence in the enzyme extract of impurities which can react stoichiometrically with the substrate can cause marked interference at low substrate concentration (473).

C. SELECTING THE OPTIMUM VOLUME FOR THE ASSAY REACTION MIXTURE

If the amount of enzyme extract and the concentrations of all other reactants are fixed, varying the volume of the reaction mixture will not affect the rate of the enzymatic reaction when it is expressed as micromoles of product formed per unit time. When the rate is expressed as the percentage conversion per unit time, however, it will decrease as the volume (and hence the amount of substrate) increases. When low levels of enzymatic activity are assayed, maximum sensitivity will be achieved when there is a minimum increase in volume as the other reactants are added to the enzyme sample. In other words, maximum sensitivity will be achieved when the concentration of enzyme in the reaction mixture is as high as possible. This has two consequences.

1. If only a limited amount of enzyme extract is available, maximum sensitivity will be achieved by carrying out the assay in the smallest possible volume. This is well illustrated by an assay of choline acetylase, which was carried out in a volume of only 1 μl, and which was sufficiently sensitive to permit the measurement of the enzymatic activity of individual cells (166).

2. If the level of enzymatic activity in the extract is low, but reasonable amounts of extract are available, no advantage will be gained by carrying out the assay in a very small volume. In fact, this will only increase the technical difficulties. Optimum conditions will be attained by taking as large a sample of enzyme extract as is appropriate for the analytical techniques used, and adding the substrate and other reactants in the minimum possible volume so that the concentration of enzyme in the reaction mixture remains as high as possible.

D. SELECTING THE OPTIMUM SUBSTRATE CONCENTRATION

When low levels of enzymatic activity are to be measured, conditions should be chosen to ensure that the rate of formation of products is as high as possible, and that blanks are at a minimum. The first condition requires high substrate concentrations so that the enzymes are saturated with substrate. However, high substrate concentrations have been shown (Section V.2.A) to magnify the effect of the three factors which cause high blanks. Blanks are minimal when the proportion of substrate converted to product per unit time is maximum. This depends on the rate of the enzymatic reaction, the concentration of substrate, and the volume of the reaction mixture. Once the optimum assay volume has been selected (Section V.2.C), the proportion of substrate converted to product, for a particular enzyme activity, will depend only on the substrate concentration and will be maximum at low substrate concentration.

The substrate concentration at which maximum sensitivity will be reached depends, therefore, on two conflicting factors and the extent to which each applies to the assay in question. Obviously optimum sensitivity will be achieved at some intermediate substrate concentration, and it has been shown (46) that, if equal weight is given to all factors, maximum sensitivity will result from using a substrate concentration equal to the Michaelis constant, K_m.

However, in most enzyme assays maximum sensitivity is not needed, and the chief requirement is usually for good kinetics, which can best be achieved by carrying out assays at substrate concentrations $> K_m$. Even when maximum sensitivity is not required, the accuracy of assays will be increased by reducing the blank readings. The use of very high substrate concentrations should, therefore, be avoided.

3. How High Should the Specific Activity of the Substrate Be?

It is often mistakenly assumed that substrates of high specific activity are required in radiometric enzyme assays in order to achieve high sensitivity. In general, however, this is not so.

When the optimum conditions for an assay have been determined (Section V.2), the substrate should be used at a specific activity sufficiently high to ensure that the product gives satisfactory counting characteristics. It is unnecessarily wasteful and expensive to use specific activities higher than are required to meet these conditions.

The amount of activity required will depend on the proportion of substrate converted to product, on the nuclide, and on the sensitivity of the counting equipment available. If a liquid scintillation counter is available, 0.02 μCi of a ^{14}C-labeled substrate, per assay sample, is sufficient to give about 300 cpm for each 1% reaction. For tritium-labeled substrates, which are counted at lower efficiency, it will be necessary to use about 0.05 μCi per assay sample to achieve the same count rate.

The specific activity necessary to give this amount of activity depends on the volume of sample taken for analysis and on the substrate concentration (which, following the recommendations in Section V.2.D, should be equal to or slightly higher than K_m). This can easily be calculated (46).

Most enzyme assays are carried out at substrate concentrations of about $10^{-3}M$, and with assay sample volumes in the range of 0.1–1.0 ml. Calculations will show, therefore, that ^{14}C-labeled substrates with specific activities in the range 0.02–2.0 mCi/mM will meet the requirements of most radiometric enzyme assays. High-specific-activity substrates are necesssary to achieve maximum sensitivity only in the assay of enzymes which have very low Michaelis constants ($<10^{-5}M$), in assays which require the use of very small samples, and in assays which result in very low ($<1\%$) conversion of substrate to product.

Because labeled substrates are now available at very high specific activities, it is often claimed that the sensitivity of radiometric enzyme assays can be increased several-hundred-fold by using such substrates. To use these substrates at the same concentration as is employed with the low-specific-activity substrates would also increase the cost of the assays several-hundred-fold. Since few workers can afford to do this, the use of high-specific-activity substrates is nearly always accompanied by a parallel decrease in substrate concentration. Unless the enzyme has a very high affinity for its substrate (i.e., a very low K_m), this decrease in substrate concentration will result in a corresponding decrease in the rate of reaction. The overall result is that, as the specific activity increases, the rate of reaction decreases, so that there is very little change in the rate of product formation, expressed as counts per minute of product per unit time.

Also, high-specific-activity compounds undergo radiation decomposition more rapidly than do the equivalent low-specific-activity compounds. If these radiation decomposition products contribute to the blank reading, the use of high-specific-activity substrates may result in a significant decrease in the sensitivity of the assays.

4. How Accurately Should the Specific Activity of the Substrate Be Known?

All methods of determining the specific activities of labeled compounds are subject to the usual errors in the measurement of both their radioactivity and their concentration. The accuracy of the quoted value of the specific activity is important if it is used directly or indirectly in the calculation of enzymatic rates.

The specific activity of most commercially available high-specific-activity materials should be known to within 5 to 10% ,which is adequate for the majority of enzyme assays. It should be noted, however, that a recent report (474) has warned that errors of up to 500% have been found in the quoted specific activities of tritium-labeled compounds obtained from some commercial sources.

However, the majority of enzyme assays use low-specific-activity substrates which are made by diluting high-specific-activity materials with large amounts of inactive substrates. In these circumstances the specific activity of the diluted material is determined by the purity of the inacitve substrate and the purity and activity of the active substrate. Errors in the specific activity of the high-specific-activity substrate will have little effect on the specific activity of the diluted material.

It is very easy to draw false conclusions from results obtained using incorrect specific activities, and users should always ascertain the reliability of the specific activity value when this may be important. A recent example of false conclusions (a large and serious isotope effect) being drawn for this reason is discussed in ref. 92.

5. How Pure Should the Substrate Be?

It would be an advantage always to work with absolutely "pure" reagents, but this ideal is unattainable. To attempt to obtain compounds of unnecessarily high purity can be very uneconomical. Labeled compounds present more purity problems than unlabeled ones because the worker has to contend with the presence of both active and inactive impurities and with the problem of radiation decomposition. Many of these problems can and should be overcome by making and storing labeled compounds, to be used as enzyme substrates, at as low a specific activity as is consistent with their intended application.

If it is impossible to purchase labeled substrates at the appropriate low specific activity, it is advisable to reduce the specific activity by adding the required amount of inactive substrate to the labeled compound immediately on receipt. Many workers purchase high-specific-activity substrates and store them, often for a long time. The labeled substrates are then diluted with unlabeled substrates immediately before use in an assay. This practice is unadvisable because not only does it cause unnecessary radiation decomposition on prolonged storage, but also it results in the use in the assay of a low-specific-activity substrate contaminated with high-specific-activity impurities. This can magnify the effect of the impurities.

Because radiation decomposition is an inescapable effect of radioactivity, even the purest labeled compound is likely to contain a very small amount of impurity. Although the rate of radiation decomposition can be markedly decreased by reducing the specific activity, it cannot be stopped completely. Some compounds which are particularly susceptible to radiation decomposition may need repurification at intervals to reduce blank readings. Simple methods of repurification of a large number of substrates have been published elsewhere (46,47).

It should be stressed that the term "satisfactory purity" is meaningless except in the context of the use for which the material is intended. For example, in an insulin fat-pad assay using 1-^{14}C-D-glucose which had undergone about 1% decomposition on storage, a "no-tissue blank," corresponding to 0.1% of the total activity added, resulted from the presence of certain radiation products (475). This represents a high blank for the fat-pad assay and severely reduced the sensitivity. The problem was overcome by producing low-specific-activity material, which gave a blank one fiftieth of the original and did not increase on storage for several months. However, the labeled glucose which was unsatisfactory for the fat-pad assay could well have been used in a glucokinase assay by the DEAE disk method. The ionic radiation decomposition products would be retained on the DEAE disk and would contribute to the blank reading, but 0.1% would be an acceptable blank in this assay. Hence glucose which is too impure to be used in one enzyme assay is pure enough for another.

6. Isotope Effects

An isotope effect may be defined as any difference in chemical or physical behavior between two compounds which differ only in the isotopic composition of one or more of their constituent elements. These effects depend on the mass ratio, and the large difference in mass between hydrogen and tritium makes large isotope effects possible with tritium-labeled compounds. Many examples have been described (476). Most reported isotope effects have been concerned with differences in reaction rate constants for two isotopically different species of the same compound. These rate differences may be classified into primary and secondary isotope effects.

Primary isotope effects are demonstrated by the difference in reaction rates of two isotopic species of the same compound, because of the formation or the breaking of isotopically different bonds in a rate-determining step. As might be expected from the large mass difference between hydrogen and tritium, quite large isotope effects have been reported for this type of reaction.

Secondary isotope effects can arise when isotopically different bonds are not formed or broken in a rate-determining step, and they should be very much smaller than primary isotope effects. It has been concluded that secondary isotope effects, even with tritium, are likely to be less than 5% and therefore will probably lie well within the limits of experimental error in enzyme assays (92).

Many allegations of secondary isotope effects have been based on the incorrect interpretation of results obtained with labeled substrates. Such spurious isotope effects are discussed in detail elsewhere (92).

In discussing isotope effects during enzyme reactions, most authors refer to the effect on the maximum velocity, V_{max}, and most assays are carried out at substrate concentrations high enough to cause saturation of the enzyme with substrate, so that V_{max} is measured. Isotope effects can also influence the binding of the substrate to the enzyme, and hence the Michaelis constant, K_m. With many enzymes, hydrogen bonds are involved in this binding, and one might predict relatively large isotope effects for these enzymes when tritium- or deuterium-labeled substrates are used. Several examples of such isotope effects have been reported for deuterium-labeled substrates. Three flavin-dependent enzymes were found to have a smaller K_m with the isotopically normal substrate, whereas three NAD- or NADP-dependent enzymes had a smaller K_m with deuteriated substrates (477). It was shown, for example, that during the reaction of yeast alcohol dehydrogenase V_{max} for isotopically normal ethanol is, as might be expected, 1.8 times as great as V_{max} for deuteriated ethanol. At the same time K_m for ethanol is 2.3 times as great as K_m for deuteriated ethanol. As a result, at high substrate concentrations, where the enzyme is saturated with substrate, there is a large isotope effect, and the deuteriated substrate reacts more slowly than the isotopically normal substrate. At low substrate concentrations, however, the enzyme is more highly saturated with the deuteriated substrate than with the isotopically normal substrate. As a result, the deuteriated substrate reacts faster than the isotopically normal substrate under these conditions.

Although the occurrence of such an isotope effect has not yet been reported for a radioisotopically labeled substrate, it remains a possibility which must not be overlooked.

The occurrence of an isotope effect does not detract from the usefulness of radiochemical enzyme assays. If the results are to be compared with those obtained by conventional methods, however, a correction factor may need to be used.

7. Problems Associated with the Handling of Small Masses and Very Dilute Solutions

An important feature of the usefulness of radiochemicals (particularly at high specific activity) is that they may be detected and measured in smaller quantities and at greater dilution than by conventional methods. The observed behavior of radiochemicals in these small quantities sometimes differs unexpectedly from the behavior of the identical nonradioactive chemicals in macroscopic amounts. This is not due solely to the effects of ionizing radiations, because under appropriate conditions very small quantities of nonradioactive substances may also show deviations from normal macroscale behavior; rather, the ease of detection of radiochemicals makes apparent abnormal behavior which would otherwise go undetected. Most of these difficulties can be overcome by the use of added nonradioactive carrier. This is achieved in most radiometric enzyme assays by diluting the labeled substrate to a suitable low specific activity. Nevertheless, the very high sensitivity of present-day counting techniques can mean that very small amounts of these low-specific-activity substrates are handled in routine assays, and these peculiarities can reappear.

In very dilute solutions, surface effects can have a very marked influence on chemical behavior, and it is possible that the quantity of solute adsorbed onto a surface from a very dilute solution may be a significant proportion of the whole. Glass vessels are usually suitable for containing very dilute solutions of organic compounds but are not always satisfactory for very dilute solutions of inorganic ions except at low pH.

^{32}P-Orthophosphate is used in many enzyme assays, and aqueous solutions of the carrier-free substrate are supplied commercially in dilute hydrochloric acid solution at pH 2 to 3. Such solutions are stable and no adsorption losses occur; however, if the user adjusts the pH to a physiological level without first adding a small amount of inactive phosphate as a hold-back carrier, adsorption will occur and severe losses will take place onto the walls of the vessel and onto any particulate matter which may be present.

It has also been shown that, during the scintillation counting of very small amounts of labeled substrates, irreversible adsorption onto the walls of the glass scintillation vials can cause low count rates (478). As the counting geometry changes from 4π (homogeneous solution) to essentially 2π (flat surface), counting losses of up to 50% occur. This effect can be prevented by adding a small amount of inactive substrate before adding the active solution to the counting vial. Alternatively, plastic vials, which do not adsorb the labeled compound, may be used.

When a compound is analyzed by paper chromatography or paper electrophoresis, a very small amount of the compound is irreversibly adsorbed by the cellulose and is left as a spot on the origin and a streak behind the main spot. The amount of material irreversibly adsorbed in this way is extremely small and cannot be detected by the usual methods; but if a high-specific-activity labeled compound is used, this small weight of material may represent a large proportion of the total weight and activity. The effect of this irreversible adsorption can be avoided by applying a spot of carrier to the origin before applying the active material, so that, in effect, the specific activity of the material is considerably reduced. The small weight of material irreversibly adsorbed onto the paper will then represent only a negligibly small proportion of the active material applied.

Similar adsorption problems can occur with denatured protein, adsorbent charcoal and alumina, and ion-exchange materials. They can similarly be avoided by adding inactive carrier material to the active sample, or by pre-washing the adsorbents or ion-exchange materials with inactive carrier.

Another possible disadvantage of working with very small amounts of material is that, when reactions are followed by precipitating the radioactive product, it is often difficult to ensure complete precipitation, and there is a possibility of relatively large losses of material during the subsequent washing. Also, in reactions which are followed by solvent extraction, adsorption, or ion-exchange methods, it is often difficult to obtain 100% recovery. In all these situations, the addition of carrier before processing will usually enable satisfactory recovery to be achieved. However, if it proves impossible to obtain satisfactory recovery in this way, the specific activity of the recovered material can be measured, and the total activity calculated from the known weight of carrier added. Alternatively, double-label techniques may be used to correct for low recovery.

Because of the high sensitivity of radiometric enzyme assays, many can be carried out on a very small scale, the reaction mixture occupying a volume of 50 μl or less. This can make mixing and sampling difficult, and evaporation from such a small volume can cause quite serious errors. These problems have been overcome by mixing the reactants at low temperature, and sealing them in a small capillary tube before initiating the reaction by immersing the tube in a thermostatted bath at the appropriate temperature. Details of this technique are given in ref. 479. McCaman (166) has also described a technique suitable for following reactions on the 1-μl scale.

8. Special Effects with Tritium-Labeled Substrates

Tritium-labeled substrates are often used in enzyme assays. Because of certain special effects which occur only with tritium-labeled compounds, however, spurious isotope effects have been inferred from rate measurements obtained by using these compounds. These special effects and ways of detecting and eliminating them are discussed in ref. 92 and are summarized in the following sections.

A. FALSE RATES DUE TO THE LABILIZATION OF TRITIUM

When tritium-labeled substrates are used, the tritium label is nearly always serving as an auxiliary label for carbon, and the assumption is made that the label is stable under the experimental conditions. If a proportion of the label is lost during the reaction or isolation of products, any measurement of the radioactivity of the product will result in an apparently low rate of reaction.

Such labilization of tritium can occur in several ways, for example, chemically or enzymatically during the reaction, chemically during the separation or isolation of products, or during storage in solution before use.

Suppliers of tritium-labeled compounds remove readily labile tritium under relatively mild conditions during manufacture. Nevertheless, because the conditions during the enzymatic reaction and during the isolation of products may be more drastic, a loss of tritium can occur from the substrate the product, or both. Some tritium-labeled compounds also slowly "leak" tritium into solution on storage. Although much information is available about the lability of tritium in labeled compounds (476,480), little attention has been given to its lability in compounds stored in aqueous solution at neutral pH. A recent publication (481) has shown that several tritium-labeled compounds slowly exchanged tritium with the solvent on prolonged storage under neutral conditions, but did not undergo serious radiation decomposition. If these compounds were used in tritium-release assays, the tritium released on storage could cause a high blank reading. Therefore solutions of tritium-labeled compounds which are to be used in such assays should first be lyophilized to remove any tritiated water formed by exchange. This exchange on storage also results in a drop in the specific activity of the labeled compound, and this too should be allowed for when the rate of reaction is calculated from the rate of tritium release. The short (12.26 years) half-life of tritium means that the normal radioactive decay process will cause the specific activity of tritium-labeled substrates to decrease at a rate of about 5% a year. Although this is unimportant for most applications, it should be allowed for when "old" tritium-labeled substrates are used in assays.

A few instances of enzymatic labilization of tritium have been reported. For example, Evans and his coworkers (482) found that tritium was lost from generally labelled L-amino acids when they were treated with D-amino acid oxidase, and that no degradation of the amino acid occurred during this treatment (482).

Chemical or enzymatic labilization of tritium during reaction can easily be detected by distilling a sample of water from the reaction mixture and determining its tritium content. If the extent of tritium loss is not too great, a correction may be applied. Because variations in the labeling pattern of generally labeled tritiated compounds can occur from batch to batch, it is better to use specifically labeled, rather than generally labeled, substrates in order to make such corrections easier and more accurate.

Alternatively, because the extent of labilization varies from position to position in the molecule, it may be possible to select a position of labeling from which losses are minimal. For example, in the reaction referred to earlier (482), tritium was lost from the α-position of the L-amino acids. The use of L-amino acids labeled in positions other than the α-position should result in a lower rate of labilization.

Tritium loss during the isolation of products may be detected by isolating a sample of the product, measuring its specific activity, and then resubmitting it to the isolation procedure and measuring its specific activity again. If labilization has occurred, the specific activity will have decreased. If the loss of tritium is significant, an alternative procedure must be found.

Where necessary, the stability of the tritium label may be checked by comparing the results with those obtained with ^{14}C-labeled substrates.

B. INTRAMOLECULAR MIGRATION OF TRITIUM, THE "NIH SHIFT"

Many enzyme reactions have been rapidly and simply assayed by measuring the rate of displacement of tritium, into solution, by the substituent. Some of these assays have been complicated by an internal migration of tritium, which results in the retention of tritium in the product instead of its release into solution. This is known as the "NIH shift," and its occurrence results in an apparently low rate of reaction. This shift, its detection, and the methods of overcoming it were discussed in Sections III.4.A and IV.1.

C. ERRORS RESULTING FROM THE USE OF SUBSTRATES IN WHICH THE POSITION OF LABELING IS NOT EXACTLY AS SPECIFIED

When a reaction is followed by measuring the rate of tritium release from a specifically labeled substrate, an apparently low rate of reaction will result if the position of labeling is not exactly as specified.

In some reactions it may also be necessary to know the stereospecificity of the tritium label. For example, in a study of the enzymatic hydroxylation of proline, it was found that the use of cis-L-proline-4-^3H resulted in the complete retention of tritium label during hydroxylation, whereas complete loss of tritium occurred with trans-L-proline-4-^3H (483).

Low rates resulting from these labeling errors may be detected, and corrected for, by isolating the product and determining its radioactivity.

This effect and the "NIH shift" can be differentiated by determining the position of labeling in the substrate, by degradation, nuclear magnetic resonance spectroscopy, or other methods. In reactions where the position of labeling is correct but the stereospecificity is wrong, no tritium will be released. This situation can easily be distinguished from the "NIH shift," which is not likely to result in complete retention of tritium.

D. LOW RESULTS DUE TO SELF-ABSORPTION LOSSES DURING THE LIQUID SCINTIL-
LATION COUNTING OF TRITIUM-LABELED COMPOUNDS

The low energy and short range of tritium β-particles make it impossible to measure the activity of tritium-labeled substances with very high counting efficiency. The counting efficiency is often less than 30%, and varies with the nature of the labeled substance and the scintillant. In order to determine the absolute activity, or even to compare the activities of samples, it is necessary to know the counting efficiency. Since this can vary from sample to sample, the efficiency is determined for each sample, either by internal or external standardization, or by the channels ratio method.

Many enzymes are assayed by measuring the incorporation of activity into an insoluble product (Section III.2). The possibility of self-absorption losses arises when such precipitates are counted, and it should not be assumed that counting efficiency determinations compensate for these losses.

The magnitude of any self-absorption losses will increase with increasing particle size of the precipitates. These losses are not likely to be appreciable with precipitates obtained by the filter disk method, which are evenly distributed as a very thin film over the surface of the fibers of the filter disk, but can be important in assays in which the macromolecules are filtered after precipitation. In assays of this type large and variable self-absorption losses can occur. For example, in experiments where the incorporation of tritiated thymidine into DNA was followed by such a method, it was shown that self-absorption losses, caused by the coprecipitation of inactive RNA with the labeled DNA, resulted in an apparently low incorporation of activity (484). Digestion of this RNA in alkali, before the precipitation of the DNA, resulted in a 50% increase in the measured activity. A duplicate experiment with ^{14}C-labeled DNA, however, showed no self-absorption losses. Although only one report of such a loss has been made, the possible occurrence of self-absorption losses should be considered in assays in which insoluble tritium-labeled compounds are determined by liquid scintillation or other methods. Their occurrence can be detected either by comparing the activity of a precipitated sample with that of a duplicate sample which has been solubilized before counting (484), or by adding varying amounts of inactive macromolecule before precipitation and counting, in order to determine at what point self-absorption losses become important. An alternative (and simpler) method is to prepare duplicate or triplicate samples in which the macromolecular product is precipitated from different volumes of reaction mixture. If self-absorption losses are insignificant, there will be a linear relationship between the counts per minute and the sample volume (and hence the weight of precipitate). If a nonlinear plot is obtained, its shape will indicate the extent of the self-absorption losses.

Many workers using liquid scintillation methods to measure the activity of precipitates have tried to correct for self-absorption losses by calibrating against a standard prepared by drying a known amount of the tritiated precursor on a disk or membrane filter. As pointed out previously this procedure can grossly overestimate self-absorption losses because the small precursor molecules are adsorbed throughout the whole disk, and self-absorption losses are caused by the support medium. When a macromolecule is filtered or precipitated on the surface of fibers, self-absorption losses are due largely to the thin layer of precipitate on the surface, and not to the more dense support medium.

Self-absorption problems can also arise when tritiated compounds are counted on paper chromatograms or electrophoretograms, or on thin-layer support media. A recent publication (485) has shown that unequal drying of the two surfaces of chromatograms can lead to the unequal distribution of solutes throughout the paper. This leads to wide variation in the magnitude of the self-absorption losses. Consistent results were obtained only when the chromatograms were dried in still air. It was also shown that the counts per minute obtained for equal amounts of different tritiated compounds depended markedly on the mobilities of the compounds in the chromatographic solvents. Compounds with high mobilities migrated nearer to the surface of the support media when the solvents were removed by evaporation. As a result, they suffered much smaller self-absorption losses than the less mobile compounds. Although semiquantitative results may be obtained when counting tritiated compounds on insoluble support media, quantitative results can be obtained only by eluting or combusting the material and then counting in homogeneous solution.

9. Miscellaneous Problems Arising during Assays by Sampling Methods

All enzyme reactions which are assayed by sampling methods need to be stopped before determining the amount of reaction, and many methods of stopping reactions have been used.

These methods include denaturation of the enzyme by heating or by the addition of protein precipitants. The presence of precipitated protein can be troublesome because it can interfere in many determinations. Heating has also been shown, in some assays, to release contaminating enzymes which degraded the products and gave inaccurate results. Precipitated protein usually needs to be removed before determining the amount of reaction, and assays can often be speeded up, therefore, by using alternative methods of stopping the reaction so that the protein is not precipitated. These include the addition of enzyme inhibitors, for example, EDTA for metal-activated enzymes, fluoride for ATPase, and organic solvents such as ethanol. Simple cooling, dilution with water, or change of pH may also serve to stop reactions. A very simple and effective method of stopping radiometric enzyme assays is to inhibit the formation of radioactive products by adding an excess of un-labeled substrate. This can be done either by adding a solution of the carrier material, or more simply by spotting an aliquot of the reaction mixture on top of a spot of carrier previously applied to the chromatogram, electro-phoretogram, or paper disk. This isotope dilution method of stopping reactions has the added advantage of avoiding many of the problems inherent in the handling of small weights of material (see Section V.7), particularly that of irreversible adsorption.

In the past, the utility of many radiometric enzyme assays has been re-duced by the use of unnecessarily complex and tedious separation procedures. Recommended methods of simplifying these procedures were discusssed in Section III. Rapid, simple, and accurate techniques are now available for the majority of radiometric enzyme assays.

10. Miscellaneous Problems Associated with the Supply, Cost, and Counting of Labeled Substrates

During the past two or three decades there has been a great increase in the development and use of enzyme assays in many fields of biochemistry. On the other hand, despite the many obvious advantages of radiometric assays, it is only in the last few years that a significant increase in the number of these assays in general use has occurred. Many factors are responsible for this, among them the following.

A large number of clinical assays (e.g., assays of serum enzymes) have not required a high degree of accuracy and sensitivity, whereas nowadays the assay of enzymes in tissue biopsy samples demands much more sensitive and accurate methods.

Twenty years ago the range of labeled enzyme substrates available was only a small fraction of the large and steadily increasing number of such compounds now marketed by the major suppliers.

Twenty years ago few laboratories had satisfactory counting equipment, whereas today most laboratories are well equipped in this regard. A wide variety of counting methods have been used; of these, the very sensitive and convenient liquid scintillation method, which has the advantage of permitting automatic counting of large numbers of samples, is most often employed. Many articles on liquid scintillation counting have been written, and a recent review (88) describes in detail the preparation of samples for counting, the selection of the most suitable scintillant "cocktail," and calibration techniques.

Other objections to radiometric enzyme assays have been based on their supposed high cost, but their high sensitivity means that only very small amounts of labeled compounds need be used. Many assays can, for example, be performed with as little as 0.01 μCi of labeled substrate per assay. Although scintillant cocktails and scintillation counting vials can contribute appreciably to the cost of radiometric assays, many biological samples can be counted in a toluene-Triton X-100 mixture (64,65,88), which is much cheaper than the customary Bray's scintillant. The cost of assays can be further reduced by using cheap, disposable scintillation vials. In many radiometric assays the products are so tightly bound to cellulose or ion-exchange papers that the vials and scintillant may be reused after the papers have been removed with tweezers after counting. Often vials and scintillant can be reused thirty or more times before the background rises significantly. This represents a marked reduction in the cost of the assays.

VI. SUMMARY: THE ADVANTAGES OF RADIOMETRIC METHODS OF ENZYME ASSAY

The chief objections to sampling methods of enzyme assay in the past have involved the time taken for each assay. The improved techniques presently used in radiometric enzyme assays have overcome these objections; relatively rapid and simple sampling and separation procedures, which can easily be performed by technical assistants, are now available for most enzyme assays.

These techniques are very versatile, offer extremely high sensitivity and specificity, and give reproducible and accurate results. They offer many advantages over conventional methods for nearly all routine determinations, other than those carried out during the study of the kinetics of reaction of highly purified enzymes, where very rapid, continuous assays may be essential.

The versatility is such that, by choosing a suitable labeled substrate, most reactions can be followed in either direction, and, by selecting an appropriate substrate, one can also avoid the interfering reactions which frequently cause errors in conventional assays. This specificity and the resulting freedom from interference make radiometric methods most suitable for assaying enzyme activity in crude extracts, and for studying the effects of activators, inhibitors, and pH variation.

Most conventional methods of assay depend on either colorimetric or spectrophotometric estimations of substrate or product, or of a secondary product formed by coupling the reaction to a suitable NAD or NADP oxidoreductase system. These methods are unsuitable, however, for the assay of crude extracts which are highly colored or contain endogenous materials that can react with the coupling enzyme system. Radiometric assays do not suffer from interference of this sort.

Because most conventional assays depend on the direct or indirect measurement of products, they cannot be used to measure enzyme activity in extracts which contain high concentrations of the product, or in studies of product inhibition where the enzyme activity must be determined in the presence of high concentrations of product. Radiometric assays, which measure only the radioactive products, are unaffected by the presence of endogenous or added inactive product.

Often a reaction may be catalyzed by two enzymes, one of high and the other of low specificity. When a highly specific enzyme is assayed in the presence of an enzyme of low specificity, the nonspecific enzyme must be inhibited by using a high concentration of one of its substrates which does not interfere in the assay of the specific enzyme. Such assays are far easier to carry out by radiometric than by conventional methods.

The sensitivity of radiometric assays enables them to be used over a wider range of substrate concentrations than is possible with other methods. This makes them very suitable for the determination of Michaelis constants, which often necessitates measurements at very low and at very high substrate concentrations. This sensitivity also facilitates the assay of low levels of enzyme activity, and of the enzymatic activity of microgram quantities of tissue, which may be obtained in needle biopsies or during a survey of enzyme distribution throughout a particular organ.

Another advantage of radiometric assays is that their sensitivity makes them suitable for use at very low substrate concentration and with very little dilution of the enzyme. This is very important in the *in vitro* determination of the *in vivo* activity of enzymes which have been inhibited by competitive inhibitors. Conventional assays, in which the enzyme is appreciably diluted and high substrate concentrations are used, cause the *in vivo* inhibition to disappear, and falsely high rates to be observed.

The development of radiometric methods of enzyme assay during the past few years has been very important in many fields. It can safely be predicted that these methods will prove to be increasingly useful, not only in general biochemistry, but also in clinical biochemistry, particularly in the study and diagnosis of diseases caused by inherited enzyme defects. They can also be expected to make a significant contribution to the pharmacological study of the effects of drugs on enzymatic activity and to the clinical monitoring of patients undergoing drug therapy.

Acknowledgments

Thanks are due to my colleagues Dr. J. R. Catch and Dr. E. A. Evans for their interest; to the director of The Radiochemical Centre, Dr. W. P. Grove, for his permission to publish this work; to the staff of The Radiochemical Centre library for unfailing assistance; and to the many workers who have discussed their enzyme assay techniques—and problems—with me.

The acquisition of much of the information in this chapter was made possible by two lecture tours of the United States. These were sponsored by the Amersham-Searle Corporation, and the assistance of this organization is gratefully acknowledged.

References

1. K. G. Oldham, *Radiochemical Methods of Enzyme Assay* (*RCC Review 9*), Radiochemical Centre, Amersham, 1968.
2. D. J. Reed, "Methodoloy of Radiotracer Enzyme Assays," in S. Rothschild, Ed., *Advances in Tracer Methodology*, Vol. 4, Plenum, New York, 1968, pp. 145–175.
3. R. E. McCaman, *Isotopes Radiation Technol.*, *3*, 328 (1966).
4. International Union of Biochemistry, *Enzyme Nomenclature: Recommendations* (*1964*) *on the Nomenclature and Classification of Enzymes, Together with Their Units and Symbols of Enzyme Kinetics*, Elsevier, Amsterdam, 1965.
5. H-U. Bergmeyer, *Methods of Enzymatic Analysis*, Academic Press, New York, 1963, p. 32.
6. International Union of Biochemistry, *Report of the Commission on Enzymes*, Pergamon Press, Oxford, 1961.
7. G. G. Guilbault, *Enzymatic Methods of Analysis*, Pergamon Press, New York, 1970.
8. M. Dixon and E. C. Webb, *Enzymes*, Longmanns, London, 1964.
9. G. I. Loutfi and D. D. Hagerman, *Acta Endocrinol.*, *54*, 122 (1967).
10. H-U. Bergmeyer, *Methods of Enzymatic Analysis*, Academic Press, New York, 1963, p. 10.
11. S. B. Rosalki and J. H. Wilkinson, *J. Clin. Pathol.*, *12*, 138 (1959).
12. J. King, *Practical Clinical Enzymology*, Van Nostrand, London, 1965, p. 125.
13. R. A. Harvey, T. Godefroy, J. Lucas-Lenard, and M. Grunberg-Manago, *European J. Biochem.*, *1*, 327 (1967).
14. F. Fonnum, *Biochem. Pharmacol.*, *17*, 2503 (1969).
15. J. Gailiusis, R. W. Rinne, and C. R. Benedict, *Biochim. Biophys. Acta*, *92*, 595 (1964).
16. R. Raunio, *Acta Chem. Scand.*, *23*, 1168 (1969).
17. A. W. Murray, *Biochem. J.*, *106*, 549 (1968).
18. R. E. McCaman, "Application of Tracers to Quantitative Histochemical and Cytochemical Studies," in S. Rothschild, Ed., *Advances in Tracer Methodology*, Vol. 4, Plenum, New York, 1968, pp. 187–202.
19. R. E. McCaman, *Life Sci.*, *4*, 2353 (1965).
20. S. Otsuka and Y. Kobayashi, *Biochem. Pharmacol.*, *13*, 995 (1964).
21. S. Burstein and R. I. Dorfman, *J. Biol. Chem.*, *238*, 1656 (1964).
22. C. C. Richardson, C. L. Schildkraut, H. V. Aposhian, and A. Kornberg, *J. Biol. Chem.*, *239*, 222 (1964).
23. A. Falaschi and A. Kornberg, *J. Biol. Chem.*, *241*, 1478 (1966).

24. G. D. Birnie and S. M. Fox, *Biochem. J.*, *104*, 239 (1967).
25. F. J. Bollum, *J. Biol. Chem.*, *234*, 2733 (1959).
26. J. S. Krakow, *J. Biol. Chem.*, *241*, 1830 (1966).
27. D. A. Smith, R. L. Ratliff, D. L. Williams, and A. M. Martinez, *J. Biol. Chem.*, *242*, 590 (1967).
28. P. H. Lloyd, B. H. Nicholson, and A. R. Peacocke, *Biochem. J.*, *104*, 999 (1967).
29. J. S. Paul, R. C. Reynolds, and P. O. Montgomery, *Nature*, *215*, 749 (1967).
30. A. Falaschi, J. Adler, and H. G. Khorana, *J. Biol. Chem.*, *238*, 3080 (1963).
31. E. A. Hyatt, *Biochim. Biophys. Acta*, *142*, 246 (1967).
32. R. J. Mans and G. D. Novelli, *Arch. Biochem. Biophys.*, *94*, 48 (1961).
33. J. M. Wilhelm and J. W. Corcoran, *Biochemistry*, *6*, 2578 (1967).
34. J. W. Davies and E. C. Cocking, *Biochem. J.*, *104*, 23 (1967).
35. J. E. Kay and A. Korner, *Biochem. J.*, *100*, 815 (1966).
36. Y. Nishizuka and F. Lippman, *Proc. Natl. Acad. Sci., U.S.*, *55*, 212 (1966).
37. T. W. Conway, *Proc. Natl. Acad. Sci., U.S.*, *51*, 1216 (1964).
38. H. De Wulf and H. G. Hers, *European J. Biochem.*, *2*, 50 (1967).
39. D. S. Feingold, E. F. Neufeld, and W. Z. Hassid, *J. Biol. Chem.*, *233*, 783 (1958).
40. J. A. Haines and P. C. Zamecnik, *Biochim. Biophys. Acta*, *146*, 227 (1967).
41. E. M. Lansford, N. M. Lee, and W. Shive, *Arch. Biochem. Biophys.*, *119*, 272 (1967).
42. M. P. Stulberg, *J. Biol. Chem.*, *242*, 1060 (1967).
43. F. J. Bollum, "Filter Paper Disc Techniques for Assaying Radioactive Macromolecules," in G. L. Cantoni and D. R. Davies, Eds., *Procedures in Nucleic Acid Research*, Harper & Row, New York, 1966, pp. 296–300.
44. R. A. Malt and W. L. Miller, *Anal. Biochem.*, *18*, 388 (1967).
45. D. M. Gill, *Intern. J. Appl. Radiation Isotopes*, *18*, 393 (1967).
46. K. G. Oldham, *Intern. J. Appl. Radiation Isotopes*, *21*, 421 (1970).
47. K. G. Oldham, *Anal. Biochem.*, *44*, 143 (1971).
48. E. P. Geiduschek and A. Daniels, *Anal. Biochem.*, *11*, 133 (1965).
49. J. F. Scott, "Membrane Filter Technique for the Assay of Charged sRNA, "in S. P. Colowick and N. O. Kaplan, Eds., *Methods in Enzymology*, Vol. XIIB, Academic Press, New York, 1968, pp. 173–176.
50. K. H. Muench and P. Berg, "Preparation of Aminoacyl Ribonucleic Acid Synthetases from *Escherichia coli*," in G. L. Cantoni and D. R. Davies, Eds., *Procedures in Nucleic Acid Research*, Harper & Row, New York, 1966, pp. 375–383.
51. J. D. Cherayil and R. M. Bock, *Biochemistry*, *4*, 1174 (1965).
52. J. D. Cherayil, A. Hampel, and R. M. Bock, "Rapid Microscale Assay for sRNA," in S. P. Colowick and N. O. Kaplan, Eds., *Methods in Enzymology*, Vol. XIIB, Academic Press, New York, 1968, pp. 166–169.
53. E. Wimmer, I. H. Maxwell, and G. M. Tener, *Biochemistry*, *7*, 2623 (1968).
54. G. Goldstein, W. L. Maddox, and I. B. Rubin, "A Semi-automated Filter Paper Disc Technique for the Determination of Transfer Ribonucleic Acid," in *Technicon Symposia 1967*, Vol. 1, Mediad, Inc., New York, 1968, pp. 47–49.
55. L. Baugnet-Mahieu, R. Goutier, and M. Semal, *J. Labelled Compounds*, *2*, 77 (1966).
56. D. J. Reed, K. Goto, and C. H. Wang, *Anal. Biochem.*, *16*, 59 (1966).
57. G. Brooker, L. J. Thomas, and M. M. Applemann, *Biochemistry*, *7*, 4177 (1968).
58. J-P. Jost, A. Hsie, S. D. Hughes, and L. Ryan, *J. Biol. Chem.*, *245*, 351 (1970).
59. D. Monard, J. Janecek, and H. V. Rickenberg, *Biochem. Biophys. Res. Commun.*, *35*, 584 (1969).
60. J. R. Steeper and C. D. Steuart, *Anal. Biochem.*, *34*, 123 (1970).
61. G. Goldstein, *Anal. Biochem.*, *20*, 477 (1967).

62. G. A. Bray, *Anal. Biochem.*, *1*, 279 (1960).
63. M. S. Patterson and R. C. Greene, *Anal. Chem.*, *37*, 854 (1965).
64. J. C. Turner, *Intern. J. Appl. Radiation Isotopes*, *19*, 557 (1968).
65. J. C. Turner, *Intern. J. Appl. Radiation Isotopes*, *20*, 499 (1969).
66. J. R. Sherman, *Anal. Biochem.*, *5*, 548 (1963).
66b. N. B. Furlong, *Anal. Biochem.*, *5*, 515 (1963).
67. T. R. Breitman, *Biochim. Biophys. Acta*, *67*, 153 (1963).
67b. D. H. Ives, J. P. Durham, and V. S. Tucker, *Anal. Biochem.*, *28*, 192 (1969).
68. E. A. Newsholme, J. Robinson, and K. Taylor, *Biochim. Biophys. Acta*, *132*, 338 (1967).
69. E. Bresnick and U. B. Thompson, *J. Biol. Chem.*, *240*, 3967 (1965).
70. J. F. Morrison and W. W. Cleland, *J. Biol. Chem.*, *241*, 673 (1966).
71. A. Orengo, *Exptl. Cell Res.*, *41*, 338 (1966).
72. D. Roberts, D. Kessel, and T. C. Hall, *Japan. Cancer Assoc. GANN, Monogr. 2*, Maruzen, Tokyo, 1967, pp. 113–126.
73. H. G. Klemperer and G. R. Haynes, *Biochem. J.*, *108*, 541 (1968).
74. J. P. Whitlock, R. Kaufman, and R. Baserga, *Cancer Res.*, *28*, 2211 (1968).
75. R. Labow, G. F. Maley, and F. Maley, *Cancer Res.*, *29*, 366 (1969).
76. J. Preiss and E. Greenberg, *Anal. Biochem.*, *18*, 464 (1967).
77. W. G. Ng, W. R. Bergren, and G. N. Donnell, *Clin. Chim. Acta*, *15*, 489 (1967).
78. S. Gabay and H. George, *Anal. Biochem.*, *21*, 111 (1967).
79. M. R. Atkinson and A. W. Murray, *Biochem. J.*, *94*, 64 (1965).
80. H. K. Miller and M. E. Balis, *Biochem. Pharmacol.*, *18*, 2225 (1969).
81. H. Cedar and J. H. Schwarz, *J. Bacteriol.*, *96*, 2043 (1968).
82. R. L. Jungas, *Proc. Natl. Acad. Sci., U.S.*, *56*, 757 (1966).
83. R. R. Bürk, *Nature*, *219*, 1272 (1968).
84. A. M. Spiegel and M. W. Bitensky, *Endocrinology*, *85*, 638 (1969).
85. G. Burke, *Endocrinology*, *86*, 346 (1970).
86. R. H. Wilson, *Biochem. J.*, *118*, 16P (1970).
87. K. G. Oldham, some unpublished experiments.
88. J. C. Turner, *Sample Preparation for Liquid Scintillation Counting (RCC Review 6)*, Radiochemical Centre, Amersham, 1967.
89. K. Randerath, *J. Chromatog.*, *10*, 235 (1963).
90. R. M. Roberts and K. C. Tovey, *Anal. Biochem.*, *34*, 582 (1970).
91. R. H. Wilson, Botany Dept., Oxford University, personal communication.
92. K. G. Oldham, *J. Labelled Compounds*, *4*, 127 (1968).
93. H. Reichard, *J. Lab. Clin. Med.*, *63*, 1061 (1964).
94. D. L. DiPetro, *Anal. Biochem.*, *6*, 305 (1963).
95. A. E. Renold, D. L. DiPietro, and A. K. Williams, in *The Structure and Metabolism of the Pancreatic Islets*, Wenner-Gren Center International Symposium Series, Vol. 3, Pergamon Press, New York, 1964, pp. 269–279.
96. R. T. Schimke, *J. Biol. Chem.*, *239*, 136 (1964).
97. N. Carulli, S. Kaihara, and H. N. Wagner, *Anal. Biochem.*, *24*, 515 (1968).
98. J. E. Ultmann and P. Fiegelson, *Ann. N. Y. Acad. Sci.*, *103*, 724 (1963).
99. J. Chiriboga and D. N. Roy, *Nature*, *193*, 684 (1962).
100. W. S. Runyan, M. A. Fardy, and R. P. Geyer, *Biochim. Biophys. Acta*, *141*, 421 (1967).
101. R. E. McCaman, *Isotopes Radiation Technol.*, *3*, 328 (1966).
102. D. R. Buhler, *Anal. Biochem.*, *4*, 413 (1962).
103. E. A. Khairallah and G. Wolf, *J. Biol. Chem.*, *242*, 32 (1967).

104. M. Lewis, G. R. Lee, G. E. Cartwright, and M. M. Wintrobe, *Biochim. Biophys. Acta*, *141*, 296 (1967).
105. R. J. Levine and D. E. Watts, *Biochem. Pharmacol.*, *16*, 993 (1967).
106. J. B. Bourke, M. Faust, and L. M. Massey, *Intern. J. Appl. Radiation Isotopes*, *18*, 619 (1967).
107. F. H. Woeller, *Anal. Biochem.*, *2*, 508 (1961).
108. P. J. Lupien, C. M. Hinse, and L. Berlinguet, *Anal. Biochem.*, *24*, 1 (1968).
109. N. Carulli, S. Kaihara, and H. N. Wagner, *Anal. Biochem.*, *26*, 334 (1968).
110. W. D. Davidson, A. D. Schwabe, and J. C. Thompson, *Anal. Biochem.*, *26*, 341 (1968).
111. F. H. DeLand and H. N. Wagner, *Radiology*, *92*, 154 (1969).
112. B. Grahn and E. Rosengren, *Brit. J. Pharmacol.*, *33*, 472 (1968).
113. C. R. Slack and M. D. Hatch, *Biochem. J.*, *103*, 660 (1967).
114. A. D. Gounaris and L. P. Hager, *J. Biol. Chem.*, *236*, 1013 (1961).
115. B. Keech and G. J. Barritt, *J. Biol. Chem.*, *242*, 1983 (1967).
116. D. O'Neal and A. W. Naylor, *Biochem. Biophys. Res. Commun.*, *31*, 322 (1968).
117. L. Peng and M. E. Jones, *Biochem. Biophys. Res. Commun.*, *34*, 335 (1969).
118. C. G. Maresh, T. H. Kwan, and S. M. Kalman, *Can. J. Biochem. Physiol.*, *47*, 61 (1969).
119. F. J. Ballard, R. W. Hanson, and G. A. Leveille, *J. Biol. Chem.*, *242*, 2746 (1967).
120. M. C. Scrutton and M. F. Utter, *J. Biol. Chem.*, *240*, 3714 (1965).
121. M. D. Hatch and C. R. Slack, *Biochem. J.*, *112*, 549 (1969).
122. M. A. Mehlman, *J. Biol. Chem.*, *243*, 3289 (1968).
123. F. P. W. Winteringham and R. W. Disney, *Nature*, *195*, 1303 (1962).
124. J. E. Casida, J. L. Engel, and Y. Nishizawa, *Biochem. Pharmacol.*, *15*, 627 (1966).
125. J. Schuberth, *Acta Chem. Scand.*, *17 suppl. 1*, S233 (1963).
126. J. E. Young, M. D. Prager, and I. C. Atkins, *Proc. Soc. Exptl. Biol. Med.*, *125*, 860 (1967).
127. N. Ishimoto and J. L. Strominger, "CDP-Ribitol: Acceptor Phosphoribitol Transferase from *Staphylococcus aureus*," in S. P. Colowick and N. O. Kaplan, Eds., *Methods in Enzymology*, Vol. 8, Academic Press, New York, 1966, pp. 423–426.
128. G. Guroff and J. Daly, *Arch. Biochem. Biophys.*, *122*, 212 (1967).
129. R. M. Roberts and K. C. Tovey, *J. Chromatog.*, *47*, 287 (1970).
130. H-P. Bär and O. Hechter, *Anal. Biochem.*, *29*, 476 (1969).
131. I. D. Algranati, N. Behrens, H. Carminatti, and E. Cabib, "Mannan Synthetase from Yeast," in S. P. Colowick and N. O. Kaplan, Eds., *Methods in Enzymology*, Vol. 8, Academic Press, New York, 1966, pp. 411–416.
132. H. Babad and W. Z. Hassid, *J. Biol Chem.*, *241*, 2672 (1966).
133. V. M. Hearn, Z. G. Smith, and W. M. Watkins, *Biochem. J.*, *109*, 315 (1968).
134. M. A. Chester and W. M. Watkins, *Biochem. Biophys. Res. Commun.*, *34*, 835 (1969).
135. F. Klink, K. Kloppstech, and H. Netter, *Z. Naturforsch.*, *21b*, 497 (1966).
136. K. Shimada and Y. Sugino, *Biochim. Biophys. Acta*, *185*, 367 (1969).
137. D. H. Duckworth and M. J. Bessman, *J. Biol. Chem.*, *242*, 2877 (1967).
138. I. R. Lehman, M. J. Bessman, E. S. Simms, and A. Kornberg, *J. Biol. Chem.*, *233*, 163 (1958).
139. R. A. Harvey, T. Godefroy, J. Lucas-Lenard, and M. Grunberg-Manago, *European J. Biochem.*, *1*, 327 (1967).
140. G. Guroff and C. A. Rhoads, *J. Neurochem.*, *16*, 1543 (1969).
141. W. L. Fangman and F. C. Neidhart, *J. Biol. Chem.*, *239*, 1839 (1964).

142. T. Sand, J. C. Siebke, and S. Laland, *Febs Letters*, *1*, 63 (1968).
143. D. M. Ireland and D. C. B. Mills, *Biochem. J.*, *99*, 283 (1966).
144. T. Nagatsu, M. Levitt, and S. Udenfriend, *J. Biol. Chem.*, *239*, 2910 (1964).
145. T. Nagatsu and K. Takeuchi, *Experientia*, *23*, 532 (1967).
146. P. Laduron and F. Belpaire, *Biochem. Pharmacol.*, *17*, 1127 (1968).
147. M. D. Prager and N. Bachynsky, *Biochem. Biophys. Res. Commun.*, *31*, 43 (1968).
148. C. M. Haskell and G. P. Canellos, *Biochem. Pharmacol.*, *18*, 2578 (1969).
149. M. Farstad, *Biochim. Biophys. Acta*, *146*, 272 (1967).
150. W. R. Metge, C. A. Owen, W. T. Foulk, and H. N. Hoffman, *J. Lab. Clin. Med.*, *64*, 335 (1964).
151. J. Daly, J. K. Inscoe, and J. Axelrod, *J. Med. Chem.*, *8*, 153 (1965).
152. J. Axelrod, J. K. Inscoe, and J. Daly, *J. Pharmacol., Exptl. Therap.*, *149*, 16 (1965).
153. J. K. Inscoe, J. Daly, and J. Axelrod, *Biochem. Pharmacol.*, *14*, 1257 (1965).
154. J. W. Daly and A. A. Manian, *Biochem. Pharmacol.*, *16*, 2131 (1967).
155. F. Fonnum, *Biochem. J.*, *100*, 479 (1966).
156. C. F. Fox and E. P. Kennedy, *Anal. Biochem.*, *18*, 286 (1967).
157. R. Cohn and S. Segal, *Biochim. Biophys. Acta*, *171*, 333 (1969).
158. W. Lovenberg, E. Jeauier, and A. Sjoerdsma, *Advan. Pharmacol.*, *6A*, 21 (1968).
159. Y. Daikuhara, T. Tsunemi, and Y. Takeda, *Biochim. Biophys. Acta*, *158*, 51 (1968).
160. B. Peterkofsky, *Arch. Biochem. Biophys.*, *128*, 637 (1968).
161. S. Hayashi and E. C. C. Lin, *Biochim. Biophys. Acta*, *94*, 479 (1965).
162. S. Hayashi and E. C. C. Lin, *J. Biol. Chem.*, *242*, 1030 (1967).
163. N. Lee and I. Bendet, *J. Biol. Chem.*, *242*, 2043 (1967).
164. M. O. Oster and N. P. Wood, *J. Bacteriol.*, *87*, 104 (1964).
165. A. Alpert, R. L. Kisliuk, and L. Shuster, *Biochem. Pharmacol.*, *15*, 465 (1966).
166. G. Buckley, S. Consolo, E. Giacobini, and R. E. McCaman, *Acta Physiol. Scand.*, *71*, 341 (1967).
167. B. Weiss and E. Costa, *Science*, *156*, 1750 (1967).
168. R. A. Eisenman, A. S. Balasubramanian, and W. Marx, *Arch. Biochem. Biophys.*, *109*, 387 (1967).
169. D. A. Kalbhen, K. Karzel, and R. Domenjoz, *Med. Pharmacol. Expt.*, *16*, 185 (1967).
170. F. Eisenberg, *J. Biol. Chem.*, *242*, 1375 (1967).
171. R. W. Schayer and M. A. Reilly, *Am. J. Physiol.*, *215*, 472 (1968).
172. R. W. Hanson and F. J. Ballard, *Biochem. J.*, *108*, 705 (1968).
173. T. I. Matula, I. J. McDonald, and S. M. Martin, *Biochem. Biophys. Res. Commun.*, *34*, 795 (1969).
174. S. Matsuhashi and J. L. Strominger, "Formation of CDP-3,6-dideoxyhexoses from CDP-D-glucose," in S. P. Colowick and N. O. Kaplan, Eds., *Methods in Enzymology*, Vol. 8, Academic Press, New York, 1966, pp. 310–316.
175. G. A. Barber, *Arch. Biochem. Biophys.*, *103*, 276 (1963).
176. T. C. Linn, *J. Biol. Chem.*, *242*, 984 (1967).
177. J. A. Sturman, *Clin. Chim. Acta*, *26*, 135 (1969).
178. M. Goulian and W. S. Beck, *J. Biol. Chem.*, *241*, 4233 (1966).
179. P. P. Saccoccia and R. P. Miech, *Mol. Pharmacol.*, *5*, 26 (1969).
180. U. A. S. Al-Khalidi, S. Nasrallah, A. K. Hachadurian, and M. M. Shamaa, *Clin. Chim. Acta*, *11*, 72 (1965).
181. R. A. Goldsby and P. G. Heytler, *Biochim. Biophys. Acta*, *97*, 162 (1965).
182. R. L. Anderson, *Biochim. Biophys. Acta*, *144*, 18 (1967).
183. R. Hakanson and C. Owman, *J. Neurochem.*, *12*, 417 (1965).
184. R. E. McCaman, M. W. McCaman, J. M. Hunt, and M. S. Smith, *J. Neurochem.*, *12*, 15 (1965).

185. Y. Kobayashi, *Biochem. Pharmacol.*, *15*, 1287 (1966).
186. R. J. Wurtman and J. Axelrod, *Biochem. Pharmacol.*, *12*, 1439 (1963).
187. J. Southgate and C. G. S. Collins, *Biochem. Pharmacol.*, *18*, 2285 (1969).
188. L. J. Ignarro and F. E. Shideman, *J. Pharmacol. Exptl. Therap.*, *159*, 29 (1968).
189. D. S. Robinson, W. Lovenberg, H. Keiser, and A. Sjoerdsma, *Biochem. Pharmacol.*, *17*, 109 (1968).
190. F. Izumi, M. Oka, H. Yoshida, and R. Imaizumi, *Biochem. Pharmacol.*, *18*, 1739 (1969).
191. J. H. Fellman, E. S. Roth, and R. F. Mollica, *Anal. Biochem.*, *30*, 339 (1969).
192. D. Aures, R. Fleming, and R. Hakanson, *J. Chromatog.*, *33*, 480 (1968).
193. Y. Kobayashi, J. Kupelian, and D. V. Maudsley, *Biochem. Pharmacol.*, *18*, 1585 (1969).
194. B. L. Hillcoat, P. F. Nixon, R. L. Blakley, *Anal. Biochem.*, *21*, 178 (1967).
195. S. P. Rothenberg, *Anal. Biochem.*, *13*, 530 (1965).
196. S. P. Rothenberg, *Nature*, *206*, 1154 (1965).
197. D. Roberts, *Biochemistry*, *5*, 3549 (1966).
198. D. Roberts and T. C. Hall, *Cancer Res.*, *29*, 166 (1969).
199. "Migration Explains Enzymic Hydroxylations," *Chem. Eng. News*, *45(8)*, 42 (1967).
200. "More Information Gathered on the Mechanism of the NIH Shift," *Chem. Eng. News*, *45(17)*, 57 (1967).
201. G. Guroff, J. Daly, B. Witkop, and S. Udenfriend, *Proceedings of the 7th International Congress of Biochemistry*, Tokyo, 1967, to be published.
202. G. Guroff and A. Abramowitz, *Anal. Biochem.*, *19*, 548 (1967).
203. J. Daly, G. Guroff, S. Udenfriend, and B. Witkop, *Arch. Biochem. Biophys.*, *122*, 218 (1967).
204. J. W. Daly, *Anal. Biochem.*, *33*, 286 (1970).
205. T. Nagatsu, M. Levitt, and S. Udenfriend, *Anal. Biochem.*, *9*, 122 (1964).
206. S. Udenfriend and C. R. Creveling, *J. Neurochem.*, *4*, 350 (1959).
207. M. Goldstein, N. Prochoroff, and S. Sirlin, *Experientia*, 592 (1965).
208. D. A. Silverman and P. Talalay, *Mol. Pharmacol.*, *3*, 90 (1967).
209. F. N. Chang and C. J. Sih, *Biochemistry*, *3*, 1551 (1964).
210. M. Tanabe, D. Yasuda, J. Tagg, and C. Mitoma, *Biochem. Pharmacol.*, *16*, 2230 (1967).
211. D. W. Russell and E. E. Conn, *Arch. Biochem. Biophys.*, *122*, 256 (1967).
212. N. Amrhein and M. H. Zenk, *Phytochemistry*, *8*, 107 (1969).
213. J. Renson, J. Daly, H. Weissbach, B. Witkop, and S. Udenfriend, *Biochem. Biophys. Res. Commun.*, *25*, 504 (1966).
214. E. Jequier, D. S. Robinson, W. Lovenberg, and A. Sjoerdsma, *Biochem. Pharmacol.*, *18*, 1071 (1969).
215. R. Hakanson and G. J. Hoffman, *Biochem. Pharmacol.*, *16*, 1677 (1967).
216. A. Tourian, J. Goddard, and T. T. Puck, *J. Cellular Physiol.*, *73*, 159 (1969).
217. R. A. Mueller, H. Thoenen, and J. Axelrod, *J. Pharmacol. Exptl. Ther.*, *169*, 74 (1969).
218. R. J. Taylor, C. S. Stubbs, and L. Ellenbogen, *Biochem. Pharmacol.*, *18*, 587 (1969).
219. J. J. Hutton, A. Marglin, B. Witkop, J. Kurtz, A. Berger, and S. Udenfriend, *Arch. Biochem. Biophys.*, *125*, 779 (1968).
220. M. Liss and A. M. Maxam, *Biochim. Biophys. Acta*, *140*, 555 (1967).
221. W. K. Paik and S. Kim, *J. Biol. Chem.*, *243*, 2108 (1968).
222. H. Kauss and W. Z. Hassid, *J. Biol. Chem.*, *242*, 1680 (1967).
223. J. Hurwitz, M. Anders, M. Gold, and I. Smith, *J. Biol. Chem.*, *240*, 1256 (1965).
224. B. C. Baguley and M. Staehelin, *Biochemistry*, *7*, 45 (1968).

225. E. S. McFarlane, *Canad. J. Microbiol.*, *15*, 189 (1969).
226. M. Gold and J. Hurwitz, *J. Biol. Chem.*, *239*, 3858 (1964).
227. K. Oda and J. Marmur, *Biochemistry*, *5*, 761 (1966).
228. B. Sheid, P. R. Srinivason, and E. Borek, *Biochemistry*, *7*, 280 (1968).
229. J. Axelrod and J. K. Lauber, *Biochem. Pharmacol.*, *17*, 828 (1968).
230. J. Daly, J. K. Inscoe, and J. Axelrod, *J. Med. Chem.*, *8*, 153 (1965).
231. D. D. Brown, R. Tomchick, and J. Axelrod, *J. Biol. Chem.*, *234*, 2949 (1959).
232. J. Axelrod and J. Daly, *Biochim. Biophys. Acta*, *159*, 472 (1968).
233. J. Axelrod, *J. Biol. Chem.*, *237*, 1657 (1962).
234. R. D. Ciaranello, R. E. Barchas, G. S. Byers, D. W. Stemmle, and J. D. Barchas, *Nature*, *221*, 368 (1969).
235. S. K. Shapiro, A. Almens, and J. F. Thomson, *J. Biol. Chem.*, *240*, 2512 (1965).
236. R. T. Taylor and H. Weissbach, *J. Biol. Chem.*, *242*, 1502 (1967).
237. J. D. Finkelstein and S. H. Mudd, *J. Biol. Chem.*, *242*, 873 (1967).
238. J. R. Guest and D. D. Woods, *Biochem. J.*, *97*, 500 (1965).
239. J. L. Botsford and L. W. Parks, *J. Bacteriol.*, *97*, 1176 (1969).
240. D. M. Shapiro, J. Eigner, and G. R. Greenberg, *Proc. Natl. Acad. Sci., U.S.*, *53*, 874 (1965).
241. J. G. Flaks and S. S. Cohen, *J. Biol. Chem.*, *234*, 1501 (1959).
242. M. I. S. Lomax and G. R. Greenberg, *J. Biol. Chem.*, *242*, 109 (1967).
243. H. O. Kammen, *Anal. Biochem.*, *17*, 553 (1966).
244. M. I. S. Lomax and G. R. Greenberg, *J. Biol. Chem.*, *242*, 1304 (1967).
245. H-S. Lin, *J. Bacteriol.*, *96*, 2054 (1968).
246. M. D. Prager, J. E. Young, and I. C. Atkins, *J. Lab. Clin. Med.*, *70*, 768 (1967).
247. M. Brin, M. Tai, A. S. Ostashever, and H. Kalinsky, *J. Nutr.*, *71*, 273 (1960).
248. B. K. Schrier and L. Shuster, *J. Neurochem.*, *14*, 977 (1967).
249. I. Diamond and E. P. Kennedy, *Anal. Biochem.*, *24*, 90 (1968).
250. D. Gallwitz, *Biochem. Biophys. Res. Commun.*, *32*, 117 (1968).
251. S. Nagai and M. Flavin, *J. Biol. Chem.*, *242*, 3884 (1967).
252. W. V. Shaw, *J. Biol. Chem.*, *242*, 687 (1967).
253. J. Bremer and K. R. Norum, *J. Biol. Chem.*, *242*, 1744 (1967).
254. L. Glaser and M. M. Burger, *J. Biol. Chem.*, *239*, 3187 (1964).
255. E. P. Camargo, C. P. Dietrich, D. Sonneborn, and J. L. Strominger, *J. Biol. Chem.*, *242*, 3121 (1967).
256. D. Shepherd and I. W. Segel, *Arch. Biochem. Biophys.*, *131*, 609 (1969).
257. R. Dedonder, "Levansucrase from *Bacillus subtilis*," in S. P. Colowick and N. O. Kaplan, Eds., *Methods in Enzymology*, Vol. VIII, Academic Press, New York, 1966, pp. 500–505.
258. H. De Wulf, W. Stalman, and H. G. Hers, *European J. Biochem.*, *6*, 545 (1968).
259. R. Piras, L. B. Rothman, and E. Cabib, *Biochemistry*, *7*, 56 (1968).
260. A. H. Gold, *Biochem. Biophys. Res. Commun.*, *31*, 361 (1968).
261. H. Vainer, *Nature*, *217*, 951 (1968).
262. H. J. Mersmann and H. L. Segal, *J. Biol. Chem.*, *244*, 1701 (1969).
263. L. M. Blatt, J. O. Scamahorn, and K. H. Kim, *Biochim. Biophys. Acta*, *177*, 553 (1969).
264. J. A. Thomas, K. K. Schlender, and J. Larner, *Anal. Biochem.*, *25*, 486 (1968).
265. D. des S. Thomas and R. G. Stanley, *Biochem. Biophys. Res. Commun.*, *30*, 292 (1968).
266. C. R. Slack, *Phytochemistry*, *5*, 397 (1966).
267. R. Roth and M. Sussman, *J. Biol. Chem.*, *243*, 5081 (1968).
268. A. D. Elbein, *J. Bacteriol.*, *96*, 1623 (1968).

269. C. Liu, B. W. Patterson, D. Lapp, and A. D. Elbein, *J. Biol. Chem.*, *244*, 3728 (1969).
270. G. S. Rao and H. Breuer, *J. Biol. Chem.*, *244*, 5521 (1969).
271. W. W. Watkins and W. Z. Hassid, *Biochem. Biophys. Res. Commun.*, *5*, 260 (1961).
272. H. Babad and W. Z. Hassid, *J. Biol. Chem.*, *239*, PC4 (1964).
273. N. J. Kuhn, *Biochem. J.*, *106*, 743 (1968).
274. D. J. Morre, L. M. Merlin, and T. W. Keenan, *Biochem. Biophys. Res. Commun.*, *37*, 813 (1969).
275. A. P. Grollman, C. Hall, and V. Ginsburg, *J. Biol. Chem.*, *240*, 975 (1965).
276. H. B. Bosmann, A. Hagopian, and E. H. Eylar, *Arch. Biochem. Biophys.*, *128*, 470 (1968).
277. A. D. Elbein, G. A. Barber, and W. Z. Hassid, "Cellulose Synthetase from Plants," in S. P. Colowick and N. O. Kaplan, Eds., *Methods in Enzymology*, Vol. VIII, Academic Press, New York, 1966, pp. 416–418.
278. P. Brennan and C. E. Ballou, *Biochem. Biophys. Res. Commun.*, *30*, 69 (1968).
279. R. K. Bretthauer, L. P. Kozak, and W. E. Irwin, *Biochem. Biophys. Res. Commun.*, *37*, 820 (1969).
280. J. M. McNab, C. L. Villemez, and P. Albersheim, *Biochem. J.*, *106*, 355 (1968).
281. M. J. Osborn and L. D'Ari, *Biochem. Biophys. Res. Commun.*, *16*, 568 (1964).
282. D. C. Collins, H. Jirku, and D. S. Layne, *J. Biol. Chem.*, *243*, 2928 (1968).
283. E. J. McGuire and S. Roseman, *J. Biol. Chem.*, *242*, 3745 (1967).
284. A. Kobata and V. Ginsburg, *J. Biol. Chem.*, *245*, 1484 (1970).
285. P. P. Saunders, B. A. Wilson, and G. F. Saunders, *J. Biol. Chem.*, *244*, 3691 (1969).
286. D. J. Rainnie, and D. G. R. Blair, *Can. J. Biochem.*, *47*, 429 (1969).
287. R. H. Lindsay, M-Y. Wong, C. J. Romine, and J. B. Hill, *Anal. Biochem.*, *24*, 506 (1968).
288. R. C. Gallo, S. Perry, and T. R. Breitman, *Biochem. Pharmacol.*, *17*, 2185 (1968).
289. H. L. A. Tarr, J. E. Roy, and M. Yamamoto, *Can. J. Biochem.*, *46*, 407 (1968).
290. P. B. Nicholls and A. W. Murray, *Plant Physiol.*, *43*, 645 (1968).
291. J. N. Santos, K. W. Hempstead, L. E. Kopp, and R. P. Miech, *J. Neurochem.*, *15*, 367 (1968).
292. A. W. Murray, P. C. L. Wong, and B. Friedrichs, *Biochem. J.*, *112*, 741 (1969).
293. B. M. Dean, R. W. E. Watts, and W. J. Westwick, *Febs Letters*, *1*, 179 (1968).
294. A. W. Murray, *Biochem. J.*, *100*, 664 (1966).
295. N. Fausto, *Biochim. Biophys. Acta*, *182*, 66 (1969).
296. L. D. Smith and R. K. Gholson, *J. Biol. Chem.*, *244*, 68 (1969).
297. D. L. Hill and L. L. Bennett, *Biochemistry*, *8*, 122 (1969).
298. B. S. Tay, R. McC. Lilley, A. W. Murray, and M. R. Atkinson, *Biochem. Pharmacol.*, *18*, 936 (1969).
299. S. Nakamura, M. Ikeda, H. Tsuji, Y. Nishizuka, and O. Hayaishi, *Biochem. Biophys. Res. Commun.*, *13*, 285 (1963).
300. P. W. Holloway and G. Popjak, *Biochem. J.*, *104*, 57 (1967).
301. F. Pan and H. Tarver, *Arch. Biochem. Biophys.*, *119*, 429 (1967).
302. G. Marino, A. M. Greco, V. Scardi, and R. Zito, *Biochem. J.*, *99*, 589 (1966).
303. A. Weinstein, G. Medes, and G. Litwack, *Anal. Biochem.*, *21*, 86 (1967).
304. R. J. Wurtman and F. Larin, *Biochem. Pharmacol.*, *17*, 817 (1968).
305. G. Litwack and J. M. Squires, *Anal. Biochem.*, *24*, 438 (1968).
306. G. N. Donnell, W. G. Ng, J. E. Hodgman, and W. R. Bergren, *Pediatrics*, *39*, 829 (1967).
307. J. S. Gulbinsky and W. W. Cleland, *Biochemistry*, *7*, 566 (1968).

308. C. Mittermayer, R. Bosselmann, and V. Bremerskov, *European J. Biochem.*, *4*, 487 (1968).
309. H. L. Gordon, T. J. Bardos, and Z. F. Chmielewicz, *Cancer Res.*, *28*, 2068 (1968).
310. H. J. Grav and R. M. S. Smellie, *Biochem. J.*, *94*, 518 (1965).
311. J. P. Durham and D. H. Ives, *J. Biol. Chem.*, *245*, 2276 (1970).
312. J. Robinson and E. A. Newsholme, *Biochem. J.*, *112*, 455 (1969).
313. J. F. Kuo and P. Greengard, *J. Biol. Chem.*, *245*, 2493 (1970).
314. O. Skold, *J. Biol. Chem.*, *235*, 3273 (1960).
315. A. Orengo, *J. Biol. Chem.*, *244*, 2204 (1969).
316. Y. Sugino and Y. Miyoshi, *J. Biol. Chem.*, *239*, 2360 (1964).
317. R. C. Nordlie and W. Arion, "Glucose-6-phosphatase," in S. P. Colowick and N. O. Kaplan, Eds., *Methods in Enzymology*, Vol. IX, Academic Press, New York, 1966, pp 619–625.
318. M. R. Stetten, *Biochim. Biophys. Acta*, *208*, 394 (1970).
319. M. C. M. Yip and W. E. Knox, *J. Biol. Chem.*, *245*, 2199 (1970).
320. S. E. Hager and M. E. Jones, *J. Biol. Chem.*, *242*, 5667 (1967).
321. V. P. Wellner, J. I. Santos, and A. Meister, *Biochemistry*, 7, 2848 (1968).
322. M. P. Argyrakis-Vomvoyannis, *Biochim. Biophys. Acta*, *166*, 593 (1968).
323. N. Fausto and J. L. Van Lancker, *J. Biol. Chem.*, *240*, 1247 (1965).
324. F. Wanka and C. L. M. Poels, *European J. Biochem.*, *9*, 478 (1969).
325. F. S. Markland and C. L. Wadkins, *J. Biol. Chem.*, *241*, 4136 (1966).
326. T. Ozaki and A. Kornberg, *J. Biol. Chem.*, *239*, 269 (1964).
327. J. S. Wiberg, *J. Biol. Chem.*, *242*, 5824 (1967).
328. R. Labow and F. Maley, *Biochem. Biophys. Res. Commun.*, *29*, 136 (1967).
329. R. J. Buccino and J. S. Roth, *Arch. Biochem. Biophys.*, *132*, 49 (1969).
330. B. Deus, H. E. C. Blum, and H. Holzer, *Anal. Biochem.*, *27*, 492 (1969).
331. A. S. Levit and G. Wolf, *Biochim. Biophys. Acta*, *178*, 262 (1969).
332. E. Ebner, D. Wolf, C. Gancedo, S. Elsasser, and H. Holzer, *European J. Biochem.*, *14*, 535 (1970).
333. R. Mertelsmann and H. Matthaei, *Biochem. Biophys. Res. Commun.*, *33*, 136 (1968).
334. P. A. Straat, P. O. P. Ts'O, and F. J. Bollum, *J. Biol. Chem.*, *243*, 5000 (1968).
335. L. A. Loeb, S. S. Agarwal, and A. M. Woodside, *Proc. Natl. Acad. Sci., U.S.*, *61*, 827 (1968).
336. C. D. Steuart, S. R. Arand, and M. J. Bessman, *J. Biol. Chem.*, *243*, 5308 (1968).
337. D. Beyersmann and G. Schramm, *Biochim. Biophys. Acta*, *159*, 64 (1968).
338. O. Westergaard and R. E. Pearlman, *Exptl. Cell Res.*, *54*, 309 (1969).
339. G. Wever and S. T. Takats, *Biochim. Biophys. Acta*, *199*, 8 (1970).
340. P. S. Fitt, F. W. Dietz, and M. Grunberg-Manago, *Biochim. Biophys. Acta*, *151*, 99 (1968).
341. K. Nakazawa, K. Ueda, T. Honjo, K. Yoshihara, Y. Nishizuka, and O. Hayaishi, *Biochem. Biophys. Res. Commun.*, *32*, 143 (1968).
342. M. E. Haines, I. R. Johnston, A. P. Mathias, and D. Ridge, *Biochem. J.*, *115*, 881 (1969).
343. L. Daneo-Moore and G. D. Shackman, *Biochim. Biophys. Acta*, *195*, 145 (1969).
344. F. Molemans, M. van Montagu, and W. Fiers, *European J. Biochem.*, *4*, 524 (1968).
345. H. G. Klemperer and G. R. Haynes, *Biochem. J.*, *104*, 537 (1967).
346. J. D. Vidre and J. D. Loerch, *Biochim. Biophys. Acta*, *159*, 551 (1968).
347. J. M. Ashworth and M. Sussman, *J. Biol. Chem.*, *242*, 1696 (1967).
348. H. Nikaido, K. Nikaido, and P. H. Makela, *J. Bacteriol.*, *91*, 1126 (1966).
349. T. Inouye, H. L. Nadler, and Y-Y. Hsia, *Clin. Chim. Acta*, *19*, 169 (1968).

RADIOMETRIC METHODS OF ENZYME ASSAY

350. C. Péaud-Lenoël and M. Axelos, *European J. Biochem.*, *4*, 561 (1968).
351. R. Kornfeld, S. Kornfeld, and V. Ginsburg, *Biochem. Biophys. Res. Commun.*, *17*, 578 (1964).
352. J. Preiss, L. Shen, E. Greenberg, and N. Gentner, *Biochemistry*, *5*, 1833 (1966).
353. M. Hirata and O. Hayaishi, *Biochim. Biophys. Acta*, *149*, 1 (1967).
354. B. Weiss, *J. Pharmacol. Exptl. Therap.*, *166*, 330 (1969).
355. J. M. Marsh, *J. Biol. Chem.*, *245*, 1596 (1970).
356. G. Brooker, L. J. Thomas, and M. M. Appleman, *Biochemistry*, *7*, 4177 (1968).
357. G. Krishna and L. Birnbaumer, *Anal. Biochem.*, *35*, 393 (1970).
358. G. Krishna, B. Weiss, and B. B. Brodie, *J. Pharmacol. Exptl. Therap.*, *163*, 379 (1968).
359. H-P. Bär and O. Hechter, *Anal. Biochem.*, *29*, 476 (1969).
360. F. F. Becker and M. W. Bitensky, *Proc. Soc. Exptl. Biol. Med.*, *130*, 983 (1969).
361. M. W. Bitensky, V. Russell, and M. Blanco, *Endocrinology*, *11*, 154 (1970).
362. G. Schultz, E. Böhme, and K. Munske, *Life Sci.*, *8*, 1323 (1969).
363. E. Böhme, *European J. Biochem.*, *14*, 422 (1970).
364. J. G. Hardman and E. W. Sutherland, *J. Biol. Chem.*, *244*, 6363 (1969).
365. P. Mattock and J. G. Jones, *Biochem., J.*, *116*, 797 (1970).
366. J. B. Adams and A. M. Edwards, *Biochim. Biophys. Acta*, *167*, 122 (1968).
367. T. Harada, S. Shimizu, Y. Nakanishi, and S. Suzuki, *J. Biol. Chem.*, *242*, 2288 (1967).
368. A. Kaplan, *Anal. Biochem.*, *33*, 218 (1970).
369. H. Chino and L. I. Gilbert, *Anal. Biochem.*, *10*, 395 (1965).
370. J. Boyer and H. Giudicelli, *Biochim. Biophys. Acta*, *202*, 219 (1970).
371. G. Brante, *Rec. Trav. Chim. Pays-Bas*, *74*, 370 (1955).
372. L. I. Gilbert, *Anal. Biochem.*, *10*, 400 (1965).
373. H. Winkler, A. D. Smith, F. Dubois, and H. van den Bosch, *Biochem. J.*, *105*, 38C (1967).
374. A. D. Smith and H. Winkler, *Biochem. J.*, *108*, 867 (1968).
375. H. Winkler, *Biochem. J.*, 38 (1967).
376. S. J. Friedberg and R. C. Greene, *J. Biol. Chem.*, *242*, 234 (1967).
377. M. Paysant, C. Soler, R. Wald, and J. Polonovski, *Bull. Soc. Chim. Biol.*, *48*, 863 (1966).
378. S. Gatt, *Biochim. Biophys. Acta*, *159*, 304 (1968).
379. L. T. Potter, *J. Pharmacol. Exptl. Therap.*, *156*, 500 (1967).
380. R. W. Disney, *Biochem. Pharmacol.*, *15*, 361 (1966).
381. K. G. Oldham, *Biochem. Pharmacol.*, *17*, 1107 (1968).
382. A. Belfield and D. M. Goldberg, *Life Sci.*, *8*, 129 (1969).
383. D. Shugar, A. Szenberg, and H. Sieraakowska, *Exptl. Cell Res.*, *13*, 424 (1957).
384. A. L. Fox and H. F. Taswell, *Exptl. Cell Res.*, *51*, 504 (1968).
385. W. F. Bridges, *J. Biol. Chem.*, *242*, 2080 (1967).
386. S. Linn and I. R. Lehman, *J. Biol. Chem.*, *241*, 2694 (1966).
387. L. Glaser, A. Melo, and R. Paul, *J. Biol. Chem.*, *242*, 1944 (1967).
388. A. Becker and J. Hurwitz, *J. Biol. Chem.*, *242*, 936 (1967).
389. N. Pfrogner, A. Bradley, and H. Fraenkel-Conrat, *Biochim. Biophys. Acta*, *142*, 105 (1967).
390. O. M. Rosen and S. M. Rosen, *Arch. Biochem. Biophys.*, *131*, 449 (1969).
391. T. Braun, O. Hechter, and H-P. Bär, *Proc. Soc. Exptl. Biol. Med.*, *132*, 233 (1969).
392. M. Vaughan and F. Murad, *Biochemistry*, *8*, 3092 (1969).
393. G. S. Levey and S. E. Epstein, *Circulation Res.*, *24*, 151 (1969).

394. D. Monard, J. Janecek, and H. V. Rickenberg, *Biochem. Biophys. Res. Commun.*, *35*, 584 (1969).
395. R. J. DeLange, R. G. Kemp, W. D. Riley, R. A. Cooper, and E. G. Krebs, *J. Biol. Chem.*, *243*, 2200 (1968).
396. M. Obinata and D. Mizuno, *Biochim. Biophys. Acta*, *155*, 98 (1968).
397. R. F. Greene, D. M. Kohl, and R. A. Flickinger, *Exptl. Cell Res.*, *48*, 685 (1967).
398. J. M. Scott and B. Spencer, *Biochem. J.*, *106*, 471 (1968).
399. R. O. Brady, J. N. Kanfer, R. M. Bradley, and D. Shapiro, *J. Clin. Invest.*, *45*, 1112 (1966).
400. D. M. Brown and N. S. Radin, *J. Neurochem.*, *16*, 501 (1969).
401. A. I. Lansing, W. E. Lynch, and I. Lieberman, *J. Biol. Chem.*, *242*, 1772 (1967).
402. P. Justice, C. Ryan, and D. Y-Y. Hsia, *Biochem. Biophys. Res. Commun.*, *39*, 301 (1970).
403. M. Wenzel, H-P. Lenk, and E. Schütte, *Z. Physiol. Chem.*, *327*, 13 (1961).
404. E. Eggermont, *European J. Biochem.*, *9*, 483 (1969).
405. H. Noll and J. Orlando, *Anal. Biochem.*, *2*, 205 (1961).
406. J. De Bersaques, *Anal. Biochem.*, *22*, 176 (1968).
407. S. Simmonds, *J. Biol. Chem.*, *241*, 2502 (1966).
408. M. K. Loken, K. D. Terrill, J. F. Marvin, and D. G. Mosser, *J. Gen. Physiol.*, *42*, 250 (1958).
409. M. Schneider, M. Ventrice, and B. Weiner, *Proc. Soc. Exptl. Biol. Med.*, *120*, 130 (1965).
410. M. von Tigerstrom and G. M. Tener, *Can. J. Biochem.*, *45*, 1067 (1967).
411. Y. Tasaka and J. Campbell, *Can. J. Biochem.*, *46*, 483 (1967).
412. P. P. K. Ho and L. Jones, *Biochim. Biophys. Acta*, *177*, 172 (1969).
413. J. H. Schwartz, J. Y. Reeves, and J. D. Broome, *Proc. Natl. Acad. Sci., U.S.*, *56*, 1516 (1966).
414. R. C. Gallo and S. Perry, *J. Clin. Invest.*, *48*, 105 (1969).
415. B. Zicha, G. B. Gerber, and J. Deroo, *Experientia*, *25*, 1039 (1969).
416. G. W. Camiener, *Biochem. Pharmacol.*, *16*, 1681 (1967).
417. F. Maley and G. F. Maley, *J. Biol. Chem.*, *235*, 2968 (1960).
418. W. H. Fleming and M. J. Bessman, *J. Biol. Chem.*, *242*, 363 (1967).
419. J. J. Scocca, S. R. Panny, and M. J. Bessman, *J. Biol. Chem.*, *244*, 3698 (1969).
420. M. Nishihara, A. Chrambach, and H. V. Aposhian, *Biochemistry*, *6*, 1877 (1967).
421. R. F. Ramaley, W. A. Bridger, R. W. Moyer, and P. D. Boyer, *J. Biol. Chem.*, *242*, 4287 (1967).
422. F. Klink, K. Kloppstech, and H. Netter, *Z. Naturforsch.*, *21*, 497 (1966).
423. Y. Nishizuka and F. Lipmann, *Proc. Natl. Acad. Sci. U.S.*, *55*, 212 (1966).
424. G. D. Birnie and S. M. Fox, *Biochem. J.*, *104*, 239 (1967).
425. G. Kahlson, E. Rosengren, and R. Thunberg, *J. Physiol.*, *169*, 467 (1963).
426. D. Aures, R. Hakanson, and C. Wiseman, *European J. Pharmacol.*, *8*, 232 (1969).
427. A. Albers, *J. Biol. Chem.*, *238*, 557 (1963).
428. P. Kraus, *Experientia*, *24*, 206 (1968).
429. P. B. Molinoff and E. A. Kravitz, *J. Neurochem.*, *15*, 391 (1968).
430. P. H. Strausbauch and E. H. Fischer, *Biochem. Biophys. Res. Commun.*, *28*, 525 (1967).
431. A. Raina and J. Janne, *Acta Chem. Scand.*, *22*, 2375 (1968).
432. D. H. Russell and S. H. Snyder, *Mol. Pharmacol.*, *5*, 253 (1969).
433. K. O. Wustrack and R. J. Levine, *Biochem. Pharmacol.*, *18*, 2465 (1969).
434. F. J. Leinweber, *Mol. Pharmacol.*, *4*, 337 (1968).
435. L. Ellenbogen, E. Markley, and R. J. Taylor, *Biochem. Pharmacol.*, *18*, 683 (1969).

436. K. A. Gumaa and P. McLean, *Biochem. J.*, *115*, 1009 (1969).
437. A. E. Pegg, D. H. Lockwood, and H. G. Williams-Ashman, *Biochem. J.*, *117*, 17 (1970).
438. A. D. Gounaris and L. P. Hager, *J. Biol. Chem.*, *236*, 1013 (1961).
439. F. J. Ballard and R. W. Hanson, *Biochem. J.*, *104*, 866 (1967).
440. L. E. Anderson and R. C. Fuller, *J. Biol. Chem.*, *244*, 3105 (1969).
441. P. Laduron and F. Belpaire, *Anal. Biochem.*, *26*, 210 (1968).
442. S. H. Snyder and J. Axelrod, *Biochem. Pharmacol.*, *13*, 805 (1964).
443. J. Vesely, A. Cihak, and F. Sorm, *Biochem. Pharmacol.*, *17*, 519 (1968).
444. J. S. Schutzbach and D. S. Feingold, *J. Biol. Chem.*, *245*, 2476 (1970).
445. D. K. Murali and A. N. Radhakrishnaa, *Biochem. Pharmacol.*, *15*, 735 (1966).
446. D. M. Shepherd and B. G. Woodcock, *Biochem. Pharmacol.*, *17*, 23 (1968).
447. W. S. Wegener, H. C. Reeves, and S. J. Ajl, *Anal. Biochem.*, *11*, 111 (1965).
448. P. Furmanski, H. C. Reeves, and S. J. Ajl, *Microchem. J.*, *13*, 279 (1968).
449. H. Nakagawa and H. Kimura, *Biochem. Biophys. Res. Commun.*, *32*, 208 (1968).
450. S. H. Mudd, J. D. Finkelstein, F. Irreverre, and L. Laster, *J. Biol. Chem.*, *240*, 4382 (1965).
451. J. G. Hollowell, M. E. Coryell, W. K. Hall, J. K. Findley, and T. G. Thevnos, *Proc. Soc. Exptl. Biol. Med.*, *129*, 327 (1968).
452. J. Koukol and E. E. Conn, *J. Biol. Chem.*, *236*, 2692 (1961).
453. C. J. Ackerman and S. Al-Mudhaffer, *Endocrinology*, *82*, 905 (1968).
454. D. S. Feingold, E. F. Neufeld, and W. Z. Hassid, *J. Biol. Chem.*, *235*, 910 (1960).
455. G. W. Jourdian and S. Roseman, *J. Biol. Chem.*, *237*, 2442 (1962).
456. B. W. Agranoff, H. Eggerer, and F. Lynen, *J. Biol. Chem.*, *235*, 326 (1960).
457. P. W. Holloway and G. Popjak, *Biochem. J.*, *106*, 835 (1968).
458. R. S. Piha, R. K. Airas, and L. I. Airi, *Suomen Kemistilehti*, *39*, 204 (1966).
459. P. D. Fullerton and L. R. Finch, *Anal. Biochem.*, *29*, 545 (1969).
460. D. J. Galton and T. R. Fraser, *Anal. Biochem.*, *28*, 59 (1969).
461. M. K. Patterson and G. Orr, *Biochem. Biophys. Res. Commun.*, *26*, 228 (1967).
462. E. Ito and J. L. Strominger, *J. Biol. Chem.*, *237*, 2696 (1962).
463. E. Ito, S. G. Nathenson, D. N. Dietzler, J. S. Anderson, and J. L. Strominger, "Formation of UDP-acetylmuramyl Peptides," in *Methods in Enzymology*, Vol. VIII, Academic Press, New York, 1966, pp. 324–337.
464. M. Cannon, *Febs Letters*, *8*, 217 (1970).
465. J. E. Allende, R. Monro, and F. Lipmann, *Proc. Natl. Acad. Sci.*, *U.S. 51*, 1211 (1964).
466. J. M. Wilhelm and J. W. Corcoran, *Biochemistry*, *6*, 2578 (1967).
467. D. M. Dawson, *J. Neurochem.*, *15*, 31 (1968).
468. P. Walter, V. Paetkau, and H. A. Lardy, *J. Biol. Chem.*, *241*, 2523 (1966).
469. W. Brech, E. Shrago, and D. Wilken, *Biochim. Biophys. Acta*, *201*, 145 (1970).
470. M. D. Greenspan and J. M. Lowenstein, *J. Biol. Chem.*, *243*, 6273 (1968).
471. I. Olsen and J. M. Merrick, *J. Bacteriol.*, *95*, 1774 (1968).
472. T. D. Thomas and R. D. Batt, *Anal. Biochem.*, *28*, 477 (1969).
473. J. B. Adams and A. M. Edwards, *Biochim. Biophys. Acta*, *167*, 122 (1968).
474. D. M. Prescott, *Science*, *167*, 1285 (1970).
475. R. J. Bayly, *Nucleonics*, *24*, No. 6, 46 (1966).
476. E. A. Evans, *Tritium and Its Compounds*, Butterworths, London, 1966.
477. B. Belleau, "The Significance of Deuterium Isotope Effects in the Formation of Bioreceptor-Substrate Complexes," in L. J. Roth, Ed., *Isotopes in Experimental Pharmacology*, University of Chicago Press, Chicago, 1965, pp. 469–481.

478. J. D. Davidson and V. T. Oliverio, "Methods for Pharmacological Studies of Methylglyoxal-bis(guanylhydrazone-^{14}C) in Man and Animals," in L. J. Roth, Ed., *Isotopes in Experimental Pharmacology*, University of Chicago Press, Chicago, 1965, pp. 345–352.
479. W. L. Porter and N. Hoban, *Anal. Chem.*, *26*, 1846 (1954).
480. E. A. Evans, H. C. Sheppard, and J. C. Turner, *J. Labelled Compounds*, *6*, 76 (1970).
481. W. R. Waterfield, J. A. Spanner, and F. G. Stanford, *Nature*, *218*, 472 (1968).
482. E. A. Evans, R. H. Green, J. A. Spanner, and W. R. Waterfield, *Nature*, *198*, 1301 (1963).
483. Y. Fujita, A. Gottlieb, B. Peterkofsky, S. Udenfriend, and B. Witkop, *J. Am. Chem. Soc.*, *86*, 4709 (1964).
484. H. S. Carr and H. S. Rosenkranz, *J. Bacteriol.*, *92*, 1840 (1966).

Polarography and Voltammetry of Nucleosides and Nucleotides and Their Parent Bases as an Analytical and Investigative Tool

PHILIP J. ELVING, JAMES E. O'REILLY,* AND CONRAD O. SCHMAKEL,†
The University of Michigan, Ann Arbor, Michigan

I. Introduction.. 291
   1. Objective and Scope of the Present Review...................... 291
   2. Significance of Electrochemical Studies of Biological Compounds.... 293
      A. Relation of Electrochemical Processes to Biological Processes.. 293
   3. Pertinent Literature... 295
II. Polarographic Techniques... 296
   1. Cyclic Voltammetry.. 297
   2. Alternating-Current Polarography.............................. 299
   3. Oscillopolarography... 299
   4. Controlled-Potential Electrolysis and Coulometry.............. 300
III. Interpretation of the Polarographic Behavior of Organic Compounds...... 301
   1. Data Obtainable: Significance and Interpretation................ 301
      A. The Half-Wave Potential................................... 301
         a. Potential Scale..................................... 303
      B. Current and Faradaic *n* Values........................... 303
      C. Reversibility Criteria..................................... 305
   2. Organic Electrode Reaction Mechanisms........................ 306
      A. General Mechanistic Path................................. 307
         a. Typical Reaction Path: Reduction of the Pyrimidine Ring 308
         b. Electroactive Sites................................. 309
         c. Effect of Substituents.............................. 310
      B. Kinetic Factors... 312
         a. Kinetics near the Electrode......................... 312
         b. Heterogeneous Electrode Kinetics................... 313
      C. Adsorption Phenomena................................... 314
         a. Adsorption Sites................................... 315
         b. Measurement of Adsorption......................... 315

* Present address: Department of Chemistry, University of Kentucky, Lexington, Kentucky 40506.

† Present address: Abbott Laboratories, Scientific Divisions, North Chicago, Illinois, 60064.

3. Correlations Involving Potentials.............................. 316
   A. Factors Involved...................................... 318
   B. Linear Free-Energy Correlations........................ 319
   C. Quantum-Mechanically Calculated Electronic Indices....... 321
      a. Relation of $E_{1/2}$ and k_i........................... 321
      b. Other Indices.................................... 323
      c. Applicability.................................... 323
   D. Miscellaneous Types of Correlations..................... 323
4. Quantitative Analysis...................................... 325
   A. Polarographic Analysis................................ 325
   B. Determination of Concentration......................... 326
   C. Titration... 327
   D. Measurement of Reaction Rates......................... 328
5. Other Areas of Application................................. 329
   A. Identification and Characterization...................... 329
   B. Determination of Chemical and Physical Properties......... 329
   C. Characterization of Chemical Reactions................... 330
   D. Association, Conformation, and Orientation............... 330
      a. Conformation and Orientation..................... 330
      b. Association...................................... 332
      c. Purine-Pyrimidine Interaction..................... 332
   E. Electrochemical Synthesis............................. 332
IV. Pyridine (Nicotinamide) Derivatives............................ 333
   1. Nomenclature... 333
   2. Redox Behavior... 334
   3. Ultraviolet Spectra...................................... 336
   4. Polarographic and Voltammetric Behavior.................... 337
   5. Nicotinamide... 337
      A. Voltammetry.. 337
         a. Direct-Current Polarography...................... 337
         b. Cyclic Voltammetry............................. 339
         c. Alternating-Current Polarography.................. 339
      B. Controlled-Potential Electrolysis...................... 340
      C. Reduction Mechanism................................ 341
   6. Model Compounds.. 344
      A. Voltammetry.. 347
         a. Wave I... 347
         b. Wave II.. 347
      B. Controlled-Potential Electrolysis...................... 348
      C. Adsorption Phenomena............................... 350
         a. Cyclic Voltammetry............................. 350
         b. Chronopotentiometry............................ 351
         c. Alternating-Current Polarography.................. 351
      D. Reduction Mechanism................................ 353
   7. Pyridine Nucleotides...................................... 354
      A. Direct-Current Polarography.......................... 357
         a. Effect of Ionic Strength.......................... 359
         b. Effect of Tetraalkylammonium Salts................ 359
         c. Interpretation of Results......................... 361
      B. Alternating-Current Polarography...................... 363
         a. Effect of Tetraalkylammonium Salts................ 364

        C. Cyclic Voltammetry..................................... 366
            a. Possible Involvement of Adenine Reduction........... 366
        D. Controlled-Potential Electrolysis........................ 368
            a. NAD⁺ (Wave I)................................ 368
            b. NAD⁺ (Wave II)............................... 370
            c. NMN⁺.. 371
            d. NADP⁺....................................... 371
        E. Reduction Mechanism................................. 372
            a. Structure of the Dimer........................... 373
    8. Biological Reduction Mechanism.............................. 377
    9. Analytical Application....................................... 379
V. Pyrimidine and Purine Derivatives................................. 380
    1. Nomenclature... 380
    2. Polarographic and Voltammetric Behavior..................... 381
    3. Pyrimidines.. 382
        A. Pyrimidine.. 383
            a. General Pyrimidine-Purine Reduction Mechanism..... 395
        B. Cytosine Series..................................... 397
            a. Reduction of Cytosine........................... 397
            b. Isocytosine................................... 399
            c. Nucleosides and Nucleotides...................... 399
        C. Hydroxypyrimidines.................................. 399
            a. 2-Hydroxypyrimidine............................ 399
            b. 4-Hydroxypyrimidine............................ 400
            c. Alloxan System................................ 400
            d. Polyhydroxypyrimidines......................... 401
        D. Uracil and Thymine Series............................. 401
        E. Azapyrimidines..................................... 401
        F. Thiopyrimidines.................................... 402
        G. Chloropyrimidines.................................. 402
        H. Oxidation of Pyrimidines............................. 403
    4. Purines: Reduction.. 405
        A. Purine... 405
        B. Adenine Series..................................... 407
            a. Adenine...................................... 407
            b. Adenine Nucleosides and Nucleotides............... 407
            c. Compounds Related to Adenine.................... 409
        C. Guanine Series..................................... 409
        D. Azapurines.. 410
        E. Hydroxypurines.................................... 411
        F. Thiopurines....................................... 411
            a. 6-Thiopurines................................. 411
            b. Purine-6-Sulfinic Acid........................... 412
            c. Purine-6-Sulfonic Acid........................... 413
    5. Purines: Oxidation.. 413
        A. Adenine.. 414
        B. Hydroxypurines.................................... 416
            a. Uric Acid..................................... 416
            b. Guanine...................................... 416
            c. Miscellaneous................................. 418

    C. Thiopurines... 418
        a. 6-Thiopurine.. 419
        b. Purine-6-Sulfinic Acid.................................. 419
    6. Correlation of Electrode Reaction Paths and Substitution........... 420
    7. Adsorption on Electrodes...................................... 421
        A. Mechanism of Adsorption.................................. 421
        B. Parent Bases.. 423
        C. Nucleosides and Nucleotides.............................. 423
    8. Applications: Analysis.. 424
        A. Control of Test Solution Composition..................... 425
        B. Determination of Pyrimidines: Reduction.................. 425
            a. Hydroxypyrimidines.................................. 426
            b. Uracil Derivatives.................................. 426
        C. Determination of Pyrimidines: Oxidation.................. 427
        D. Determination of Purines: Reduction...................... 427
            a. Purine... 427
            b. Adenine.. 428
            c. Hypoxanthine... 429
            d. Thiopurines... 429
        E. Determination of Purines: Oxidation...................... 430
            a. Hydroxypurines....................................... 430
            b. Thiopurines... 431
        F. Oscillopolarography....................................... 432
        G. Analysis of Mixtures..................................... 433
    9. Applications: Correlations Involving $E_{1/2}$..................... 434
        A. Correlation with pK_a.................................. 435
        B. Correlations with Linear Free-Energy Parameters.......... 435
        C. Correlations with Molecular Orbital Parameters............ 437
    10. Applications: Miscellaneous.................................. 438
        A. Chemical Kinetics.. 438
        B. Synthesis: Macroscale Electrolysis....................... 439
        C. Association, Conformation, and Orientation............... 439
VI. Flavin Nucleotides... 440
    1. Nomenclature... 440
    2. Redox Activity... 441
    3. Polarographic and Voltammetric Behavior...................... 442
        A. Riboflavin and Related Compounds........................ 442
            a. Adsorption of Riboflavin and Leucoriboflavin......... 444
            b. Catalytic Wave...................................... 446
            c. Behavior in Dimethyl Sulfoxide...................... 446
        B. Flavin Nucleotides....................................... 447
            a. Adsorption Prewave.................................. 448
            b. Adsorption Postwave................................. 449
    4. Electrochemically Calculated Molecular Areas.................. 450
    5. Complexation.. 450
    6. Biological Fuel Cells.. 450
    7. Analytical Application....................................... 451
    Acknowledgments... 454
    References.. 454

I. INTRODUCTION

1. Objective and Scope of the Present Review

The enormous growth in recent years in biochemical and medical research, for example, in the nucleic acid area, has created a need for new approaches to the study of the structure, chemical and biological reactivity, physicochemical properties, and other aspects of biologically important compounds.

The value of electrical measurements in elucidating and utilizing the oxidation-reduction behavior of organic compounds has long been recognized. However, since relatively few organic redox systems behave reversibly, valid potentiometric data have been obtained for only a limited number of compounds, for example, the work of Michaelis, Clark, and their collaborators, which was largely centered in the 1920s and early 1930s. Since then, investigation of the redox behavior of organic compounds has been based largely on polarography and derived techniques; half-wave potentials so determined are frequently the only energetic data readily obtainable by electrochemical measurement.

The need, for example, for reliable potential data is emphasized by the possible use of such data for correlation with quantum-mechanically calculated and other theoretical and experimental parameters. These, in turn, may be correlatable with biological phenomena.

The present review attempts a comprehensive and critical survey of the electrochemical behavior of nucleosides and nucleotides and their parent bases, which are heterocyclic nitrogen-containing compounds, and of key related model compounds. Emphasis is placed on elucidation of the pathways involved in their oxidation and reduction at solution-electrode interfaces and on correlation of these pathways and the associated energetic parameters with biologically relevant physical and chemical parameters and processes, from the viewpoints of shedding further light on biological phenomena and of providing additional methodology for their investigation.

One reason for the relatively small use made of polarographic techniques by chemists interested in biological compounds is their lack of sufficient theoretical background in electrochemistry. Although the latter is no more complicated, even if not yet as unified, as the theoretical aspects of radiant energy absorption, radiant energy absorption is generally taught in undergraduate chemistry courses as background for an understanding of chemical binding and structure. There is, though, the complication that electrode reactions are per se heterogeneous reactions, and, by and large, chemists are not oriented toward such reactions (e.g., the relatively minor attention given in undergraduate and graduate chemistry courses to such surface phenomena as adsorption and heterogeneous kinetics). It is hoped that the present review may serve in small part to remedy this situation.

Attention is focused on behavior in aqueous solution, since biological processes occur in such media. However, data obtained in nonaqueous media will be considered in cases where such data amplify the information obtained in aqueous media. Consequently, the bulk of the present discussion deals with electrochemical reduction, since, until relatively recently, systematic study of the electrochemistry of organic compounds in aqueous media was confined largely to their reduction at mercury electrodes, except for the rather limited number which were sufficiently readily oxidized and/or sufficiently reversible to be studied polarographically or potentiometrically at platinum electrodes. Study of electrochemical oxidation was limited by lack of an electrode which would not itself be oxidized at the relatively positive potentials necessary for investigating organic compounds, whose oxidation generally involves a high activation overpotential. This need has been met in large part by the development of a variety of carbon-based electrodes (1,2).

Where possible, account has been taken of the effect of variation in experimental conditions (e.g., pH, ionic strength, specific solution components, and temperature), and an attempt made to elucidate the redox reaction mechanisms for the individual compounds and related series of compounds. Such efforts optimally include, inter alia, specification of reaction sites in the molecule, evaluation of the role of adsorption in relation to electron-transfer processes and determination of adsorption sites in the molecule, and explication of the sequence of charge transfer, protonation, and intervening nonelectron-transfer chemical reactions in the overall redox reaction. In many instances, interpretative comments have been supplied on the data and explanations given in the literature; at times, the fragmentary nature of the latter have caused the present comments to be expressed essentially as questions.

Because of the differing natures of the published papers on the polarography of pyridine and flavin derivatives, on the one hand, and on the polarography of purine and pyrimidine derivatives, on the other hand, it has not been possible to organize the reviews of the electrochemical behavior of the two groups of compounds in exactly the same fashion or to treat the two groups equivalently; for example, the literature on the use of polarography for determining purines and pyrimidines is quite extensive, whereas that on nicotinamide derivatives is rather sparse. Furthermore, there is much less agreement concerning the basic mechanism for the electrolytic reduction of nicotinamides than concerning that for purines and pyrimidines.

This book provides a comprehensive review of the literature through the middle of 1970. Selected references through 1972 are also included.

2. Significance of Electrochemical Studies
of Biological Compounds

In the concluding chapter of *Quantum Biochemistry*, Pullman and Pullman (3) emphasize the significance of that aspect of biochemistry which "concerns the obviously essential importance of molecular systems with mobile electrons, and therefore of electronic delocalization, in the phenomena of life." This importance results from the chemical behavior of the key elements (hydrogen, carbon, nitrogen, oxygen) in living systems, which means that "all the essential biochemical substances which are related to or perform the fundamental functions of the living matter are constituted of completely or, at least, partially conjugated systems."

It is precisely the chemical configurations specified in the preceding paragraph which are involved in electrochemical oxidation and reduction. Since energy transfer and other processes in biological systems are frequently intimately associated with actual electron-transfer processes, information which can be obtained on the nature of the latter processes, for example, in respect to the energy levels (electrical potentials) and the reaction pathways involved, will be helpful in understanding, controlling, and utilizing such processes in biological and clinical situations. In general, polarographic techniques can yield information about the energetics, kinetics, and mechanisms of electron-transfer processes, the nature and kinetics of accompanying non-electron-transfer reactions, the accessibility of potentially oxidizable and reducible groups in the molecule, the nature of the molecule in solution, and its surface activity and behavior in respect to adsorption at electrically charged interfaces.

A. RELATION OF ELECTROCHEMICAL PROCESSES TO BIOLOGICAL PROCESSES

Although biological redox systems may be of considerable complexity, because, for example, of enzyme interactions, much potentially useful information can be obtained from examination by electrochemical techniques of the redox, solution, and adsorption behavior of compounds of interest in aqueous solution with respect to the nature of the factors controlling the electron-transfer process under conditions which approach those prevailing in a biological situation (e.g., in respect to reactions at interfaces in living cells). In particular, polarographic techniques offer the possibility of studying some aspects of the dynamics of the structure and behavior of such biological species as ATP, the nucleic acids, and important coenzymes.

Postulation of the degree of correlation that may exist between electrolytic and biological processes is still speculative; nevertheless, there is a similarity of conditions (cf. below) under which the two kinds of processes occur. Hopefully, knowledge gained in the investigation of the electrochemical behavior of compounds of biological interest may allow evaluation of the extent to which correlations can be made. The fact that analogous reaction paths have to be postulated to explain both the enzymatic and the electrochemical oxidation of two purines, that is, adenine (4) and uric acid (5), which are key compounds in biological processes, reflects a parallelism between the two types of processes. This supports the notion that study of the electrolytic mechanistic pathways for the oxidation and reduction of biologically significant compounds may be of service in elucidating their behavior in biological electron-transfer processes in respect to sites of oxidation and of reduction, reaction paths, experimental conditions, effect of substituents, and adsorption at the electron-transfer interface.

It is pertinent to note in this connection that, as mentioned, electrolytic reductions and oxidation often occur under conditions resembling those of enzymatic and other biological transformations:

1. Both types of processes involve heterogeneous electron-transfer reactions, in which the nature of the electrical double layer and the phenomena in the layer are important factors.

2. Both may involve previous occurrence of adsorption and/or formation of an adduct and may be accompanied by non-electron-transfer chemical reactions.

3. In both, there are generally stereochemical requirements that must be met.

4. Both occur in dilute aqueous solution of generally similar pH range with electrolyte compositions of comparable ionic strength.

5. The temperature range involved is also more or less comparable between electrochemical studies at 25° and biological reactions at animal body temperature.

6. The operative mass transport processes of convection and diffusion are not too dissimilar in electrolytic and biological processes.

To the extent that there is a similarity of conditions under which the two kinds of processes occur, knowledge gained in the investigation of electrochemical behavior may help to clarify and explain enzymatic and other biological processes, and may even lead to the preparation of compounds of interesting biological activity. One function of the present review is to provide background for a critical evaluation of the extent to which such correlations and extrapolations can be made.

In order to fully utilize the electrochemical information available, it is necessary to understand qualitatively and, where possible, quantitatively specific electrode-substrate phenomena which may not be present in biological systems, and which could, unless taken into account, invalidate correlations of biological and electrochemical data. Consequently, it will be necessary in the future to enlarge our knowledge of the extent of involvement of such phenomena as adsorption-desorption on the electrode surface and substrate-electrode reaction. On the other hand, such information on adsorbent-adsorbate interaction may be applicable in biological situations.

One area of reasonably valid immediate application of electrochemical data to the interpretation of biological and clinical phenomena would be in the study of solution reactions, which do not involve the electrode surface itself but which use the electroactivity of the compound (as measured by its potential and current) as an index to its nature, for example, investigation of the blocking or alteration of oxidizable and reducible sites in a molecule by solution species due to the formation of Lewis acid-base adducts, charge-transfer complexes, ion association, and other interrelated phenomena. Thus Palecek in Czechoslovakia (6) and Berg et al. in East Germany (7,8) have successfully applied polarographic methods to the study of native DNA and its polymeric degradation products, conformation changes in DNA, genetic relationships in bacterial DNA, and the estimation of de-natured DNA.

3. Pertinent Literature

In addition to the many references on specific compounds, techniques, and matters of methodology and experiment scattered throughout this chapter, attention should be called to the fact that desired information—if available in the extensive literature on organic electrochemistry under controlled potentiostatic and galvanostatic conditions—can be conveniently located in the 1969 bibliography on polarography covering the period 1922–67 (9) and in the annual bibliographies issued by the Prague and Padua polarographic centers (10,11). The current literature is summarized in biennial reviews (April of even-numbered years) in *Analytical Chemistry* by Bard, Hume, Nicolson, Pietrzyk, Reinmuth, Stock, Wawzonek, and others. A helpful listing of books and review chapters and papers will be found on pp. 254–260 of Zuman, *Organic Polarographic Analysis* (12).

In the specific area of the present chapter, pertinent references will be found in the reviews of the polarography of purines and pyrimidines by Elving et al. (13), and of nucleosides and nucleotides by Janik and Elving (14). Particular reference should be made to Clark's excellent book, *Oxida-tion-Reduction Potentials of Organic Systems* (15), Volke's reviews of the polarog-raphy of heterocyclic compounds (16,17), Vetterl's papers on the adsorption of bases, nucleosides, and nucleotides on mercury electrodes (18,19), and the Brezina and Zuman book entitled *Polarography in Medicine, Biochemistry and Pharmacy* (20). Müller (21) has reviewed the catalytic hydrogen polaro-graphic wave given by proteins, amino acids, and some other compounds (the Brdička reaction); Homolka (22) in a Czechoslovak book has described the use of polarography in diagnosing malfunctions, including cancer of various types; and Palecek (6) has reviewed polarographic techniques in nucleic acid research in a volume which also contains a useful summary by Pullman and Pullman of quantum-mechanical calculations (methodology and results) on nucleic acid constituents (23). Underwood and Burnett have reviewed the electrochemistry of biological compounds (23a).

A useful auxiliary reference in considering the electrochemical behavior of purine and pyrimidine derivatives, especially in regard to variation with pH, is the review by Izatt, Christensen, and Rytting (23b). It deals with the sites and thermodynamic quantities, including pK_a values, associated with proton and metal ion interaction of RNA, DNA, and their constituent bases, nucleosides, and nucleotides.

II. POLAROGRAPHIC TECHNIQUES

No attempt will be made to review the theory, methodology, and instrumentation of the various polarographically related techniques which have been and are being used to investigate the electrochemical behavior of nucleosides and nucleotides and their parent bases, except to indicate the type of information obtained and its interpretation. Section III of this review covers the latter aspect.

It is assumed that the reader is acquainted with the basic principles of polarography at the dropping mercury electrode (D.M.E.). Reviews of varying degrees of comprehensiveness in regard to the theoretical and practical fundamentals are readily available (e.g., refs. 24–28). The terms *polarography* and *voltammetry* are often used interchangeably, even though some electroanalytical chemists would prefer to define them as referring to techniques based on current-potential curves obtained under essentially constant-potential conditions at the D.M.E. and at constant-area electrodes, respectively. However, in the present chapter these two words, when used without a modifier, will with their adjectival forms generally refer to techniques which have developed from polarography as originally described by Heyrovsky.

A few relevant aspects of current instrumentation should be noted. Probably the most significant development has been the increasing availability of operational amplifier-based systems utilizing solid-state components for the convenient application of a variety of electrochemical techniques (29,30). Other developments of particular relevance include (a) three-electrode configurations, usually involving operational amplifier control systems, which have largely minimized the problems associated with the large iR drop experienced on the polarography of organic solvent solutions of high resistance; (b) reliable indicating electrodes based on

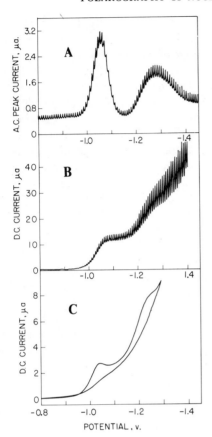

Figure 1. Response of purine in pH 4.1 aqueous acetate buffer solution (ionic strength: 0.5M). (a) Alternating-current polarography at D.M.E.; dc scan rate, 2 mV/sec; ac amplitude, 10 mV peak-to-peak; frequency, 50 Hz; 2mM purine. (b) Direct-current polarography at D.M.E.; scan or polarization rate, 2mV/sec; 2mM purine. (c) Cyclic voltammetry at H.M.D.E.; scan rate 20 mV/sec, 0.5mM purine.

graphite, which have extended the potential range available for studying electrochemical oxidation; (c) perturbation techniques, such as cyclic and potential-step voltammetry, which may permit separation of the electron-transfer step from subsequent reactions of the reduced or oxidized species; (d) alternating-current polarography with special reference to the phase-selective approach; and (e) optical and magnetic resonance techniques for identifying electrode reaction products and intermediates.

Comparative polarograms produced by dc polarography, ac polarography, and cyclic voltammetry are illustrated in Figures 1 and 15.

1. Cyclic Voltammetry

Cyclic voltammetry involves application of a triangular potential sweep to a constant-area electrode such as a hanging mercury drop, whose potential is, consequently, rapidly varied from, for example, 0 to −2 V and then back to 0 V, followed by repetition of this cycle as desired; usually the current is continuously measured. The rate of potential change (polarization, sweep, or scan rate), dE/dt or v, is usually in the range of 0.02 to 20 V/sec.

If the original electroactive species produces a product (or a chemically altered product of the one formed on the original reduction or oxidation) which can be electrolytically converted back to the original species (or some other species), a composite cathodic-anodic pattern may result because of the redox product of the original species generated during one part of the triangular sweep being electrolyzed back during the return portion of the sweep. For example, in some cases, especially in nonaqueous media, a pattern can be observed for the $1e$ production of a free radical and for its electrolytic conversion back to the original product (Figure 2).

The value of cyclic voltammetry in mechanistic studies of electrode

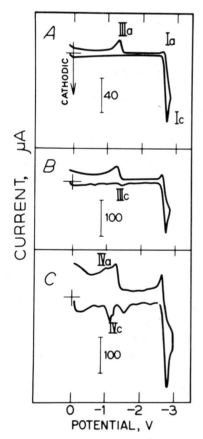

Figure 2. Steady-state cyclic voltammograms of pyrimidine (1.03mM) in acetonitrile. Scan rates: (a) 3 V/sec; (b) 15 V/sec; (c) 60 V/sec. Peaks are identified in O'Reilly and Elving (249).

processes has been amply demonstrated; when used over properly selected ranges of initial and final potentials and with a wide range of scan rates, it facilitates identification of reversible redox couples and frequently of electro-active intermediates formed chemically or electrolytically. Papers on the theory of stationary electrode polarography by Shain, Nicholson, and others (e.g., refs. 31–47) have provided possible means for quantitatively resolving the electrochemical and associated chemical steps in electrode processes, including the estimation of rate constants for intervening and accompanying chemical reactions.

2. Alternating-Current Polarography

Increasing use is being made in the investigation of organic electrode processes of alternating-current (ac) polarography and tensammetry (48–50a), in which a sinusoidal wave alternating potential (about 10 to 500 Hz) of a few millivolts amplitude is superimposed on the conventional dc potential used in D.M.E. polarography. Emphasis is being increasingly placed on the utility of this approach in obtaining information regarding the reversibility of the energy-controlling electron-transfer step, the presence of coupled chemical steps, and adsorption of reactant and product on the electrode and other phenomena occurring in the double layer. In some cases, measurement of rate constants for the charge-transfer and chemical reactions involved may be possible.

A recent factor of importance is the observation of ac polarographic waves for compounds which were not expected to undergo reversible electrode processes, as in Figure 1, where it is apparent that the second wave is better defined on ac than on dc polarography. An even more important development is the availability of instrumentation which allows separate measurement of the current component in phase with the applied alternating voltage, which is primarily associated with faradaic processes and their kinetic aspects, and the quadrature or 90° out-of-phase current component, which is associated with capacitive factors such as alteration of the electrode-solution interface due to adsorption as well as with faradaic processes.

The theory of ac polarography for irreversible systems has been presented by Timmer et al. (51,52) and by Smith and McCord (53).

The interpretation of ac polarographic curves relative to elucidation of the individual steps in the electrode process is explained in the discussion of the results for nicotinamide model compounds (Section IV).

3. Oscillopolarography

Oscillographic polarography (oscillopolarography), in which the D.M.E. is polarized with a constant-magnitude alternating current, generally about

50 Hz, and the variation of dE/dt with E is recorded (54), has been a useful tool, primarily in the hands of Czechoslovak investigators, for the study of nucleic acids and their constituents, for example, for the estimation of the guanine plus cytosine content of DNA (55), for the detection and estimation of denatured DNA (56), and for following the course of thermal denaturation of DNA (56,57) and the genetic relationships of bacterial DNA (58).

Characteristic electrochemical phenomena (e.g., faradaic processes and variation in double-layer capacity due to adsorption-desorption processes) manifest themselves as notches, commonly referred to as incisions or indentations, in the roughly circular plot usually obtained. The qualitative characteristic (relative position on the potential axis of an incision) is usually expressed as the Q value, for which Q = (linear distance of incision peak from potential of anodic mercury dissolution)/(linear distance between latter potential and that of background electrolyte discharge).

4. Controlled-Potential Electrolysis and Coulometry

Frequently the relatively old technique of electrolysis at controlled electrode potential (59–61) is still essential in elucidating organic electrode processes, since such electrolyses will usually allow determination of the faradaic n value (number of electrons transferred per molecule oxidized or reduced) and preparation of sufficient product or products for identification and characterization, as well as for calculation of a material balance in terms of compound electrolyzed and products produced. These latter objectives can generally be met by examination of the electrolyzed test solution by separation processes such as paper, thin-layer, column adsorption, and ion-exchange chromatography, lyophilization, and extraction, combined with measurement processes which give both qualitative and quantitative data (e.g., ultraviolet and infrared absorption spectrophotometry, gas chromatography, and polarography).

It is frequently informative to follow the course of electrolysis by monitoring the changes with time in the nature and concentration of the species present by suitable analytical techniques, such as repeated scanning of the ultraviolet absorption of the test solution or repeated papergramming of portions of it. In some cases, for example, in studies of the NAD system, it is helpful to determine the activity of the electrolyzed solution enzymatically and to carry the test solution through cycles of alternate reduction and oxidation with enzymatic monitoring at each stage.

The products of transitory electrolysis at microelectrodes such as the D.M.E. and of exhaustive electrolysis at large electrodes are generally similar except when chemical rate phenomena can cause differences. Controlled-potential electrolyses have permitted the elucidation of complex electrode processes with unexpected products, for example, the probable presence of dicarbonium-ion intermediates in the oxidation of uric acid (5) and adenine (4).

Two increasingly important applications of controlled-potential electrolysis are in the generation of free radicals for study by electron spin resonance and in the development of electrolytic methods of organic synthesis. Thus the presence and characteristics of a free-radical intermediate of sufficiently long life can at times be advantageously shown by running the solution from an electrolysis cell directly into the cavity of an electron spin resonance spectrometer.

The results of controlled-potential electrolyses are subsequently discussed in connection with the polarography of the compounds involved, since the potentials at which the electrolyses were made were almost invariably selected on the basis of the polarographic data.

Other polarographically derived electrochemical techniques will be briefly explained where they are first introduced.

III. INTERPRETATION OF THE POLAROGRAPHIC BEHAVIOR OF ORGANIC COMPOUNDS

As previously indicated, no attempt is being made in the present chapter to examine thoroughly the underlying theory, the interpretation, or the utilization of polarographic data. The following discussion is intended only to emphasize the principal factors involved in utilizing polarographically obtained numbers for producing information concerning nucleotide-related species and to include adequate references to enable an interested reader to pursue a subject in suitable detail.

1. Data Obtainable: Significance and Interpretation

A. THE HALF-WAVE POTENTIAL

For the purposes of this review, the most important characteristic of a compound, when examined electrochemically, is the half-wave potential, $E_{1/2}$, which in the case of a thermodynamically reversible redox half-reaction system would correspond to the formal potential, $E_c{}^0$ or E_0', for the specific experimental conditions (corrected for diffusion coefficient differences), but in the case of an irreversible system would correspond to the formal potential plus terms involving the diffusion coefficients, the activation energy for the heterogeneous electron-transfer process, certain experimental conditions, and perhaps the rate constants for chemical steps in the overall electrode reaction process (cf. refs. 25, 62, and 63 for a summary of the nature of $E_{1/2}$ and the factors which enter into its measured value).

The ease of electrochemical reduction or oxidation of a given functional group, as measured by its $E_{1/2}$, may be markedly affected by other substituents on the molecule and stereochemical factors, since the ease of introducing or removing an electron into or from a functional group is obviously dependent on the electron density at the reactive site, which is affected to a greater or lesser degree by the whole molecular constitution and configuration. Thus the $E_{1/2}$ of a functional group allows to some extent the specific characterization of a molecule in terms of its functional group, for example, determination of the amounts of cis and trans isomers in a mixture or of aromatic carbonyl groups in the presence of aliphatic ones.

The most common experimental factor influencing the $E_{1/2}$ of a given compound is the pH of the solution. This is due primarily to the facts that hydrogen ions may participate in the energy-controlling step of the half-reaction for the reduction or oxidation of the electroactive function and that the pH may control the chemical nature of the electroactive species, as in an acid-base equilibrium (cf. ref. 63 for a discussion of the way in which pH may affect the $E_{1/2}$ of organic compounds). The change in $E_{1/2}$ with pH is often used as a basis for drawing inferences concerning the role of hydrogen ions in the electrode reaction. In the case of many reversible and irreversible organic electrode processes of the type

$$Ox + ne + pH^+ = Red \qquad [1]$$

the variation of $E_{1/2}$ with pH can be described by the relationship at 25°;

$$\frac{d(E_{1/2})}{d(\text{pH})} = \frac{-0.059p}{\alpha n_a} \qquad [2]$$

where αn_a can be replaced by n in the case of a reversible half-reaction (symbols are subsequently defined). Consequently, the variation of $E_{1/2}$ with pH allows ready approximation of p (cf. refs. 25 and 63); for example, a shift of 60 mV per pH unit change is often taken as an indication that the numbers of electrons (n) and of protons (p) involved in the electrode reaction are identical.

In general, where reference is made to such behavior as "the wave is strongly pH dependent," the phenomenological fact will be that the $E_{1/2}$ of the wave shifts considerably in magnitude with change in pH (e.g., 60 mV per unit pH change).

In addition to the pH, the background electrolyte and other solution components, such as surfactants, may markedly affect the observed polarographic behavior, especially $E_{1/2}$, of organic compounds through, among others, the following factors and phenomena: (a) composition, structure, and potential gradient in the double layer; (b) competitive adsorption on the electrode of electroactive species, intermediates and products, and other solution components; (c) effect on activities of electroactive species, intermediates and products of ionic strength, equilibrium constants, and so on; (d) solvation of such species; (e) their complexation, including ion pairing, protonation, and other Lewis acid-base adduct equilibria, and charge-transfer complex formation; (f) proton activity and concentrational stability (e.g., buffering); and (g) kinetics of the various equilibria listed as well as of others involving the electroactive species, for example, its formation from a more stable solution component.

a. Potential Scale. The signs of the potentials for the half-reactions or half-cell reactions discussed are determined by the fact that the equations for the reactions are formulated as reductions with the usual thermodynamic conventions, that is, as

$$Ox + ne = Red \qquad [3]$$

with a positive potential indicating an equilibrium to the right, as compared to the hydrogen-ion reduction half-reaction at standard conditions:

$$H^+ + e = 0.5H_2 \qquad [4]$$

The numerical values of all potentials cited in this chapter, unless otherwise explicitly indicated, are referred to the saturated calomel electrode (S.C.E.) as reference electrode, in accordance with the well-nigh universal practice in polarography and voltammetry; this electrode has a potential of 0.242 V against the normal hydrogen electrode (N.H.E.) at 25°C. Consequently, potentials, E, can be converted from one reference basis to the other by the following relation:

$$(E \text{ vs. N.H.E.}) = (E \text{ vs. S.C.E.}) + 0.242 \qquad [5]$$

Thus $E_{1/2}$ for the reduction of Na(I) to the metal amalgam is about -2.1 V versus S.C.E.

B. CURRENT AND FARADAIC n VALUES

The equations for the faradaic current obtained at various types of polarographic electrodes usually contain (a) a numerical term involving geometric, material, and other fixed or temperature-dependent constants, for example, π and the density of mercury; (b) a group of terms involved in defining the electrode area and other experimentally imposed conditions; and (c) a group of terms characteristic of the electroactive species and of the specific electrode reaction giving rise to the current. Thus the average limiting current at the D.M.E. for a diffusion-controlled electrode process at 25°C is

$$i_l = 607nD^{1/2}Cm^{2/3}t^{1/6} \tag{6}$$

and the peak current at 25°C for an essentially irreversible electrode process in stationary electrode voltammetry is defined by

$$i_p = 2.99 \times 10^5 n(\alpha n_a)^{1/2}AD^{1/2}Cv^{1/2} \tag{7}$$

where i_l and i_p are in microamperes, n is the faradaic number of electrons transferred per molecule of reactant electrolyzed, C is the concentration of the electroactive species in millimoles per liter, D is its diffusion coefficient in square centimeters per second, m is the flow of mercury in milligrams per second, t is the lifetime of mercury drop in seconds, α is the transfer coefficient and n_a is the number of electrons involved in the rate-determining step of the electrode process, A is the electrode area in square centimeters, and v is the scan or polarization rate, dE/dt, in volts per second.

It is immediately apparent how faradaic n values can be approximated from [6] by a reasonable estimation of D or by comparison with the limiting current for a compound of known n and expected similar D.

Deviations from the equations expected to be operative can often be used, for example, to determine whether the faradaic current is controlled by factors other than diffusion, as expressed by [6] (2), or whether the observed current is due to an electrochemical step followed by a chemical step to produce a species electroactive at the applied potential, which results in a second electrochemical step (a so-called ECE process) that would be indicated by the nonlinear variation of the peak current with concentration and with the square root of scan rate (see [7]).

Thus a common diagnostic test used to determine the nature of the process controlling the current at the limiting current portion of the wave obtained at the D.M.E. is based on the variation of the limiting current with change in the hydrostatic mercury height, h, in the capillary, which varies m and t of [6]. Theory predicts that the current will be proportional to $h^{1/2}$ for a diffusion-controlled or limited electrode process and to h^0 for a kinetic-controlled process, that is, one limited by the rate of a chemical reaction which generates the electroactive species; where adsorption is involved, the limiting current should vary directly with h. Temperature variation of the limiting current is also used for the same purpose: a temperature coefficient of about $+2\%$ per degree indicates a diffusion process; one of about $+8$ to $+10\%$, kinetic control; and a negative temperature effect, adsorption.

Since n_a is commonly 1 and α generally varies between 0.3 and 0.7, the term $(\alpha n)^{1/2}$ in [7] is usually between 0.55 and 0.84.

C. REVERSIBILITY CRITERIA

Determination of the degree of thermodynamic reversibility of an electrode reaction is often important in attempting to apply certain theoretically derived equations and arguments for evaluating the nature of the reaction from the experimentally observed behavior, as well as in certain types of possible correlations involving $E_{1/2}$ values (cf. Section III.3). Unfortunately, in the case of electrode reactions the term "reversibility" is often used rather loosely; for example, if the cathodic and anodic peak potentials for a redox couple on cyclic voltammetry are separated by about the theoretically expected potential difference for a reversible couple, the couple is considered reversible. Terms such as "nearly reversible," "quasi-reversible," and "high (low) degree of reversibility" sprinkle the literature. Essentially, deviation from reversibility is indicated by the presence of an activation energy in the form of an overpotential.

A ready quantitative assessment of reversibility is provided by ac polarography. Thus, as mentioned, for a reversible electron-transfer process, $E_{1/2} = E_0'$, the formal electrode potential (due allowance being made for the ratio of the square roots of the diffusion coefficients of the oxidized and reduced species). Then $E_{1/2}$ is simply related to the standard free energy of the process, ΔG°. Any deviation of $E_{1/2}$ from E_0' can be expressed as

$$E_{1/2} = E_0' + \eta \qquad [8]$$

where η is the overpotential, that is, the extra energy required to cause the reaction to proceed at a reasonable rate which, accordingly, is a measure of departure from reversibility. Comparison of the ac polarographic currents for two related compounds, involving the same number of electrons in the rate-determining step and in any subsequent steps contributing to the overall ac current, should give a measure of the departure from reversibility for these compounds, that is, the extent to which $E_{1/2}$ would be expected to differ from E_0' [or from quantum-mechanically calculated LEMO (lowest empty molecular orbital) values or other parameters for which reversibility is assumed; cf. the subsequent discussion].

A reversibility factor can be calculated by the relation

$$\text{Reversibility factor} = (\Delta i_s)_{\text{expt}} / (\Delta i_s)_{\text{theory}} \qquad [9]$$

where Δi_s is the total faradaic alternating current (A) on ac polarography, measured experimentally (expt) or calculated from the following theoretical equation (theory):

$$(\Delta i_s)_{\text{theory}} = \frac{n^2 F^2 V A \omega^{1/2} D_0^{1/2} C}{4RT} \qquad [10]$$

where n is the number of electrons in the reaction producing the ac wave, F the faraday, V the amplitude of the applied alternating voltage (V), A the maximum area of the mercury drop (cm²), ω the angular frequency (sec⁻¹), D_0 the diffusion coefficient of the electroactive compound involved (cm²/sec), C its concentration (mM/cm³), R the gas constant, and T the absolute temperature.

Slopes of plots of log $[i/(i_d - i)]$ versus E, where i is the current at potential E for a D.M.E. polarogram and i_d is the diffusion-controlled limiting current, are frequently used to evaluate the reversibility of the process producing a polarographic wave. A value of about 0.059 V at 25°C (e.g., 55 to 65 mV) is taken to indicate a one-electron (1e) reversible process; a value of about 30 mV would then indicate a reversible 2e process. Values appreciably greater than 60 mV are usually taken to indicate irreversible electrode processes (cf. refs. 24–28). These plots are often referred to as log current or log current-potential plots. A similar criterion, commonly referred to as the wave slope, is based on using only the potentials at $0.25i_d$, $E_{1/4}$, and at $0.75i_d$, $E_{3/4}$, and the resulting equation at 25°C:

$$\text{Slope} = E_{1/4} - E_{3/4} = \frac{0.056}{n} \qquad [11]$$

which would obviously give a value of 56 mV for a 1e reversible process.

Unfortunately, the difficulty of measuring the slope to +5 mV for many of the polarogaphic waves obtained for organic compounds often makes this test of only limited diagnostic value. Not infrequently, values of the slope which do not fit an integral value of n to a reasonable approximation are taken to indicate an irreversible wave and are fitted to the expression

$$\text{Slope} = \frac{0.059}{\alpha n_a} \quad \text{or} \quad \frac{0.056}{\alpha n_a} \qquad [12]$$

2. Organic Electrode Reaction Mechanisms

Current views regarding the mechanisms of organic electrode processes have been perceptively summarized by Perrin (62), who emphasizes, in common with others, that, because an electrochemical reaction is a heterogeneous one, a completely satisfactory description of its mechanism must await a detailed understanding of phenomena in the electrical double layer and the surrounding solvent layer. At present, descriptions of organic electrode processes frequently still consist only of a specification of the electrode reaction products (at times of questionable identification) with, in some cases, more or less speculative postulation of reaction intermediates. Only in relatively few cases is the experimental evidence sufficiently detailed to permit postulation of structures for the transition states involved.

More extensive application of electrochemical investigative techniques of increased information output, supplemented by optical and magnetic examination of solutions during and after electrolysis, in systematic studies of specific organic functional systems in a variety of compounds should aid in defining such important factors in redox reactions as the extent and manner of adsorption and orientation of reactant and products at the electron-exchange surface; the steps in the overall redox reaction, including the reversibility of each step and the presence of preceding, accompanying, and following chemical reactions; the role of free-radical and other short- and long-lived intermediate species; and the changes, if any, produced on the electrode surface during the redox reaction (e.g., film formation). It should then be possible, for example, to establish more firmly the sequence of electron-transfer and non-electron-transfer steps in a redox process.

An important aspect of mechanism elucidation is the employment of chemical effects, such as the use of aprotic solvents to isolate proton-dependent from proton-independent reactions and the addition of inter-mediate traps (e.g., nucleophiles for cation radicals, and proton sources for anion radicals). The latter is exemplified by the use of pyridine to form adducts with carbonium ions (64,65).

A. GENERAL MECHANISTIC PATH

Electrochemical reduction is illustrated, since reduction, as previously discussed, has been investigated much more extensively than oxidation.

The fundamental process in an organic electrochemical reduction is bond rupture, which requires only one electron to produce a free-radical species; addition of a second electron completes rupture of the bond to give a carb-anion. As a consequence, from the mechanistic viewpoint, only $1e$ or $2e$ processes need to be considered, even though polarographic waves due to multiple-electron processes are frequently observed as a result of the in-stability at the potential of its formation of the species produced in the first electron-transfer step, perhaps following an intermediate chemical reaction, and, consequently, its immediate further reduction.

On this basis, a general mechanistic pattern has been outlined (66) for organic electrode reactions, which serves not only to rationalize the general course of such processes but also to explain changes in mechanism with experimental conditions or between members of a homologous series. The pattern (Figure 3) assumes a generalized reaction site, $R:X$, where R repre-sents the reactive carbon center and X represents another carbon, oxygen, nitrogen, halogen, or other atom; there may be more than one bond between R and X, as in a $C=N$ linkage.

In the primary step of the electrode process, the reaction site accepts a single electron to form the electrode-activated complex, which can either

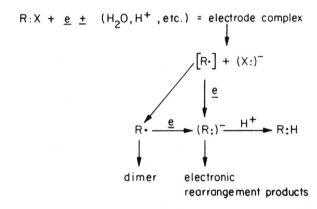

Figure 3. Generalized mechanistic path for an organic electrode reaction, in which R:X represents a generalization reaction site; the individual steps in the path are discussed in the text.   Allowance may have to be made in the above formation for (1) the participation of protons, solvent, other solution constitutuents, and the electrode surface in various steps; (2) the occurrence of chemical reactions preceding, accompanying, and following charge-transfer processes; and (3) resulting modification of the species shown.

revert to the original species or dissociate to give a free-radical precursor and an anionic species; if a multiple bond was originally present between R and X, these two species could form a single free-radical anion. The exact nature of these species will be modified by the extent of the participation of protons, solvent molecules, and other solution constituents or even the electrode surface (participation of these species in subsequent steps is not explicitly indicated but may be involved).

The free-radical precursor can either immediately on formation accept a second electron and be reduced to a carbanion or can exist as a stable free-radical species, which eventually either dimerizes or, at more negative applied potential, is further reduced to the equivalent of a carbanion. The charge on the carbanion, formed by either path, is usually neutralized, either by acceptance of a proton from the solution or by electronic rearrangement with the charge being transferred to another part of the molecule where it can be suitably handled.

Chemical reactions preceding, accompanying, or following charge transfer may (and often do) play significant roles in the overall process, as subsequently indicated.

a. Typical Reaction Path: Reduction of the Pyrimidine Ring.
The complex pattern observed for the polarographic reduction of pyrimidine, in which five waves appear (Figure 4), can be explained (67,68) on the basis

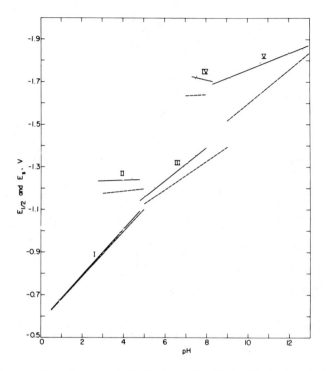

Figure 4. Variation of E_s (solid line) and $E_{1/2}$ (dashed line) with pH for pyrimidine (68,133).

of the general electrode reaction mechanism outlined (Figure 5): Wave I results from $1e$ reduction of the 3,4 bond with the simultaneous acquisition of a proton to form a free radical, which may dimerize to 4,4'-bipyrimidine or be reduced at more negative potential in a further $1e$ process (wave II) to 3,4-dihydropyrimidine. The strong pH dependence of wave I and the slight pH dependence of wave II result in their merging at about pH 5 to form pH-dependent $2e$ wave III. Essentially, pH-independent wave IV, which appears at higher pH, represents a $2e$ reduction at still more negative potential of the dihydropyrimidine to a tetrahydropyrimidine. The difference in pH dependence between waves III and IV causes their merging at about pH 9 to form pH-dependent $4e$ wave V.

b. Electroactive Sites. Reduction of nicotinamide derivatives involves the pyridine ring; that of flavin derivatives, the isoalloxazine moiety. Purine and pyrimidine derivative reduction essentially follows the course outlined for pyrimidine itself with possible modification as discussed in the sections

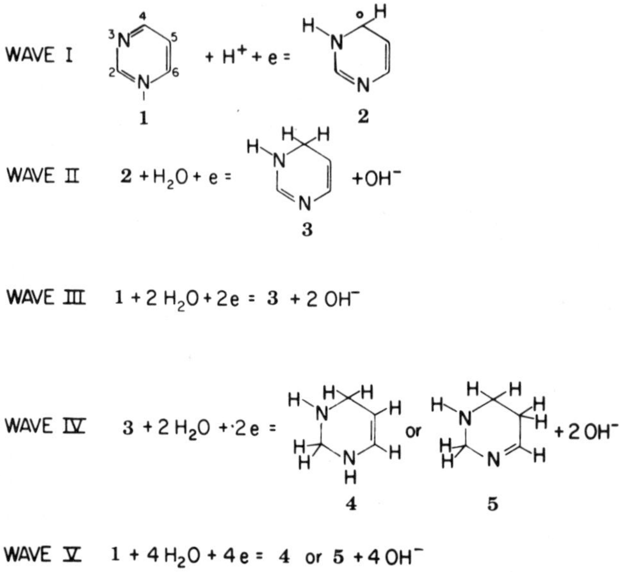

Figure 5. Reactions producing the five polarographic waves observed (cf. Figure 4) in the electrochemical reduction of pyrimidine (compound 1).

dealing with these compounds. It is probably superfluous to note that the presence of more readily reducible substituent groups on the heterocyclic base, for example, the nitro group, will result in prior reduction of such substituent.

Generalizations concerning oxidation sites are not feasible because of the paucity of data; the only two purines whose electrochemical oxidation has been thoroughly investigated are apparently attacked on oxidation at the double bond common to both rings.

c. Effect of Substituents. Although electrochemical oxidation and reduction in nucleosides and nucleotides occur primarily in the heterocyclic ring moiety, the sugar and sugar-phosphate groups (e.g., ribose and ribosophosphate) influence diffusion, association, conformation, adsorption, and electron density at reactive sites; the last effect will generally manifest itself in potential shifts and possibly in alteration of the reaction path for the overall redox process. More detailed study is necessary for estimation of the influence of the sugar and sugar-phosphate moieties on the redox behavior of the parent base and for meaningful quantitative correlations of potentials with structure, reactivity indices, and biological phenomena; in particular, as previously mentioned, only negligible data are available for oxidation processes.

Since the electrochemical behavior of polymeric nucleic acids and synthetic polynucleotides is also determined essentially by that of the constituent bases, detailed knowledge of the behavior of the monomeric units (e.g., nucleosides and nucleotides of purines and pyrimidines, including di- and oligonucleoside phosphates) should facilitate correct interpretation of results obtained with polymers. This should, in turn, allow estimation of the effects of polymer secondary structure and ribose and ribosophosphate moieties on the behavior of the parent bases, including the nature of the species in solution and on adsorption.

The effects of specific substituent groups and substitution at specific sites on the various heterocyclic rings are subsequently discussed for each parent base.

The effect of substitution can be illustrated by the variation of $E_{1/2}$ values for the reduction of two interrelated series of compounds: a group of 6-substituted purines (69) and one of adenine nucleosides and nucleotides (70). The 6-substituted purines were investigated because the naturally occurring purines are, with few exceptions, substituted in that position, for example, the major purine constituents of nucleic acids, adenine and guanine, as well as minor constituents such as 2-methyladenine, 6-methylaminopurine, and 6-dimethylaminopurine. The nucleoside-nucleotide group was investigated to determine the effect of the ribose and ribosophosphate groups.

Substitution in the 6-position of purine with simple electron-repelling groups decreases the ease of reducibility of the parent purine, with the magnitude of the effect increasing with increasing electron repellency (basicity) of the constituent. The effect of the amino group can be counterbalanced by electron-withdrawing substituents on that group.

The net effect of addition of ribose and ribosophosphate groups is more complex than in the case of simple substituents such as alkyl, amino, alkylamino, arylamino, and hydroxyl, which decrease the ease of reducibility of the parent base (in addition to substituent inductive effect, other effects may be involved, e.g., saturation of the ring as a result of tautomeric shifts). The complexity of the situation is due, at least in part, to the conflicting influences of (a) an electron-withdrawing effect due to substitution by ribose or ribosophosphate, which increases the reducibility of the purine ring, and (b) adsorption and intermolecular association, which can be expected to decrease the ease of reducibility.

An important factor in decreasing reducibility in nucleotides is the presence of negatively charged phosphate groups, which results in electrostatic repulsion of the compound from the similarly charged electrode surface and in weakening electron withdrawal by ribose; this effect is negligible at low pH, where the phosphate group is screened by protons. The condensed structure which ATP assumes in aqueous media is another reducibility-decreasing factor. The molecule is so folded that the two terminal dissociated phosphate groups approach the protonated N(1) and amino groups to form a closely packed zwitterion; the N(1)=C(6)—NH$_2$ reduction site is thus surrounded by a protective cloud of negatively charged phosphate groups. A similar situation may occur with ADP.

B. KINETIC FACTORS

In connection with the deduction and specification of the mechanisms of electrode reactions of biological compounds, two types of kinetic phenomena are of immediate interest. One type involves rapid solution reactions, which can occur at or near the electrode surface as a result of the disturbance of an equilibrium by electrolysis of one component and the resulting re-establishment of equilibrium, for example, recombination of hydrogen ion and an acid anion due to electrolysis at the applied potential of only the more readily reduced acid. The other type consists of heterogeneous reactions involving electron transfer across the solution-electrode interface.

a. **Kinetics near the Electrode.** New techniques and methodologies are constantly being devised for the study of the mechanisms and kinetics of chemical reactions, which may precede, accompany, or follow electron-transfer processes and which may be studied by their effects on the observed polarographic currents. The classical example is the reduction of the protonated form of a species, in which the current may be controlled not only by the bulk concentration and diffusion coefficient of the unprotonated species and by the rate of the charge-transfer process, but also by the rate of protonation of the unprotonated species in the layer next to the electrode as the more readily reduced protonated species is reduced:

$$H^+ + A^- \rightleftharpoons HA \qquad [13]$$

The basic theory of such polarographic kinetic currents at the D.M.E. has been reviewed by Brdička et al. (71). Determination of the kinetics of fast reactions in solution by electrochemical methods has been recently reviewed by Bewick et al. (72); Ashley and Reilley (73) have reviewed chemical kinetics in electrochemical processes.

The development of the theory of observed current-potential curves for constant-area electrodes (cyclic voltammetry) under the condition of rapid polarization (high values of dE/dt) has included extensive work on the measurement of the rates of chemical reactions accompanying electron-transfer processes (cf. Section II.1).

A nomenclature has evolved in which E refers to an electron-transfer reaction and C to a chemical reaction. Thus an ECE process consists of a chemical reaction intervening between two electron-transfer steps, for example, the second electron transfer involves the rearránged or reacted product of the first one.

One kinetic process of special interest in connection with the compounds covered in this chapter is the catalytic reduction of hydrogen ion, which results from the large activation energy required for hydrogen-ion reduction at most electrodes, especially those of mercury, and the comparatively less difficult reduction of the adducts formed by hydrogen ion with heterocyclic nitrogen, sulfur, and other species. On reduction, the latter adducts evolve hydrogen and regenerate the adduct-forming species, which can again complex with hydrogen ion, resulting in a typical catalytic cyclic process. In some instances, both the oxidized and reduced forms of a compound seem to catalyze hydrogen-ion reduction (e.g., adenine); in other cases, only one form (e.g., the reduced form of purine) has this capability.

Since all nucleosides and nucleotides contain heterocyclic rings with nitrogen in them, all reduction waves, especially in acidic media, need to be carefully examined for the concomitant presence of a hydrogen-ion reduction process, for example, the anomalous reduction wave observed for 6-substituted purines including adenine (74). In the case of many compounds the behavior patterns involved are not yet clear, and further work is needed.

In solution of pH 5 or less, the catalytic hydrogen wave frequently appears as the solution decomposition wave, thus limiting the potential range available and preventing the observation of reduction waves which occur at more negative potential than the hydrogen evolution.

b. Heterogeneous Electrode Kinetics. Polarographically measured kinetic data on organic species involving the determination of heterogeneous rate constants and transfer coefficients for electrode reactions have been relatively scanty. The difficulties in applying the Koutecky approach to organic compounds have been discussed (75–78). Alternating-current polarography can be used to investigate electrode reaction reversibility, including the kinetics of charge transfer. A variety of other electrochemical techniques (e.g., pulse polarography) can be employed to determine the kinetic parameters for both charge-transfer and coupled chemical reactions in electrode processes of the so-called "quasi-reversible" and "totally irreversible" types. Recent developments in such electrochemical relaxation and other cyclic techniques may result in a revival of interest in the accumulation of adequate data on the subject; Reinmuth (79) has reviewed the application of electrochemical relaxation techniques to the study of electro-chemical reactions [cf. use of nonlinear relaxation methods to study very fast electrode processes (80) and review of the study of fast electrode processes by relaxation methods (81)].

The standard reference on electrochemical kinetics is Vetter's book (82), which is available in English translation (83,84). Damaskin's brief book (85) is a helpful summary of the principles underlying the methods used to investigate the kinetics of fast electrode processes. The effect of adsorption of organic compounds on their electrochemical kinetics has been considered (86).

The present state of the art does not permit quantitative evaluation of kinetic parameters or detailed specification of the various steps in a reduction mechanism such as those of the nucleic acid bases. For example, there have been no theoretical nor (until recently) experimental treatments of a system as complicated as pyrimidine, which has five discrete reduction waves and numerous intermediate chemical reactions in aqueous solution. The investigation of pyrimidine (68) has shown that application of ac polarography can aid in the understanding of such a complicated, quasi-reversible organic reduction series by yielding interpretable results. The qualitative correspondence of the results with theoretical predictions for irreversible waves (effects of drop time, frequency, applied voltage, and concentration) indicates the validity of available theory and the possibility that further theoretical and experimental work on even more complicated reaction mechanisms may be fruitful.

Use of phase-selective ac polarography, including phase angle measurements under a variety of conditions, should yield more easily interpretable results (cf. the subsequent discussion of the complicated patterns characteristic of NAD, as well as of the nucleic acid bases and their derivatives).

C. ADSORPTION PHENOMENA

Adsorption on electrodes is a very complex subject and is only partially understood, despite the large amount of work done in the area in recent years. It is probably a safe generalization to say that *all* substances, when present in solution in sufficiently high concentration, produce the measurable effects of adsorption on the electrode surface, that is, the displacement of solvent molecules or background electrolyte ions from the electrode surface.

Adsorption of electroactive species, intermediates, and products at the electrode surface—actually at the solution-electrode interface—is an important factor in electrochemical measurements, since it could seriously alter the potential which would be observed in its absence and which would be the potential desired, for example, for comparison with calculated energies for the electron-transfer step. Adsorption of a depolarizer, for instance, frequently inhibits the electron-transfer process and consequently shifts the polarographic wave to a more negative potential (since adsorbed molecules are in a lower free-energy state than those in solution, they are more difficultly reduced).

More importantly, study of such adsorption could yield information of value, for example, in reference to the behavior of nucleic acids at electrically charged boundaries in the living cell, as at the surfaces of ribosomes and the nuclear membrane, and the behavior of coenzymes such as NAD at charged interfaces in respect to electron transfer, surface activity, and association in the adsorbed layer.

Recent studies of 6-substituted purines (69) and the adenine nucleoside-nucleotide sequence (70) have shown how complicated the adsorption patterns of these compounds are and how adsorption influences their observed behavior.

a. Adsorption Sites. Adsorption of nucleotide-related compounds on mercury and other electrodes can be due to a variety of causes and potential adsorption sites in the compounds, for example, π-bonding involving aromatic rings, specific bonding connected to individual nitrogen atoms, C—N double bonds and oxygenated sites, and chemical binding involving negatively charged phosphate groups and metal complex formation. Adenine possesses the steric configuration necessary for metal chelation; the N(7) and the extracyclic heteroatom attached to C(6) provide favorable sites for formation of a five-membered ring. Actually, divalent metals chemically related to mercury (copper, zinc, and nickel) do form metal complexes with purines; formation constants for such complexes have been reported (87,88), as has the reaction of mercury(II) with purines, including adenine, in aqueous solution (89).

b. Measurement of Adsorption. Alternating-current polarography, measurement of the differential capacity of the electrical double layer (DL) at the solution-electrode interface, current-time curves at fixed potential, and electrocapillary curves are particularly well suited for identifying and measuring adsorption at electrodes. For example, adsorption of the molecules of a surface-active organic compound at the interface generally lowers the magnitude of the differential capacity of the DL and, correspondingly, the intensity of the alternating current which is involved in the charging of the DL. The potential at which the differential capacity is most reduced, compared to that in the absence of the surface-active species, defines the potential range in which the greatest portion of surface-active molecules is adsorbed. Potentials corresponding to capacity or current maxima are those where the most rapid adsorption or desorption occurs. Appearance of a sharply defined pit or well on the current- or capacity-potential curve is usually identified with the occurrence of association between the adsorbed molecules. Thus the potential region around E^0 in Figure 6, where the decrease of alternating

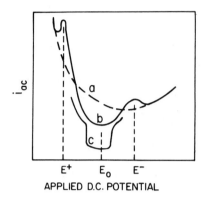

Figure 6. Dependence of alternating current, i_{ac}, on potential of the electrode, E. (a) Alternating-current polarogram of background electrolyte in the absence of specific adsorption. (b) Alternating-current polarogram of the same solution containing some surface-active substance. (c) Alternating-current polarogram in the case that intermolecular association of adsorbed species takes place on the electrode surface. Taken from Vetterl (18), with some modification.

current (i_{ac}) relative to the base line (curve a) is the greatest, is the range where the greatest amount of adsorption occurs. At potentials on either side of the adsorption minimum, peaks in the alternating current, called the anodic (E^+) and cathodic (E^-) desorption maxima, occur where the fast adsorption-desorption process results in an increase in i_{ac} across the electrode-solution interface. Association among the adsorbed molecules on the electrode surface, as mentioned, often results in a well-defined minimum or pit in the ac polarogram (90), for example, curve c.

There is an abundant literature on adsorption on electrodes, due (at least in part) to the controversial nature of the theories proposed and the great difficulty involved in doing reproducible experiments. Much of the work in the area is well summarized in Delahay's(91) and Gileadi's (92) books and the review papers of Frumkin and Damaskin (93), as well as in the extensive studies of Gierst, Parsons, Mairanovskii, and others (e.g., refs. 94 and 95). Several treatments of the subject of ac polarography and adsorption at electrode surfaces are available, including the monograph by Breyer and Bauer (48) and the chapter by Smith (49). The general area of adsorption at electrode-solution interfaces has been critically reviewed by Bauer et al. (138). There are lucid reviews of the theory of the electrical DL by Mohilner (96), of adsorption in polarography by Reilley and Stumm (97), of the adsorption of organic compounds at the electrode-solution interface, largely as observed by electrocapillary and differential capacity methods, by Frumkin and Damaskin (93), of the theory of charge effects in the DL by Barlow and MacDonald (98), and of DL structure and influence on electrode reaction rates by Parsons (99).

3. Correlations Involving Potentials

An increasingly important reason for investigating the electrochemical behavior of biologically important compounds is the development of correlations between (a) the mechanistic paths involved and the characteristic potentials associated with them, and (b) chemical, biological, and clinical activity.

Correlation of polarographic $E_{1/2}$ values with numerical structural and reactivity characteristics is generally based on the implicit or explicit postulation that the characteristic $E_{1/2}$ of a compound is a function of electron density and other factors which, in turn, are also related in a relatively simple way to some biological, physical, or chemical property. The frequently resulting linear relationship between $E_{1/2}$ for a series of more or less closely related compounds and a suitably selected mathematical function of the values of the given property for that series of compounds permits (a) prediction of the magnitude of the property for a compound not originally included in the series from its readily measured $E_{1/2}$, and (b) the rapid comparative evaluation of a property for the group of compounds, based on comparison of their $E_{1/2}$ values.

The large variety of experimental and theoretically calculated properties and phenomena which have been compared to polarographically determined potentials include structural summation characteristics, quantum-mechanically calculated parameters, photoionization potentials, degree of carcinogenesis, fungicidal activity, transition energies of charge-transfer complex formation, wavelengths of spectrophotometric absorption maxima, nuclear magnetic resonance data, and antioxidant ability. The most extensively used correlations have involved various forms of the Hammett sigma-rho equation based on polar substituent quantities and the Taft modification; perhaps the potentially most significant have involved quantum-mechanical electronic indices. As Gleicher and Gleicher (100) have stated, "One of the most successful applications of the Hückel molecular orbital theory has been in the correlation of polarographic oxidation and reduction potentials."

The need for reliable experimental measures of electron affinity, such as polarographic half-wave potentials, has been emphasized by quantum chemists concerned with evaluation of theoretically based calculations of the properties of biologically related compounds. Thus calculations of the ionization potentials (electron-donor properties) and electron affinities (electron-acceptor properties) for the four nucleic acid bases (guanine, adenine, cytosine, and uracil) by semiempirical self-consistent field and Hückel molecular orbital approximation methods were evaluated by comparing them with oxidation potentials determined at the stationary graphite electrode and reduction potentials determined at the D.M.E. (101).

The principal advantages in the use of polarographic half-wave potentials in linear free-energy and other types of correlations (e.g., in verifying assumptions regarding the structure of a molecule or in evaluating a theoretical approach to calculating the energy level for adding an electron to the compound) are the ease with which such potentials can be measured, as compared, for example, to rate or equilibrium constants, and the relative precision which is obtainable.

A. FACTORS INVOLVED

In examining the correlation of $E_{1/2}$ with experimental and calculated parameters, particularly the latter, consideration must be given to three frequently overlooked but important factors. First, the reliability of the correlations depends not only on the validity of the mathematical approaches and on the accuracy of the experimental measurements, but also on the calculations and measurements having been made on identical molecular species. This is related to the solvation energy problem (cf. the next paragraph) and to the need for determining the sites at which electron transfers occur, based on the electrochemical redox reaction pathway, and their exact chemical nature in order to assure that theoretical and experimental data are being compared for the same molecular species.

Second, account must be taken of the solvation energy contribution to the $E_{1/2}$, when the theoretical calculations are based on an idealized gas-phase molecule. In many past correlations there has been an implicit or explicit assumption that the solvation energy term is constant, is inconsequential, or varies in a regular fashion for the series of compounds studied. Although this assumption may be essentially valid for a series of large aromatic hydrocarbons and their $1e$ oxidation or reduction products in an "inert" organic solvent, its validity would have to be carefully examined, for example, for a series consisting of a purine and its nucleosides and nucleotides in aqueous medium because of the variation in nature, number, and strength of hydrogen-bonding donor and acceptor groups in the different compounds and in their redox products.

Finally, account must be taken of the influence of the degree of reversibility of the electron-transfer process on the observed potential; whereas the potential observed for a "completely reversible" electrode process, as previously discussed, is directly relatable to the standard free energy of the process, that observed for one which is not so reversible contains a component due to the necessary energy of activation, as well as, perhaps, components due to adsorption and other accompanying phenomena (e.g., ref. 63). Again, the assumption is usually made that, although irreversibility and adsorption are complicated factors, they may be expected to operate uniformly in a series of closely related derivatives; the validity of this assumption is usually judged on the basis of the consistency of the correlation (e.g., ref.

102). Thus $E_{1/2}$ values for overall irreversible polarographic reductions have frequently been correlated with molecular orbital and free-energy indices (e.g., refs. 103 and 104); the usefulness of such correlations has been emphasized by, for example, Zahradnik and Parkanyi (105).

In general, correlations of calculated molecular orbital (MO) and structure reactivity indices with experimental electron affinity values (e.g., $E_{1/2}$ for reduction) and other properties, such as ultraviolet absorption bands, must, for the time being, be considered as semiempirical, since there are no exact methods for calculation of MO and other parameters which would involve all possible intra- and intermolecular effects. In any event, comparison of calculated values with $E_{1/2}$ for a series of related compounds may provide information on (a) whether effects of the types mentioned operate with all individual derivatives to comparatively the same extent, and (b) whether differences in $E_{1/2}$ due to structural changes have a counterpart in differences in MO and other indices, and vice versa. This approach is related to one of the applications of $E_{1/2}$-property relations as yet only slightly exploited: the use of deviations from the expected linear relationship to shed light on the natures of the property and electrochemical process; for example, deviations from expected correlations for electrochemical carbon-halogen bond fission in aliphatic compounds could be related to adsorption on the mercury electrode with resulting polarization of the molecules and other stabilizing effects (106). Analysis of the deviations encountered should lead to refinement and sophistication of the postulated relationships, resulting in ones which better fit the experimental data and therefore involve more useful models.

One group of compounds whose study may be particularly rewarding in terms of the relation of their biological activity to the experimental electrochemical potentials as a measure of electron distribution in the molecules consists of the purine and pyrimidine derivatives of clinical interest in cancer chemotherapy because of their being antimetabolites. Another fruitful area of correlation will involve elucidation of the factors determining whether biological and electrochemical redox processes involve $1e$ or $2e$ transfers in the energy-controlling steps, for example, in overall $2e$ processes involving coenzymes.

B. LINEAR FREE-ENERGY CORRELATIONS

The physical basis for using half-wave potentials in linear free-energy relations such as the Hammett-Taft equations is the fact that $E_{1/2}$ is a simple function of the logarithm of (a) the heterogeneous rate constant for irreversible electrode processes or (b) the equilibrium constant for reversible electrode processes. Hence the basic Hammett equation for two members of a related series of compounds involving the same electroactive group can be expressed as

$$(E_{1/2})_i - (E_{1/2})_j = \rho(\sigma_i - \sigma_j)$$ [14]

where ρ is a proportionality constant for the series, σ is a substituent constant summing up the effects of the substituents on a compound by which it differs from the selected reference compound, and one of the compounds (usually the one least substituted in respect to the substituents being examined) may be selected as the reference compound of $\sigma = 0$. A linear $E_{1/2}$ versus σ plot for a series indicates that a polar substituent effect predominates on electroreduction and that the σ values used correctly assess the factors determining $E_{1/2}$.

Zuman (107) has reviewed the correlation of half-wave potentials based on linear free-energy relationships of the Hammett-Taft variety in a book which examines exhaustively the effect of substituents on polarographic behavior on the basis of polar, resonance, and steric phenomena, and the consequent applicability of polarography to the study of mechanisms and structural effects in organic chemical and electrochemical reactions. Perrin (62) has reviewed the subject from the viewpoint of the physical organic chemist.

The nature of the information obtainable from even a simple series of compounds can be illustrated by a brief summary of some of the results of a polarographic investigation of the reduction of 6-substituted purines, in which the degree of correlation of $E_{1/2}$ with a variety of experimental and theoretical parameters, including those most commonly used in $E_{1/2}$ correlations, was examined (69).

Correlations of $E_{1/2}$ with the polar substituent constants for the 6-alkyl-aminopurines and with the total polar substituent constants for all 6-substituted purines examined (Figure 25) indicate either that no mesomeric and steric effects due to the substituents are involved or that they are transmitted by an inductive mechanism and, furthermore, that none of the substituents exerts any specific effect different from the effects exerted by the other substituents. The positive ρ values found indicate that the potential-determining step involves a nucleophilic mechanism, with an electron being the most probable nucleophilic agent. Thus any primary electrophilic attack (e.g., protonation before electron transfer) has no significant role in the potential-determining step. Since the protonated form is the species believed to undergo electroreduction, the protonation must be a very rapid reaction and hence not potential determining. Furthermore, since the value of ρ depends on the charged state of the depolarizer and on the charge-transfer kinetic transfer coefficient, the similar values of ρ obtained at pH 2.5 and 4.0 indicate that the same species is reduced at both pH values by the same mechanism.

C. QUANTUM-MECHANICALLY CALCULATED ELECTRONIC INDICES

Correlations of molecular orbital energy level calculations with polarographic data have commonly involved the energies for adding an electron to the lowest empty or unoccupied MO (LEMO) or removing one from the highest occupied MO (HOMO), based on calculation of the energies of the molecular orbitals of the mobile or π-electrons. The theory of linear combination of atomic orbitals (LCAO) gives the latter in the form

$$E_i = \alpha + \beta k_i \qquad [15]$$

where α is a coulomb integral, β is a resonance integral, and k_i is the number calculated theoretically. Positive k_i values are associated with occupied (bonding) orbitals in the ground state of the molecule; negative k_i values are associated with unoccupied (antibonding) orbitals, which are usually occupied only in the excited state. For a homologous series, the smaller the absolute value of LEMO k_i, the greater is the electronegativity and, consequently, the greater the electron-acceptor properties of the molecule.

The various MO parameters and the methods for their computation, including the application to biological molecules, are discussed, *inter alia*, by Pullman and Pullman in *Quantum Biochemistry* (3) and in Vol 9 of *Progress in Nucleic Acid Research and Molecular Biology* (23), and by Streitwieser in *Molecular Orbital Theory for Organic Chemists* (108).

a. Relation of $E_{1/2}$ and k_i. For a reduction process in solution,

$$\text{Ox} + ne \rightleftharpoons \text{Red} \qquad [16]$$

it can be shown that, for a "reversible" polarographic process,

$$E_{1/2} = \frac{-\Delta H^\circ}{nF} + \frac{T \, \Delta S^\circ}{nF} - \frac{RT}{nF} \ln \frac{f_{\text{Red}}}{f_{\text{Ox}}} - \frac{RT}{nF} \ln \frac{D_{\text{Ox}}^{1/2}}{D_{\text{Red}}^{1/2}} \qquad [17]$$

where the symbols have their usual thermodynamic and electrochemical significance. By considering the reactions,

$$\text{Ox (g)} \quad + ne \rightleftharpoons \text{Red (g)} \qquad \text{EA} \qquad [18]$$
$$\text{Ox (soln)} + ne \rightleftharpoons \text{Red (soln)} \quad \text{EA} + (\text{SE}_{\text{Ox}} - \text{SE}_{\text{Red}}) \qquad [19]$$

where EA is the electron affinity of the Ox species, and SE is the solvation energy of the oxidized and reduced species as designated, and using the approach of Pysh and Yang (102), $\Delta H°$ in [17] can be replaced to give

$$E_{1/2} = \text{EA} + (\text{SE}_{\text{Ox}} - \text{SE}_{\text{Red}}) + \frac{T\,\Delta S°}{nF} - \frac{RT}{nF}\ln\frac{f_{\text{Red}}D_{\text{Ox}}^{1/2}}{f_{\text{Ox}}D_{\text{Red}}^{1/2}} \quad [20]$$

This expression is similar to Eq. 1 of Pysh and Yang (102), although the latter refers to an oxidation process and contains minor error. Combining [15] and [20], where $E_i = \text{EA}$, and neglecting the small entropy and logarithmic terms yields

$$E_{1/2} = \beta k_i + (\text{SE}_{\text{Ox}} - \text{SE}_{\text{Red}}) + \text{constant} \quad [21]$$

Then plots of $E_{1/2}$ versus the LEMO energy (usually, k_i expressed in units of β) should be linear, provided that, for the series of compounds studied, the solvation energy terms between members of the series are constant or vary in a regular fashion. It is unlikely that the latter would be true, for example, for sequences of bases and their nucleosides and nucleotides in aqueous solution, as earlier noted. However, it is not unlikely that the rate-controlling step for the reduction of each compound might involve only a single electron and that the difference in solvation energies of the reactant and initial product in the rate-controlling step for the first electron transfer for a given compound would be small, if not negligible.

Many purines and possibly pyrimidines appear to be reduced only in the protonated state (e.g., adenine), whereas others (e.g., purine and pyrimidine themselves) are reduced in an unprotonated form. This, of course, does not affect the initial or overall electron consumption, although significant solvation energy differences may be involved.

For the energy of the LEMO to be simply related to $E_{1/2}$ for a reduction, the process must be reversible, as noted. However, $E_{1/2}$ values for overall irreversible polarographic reductions have frequently been correlated with MO and free-energy indices, as was also mentioned.

A priori, the optimum approach would seem to involve the use of electrochemical data based on initial $1e$ processes, since the MO data apply to such processes. Many organic compounds which give multiple-electron polarographic waves in proton-available media (e.g., aqueous solution) give initial $1e$ waves in nonaqueous media of low proton availability; this fact undoubtedly accounts for the commonly better correlations based on $E_{1/2}$ data obtained in nonaqueous media. However, data obtained for multielectron waves in aqueous and nonaqueous media have been used (e.g., refs. 103 and 104). In the case of biologically important molecules, it is preferable to examine data obtained under conditions similar to physiological ones, that is, in aqueous media of normal pH range, temperature, and other parameters.

b. Other Indices. In addition to correlation of $E_{1/2}$ with LEMO and HOMO values, correlations have been investigated between $E_{1/2}$ and (a) bond orders for selected pairs of atoms in the molecules and (b) net charges on certain atoms in the molecules, in order to determine whether such correlations could be used to identify the electroactive sites in the molecules (69).

The *bond order* may be physically associated with the binding power of a bond as the MO analog of the resonance theory concept of double-bond character (3,108). The *net charge* at an atom represents the difference between the charge that an atom would carry in the absence of delocalization and its actual calculated electronic charge (3). The latter, that is, the total electron density at an atom, is the sum of electron densities contributed by each electron in each MO and is a convenient value, even though it has only approximate physical significance (108).

c. Applicability. Since MO calculations have been correlated with various types of chemical and biochemical activity for a variety of molecules (3), a successful correlation with polarographic data will, in turn, facilitate correlation of biochemical and related medical activity with experimental numbers, that is, potentials, which are readily obtained in solution media simulating biological conditions. It is clear, however, that, in order validly to correlate theoretically calculated data with redox potentials, the effects of adsorption, electron-transfer reversibility, and solvation energy, as earlier discussed, must be considered, since these seriously alter the potentials associated with redox processes and perhaps even the mechanistic route.

Correlations of the types discussed can be illustrated by data for the reduction of the 6-substituted purines (69), already considered in the case of linear free-energy correlations. When the ease of polarographic reduction $(E_{1/2})$ is compared with electron-acceptor properties, a fairly linear relationship is obtained between $E_{1/2}$ and the LEMO k_i coefficient (Figure 26); ATP (adenosine triphosphate) fits the same correlation, but AMP (adenosine monophosphate) does not (70). Substitution in the 6-position influences chiefly the $N(1)$—$C(6)$ bond order and the electronic charge distribution on $N(1)$ and $N(3)$, which, in turn, largely determine the reducibility; $E_{1/2}$ correlates with these, but not with the $C(2)$—$N(3)$ bond order and the charges on $C(6)$ and $C(2)$ (Figure 27).

D. MISCELLANEOUS TYPES OF CORRELATIONS

Half-wave potential data for organic compounds, particularly aromatic hydrocarbons and similar polyconjugated compounds, have been compared with a variety of experimental and calculated data, as earlier mentioned. Thus Zahradnik and Parkanyi (105), who reviewed empirical correlations of $E_{1/2}$ data with Hückel MO characteristics, also compared these relationships with analogous dependences for electronic spectra and ionization potentials. Perrin (62) has provided a succinct general summary of $E_{1/2}$ correlations.

In the case of the 6-substituted purines, for which various correlations have been indicated, it was found that apparently only the protonated form is polarographically reducible (69). Consequently, correlation of $E_{1/2}$ with the acidic dissociation constant, pK_a, for the acid-base system involving the uncharged purine, P, and its conjugate acid:

$$HP^+ \rightleftharpoons H^+ + P \qquad [22]$$

might be expected on the bases (a) that the initial potential-determining reduction step is associated with N(1); (b) that adenine (and presumably other 6-substituted purines) is protonated first at N(1), that is, at the most basic nitrogen; and (c) that the electron density at N(1) would affect both $E_{1/2}$ and pK_a; $E_{1/2}$ does become more negative as the net negative charge on N(1) increases. Although the electronic density on the ring nitrogen cannot be taken alone as a measure of its basicity, the reducibility should, as an initial approximation, be inversely proportional to the basicity represented by pK_a. Actually, six of the seven purines for which pK_a data are available do give a reasonably linear $E_{1/2}$ versus pK_a plot.

It seems likely, as well as highly desirable, that an increasingly important aspect of the study of organic electrode processes will be the correlation of polarographic behavior with environment, structure, and reactivity, and that a major area of effort will involve examination of the degree of correlation of mechanistic paths, the potentials associated with these and other data for the electrochemical oxidation and reduction of biologically significant compounds, with their chemical, physical, biological, and clinical activity. Thus correlations between (a) potentials and reaction patterns and (b) interfacial factors and behavior, for example, adsorption, will be sought because of the obvious biological significance of adsorption at charged interfaces.

In general, the relationships between biological and electrochemical redox processes, in reference to mechanisms, energy requirements, adsorption and association phenomena, and related factors, will be carefully explored, particularly in respect to the extent to which it may be possible to obtain information on biological electron-transfer processes by observing these or related processes by electrochemical techniques. Examples of this approach are given throughout the present review, for example, in Section III.5.D on association and orientation at the solution-electrode interface.

4. Quantitative Analysis

A. POLAROGRAPHIC ANALYSIS

The applicability of voltammetric or polarographic techniques to quantitative organic analysis, based on the electrolytic oxidation or reduction of organic functional groups or compounds, is well covered in Zuman's book, *Organic Polarographic Analysis* (12), and in the comprehensive, well-documented reviews on various aspects of the subject areas published every 2 years in *Analytical Chemistry*, as well as in two reviews (109,110) by Elving on the basic methodology of polarographic analysis, which has not changed appreciably since these were published in 1954 and 1962. A more detailed and more critical consideration of the applicability of polarographic techniques to all aspects of organic analysis is in press (111). The methodological aspects of polarographic or voltammetric analysis are covered in the references just cited, as well as in the standard works on polarography (e.g., refs. 20, 24–28, 61, 107, 112) mentioned earlier.

The applications of polarography to the determination of specific heterocyclic bases and their nucleosides and nucleotides are covered in the subsequent sections dealing with each class of compounds. In this section, only the general applicability of the approach will be indicated.

Polarographic determination of biologically important compounds, pharmaceuticals, and other substances is probably not a widely used or recognized technique in laboratories, except perhaps in Czechoslovakia. This accounts for the fact that actual published polarographic procedures are fairly scarce.

The general situation seems to be that the organic functionalities which are susceptible to electrochemical reduction are sufficiently activated by adjacent structure so as to produce chromophores of high absorptivity in the ultraviolet and/or chemical reactivity, which readily allows formation of such chromophores. Furthermore, polarography is less versatile than absorption spectrophotometry in respect to responsive functions in organic compounds; however, this very restriction tends to build a certain amount of specificity into polarographic measurements. Nevertheless, polarography often offers some advantage if moderately dilute aqueous solutions are to be analyzed, if thermally unstable material is to be determined, or if the number of samples to be analyzed does not warrant extensive calibration work. The actual measurement time may be very short, especially if the wave is at potentials more negative than that of the oxygen wave, as is frequently the case, and if it is sufficiently well formed so that its height can be determined by measuring the currents at only two potentials, one preceding the wave and one on the current plateau.

From the convenience standpoint, it must be emphasized that most polarographic measurements of concentration, which constitute the bulk of normal quantitative analysis, can be made with simple dc polarographs, the commercial strip-chart recording models of which retail for under $2000.

B. DETERMINATION OF CONCENTRATION

Conventional polarography with the D.M.E. is still the most commonly used polarographic method for the determination of concentration, usually on a known weight of sample dissolved to a known volume, which permits the calculation of percentage composition by weight after suitable calibration.

In general, the objective in voltammetric analysis is to obtain a single, diffusion-controlled polarographic wave, free from interference, whose height is more or less easily relatable to the concentration of the electroactive species. Diffusion-controlled waves are preferred because they are least influenced by small factors that can have a very large effect on kinetic or catalytic waves. The polarographic determination of a single species in, for example, synthetic plant samples can routinely achieve an accuracy of 1 to 3% relative; linear scan or cyclic voltammetry can also approach this limit. In any accurate determination, care must be taken that polarographic variables, such as temperature, pH, buffer composition and concentration, concentration of electroactive species, existence of interfering substances, and presence of maximum suppressors, are suitably accounted for; often, any one of these factors can be critical.

Temperature should be controlled to within 0.2°C; diffusion currents at the D.M.E. usually have a temperature coefficient of 1 to 2% per degree, and kinetic and catalytic waves often have much larger temperature coefficients. Since most organic compounds exhibit a shift of $E_{1/2}$ with pH and show polarographic activity only within a specific pH range, the reason for pH control is obvious; pH may well be the single most critical factor in organic polarography (63). In addition to pH, actual buffer composition and concentration have an effect on polarographic behavior.

Generally, it is desirable to keep the concentration of the electroactive species low (0.05 to 1mM), where the polarographic limiting current is often linear with concentration or easily relatable to concentration by means of calibration curves. Often, there is an ideal concentration range that should be sought. Interfering compounds, generally other electroactive species with similar half-wave potentials, must be accounted for or eliminated; this can often be achieved by simply changing the pH or background electrolyte. Maximum suppressors are sometimes necessary for accurate current measurement when a wave is distorted by a maximum. The usual effect of a maximum suppressor is to shift the $E_{1/2}$ to slightly more negative potential and to lower the current; the minimum amount of maximum suppressor necessary to suppress a maximum is recommended. Often the effect of a maximum may be minimized or even eliminated by lowering the concentration of the electroactive species.

Although it may seem complicated or excessively cumbersome to deal with many of these considerations, these factors are largely electrochemical analogs of factors that must be considered in using any analytical technique, for example, spectrophotometry.

Some use is being made of graphite electrodes for concentration measurement, especially in connection with the oxidation of nonreducible organic compounds; for example, the oxidizability of purines as compared to the nonoxidizability of naturally occurring pyrimidines offers obvious analytical possibilities, as does the difference in ease of oxidation of such compounds as guanine and adenine, and xanthine and hypoxanthine in $2M$ H_2SO_4 solution (113). Little or no significant use for quantitative analysis is being made of such phenomena as the one-electron electrochemical reduction and oxidation of aromatic heterocyclic rings in nonaqueous media.

The analytical implications of the ac polarographic behavior of purine, pyrimidine, and nicotinamide derivatives are obvious; it would be a relatively simple matter, for instance, to determine these compounds in the presence of other compounds which do not possess any ac polarographic reversibility. The sensitivity is quite high, with a lower limit of 0.1 mM or less, depending on the sensitivity of the current-measuring system used.

There are many other possibilities; for example, the measurement of NAD by polarography or coulometry seems to be both specific and precise, and, in the former case, quite rapid. Cyclic voltammetry of pyrimidine, cytosine, purine, and adenine (114) indicates a relative accuracy of 2 to 3% for the determination of concentration, the time required being even less than that for normal polarography.

Of particular interest in respect to future utility is the use of polarographic techniques for trace analysis (cf. the discussion of the latter topic in ref. 111). Webb and Elving (115,115a) have obtained well-defined dc and ac polarographic waves for adenine and cytosine dinucleoside and oligonucleoside phosphates at the 2×10^{-5} M level and lower in aqueous media.

C. TITRATION

Rather extensive use is being made in organic analysis of titration methods involving (a) polarographic detection of the equivalence point (amperometric titration) and (b) electrochemical generation of titrant (coulometric titration) with amperometric or related equivalence point detection (e.g., use of "dead-stop" or other polarized electrode pair). Measurement of compounds by direct electrolysis, usually at controlled potential (coulometry), in which the titrant is the passage of electrons, finds some application.

The simplicity of the *amperometric titration* apparatus and technique makes the use of this method quite advantageous. Amperometric titration can usually be applied to any reaction involving an organic compound which is reproducible and which proceeds rapidly enough; substances which react slowly with the reagent can at times be determined by addition of an excess of reagent and titrimetric measurement of the unreacted reagent. Polarographically inert organic compounds can be determined by using electroactive titrants. Thus the oxidation of tetraphenyl borate at graphite, which was originally applied for the amperometric titration of potassium with tetraphenylborate (116), has been applied to the titration of a wide variety of amines (117).

The amperometric titration of organic compounds is summarized by Stock (118).

Coulometry involves selective electrolytic oxidation or reduction of the desired constituent at an electrode at controlled potential and measurement of the amount of the constituent reacted via measurement of the total current flow with a suitable coulometer. The coulometric determination of NAD has been mentioned.

Coulometric titration methods, usually based on constant-current electrolytic generation of titrant, with the time required for reaction being measured, are more generally applicable to organic analysis, if only because of the vast existing literature on titration of organic compounds, as indicated by Ashworth's compilation on this subject (119). However, actual applications have been largely restricted to silver precipitation and bromination reactions.

Coulometric techniques are reviewed by DeFord and Miller (120).

D. MEASUREMENT OF REACTION RATES

Polarography has long been used in homogeneous solution kinetic studies involving organic compounds as a means of measuring changes in concentrations of reactants and/or products, either by repeated scanning of the necessary potential range, if the reaction was slow enough or the potential sweep technique used was rapid enough, or by following concentration change amperometrically at a fixed applied potential. By 1950, the volume of literature on polarographic measurement in chemical kinetics was sufficient to justify a 167-page review by Semerano (121), covering isomerization, elimination, addition, autoxidation, polymerization, dissociation, decomposition, and other types of reactions. More recently, Zuman (122) reviewed the kinetic investigation of homogeneous organic reactions by polarographic measurements. The use of electrochemical methods based on polarography for the investigation of the rates of fast reactions is well reviewed by Strehlow (123).

5. Other Areas of Application

A. IDENTIFICATION AND CHARACTERIZATION

The use of polarography for qualitative analysis in the strict sense of identifying organic compounds is limited. Normally, the $E_{1/2}$ of a compound is regarded as its identifying characteristic and has been so used when the expected population of compounds is small in number and more or less known in regard to type. This restricted application is related to the nature of the polarographic half-wave potential. The $E_{1/2}$ for a typically irreversible organic electrode process cannot usually be measured more precisely than ± 0.01 to ± 0.02 V. Although this precision may seem reasonable on the basis of the about 0.2 to -2.0 V range available at mercury electrodes, it is not adequate when compared to the more restricted range of a few tenths of a volt in which compounds of the type under consideration might be reduced.

Occasionally, several characteristics of a polarographic pattern (e.g., number of waves, relative magnitudes, $E_{1/2}$ values, and effect of pH) allow the identification of a pure compound or its recognition in a mixture. In addition, many electroactive groups are markedly affected in respect to their ease of reduction by substituents on the molecular skeleton, for example, the variation in $E_{1/2}$ of pyrimidines and purines (124). The effect of substituents and structure on ease of reduction is covered by Perrin (62) and Zuman (12,122).

The straightforward polarographic characterization of reaction products in complex mixtures is seen in the recognition of parabanic acid and alloxan among the products of the electrolytic oxidation of uric acid (5). The utility of polarographic characterization of organic compounds is further illustrated by Abrahamson's studies of heterocyclic compounds (125,126) and by Olson's studies of the structure and chemical binding of metal complexes with organic ligands (127,128).

B. DETERMINATION OF CHEMICAL AND PHYSICAL PROPERTIES

The use of polarography to determine oxidation states, redox potentials, thermodynamic quantities related to potentials in the case of reversible systems, equilibrium constants based on concentration measurements, and so forth is apparent from the previous discussions, for example, those on determination of structure, reaction paths, and correlations of $E_{1/2}$.

C. CHARACTERIZATION OF CHEMICAL REACTIONS

The polarographic investigation of reactions is exemplified by that (129) of the alloxan-alloxantin-dialuric acid system, which constitutes a reversible redox system of the semiquinone type:

$$\text{Alloxantin} \rightleftharpoons \text{alloxan} + \text{dialuric acid} \qquad [23]$$

Investigation of the oxidation and reduction of the three compounds by D.M.E. polarography, coulometry, and controlled-potential electrolysis served to rectify a number of incorrect statements in the literature concerning the system and the individual compounds. In particular, it was established that, contrary to reports of the earlier potentiometric studies, alloxantin is only very slightly dissociated (cf. Section V.3.C.c).

Berg and Schütz (129a) have described the polarographic determination of the formation constants for complexes of antibiotics with DNA.

D. ASSOCIATION, CONFORMATION, AND ORIENTATION

Investigation of the polarographic reduction of the adenine nucleoside-nucleotide sequence (70) indicated that the variation of current and potential with concentration could be interpreted in terms of association and preferential orientation and conformation of the species diffusing to the electrode. In addition, purine-pyrimidine association was polarographically detected at the millimolar and lower concentration level (70,115). Because of the possible value of the electrochemical technique in studying these phenomena, the approaches involved are briefly described. Typical of what can be done is the application of polarography to investigate the conformation of poly-nucleotides, for example, poly(adenylic acid) (129b,129c).

a. **Conformation and Orientation.** The Stokes-Einstein equation defines the diffusion coefficient, D_0, of a spherical particle at infinite dilution, in terms of the apparent molecular volume, V_m (molecular weight/density), of the solid compound:

$$D_0 = \frac{k}{(V_m)^{1/3}} \text{ cm}^2/\text{sec} \qquad [24]$$

where k is a numerical constant (3.32×10^{-5} for aqueous solution at 25°). The coefficient D_0 can be used to calculate the limiting diffusion current constant at zero concentration, I_0, from the Ilkovic equation for average currents:

$$I = 607nD^{1/2} \qquad [25]$$

where n is the number of electrons transferred per molecule in the faradaic electrode reaction. Experimental D values can be calculated by [25] from diffusion current constants based on polarographic diffusion-controlled limiting currents, i_d:

$$I = \frac{i_d}{Cm^{2/3}t^{1/6}} \qquad [26]$$

The relative magnitudes of I for the adenine nucleoside-nucleotide series generally correspond to those expected on the basis of the calculated V_m values; calculated D_0 values are comparable with experimental diffusion coefficients for compounds of similar molecular weight (e.g., sucrose and FMN). However, experimental I_0 values, being 1.7 to 2 times the calculated I_0 values, produce diffusion coefficients which are 2.8 to 4 times the calculated D_0 values. Thus the effective barrier to diffusional mass transport for the series seems to correspond to an effective molecular volume which is only a fraction of that expected for a spherical type of molecule.

The apparent anomaly can be considered and rationalized—at least, in part—on the basis of the preferred conformation of the nucleosides and nucleotides in solution. Support for this approach is provided by the decrease of I with increasing depolarizer concentration, which is much greater than normally encountered.

The preferred conformation of purine nucleosides and nucleotides in solution is *anti* (130–132), resulting in a relatively planar arrangement of the molecule. The latter conformation is characterized by the $C(1')$—$O(1')$ bond of the sugar moiety being rotated about the glycosidic $C(1')$—$N(9)$ bond to a position which is about 6 to 20° anticlockwise, when viewed along $C(1')$—$N(9)$ from the purine ring plane. The angle between the purine and sugar ring planes can be expected to be similar to that in the solid state, that is, between 55° and nearly 90°.

Consequently, the considerably larger experimental diffusion coefficients as compared to the calculated ones suggest that the average preferred orientation of the diffusing adenine species, which approaches the electrode, is with the purine ring plane perpendicular to the surface and with the protonated reduction side facing the surface; that is, the effective barrier to diffusional transport of the species is the minimal barrier, being the cross-sectional areal bulk of the planar purine moiety plus the portions of the ribose or ribosophosphate moieties which protrude from the purine ring plane. The fact that the molecular volume of ATP, as calculated from polarographic current data, deviates least in the series from the volume calculated by using [24], indicates that ATP exhibits a structure which is the closest to spherical, for example, a folded structure, as previously discussed.

b. Association. Nuclear magnetic resonance and vapor pressure measurements (largely between pH 5 and 7) indicate association of all purines and their nucleosides and nucleotides in aqueous solution; the extent of association increases with concentration. The mode of the association is essentially that of vertical stacks with hydrophobic interaction of bases. Since the standard free energy for this association is of the order of the thermal energy, the stacks must break and reform rapidly, and higher stability can be expected at lower temperatures. Nuclear magnetic resonance measurements do indicate a greater degree of association at lower temperature; correspondingly, the decrease of the diffusion current constant with increasing concentration is about 2 to 3 times greater at 1.5° than at 25°, indicating the considerably greater association at the lower temperature. The decreasing reversibility of the electrode process with increasing concentration is in agreement with additional energy being involved in the electron-transfer process with increasing concentration, as would result from increased association and/or adsorption.

c. Purine-Pyrimidine Interaction. The detection of purine-pyrimidine association is illustrated by a preliminary polarographic examination of adenosine-guanosine mixtures (guanosine is not reducible under normal polarographic conditions) to see whether the experimental diffusion current constant of adenosine would decrease because of its association with guanosine. No such decrease or noticeable shift in $E_{1/2}$ was observed on adding up to 0.5mM guanosine to a 0.12mM adenosine solution. However, an apparent capacity wave appeared at 150 to 250 mV more positive than $E_{1/2}$ of the adenosine reduction wave; no such wave was observed with either compound alone. The height of the capacity wave increased and its potential became more negative up to a ratio of 2 guanosine to 1 adenosine; above this ratio, there was no further change. Such a wave indicates sudden changes in the double-layer capacity, such as might arise from dissociation and/or desorption of associated adenosine-guanosine.

E. ELECTROCHEMICAL SYNTHESIS

Preparative organic electrochemistry, especially when used at controlled electrode potential, can have the advantages of (a) precise control of the intensity of reactivity of the reagent, that is, the electron; (b) consequent possibly high selectivity of reaction, resulting in a high yield of pure product; and (c) ease of product recovery due to the presence of minimum amounts of side products and absence of excess reagents. In addition, electrolytic processes are readily automated, are applicable on a micro- to macroscale, and can use polarography to select optimum conditions.

Two important areas of application of electrosynthesis are (*a*) the generation of free-radical species for study by electron spin resonance, and (*b*) the elucidation of the nature of organic electrode processes by controlled-potential electrolysis and coulometry (e.g., the use of controlled-potential electrolysis to prepare dihydro- and tetrahydropurines (133)).

Wawzonek (134) has reviewed important developments in preparative electrolytic organic chemistry, supplementing the reviews of the subject by Swann (135) and Popp and Schultz (136). Zuman (137) has reviewed preparative methods used in the elucidation of organic electrode processes.

IV. PYRIDINE (NICOTINAMIDE) DERIVATIVES

A wealth of physical and chemical knowledge related to the pyridine nucleotide coenzymes and model compounds has been gathered during the last two decades. A number of general reviews are of particular interest (139–143), but only one (14) is concerned primarily with electrochemical results. The present review brings the coverage of the electrochemical behavior of these compounds up to date and interprets this behavior, bringing to bear known physical and chemical properties where necessary.

1. Nomenclature

The essential structural component of the pyridine nucleotides is nicotinamide (3-carbamoylpyridine), which is itself widely distributed in plant and animal tissues and is an essential dietary factor (i.e., a vitamin) for nearly all mammalian systems. Its absence, along with that of nicotinic acid and other members of the B complex, results in a deficiency disease called pellagra. The "essential" role of nicotinamide undoubtedly stems from its importance in the production of certain types of pyridine nucleotides under physiological conditions. Two such nucleotides—nicotinamide-adenine dinucleotide (NAD^+) and nicotinamide-adenine dinucleotide phosphate ($NADP^+$) (Figure 7)—are hydrogen-transferring coenzymes for a class of enzymes known as dehydrogenases, which catalyze oxidation-reduction reactions; a third nucleotide, nicotinamide mononucleotide (NMN^+), is less well known.

The abbreviations NAD^+ and $NADP^+$ were proposed in 1961 by the Commission on Enzymes of the International Union of Biochemistry. Much of the current literature, however, refers to NAD^+ as diphosphopyridine nucleotide (DPN^+) or coenzyme I, and to $NADP^+$ as triphosphopyridine nucleotide (TPN^+) or coenzyme II. The abbreviations DPN^+ and TPN^+ were originally proposed by Warburg (144). Although it appears that the transition from the general use of DPN^+ and TPN^+ to NAD^+ and $NADP^+$ is

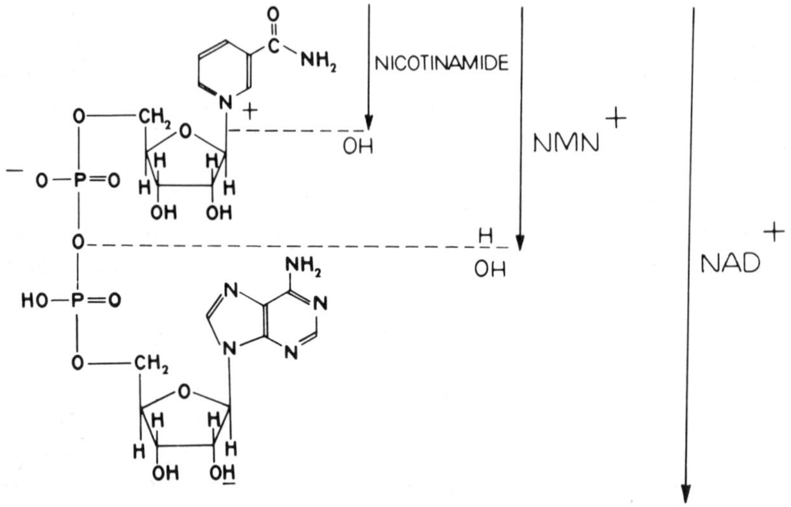

Figure 7. Formulas for NAD^+ (nicotinamide-adenine dinucleotide or coenzyme I) and derived nucleotide derivatives. In $NADP^+$ [nicotinamide-adenine dinucleotide phosphate, triphosphopyridine nucleotide (TPN^+), or coenzyme II], the underlined H is replaced by an $-PO(OH)_2$ group. Configuration, dissociation equilibria, and tautomeric shifts are not shown.

not going to take place (143), the more recent abbreviations will nevertheless be used in this review.

The abbreviation NADH is taken to mean 1,4-NADH; in the case of other isomers arising from the reduction of NAD^+, the reduction site is specified.

2. Redox Behavior

As mentioned, NAD^+ and $NADP^+$ are coenzymes for a large number of dehydrogenases (E) which catalyze reversible redox processes in mammalian biological systems, involving the transfer of hydrogen between substrate and coenzyme:

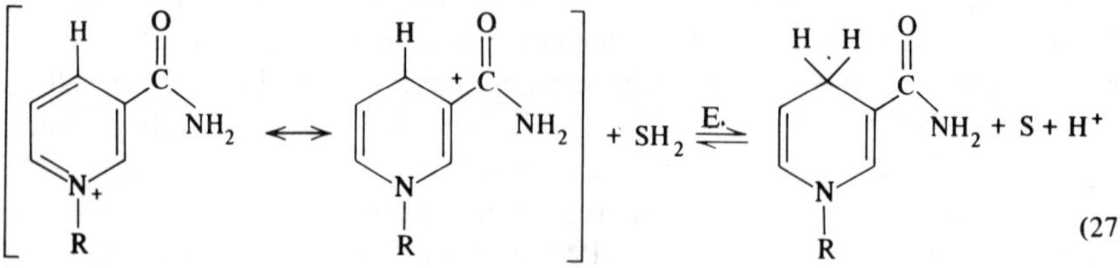

$$(27)$$

(where S represents the oxidized substrate and SH_2 the reduced substrate). Alcohol dehydrogenase (ADH), for example, catalyzes the oxidation of ethyl alcohol to acetaldehyde:

$$CH_3CH_2OH + NAD^+ \underset{}{\overset{ADH}{\rightleftharpoons}} CH_3CHO + NADH + H^+ \qquad [28]$$

where NADH and NADPH are the reduced forms of NAD^+ and $NADP^+$, respectively. The coenzyme NAD^+ is simultaneously reduced to NADH with the transfer of a hydride ion from the alcohol to C(4) of the pyridine ring (145,146). The existence of the reversible redox behavior shown by the pyridine nucleotides under physiological conditions has prompted extensive study of the reaction by electrochemists, using polarographic techniques, as subsequently discussed.

In addition to the use of enzymatic methods, NAD^+ and a number of model compounds (e.g., nicotinamide with various substituents on the ring nitrogen) have been reduced by a variety of other methods.

At least two fundamental types of reduced structure are possible for 1-substituted-3-carbamoylpyridinium ions: a dimeric structure produced on one-electron (1e) reduction, and a dihydropyridine species produced on 2e reduction; a number of isomers are possible for each structure type. (The term dihydropyridine, as used throughout the present section, refers to the dihydro-3-nicotinamide; cleavage of the amide group, where it may occur, is expressly indicated.) The following three isomeric dihydropyridines are generally possible:

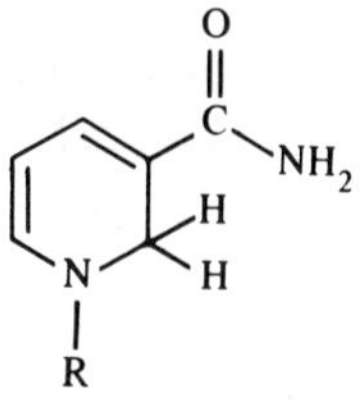

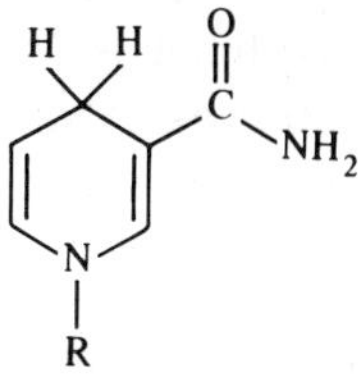

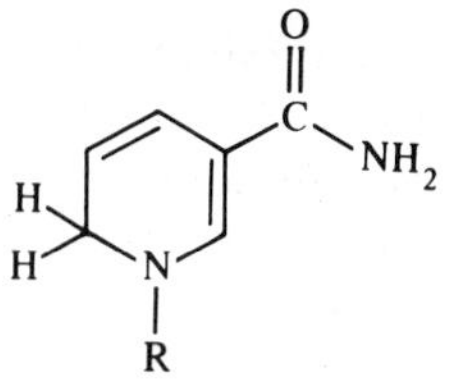

In slightly alkaline solutions, dithionite reduces NAD^+ and its model compounds exclusively via a 2e process to 1,4-dihydropyridines (147–150); the NADH thus formed is enzymatically active. Nuclear magnetic resonance data substantiate that the dithionite reduction of 1-methylnicotinamide results in hydrogen being added to the 4-position (151,152). Borohydride reduction, on the other hand, produces three different 2e products; for example, at least 92% of the NAD^+ reduced could be accounted for as three isomers: 1,4-NADH (29%), 1,6-NADH (28%), and 1,2-NADH (35%) (153,154). Reduction of the model compound, 1-dichlorobenzylnicotinamide, with NaBH$_4$ also leads to three isomers, but with only a small amount of the 1,2-isomer unless the 4- and 6-positions of the nicotinamide moiety are substituted (155,156).

Zinc reduction of NAD$^+$ models such as 1-n-propyl-, 1-benzyl-, and 1-dichlorobenzylnicotinamide gives rise to 1e products (157,158). Dimers can also be formed as a result of irradiation by X-rays (205,217,218,434). When NAD$^+$ and 1-methylnicotinamide are irradiated in the presence of an excess of an organic substance such as ethanol or formate, reduction occurs via the formation of pyridinyl radicals, which then dimerize.

3. Ultraviolet Spectra

Before reduction, most 1-substituted 3-carbamoylpyridinium ions show a single absorption band at about 265 nm. On reduction, this characteristic band is lost and one or more new bands appear, whose wavelength(s) depend on the particular reduction site in the pyridine ring; in the case of the coenzymes, an additional absorption band at 259 nm due to adenine is present in both the oxidized and the reduced forms. Thus NAD$^+$ has a single band at 259 nm ($\epsilon = 17{,}800$) arising from absorption by both the pyridine and adenine moieties; on conversion to NADH, absorption due to the pyridine moiety shifts to a longer wavelength, and two absorption bands are seen at 259 nm ($\epsilon = 14{,}400$) and 338 nm ($\epsilon = 6{,}220$).

The three dihydropyridine species can be distinguished by their NMR spectra (151,152,159,160), but much more emphasis has been placed on their characteristic ultraviolet absorption spectra (153,154,156,157).

Above 240 nm, 1,2-dihydropyridine has a single absorption band at about 410 nm, 1,4-dihydropyridine absorbs in the region of 350 nm, and 1,6-dihydropyridine exhibits two absorption bands at about 265 and 360 nm. In an attempt to account qualitatively for these spectra, it has been suggested (142,157) that the single-peak spectra of the 1,2- and 1,4-dihydropyridines are attributable to a straight conjugated system of double bonds, while the double-peak spectrum of the 1,6-isomer is due to the presence of the crossed conjugated system. The chain in the 1,2-isomer is longer by one double bond, which corresponds to a shift of 50 to 70 nm toward longer wavelength (142,157) and causes its solutions to appear bright yellow (153,154).

Recently, the dihydropyridine spectra have been interpreted on the basis of predictions obtained from molecular orbital calculations on the three isomers of reduced 1-methylnicotinamide (161). This has justified on theoretical grounds the experimental conclusions outlined in the previous paragraph.

4. Polarographic and Voltammetric Behavior

In order to present a coherent and logical discussion of the electrochemical behavior of the pyridine-based nucleosides and nucleotides, it seems desirable to discuss first the behavior of nicotinamide, next that of other model compounds which have been investigated, and then that of the pyridine nucleotides, with a final review of the electrochemical reduction mechanism for the system.

5. Nicotinamide

In reviewing the electrochemistry of pyridine nucleotides, it is important to have available a well-defined picture of the electrochemical reduction of nicotinamide itself. Although the electronic structures of nicotinamide and the 1-substituted 3-carbamoylpyridinium ion are considerably different, the electrochemical theory developed for each structure type might be made appreciably more sound if the electrochemical results could be correlated with known chemical differences.

An understanding of the behavior of nicotinamide is also of interest because it is a decomposition product in slightly alkaline solution of the three pyridine nucleotides, NAD^+, $NADP^+$, and NMN^+; thus, during macroscale electrolysis of the nucleotides, small amounts of nicotinamide are ultimately formed in either the normal or the reduced state, depending on the applied potential. Knowledge of the electrochemical and spectral properties of nicotinamide and its reduced forms would be helpful, therefore, in characterizing the reduction products of the nucleotides.

A. VOLTAMMETRY

a. Direct-Current Polarography. A great quantity of fundamental polarographic data has been reported for nicotinamide (162–168). In acidic media, this compound yields two ill-defined waves, whose half-wave potentials are extremely close to one another (20,162,168). Wave II, which is less well-defined than wave I (168), occurs just before a catalytic wave (162,168). With decreasing pH, the catalyzed wave appears at pH 7.2, increases in height as the solution is made more acidic, and finally coalesces with the normal hydrogen discharge wave at pH 5.5 (162); the catalytic wave is seen even at nicotinamide concentrations as low as $5 \times 10^{-5}M$. Since the wave occurs at a more negative potential than the two nicotinamide waves, it has been suggested that the catalyst is a protonated form of the dihydro reduction product (162).

The two nicotinamide waves are of equal height between pH 3.15 and

7.00, but are sensitive to experimental conditions because of their close proximity to the catalytic wave. In the same pH range, $E_{1/2}$ for both waves becomes more negative at about 50 to 80 mV/pH unit in a nonlinear fashion (164,168); the relative separation of the two waves remains constant at about 0.11 V. Between pH 7 and 8, the catalytic wave disappears. The single $2e$ wave seen at pH 8.0 has an inflection and appears to consist of two extremely close $1e$ waves; the wave height is the same as the sum of the two waves seen at pH 7. With increasing pH (pH 8 to 12 in the absence of tetraalkylammonium salts), the inflection disappears, $E_{1/2}$ shifts to more negative potential at 39 mV/pH unit, the wave slope approaches a more reversible value (i.e., $E_{1/4} - E_{3/4}$ shifts from 120 to 34 mV), and the diffusion current constant remains constant at 3.78 (168).

The single $2e$ wave seen above pH 8 (e.g., in pH 9.4 carbonate buffer) is

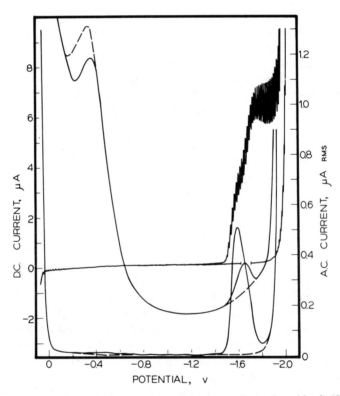

Figure 8. Direct- and alternating-current polarograms of nicotinamide (1.27mM) in Et$_4$NCl-carbonate buffer (pH 9.37). For ac polarograms: controlled drop time of 3.0 sec; $f = 50$ Hz; $\Delta E = 3.54$ mV rms; lower curves are for in-phase component and upper curves for quadrature components; dashed line is background electrolyte base current.

markedly altered when a tetraalkylammonium cation is present (Figure 8); a wave is re-established which has an inflection on its rising portion, indicating the presence of two separate $1e$ waves (168). The limiting portion of the total current is diffusion controlled and is proportional to concentration; the appearance of two closely adjacent $1e$ waves is much more pronounced at $1.27 \mathrm{m}M$ than at $0.13 \mathrm{m}M$. An I of 3.44, corresponding to the limiting portion of the total current, is slightly lower than the value in solutions without the tetraethyl salt because of a viscosity effect. Despite some error due to the small potential separation of waves I and II, both appear to have reversible $1e$ transfer steps (at 1.27 mM nicotinamide), since their wave slopes, measured as $E_{1/4} - E_{3/4}$, are 50 and 61 mV, respectively (the value for a reversible $1e$ cathodic wave at 25°C is 56.4 mV).

b. Cyclic Voltammetry. The behavior of nicotinamide under cyclic voltammetric conditions at the H.M.D.E. is consistent with the dc polarographic results. In acid solutions, a sweep to more negative potential gives rise to two cathodic peaks, which shift to more negative potential with increasing pH and merge into a single peak in slightly alkaline solution (tetraalkylammonium salts absent)'. In the presence of Et$_4$NCl, the single cathodic peak splits into two closely spaced peaks (Figure 9). The first peak corresponds to a $1e$ reduction of nicotinamide to a free radical, which then dimerizes; the second peak is due not to reduction of the dimer, but to a direct $2e$ reduction of nicotinamide in the partially depleted layer at the electrode surface. The results seen on the reverse sweep depend on the switching potential (Figure 9). When the potential sweep is reversed immediately after the first cathodic peak, a single anodic peak (Ia) is seen, which is due to oxidation of the dimeric $1e$ product; when the sweep is reversed after the second peak, an additional anodic peak (IIa) appears at more positive potential because of the oxidation of the $2e$ reduction product. Both of the oxidation processes producing peaks Ia and IIa yield a proton or protons during electron transfer, since the peak potentials become more positive with decreasing pH, that is, oxidation becomes more difficult (168).

c. Alternating-Current Polarography. The in-phase and quadrature components of the total ac current in the presence and absence of nicotinamide are shown in Figure 8. Nicotinamide exhibits a single ac polarographic peak (in-phase component, -1.59 V; quadrature component, -1.66 V) due to the first $1e$ addition to the molecule (168). The compound is only slightly surface active even at relatively high concentration (1.27 mM); in the potential region immediately before the first reduction wave, the differential DL capacity is identical to that of the supporting electrolyte alone, showing an absence of adsorption phenomena involving nicotinamide. In the potential region where faradaic processes occur, the absence or presence

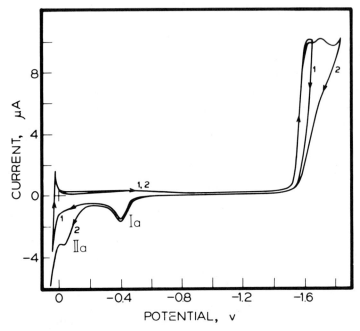

Figure 9. Cyclic voltammogram of nicotinamide (1.27mM) in Et₄NCl-carbonate buffer (pH 9.37) at the H.M.D.E. Potential sweep rate = 100 mV/sec; electrode area = 0.03 cm².

of adsorption phenomena involving the reduction products cannot be ascertained with certainty because of the close proximity to background discharge. Significant adsorption of reduction products is considered unlikely, however, because the quadrature component in this potential region shows no capacitance depressions or anomalous peaks which might be attributed to adsorption-desorption (tensammetric) phenomena. Moreover, the nicotinamide reduction products are uncharged; and, while they might possibly be adsorbed at potentials around the electrocapillary maximum (ecm), adsorption at more negative potential should become increasingly more difficult because of the considerable negative charge on the electrode surface—that is, polar solvent molecules and cations of the supporting electrolyte would be expected to displace uncharged surfactants. This would be particularly true of the tetraethylammonium ion, which is not only positively charged but also highly surface active (168).

B. CONTROLLED-POTENTIAL ELECTROLYSIS

Electrolyses of nicotinamide were conducted under conditions which enhanced the formation of two waves, that is, using pH 10 buffers containing the tetraethylammonium ion (168). Electrolysis at a potential slightly more negative than $E_{1/2}$ of the first wave, in order to minimize formation of the second wave product, gave a faradaic n value of 1.09. The two nicotinamide cathodic waves were absent on a polarogram of the completely electrolyzed solution, but two anodic waves appeared, the first of which ($E_{1/2} = -0.45$ V) had a limiting current about three times that of the second ($E_{1/2} = -0.11$ V); these waves correspond to the anodic peaks observed on cyclic voltammetry. Thus, because of the small potential separation of the two reduction processes, some of the $2e$ reduction product is also obtained.

The electrolyzed solution had absorption maxima at 266 and 346 nm; when the pH was brought to 7, the $1e$ product decomposed, causing the disappearance of these maxima and the appearance of a band at 280 nm due to an acid-catalyzed decomposition product. Exposure of the electrolyzed solution to air caused a 91% reoxidation of the reduced material back to nicotinamide; the remaining 9% decomposed to the acid-catalyzed product, as shown by the presence of a small 280-nm band.

Electrolysis on the limiting portion of wave II yields a faradaic n value of 1.82. The completely electrolyzed solution had absorption maxima at 265 and 346 nm and exhibited the two anodic polarographic waves seen after electrolysis at $E_{1/2}$ of wave I. However, the height of the first anodic wave was only about 6% that of the second; hence, mostly the $2e$ reduction product was obtained.

Exposure of the electrolyzed solution to air produced a 95% decrease in absorption at 346 nm and a new band at 272 nm. This can best be explained by assuming that the $2e$ reduction product dissappeared via two paths similar to the $1e$ reduction product and that the 272-nm band is a combination of absorption bands due to nicotinamide ($\lambda_{max} = 261$ nm) and an acid-catalyzed decomposition product ($\lambda_{max} = 280$ nm). Polarographic examination of the air-exposed solution indicated that about 51% of the original nicotinamide had been regenerated; a broad wave characteristic of hydrogen peroxide was also observed.

C. REDUCTION MECHANISM

The reaction scheme outlined in Figure 10 seems to fit best the electrochemical, chemical, and spectral data for nicotinamide (1) over the pH range studied. At pH exceeding the pK_a of nicotinamide (3.3 at 20°) (168a), the initial reduction can be regarded as uptake of an electron and a proton to form a neutral free radical (2) (source of polarographic wave I; in Et₄NCl-carbonate buffer, the process appears to be reversible). The electron-transfer reaction is followed by irreversible dimerization to the 6,6′-dimer (3). At the potential of wave II, nicotinamide is reduced to 1,6-dihydropyridine (4), which is equivalent to a further $1e$ reduction of the free radical; the dimer is

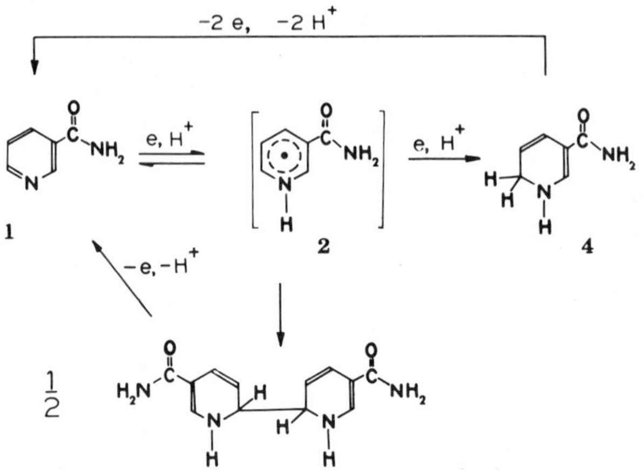

Figure 10. Interpretation of the electrochemical behavior observed for nicotinamide (**1**) and its reduction products.

not further reduced at this potential. At a potential considerably more positive than that of the first reduction wave, the dimer can be oxidized back to nicotinamide. At still more positive potential, the 1,6-dihydro species can also be oxidized to nicotinamide.

The pH dependence of both the two electrochemical reductions and the two electrochemical oxidations is compatible with the proposed reduction scheme. Below pH 3.3, the wave I process should be independent of pH, because most of the nicotinamide will be protonated at the 1-position of the pyridine ring. Above pH 3.3, the energy-controlling step in the reduction of nicotinamide involves the simultaneous addition of an electron and a proton to form a neutral free-radical species:

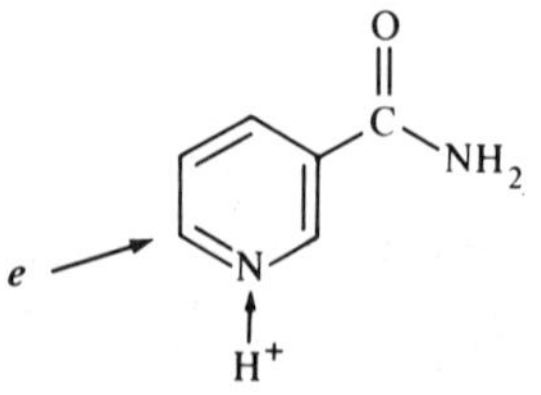

(29)

In alkaline solution, the proton is abstracted from a water molecule. Addition of a proton to the ring nitrogen facilitates reduction by decreasing the electron density in the ring. Below pH 8, further reduction of the neutral free-radical species requires a more negative potential; in alkaline solution, however, the radical, once formed, is more easily reduced than the original electroactive species. Thus, at pH 12, the two electrons add almost simultaneously to the nicotinamide nucleus, giving rise to a polarographic wave which has a slope only slightly less reversible than that expected for a reversible $2e$ wave.

In Et$_4$NCl-carbonate buffer, an analysis of the rising portion of the first dc wave suggests that the electrode process occurs reversibly. This contrasts with the fact that the alternating current for the first faradaic process represents less than 3% of that expected for a reversible $1e$ process under the experimental conditions employed. The apparent discrepancy is accounted for by noting that the ac method, as it was used in the case of nicotinamide (168), tests the reversibility of the overall electrochemical process, including the irreversible dimerization reaction which occurs after electron transfer. The period of the alternating voltage used (20 msec) is much greater than the half-life which would be expected for a nicotinamide radical (cyclic voltammetry indicates that the radical has a half-life of less than 2 msec); therefore most of the nicotinamide radicals dimerize before they can be reoxidized.

Postulation of the 6,6'-dimer and the 1,6-dihydro species reflects the two absorption bands observed for these compounds in the ultraviolet region above 240 nm, as previously discussed. Absorption bands have been observed (156,158,169) at about 265 nm and 350 nm for 1-alkyl-substituted pyridine rings, which are either the fully reduced 1,6-dihydro species or the partially reduced 6,6'-dimeric structures.

The two reduction products decompose in the presence of a proton donor such as $H_2PO_4^-$ at pH 7 to give a single absorption band at about 280 nm; the same reaction occurs more slowly in pH 10 carbonate buffers. The decomposition involves hydrolysis of the 4,5 double bond and can be regarded as an instance of enamine nucleophilicity (140,169):

(30)

This reaction accounts for the loss of the 266- and 346-nm absorption bands and for the concomitant increase in absorption at 280 nm; the latter absorption is due to the $-NH-\overset{|}{C}=\overset{|}{C}-\overset{|}{C}=O$ chromophore (170). Both 1,4- and

1,6-NADH decompose under similar conditions, producing an absorption band in the vicinity of 280 to 290 nm due to a similar chromophore (154, 171,172).

The two nicotinamide reduction products decompose, as indicated, even in solutions as alkaline as pH 10. Under these conditions, the compounds are also directly oxidized on exposure to air by molecular oxygen to form nicotinamide and hydrogen peroxide. For the 1,6-dihydro species, the two pathways appear to be of equal importance since only half of the original nicotinamide is recovered. The dimer, on the other hand, appears to be more easily oxidized by oxygen than the 1,6-dihydro species, as is consistent with their electrochemical oxidation potentials, and most of the original nicotinamide is recovered.

Recently, a single 2e reduction wave was reported (172a) for quinoline in DMF, which led to a 1,6- or 1,4-dihydro isomer. Although two distinct 1e waves were not seen, the proposed mechanism is similar to that outlined for nicotinamide.

6. Model Compounds

The electrochemical behavior of 1-substituted 3-carbamoylpyridinium ions is in many ways very similar to that observed for nicotinamide. With the 1-substituted compounds, however, the first reduction step is independent of pH and is shifted to more positive potential, as would be expected for a positively charged nicotinamide nucleus; thus two well-defined 1e polarographic waves appear in slightly alkaline solution, rather than the single 2e wave found under most conditions for nicotinamide. The polarographic behavior of the following NAD$^+$ model compounds, where R represents the N(1) substituent, has been described:

$$R = —CH_3 \ (164,168,173–178,178a)$$
$$R = —C_2H_5 \ (174)$$
$$R = —C_3H_7 \ (174)$$
$$R = —C_4H_9 \ (174)$$
$$R = —CH_2C_6H_5 \ (174,179)$$
$$R = —CH_2CH_2SO_3^- \ (179)$$
$$R = —CH_2C_6H_4SO_3^- \ (179)$$
$$R = \text{D-glucopyranosidyl} \ (20,180)$$
$$R = \text{D-glucopyranosidyl tetraacetate} \ (20,176,180)$$

Tables I and II summarize the data for these model compounds; other model compounds are included in some of the references listed. Other electrochemical studies of nicotinamide derivatives have recently been reported (180a–180c). Electrochemical studies have also been made on compounds of related interest, such as 1,1′-ethylenebis(3-carbamoylpyridinium bromide) (169) and a number of bipyridylium herbicides, such as Diquot and Paraquot (181–186).

Most of the model compounds exhibit two diffusion-controlled reduction waves of about equal height (164,168,173,175,179). In general, $E_{1/2}$ of wave I is independent of pH, background electrolyte composition, and concentration of model compound; $E_{1/2}$ for wave II varies with pH, but details regarding the manner of variation are commonly not given. Only in the case of the 1-methyl-3-carbamoylpyridinium ion have extensive data

TABLE I

Half-Wave Potentials at D. M. E. of NAD Model Compounds:
1-Methylnicotinamide Salts

| Anion[a] | Medium[b] | pH | $E_{1/2}$ (V)[c] | | Ref. |
			Wave I	Wave II	
Cl	23, 24, 31	1.0–7.3	1.0–1.03	[d]	164
	28, 45, 46	8.9–12	1.02–1.03	1.58–1.64[e]	164
	25, 27, 37	4–7	1.10–1.12	—	173
	27, 30	7.5–8.8	1.11	1.68–1.70	173
	37	9	1.12	—	173
	27	9	1.10	1.72	173
	34	9.9	1.09	1.78	173
	35	9.7	1.08	1.65	174
	47	7.43	1.02	[d]	168
	48	9.37	1.03	1.59	168
	53	11.13	1.03	1.59	168
	54	12.83	1.04	1.60	168
Br	11	Acidic	1.09	—	176
	11	Alkaline	1.09	1.66	176
	35	9.7	1.07	1.65	174
I	22	2.0	0.9	[d]	175
	8	5.2	0.94	1.32	175
	8	7.0	0.94	1.5	175
	33	9.0	0.94	1.57	175
	35	9.7	1.07	1.65	174

[a] Anion of salt added.

[b] Media are listed in Table IV.

[c] A dash indicates no wave was observed or reported.

[d] Second wave partially or completely obscured because of deposition of hydrogen from buffer.

[e] Potential varied irregularly over the pH range with no clear trend.

TABLE II

Half-Wave Potentials at D. M. E. of NAD Model Compounds:
1-Substituted Nicotinamide Salts

N-Substituent	Anion[a]	Medium[b]	pH[c]	$-E_{1/2}$ (V)[d] Wave I	Wave II	Ref.
—C$_2$H$_5$	Cl	35	9.65	1.065	1.650	174
	Br	35	9.65	1.070	1.654	174
	I	35	9.65	1.060	1.650	174
—C$_3$H$_7$	Cl	35	9.65	1.055	1.650	174
	Br	35	9.65	1.055	1.655	174
	I	35	9.65	1.055	1.653	174
—C$_4$H$_9$	Cl	35	9.65	1.048	1.655	174
	Br	35	9.65	1.050	1.655	174
	I	35	9.65	1.05	1.655	174
—CH$_2$C$_6$H$_5$	Cl	35	9.65	1.003	1.673	174
	Br	35	9.65	1.00	1.65	174
	I	35	9.65	—	1.62	174
—CH$_2$C$_6$H$_5$	Cl	40	7–9	1.00	—	179
—CH$_2$C$_6$H$_4$SO$_3$	Cl	40	7–9	0.96	—	179
—CH$_2$CH$_2$SO$_3$	Cl	40	7–9	1.02	—	179
D-Glucopyranosidyl	Br	41		0.871	—	180
D-Glucopyranosidyl tetraacetate	Br	41		0.991	—	180
	Br	11		e		176

[a] Anion of salt added.

[b] Media are listed in Table IV.

[c] A blank indicates that the pH was not specified.

[d] A dash indicates that no wave was observed or reported.

[e] One wave was observed in acidic solution and two waves in alkaline solution; numerical data were not given.

been collected in the areas of dc and ac polarography, coulometry, controlled potential-electrolysis, and ultraviolet spectrophotometry. Early polarographic investigators postulated the following electrode reaction:

$$\text{(31)}$$

More recent electrochemical work, although in accord with this mechanism, has attempted to determine the isomeric form of the reduction products, relying heavily on ultraviolet spectral arguments.

A. VOLTAMMETRY

a. Wave I. Most investigators report $E_{1/2}$ for wave I of the 1-methyl-3-carbamoylpyridinium ion to be independent of pH and the supporting electrolyte, but disagree as to the exact potential; for example, under similar conditions $E_{1/2}$ has been reported over a range from -0.94 to -1.11 V (164,168,173–175), although in two investigations a value of -1.03 V was obtained (164,168). The value of $E_{1/2}$ has been reported to be independent of the concentration of the electroactive species (82,33) and to shift more positive by 80 mV for an increase in concentration from 0.001 to 1.00mM (168). Also, $E_{1/2}$ becomes more positive with increasing drop time (168). The significance of the positive shifts in $E_{1/2}$ with increasing concentration and drop time is obscured by the presence of adsorption phenomena involving the reduction product (cf. the subsequent discussion). In the absence of adsorption effects, such shifts (for a reversible electrode process) have been interpreted as being indicative of an irreversible dimerization subsequent to charge transfer (187).

Linear plots of log $[i/(i_d - i)]$ versus E were reported to have slopes close to 59 mV, as expected for a reversible 1e redox couple (164,168); additional evidence for a reversible charge transfer has been obtained using fast-scan cyclic voltammetry (188,189). However, the overall electrochemical process is irreversible because of a rapid, irreversible chemical reaction subsequent to electron transfer (164,168,173,188). Under cyclic voltammetric conditions, a negative potential scan followed by a rapid reverse potential sweep discloses the reversible reoxidation of a transient radical intermediate, as well as the oxidation of the chemical product at a much more positive potential. Similar behavior has also been observed for the 1-ethyl-, 1-n-propyl-, and 1-benzyl-3-carbamoylpyridinium chlorides. Adsorption effects at the electrode surface complicated the theoretical treatment of data and led to uncertainty in the calculated dimerization rate constants. For the 1-methyl compound, a rate constant of 2.24×10^{-2} l/mole-sec was obtained (188), which is in poor agreement with the value of 6.9×10^7 l/mole-sec determined by pulse radiolysis (205).

b. Wave II. Above pH 8, the second 1-methylnicotinamide wave is well defined, diffusion controlled, and equal in height to the first wave (164,168, 173). Below pH 8, the wave is either completely or partially obscured by catalytic hydrogen evolution (164,168). Both the 1e and 2e reduction products, as well as their acid-catalyzed decomposition products, act as catalysts for hydrogen evolution (cf. the subsequent discussion).

In slightly alkaline solution $E_{1/2}$ of wave II becomes more negative with increasing pH and, in addition, varies with the background electrolyte (164,173,175); in one study, only a very slight pH dependence was reported (168). As in the case of wave I, reported $E_{1/2}$ values vary considerably, even when approximately the same experimental conditions are employed; for example, $E_{1/2}$ ranges at pH 9 from -1.57 to -1.72 V (164,168,173–175). Although all investigators claim to report their potentials versus the S.C.E., it is likely that the various reference electrodes differed in potential, since in all cases $\Delta E_{1/2}$ between waves I and II is constant at 0.60 ± 0.03 V. The results of two investigations (164,173) show $E_{1/2}$ of wave II to be independent of the concentration of the electroactive species in buffered solutions, but in another study (168) $E_{1/2}$ became gradually more negative with increasing concentration of electroactive species. In the latter case, the negative shift of wave II with increasing depolarizer concentration was interpreted not as being due to a pH effect at the electrode surface, but as resulting from competition between two reactions, in one of which the radicals formed in the wave I process dimerize whereas in the second they are reduced to form wave II. An increase in the depolarizer concentration would raise the dimerization rate more rapidly than the rate of the electrode process; thus, in order to obtain wave II, it would be necessary to raise the electrochemical reduction rate of the radicals, that is, the cathodic potential of the electrode (95).

The log current-potential plots for wave II are straight lines, whose slopes vary from 53 to 99 mV; the second wave is therefore considered to be irreversible under dc polarographic conditions (164,168,173). In basic solution (pH 12), hydrolysis of the amide group occurs, causing the disappearance of the polarographic pattern of 1-methylnicotinamide and the appearance of a new wave in the potential range from -1.3 to -1.5 V due to trigonelline (1-methyl-3-carboxylpyridinium chloride) (164,168,173).

B. CONTROLLED-POTENTIAL ELECTROLYSIS

A number of macroscale electrolyses of 1-methyl-3-carbamoylpyridinium ion have been attempted (168,173,175,177,178,190). Under controlled-potential conditions, exhaustive electrolysis on the first wave yields a faradaic n value of 1 and on the second wave a value of 2 (168,173). Direct-current polarograms of the initial and final electrolysis solutions (Figure 11) show the absence of the characteristic current-potential curve of the initial electroactive species in the electrolyzed solutions.

Solutions electrolyzed at the wave I potential exhibit a diffusion-controlled anodic wave at a potential considerably more positive than the original cathodic wave (168); the initial electroactive species can be regenerated by

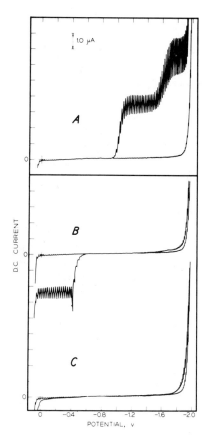

Figure 11. Electrolysis of 1 methyl-3-carba-moylpyridinium chloride (1.20mM) in KCl-carbonate buffer (pH 10). Direct-current polarograms: (A) before electrolysis; (B) after electrolysis at −1.25 V; (C) After electrolysis at −1.80 V. Background polarograms shown in all cases.

electrolysis at the anodic wave potential or by molecular oxygen. The first wave-electrolyzed solution has absorption maxima at 277 and 358 nm, suggesting the presence of the 6,6′-dimeric structure (168). The electrolysis product slowly decomposes at pH 10 via hydrolysis of the 4,5 double bond of the partially reduced pyridine ring in a reaction similar to that found for the nicotinamide reduction products (168). The rate of this reaction increases with decreasing pH; the half-life for a $2.4 \times 10^{-5}M$ solution of the dimer in pH 7.36 McIlvaine buffer is 5.1 min at 25°C. During decomposition, the anodic polarographic wave and the two characteristic ultraviolet absorption bands are lost, with a concomitant increase in absorption at 296 nm due to the acid-catalyzed decomposition product (168).

The investigators (173) who first isolated what was thought to be the 6,6′-dimer may actually have isolated the dimeric acid-catalyzed decomposition product, since electrolyses were carried out in pH 7 to 8.1 phosphate buffers. The isolated product exhibited a single absorption band at 298 nm, and no anodic wave was mentioned when a polarogram of the first-wave-electrolyzed solution was taken (173). Instead, a cathodic wave was seen at -1.65 to -1.70 V (depending on pH), which was attributed to the further reduction of the 4,5 double bonds of the dimer (2e per bond) (173). In a more recent study (168), the initial dimeric product was found to be electrochemically nonreducible.

The first investigators to isolate a 2e product, produced by electrolysis at a potential on the limiting portion of wave II of 1-methyl-3-carbamoyl-pyridinium ion in pH 7.8 to 9.2 phosphate buffers, showed it to be a 1,4-dihydropyridine species by comparing its ultraviolet and infrared spectra with those of authentic material prepared by dithionite reduction (173).

The 2e product of electrolysis on the second wave of the 1-methyl-3-carbamoylpyridinium ion was not oxidizable at the D.M.E. (Figure 11) (168). However, the product could be readily oxidized to the initial electroactive species at a pyrolytic graphite electrode (P.G.E.) at -0.04 V and by molecular oxygen. As in the case of the dimer, the dihydropyridine had absorption maxima at 270 and 358 nm, which suggested formation of the 1,6-isomer. Below pH 8, the 2e product rapidly hydrolyzed to a product with a single absorption band at 298 nm (168).

Recently, the mechanism of electrochemical oxidation of 1-methyl-1,4-dihydronicotinamide and other NADH analogs in acetonitrile (AN) was elucidated (191). In both basic and unbuffered solutions, the dihydronicotinamides were oxidized to the corresponding pyridinium salts (RN$^+$). In basic AN, RN$^+$ was formed quantitatively. In unbuffered AN, an acid decomposition reaction involving RNH accompanied the electrode reaction, and the RN$^+$ yield was only 40%. In unbuffered media, the oxidation process was resolvable into at least two steps, the first of which involved loss of a single electron; this was followed by one or more steps involving proton transfer and the loss of an additional electron to give RN$^+$. These results clearly excluded the oxidation of the 1,4-dihydronicotinamides by a 2e mechanism in AN (191).

C. ADSORPTION PHENOMENA

a. Cyclic Voltammetry. Until recently, little direct evidence was available regarding the surface activity of either 1-methylnicotinamide or its reduction products at the electrode-solution interface. In a cyclic voltammetric study of pyridine nucleotides and nicotinamide model compounds (188,189), the first cathodic process for 1-ethyl-, 1-propyl-, and 1-benzyl-nicotinamides was observed to consist of two peaks very close together. The process was described as following an AR, SR mechanism, in which adsorbed material is reduced first and then soluble material is reduced

(192,193). In the case of 1-methylnicotinamide, only the second of the two cathodic peaks was observed, and it was assumed that adsorption of the compound was much less pronounced.

b. Chronopotentiometry. Stronger evidence for the AR, SR mechanism was obtained by examining 1-methyl-, 1-propyl-, and 1-benzyl-3-carbamoylpyridinium chloride in pH 8 buffered solutions with current-reversal chronopotentiometry at the hanging mercury drop electrode (H.M.D.E.) (194); each mercury drop was allowed to equilibrate for 10 min to establish adsorption equilibrium. In all cases, the data indicated strong adsorption of the dimeric 1e reduction product. No evidence for adsorption of the methyl reactants was found. The propyl and benzyl reactants were considered to be so strongly adsorbed as to be more easily reduced than their freely diffusing counterpart.

c. Alternating-Current Polarography. Phase-selective ac polarography has been employed to examine the adsorbability of the 1e and 2e reduction products in the potential regions where these compounds are formed (168). The in-phase and quadrature components of the total alternating current for the initial and final electrolysis solutions of 1-methyl-3-carbamoylpyridinium ion are shown in Figure 12; the conditions are the same as those for the dc polarograms in Figure 11.

In the potential region before the first reduction step (Figure 12a), the differential DL capacity is altered only slightly because of the oxidized species, even though the latter is present at a relatively high concentration (1.2mM). There does appear to be an increasing trend toward greater adsorption with increasing negative potential, which would not be unexpected for a positively charged species at potentials more negative than the ecm; however, little can be said concerning the adsorbability of 1-methyl-3-carbamoylpyridinium ion beyond -0.90 V because of its reducibility.

In the potential region of the first reduction step (Figure 12a), adsorption effects and changes in the differential DL capacity, relative to the supporting electrolyte alone, are due to the surface activity of the dimeric 1e reduction product. The faradaic peak (in-phase component) at -1.07 V corresponds to the dc wave I process. The quadrature response in the potential region of wave I is complicated by the fact that there occur simultaneously a decrease in the capacity current due to strong adsorption of the dimeric product and an increase in the faradaic current due to the first reduction step. A tensammetric peak at slightly more negative potential (1.43 V) is due to desorption of the dimeric product; the peak is primarily capacitive in nature but has a resistive in-phase component at -1.41 V because the adsorption-desorption process is diffusion controlled. The fact that the dc polarogram does not

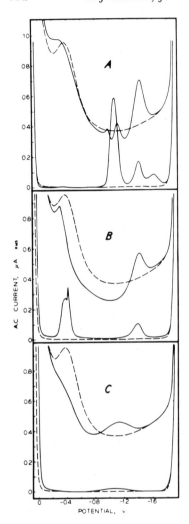

Figure 12. Electrolysis of 1-methyl-3-carbamoyl-pyridinium chloride (1.20mM) in KCl-carbonate buffer (pH 10). Alternating-current polarograms: (A) before electrolysis; (B) after electrolysis at −1.25 V; (C) after electrolysis at −1.80 V. Lower curves are for in-phase component and upper curves for quadrature components; dashed line is background electrolyte base current.

exhibit a wave at −1.4 V is proof that the ac peak arises from adsorption-desorption phenomena. An increase in the concentration of the electroactive species, the ionic strength of the supporting electrolyte, or both decreases the differential DL capacity in the vicinity of −1.0 V, increases the height of the tensammetric peak, and shifts the peak in a more negative direction. This behavior results from an increase in the surface activity of the dimeric 1e reduction product.

In the potential region of the second polarographic step (Figure 12a), all adsorption effects, if any, can be attributed to the 2e reduction product. The faradaic peak (in-phase component) at -1.62 V corresponds to the dc wave II process; no quadrature component is observed. In addition, the differential DL capacity is identical to that for the supporting electrolyte alone, showing that the $2e$ reduction product is not adsorbed in the potential region where it is formed.

The ac polarographic behavior of the $1e$ product itself is shown in Figure 12b. Faradaic peaks due to reduction of the original oxidized species are absent, but both the in-phase and quadrature components of the tensammetric process involving dimer desorption at -1.4 V remain. This is added proof supporting the interpretation already discussed regarding the ac polarographic pattern of the oxidized species (Figure 12a). Actually, the dimeric product is strongly adsorbed over a rather wide potential region both positive and negative to the ecm, which reflects the fact that the product is uncharged. The faradaic peak at -0.30 V (in-phase component) corresponds to the dc anodic wave and exhibits a maximum of the first kind (25), as does the dc wave. Finally, it is interesting to note that the strong adsorption of the dimer has relatively little effect on the dc polarographic pattern, for example, the surprisingly small effect on current-potential plots, the absence of an adsorption prewave, and the absence of anomalous behavior involving the limiting current of wave I with increasing depolarizer concentration.

In the ac polarographic pattern of the $2e$ product (Figure 12c), faradaic peaks due to reduction of the original oxidized species are absent and no faradaic behavior due to the $2e$ product is observed; this is in agreement with the dc result. The absence of the tensammetric process at -1.4 V further supports the idea that this peak is due to the dimer desorption process. The $2e$ product is rather weakly adsorbed compared to the $1e$ product. The presence of the $2e$ product most strongly affects the differential DL capacity in the potential region on either side of the ecm. At more negative potential, it is involved in a rather broad desorption hump centering at -1.12 V. Beginning at about -1.5 V, the differential DL capacity is identical to that of the supporting electrolyte alone. Thus the $2e$ reduction product is not adsorbed in the potential region of its formation, a finding which supports the conclusion reached after considering the ac results for the initial electrolysis solution.

D. REDUCTION MECHANISM

The reaction scheme outlined in Figure 13 seems to fit best the electrochemical, chemical, and spectral data for reduction of the 1-methyl-3-carbamoylpyridinium ion (5). The initial reduction can be regarded as the reversible uptake of an electron to form a neutral free radical (6) (source of

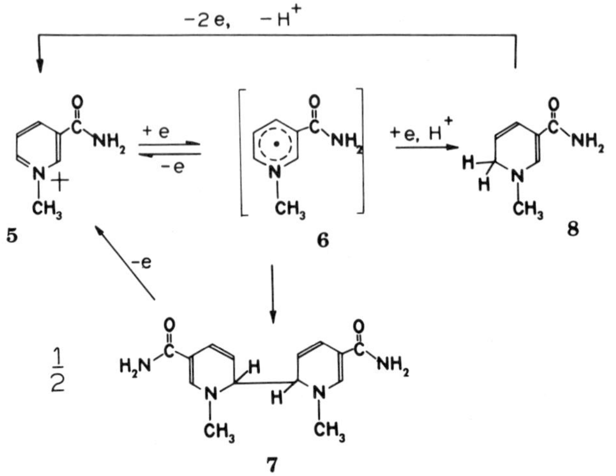

Figure 13. Interpretation of the electrochemical behavior observed for
1-methylnicotinamide (5) and its reduction products.

polarographic wave I). The electron-transfer reaction is followed by an
irreversible dimerization to form the 6,6'-dimer (7). At the potential of
wave II, the 1-methyl-3-carbamoylpyridinium ion is reduced to either a
1,4- or a 1,6-dihydropyridine (8) or, perhaps, to both; the effect of experimental conditions on the relative amounts of the two isomers formed is
unclear. In the potential region of the second wave, the dimer is not further
reduced. At a potential considerably more positive than that of the first
reduction wave, the dimer can be oxidized to the original starting material,
the 1-methyl-3-carbamoylpyridinium ion. At still more positive potential,
the $2e$ products (8) can also be oxidized to the latter ion.

7. Pyridine Nucleotides

The majority of papers concerned with the electrochemistry of the
pyridine nucleotides have dealt primarily with NAD^+. Relatively few
reports have dealt with the behavior of $NADP^+$ and NMN^+ (168,179,
195–197b).

The electrochemistry of NMN^+ is of interest because in many respects it
is a more realistic model compound for NAD^+ and $NADP^+$ than 1-methyl-
nicotinamide. The electronic structure of the pyridine ring of NMN^+ would
be expected to be more closely related to that of the coenzymes because of
the similar inductive effects of the N-substituents. In addition, NMN^+, like
the coenzymes, has a bulky N-substituent, which presumably plays a role in

TABLE III
Half-Wave Potentials at D. M. E. of NAD$^+$ and Related Pyridine Nucleotides

Nucleotide	Medium[a]	pH	$-E_{1/2}$ (V)[b]		Ref.
			Wave I	Wave II	
NAD$^+$	38	3.2–7.1	0.91–0.92	—	196
	38	7.3–8.7	0.92–0.96	—	196
	26	4–9	0.93	—	195
	47	1–8	0.88–0.90	—	168
	48	9.3	0.88	1.61	168
	49	9.3	0.92	[e]	168
	40	7–9	0.91	—	179
	8	7.4	0.93	—	346
	39	10.3–10.6	0.98	—	239
	37	5.0	0.97	1.38	196
	50	9.6	1.03	1.62	168
	36	9.1	1.12	1.72	195
	41	4–8	1.09	—	20
	42	4.2–6.6	1.09	—	180
	43	8.3	1.10	—	180
	44	10.4	1.18	—	180
	[e]	10.6	0.98	—	278
NADP$^+$	51	5–7	0.92	—	197
	47	1–8	0.89–0.90	—	168
	52	9	0.92	1.70	197
	39	10.3–10.6	1.23	—	239
	50	9.6	1.06	1.61	168
NMN$^+$	47	1–5	0.89–0.91	—	168
	48	9.3	0.96	1.63	168
	49	9.3	1.03	[c]	168
	39	10.3–10.6	1.14	—	239
	50	9.6	1.08	1.64	168
α—NAD$^+$	50	9.6	1.07	1.61	168
NHD^{+d}	50	9.6	1.05	1.60	168

[a] Media are listed in Table IV.
[b] A dash indicates that no wave was observed or reported.
[c] Second wave ill defined in slightly alkaline solutions of low ionic strength.
[d] Nicotinamide-hypoxanthine dinucleotide.
[e] Medium not specified.

TABLE IV

Media and Buffer Systems[a] Specified in Tables I to III, VI to VIII, and X

1	$2M$ Sulfuric acid
2	KCl or NaCl plus HCl
3	McIlvaine buffer: Na_2HPO_4 + citric acid + KCl[b]
4	Acetic acid + sodium acetate buffer
5	Potassium chloride
6	Ammonia + ammonium chloride buffer
7	NaOH + KCl
8	Phosphate buffer
9	Perchloric acid
10	K_2CO_3 + KOH buffer
11	Britton-Robinson[c] buffer
12	$1M$ Acetic acid
13	$0.05M$ Perchloric acid
14	$0.1M$ Perchloric acid
15	Britton-Robinson buffer + $0.3M$ ammonium formate
16	Britton-Robinson buffer + $1.0M$ ammonium formate
17	$0.08M$ Acetic, phosphoric, or boric acid plus varying amounts of $0.4M$ NaOH
18	Citrate buffer
19	$0.1M$ Sodium hydroxide
20	$1M$ Sulfuric acid
21	Carbonate buffer
22	Hydrochloric acid
23	$0.1M$ Hydrochloric acid
24	$0.2M$ Acetate buffer
25	$0.02M$ Acetate and citrate buffers
26	$0.1M$ Acetate, citrate, phosphate, and pyrophosphate buffers
27	$0.02M$ Phosphate buffer
28	$0.1M$ Phosphate buffer
29	$0.1M$ Phosphate-citrate buffer
30	$0.02M$ Pyrophosphate buffer

determining the isomeric form of the first-wave reduction product, for example, a 4,4'- as opposed to a 6,6'-dimer. Moreover, NMN⁺ enjoys the same freedom from the adenine moiety as does 1-methylnicotinamide; the ultraviolet spectra are, consequently, characteristic only of the pyridine ring, and adsorption phenomena at the electrode-solution interface due to the adenine moiety are eliminated (70).

In general, the pyridine moiety of all three nucleotides gives rise to the same fundamental polarographic behavior, which is, however, moderately sensitive to experimental conditions (Table III). In the case of NAD⁺, for example, the results of different investigations vary in regard to the potential of the first polarographic wave and the existence of a second wave. In most cases for aqueous solutions, these discrepanceis can be accounted for

TABLE IV (*continued*)

31	0.01*M* Phthalate buffer
32	0.3*M* Tetramethylammonium chloride (TMAC)
33	Glycine buffer
34	0.02*M* Ammonia
35	Tetramethylammonium borate buffer
36	0.1*M* Tetra-*n*-butylammonium carbonate buffer
37	0.02*M* Tris buffer
38	0.1*M* Tris buffer
39	0.5*M* Tris buffer
40	Tris and phenol sulfonate buffers
41	50% Dioxane with 0.1*M* tetra-*n*-butylammonium iodide (TBAI)
42	Citrate buffer in 50% dioxane with 0.1*M* TBAI
43	Phosphate buffer in 50% dioxane with 0.1*M* TBAI
44	Borate buffer in 50% dioxane with 0.1*M* TBAI
45	0.1*M* Ammonia buffer
46	0.1*M* Borate buffer
47	0.5*M* Buffers: HCl + KCl, acetate, and McIlvaine
48	0.4*M* KCl; 0.1*M* carbonate buffer
49	0.1*M* Carbonate buffer; 0.0008% Triton X-100
50	0.4*M* Tetraethylammonium chloride (TEAC); 0.1*M* carbonate buffer
51	0.2*M* Acetate, citrate, and phosphate buffers
52	0.2*M* Pyrophosphate buffer
53	0.3*M* Phosphate buffer
54	0.4*M* KCl + 0.1*M* KOH

[a] Concentrations of electrolyte or principal buffer component are given in cases where they could be located in the original paper. Cations, such as quaternary ammonium ions, are indicated where they may have had a significant effect (e.g., on surface activity).
[b] See ref. 354.
[c] See refs. 348 and 349.

by considering the differences in the experimental conditions employed (e.g., pH, the presence or absence of surface-active substances, and ionic strength). Certain aspects of the electrochemistry of each of the three nucleotides are virtually identical, and in the following sections no attempt is made, therefore, to discuss these topics separately. In cases where differences arise, appropriate comparisons are made.

A. DIRECT-CURRENT POLAROGRAPHY

The pyridine nucleotides generally exhibit a single cathodic wave in buffered solution of pH less than 6 (tetraalkylammonium salts absent) (168,179,195–197). In buffered solution of pH 1, the wave is ill defined with a limiting current having kinetic character and occurs on the side of background discharge, the latter being 0.2 to 0.3 V more positive than that of the supporting electrolyte alone (168). With increasing pH, the wave decreases in height, reaching a limiting value at pH 4 to 5, and becomes diffusion controlled (168,195,197). Above pH 5, the i_d/C ratio for all the compounds remains constant up to about pH 10 to 11, where it begins to decrease because of alkaline hydrolysis (168,198).

In the case of NAD+ and NADP+, $E_{1/2}$ is virtually independent of pH from pH 1 to 10 (168,197,198); the positive shift of background discharge is relatively constant up to about pH 6 but then becomes gradually less pronounced, until at pH 9 background discharge occurs at a potential identical to that of the supporting electrolyte alone (168). In buffered solution above pH 9 ($\mu \sim 0.1M$), virtually only the first wave process is observed; a drawn-out shoulder appears on the side of background discharge corresponding to the wave II process (the second $1e$ addition to the pyridine ring) but does not develop into a wave (168,195,197).

In the case of NMN+ (168), the single wave seen in solutions of low pH is independent of pH up to about pH 5, becomes gradually about 80 mV more negative in the pH range of 5.0 to 7.5, and remains constant thereafter up to pH 12. In the transition range, NMN+ would be expected to have a secondary phosphate dissociation; for example, pK_a for the corresponding dissociation of AMP is between 6.0 and 6.7 (199). At pH 6, NMN+ exhibits a second more negative wave ($E_{1/2} = -1.58$ V), whose limiting current is kinetically controlled and much greater than that of wave I; the current results from two simultaneously occurring electrode processes: formation of the $2e$ product, and catalytic hydrogen evolution due to reduction of the protonated form of the product [nicotinamide (162) and 1-methylnicotinamide (164,168) also show this behavior]. With decreasing pH, the catalytic wave height increases rapidly, accounting for the positive shift in background discharge, that is, the rising portion of the catalytic wave serves as discharge (168). With increasing pH, the catalytic wave rapidly decreases, and in buffer solutions above pH 9 ($\mu \sim 0.1M$) the behavior found for NMN+ is identical to that seen for NAD+ and NADP+, that is, essentially only wave I is observed with an ill-defined shoulder on background discharge representing the wave II process.

The positive background shifts of NAD+ and NADP+ are also due to catalytic hydrogen evolution; in the case of these compounds, it is likely that the adenine moiety also contributes to the observed current. In buffered solutions of pH 6 to 8, for example, the total polarographic current in the potential region near discharge can result from three electrode processes that occur essentially simultaneously: formation of the $2e$ product, catalytic hydrogen reduction of the latter protonated product, and catalytic hydrogen reduction of the protonated adenine moiety (168). The presence of the adenine moiety as part of the structure of NAD^+ and $NADP^+$ consequently complicates interpretation of the catalytic wave due to the reduced pyridine moiety. The catalytic wave found for NAD^+ and $NADP^+$ does not develop into a well-defined wave as in the case of NMN^+, but gradually merges with background discharge (a process which is best seen at low polarographic sensitity). Under certain conditions, however [e.g., in solutions of low buffer capacity (pH 7)], the catalytic wave found for NAD^+ does exhibit a kinetically controlled limiting portion (168). In solutions of pH less than 6, direct reduction of adenine may also be involved (cf. the subsequent discussion on cyclic voltammetry).

a. Effect of Ionic Strength. Above pH 9 ($\mu \sim 0.1M$), the second $1e$ reduction step of all three nucleotides is ill defined, as previously discussed. However, in solutions of ionic strength greater than about $0.5M$ (168) or containing tetraalkylammonium salts (168,195–198), two waves are observed for each of the three nucleotides (168,189,195–197,200–202) (Table III). In slightly alkaline solution (tetraalkylammonium salts absent), an increase in ionic strength from 0.1 to $2.0M$ ($0.1M$ carbonate buffer plus KCl; pH 9.30) causes the second wave of both NMN^+ and NAD^+ to grow in and to become much better defined with a limiting current about the same as that of the first wave (168). During this process, $E_{1/2}$ of NMN^+ wave II shifts from -1.73 to -1.56 V (170 mV). An equivalent shift is also seen for the second NAD^+ wave, but in this case a third wave appears with increasing ionic strength between waves I and II ($E_{1/2} = -1.38$ V at $\mu = 0.5M$; Figure 14). The third wave is related to adsorption phenomena involving the wave I product (cf. the subsequent ac polarography discussion). The same increase in ionic strength also causes $E_{1/2}$ of wave I of both compounds to become more positive: from -1.03 to 0.93 V (100 mV) for NMN^+ and from -0.92 to -0.89 V (30 mV) for NAD^+. The addition of Triton X-100, on the other hand, at a level of 0.001%, shifts $E_{1/2}$ of NAD^+ wave I in a more positive direction by 10 to 20 mV at all ionic strengths.

b. Effect of Tetraalkylammonium Salts. The presence of tetra-alkylammonium salts also promotes formation of the second wave. In tetra-n-butylammonium carbonate buffers, for example, where the more negative hydrogen overvoltage allows a broader plateau to be observed for wave II (189), NAD^+ (195,198) and $NADP^+$ (189,197) exhibit essentially the same polarographic behavior; the two waves seen are of about equal height. Wave I is diffusion controlled, but wave II exhibits kinetic character. In the case of NAD^+, log plots for wave I have slopes corresponding to a reversible $1e$ process; slopes for wave II indicate an irreversible electrode process. In addition, $E_{1/2}$ of wave II is independent of pH from pH 6 to 10. In dilute NAD^+ solutions ($\sim0.1mM$), a small concentration-independent prewave is

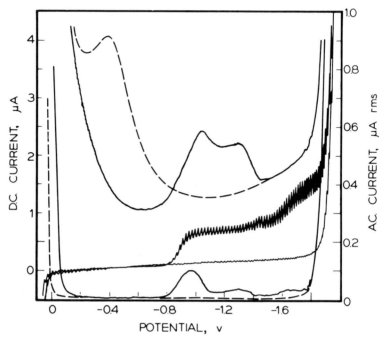

Figure 14. Direct- and alternating-current polarograms of NAD$^+$ (0.32mM) in KCl-carbonate buffer (pH 9.30). For AC polarograms: controlled drop time of 3.0 sec; $f = 50$ Hz; $\Delta E = 3.54$ mV rms; lower curves are for in-phase component and upper curves for quadrature components; dashed line is background electrolyte base current.

observed about 220 mV more positive than wave I; this was not studied in detail but was thought to be due to adsorption (195,198) (presumably involving the wave I reduction product, although this was not stated).

The features of the second wave of the three nucleotides are also enhanced by the presence of the tetraethylammonium ion (168). However, the appearance of wave II is not due to a simple shifting of background discharge to more negative potential. In solutions maintained at an ionic strength of 0.1M (0.05M potassium carbonate buffer of pH 9.90 plus KCl and Et$_4$NCl), NAD$^+$ wave II appears, grows, and becomes well defined as the Et$_4$NCl concentration increases from zero to 0.05M [wave II $E_{1/2}$ shifts from -1.83 to -1.61 V (220 mV)]; no change in the background discharge potential is observed. In addition, $E_{1/2}$ of wave I shifts from -0.91 to -0.98 V. When 0.4M Et$_4$NCl is exchanged for 0.4M KCl in solutions which are also 0.1M in carbonate buffer, $E_{1/2}$ for wave I shifts from -0.89 to -1.03 V. The behavior is essentially the same for all three nucleotides (Table III; Figures 15 and

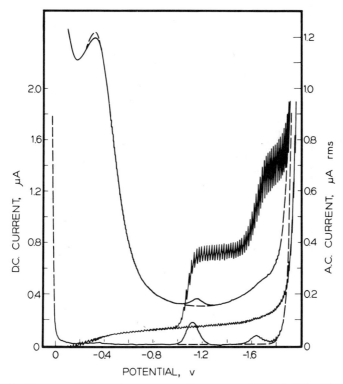

Figure 15. Direct- and alternating-current polarograms of NMN⁺ (0.3mM) in Et₄NCl-carbonate buffers (pH 9.63). For ac polarograms: controlled drop time of 3.0 sec; $f = 50$ Hz; $\Delta E = 3.54$ mV rms; lower curves are for in-phase component and upper curves for quadrature components; dashed line is background electrolyte base current.

16). In the presence of sufficient Et₄NCl, the two waves are of about equal height and diffusion controlled; each wave is due to a 1e process. In the case of NMN⁺, the slope of wave I is that for a reversible 1e cathodic wave. The wave I slopes for NAD⁺ and NADP⁺ are less reversible, being close to 70 mV. On the basis of wave slopes, wave II in all cases involves an irreversible electrode process.

c. Interpretation of Results. The effects of ionic strength and tetra-alkylammonium salts on the polarographic pattern of the pyridine nucleotides appear to be related to changes in the DL structure and probably also involve the two charge centers of the electroactive species themselves. The effect of ionic strength in the case of NMN⁺, for example, suggests that an intramolecular complex between the negatively charged phosphate group

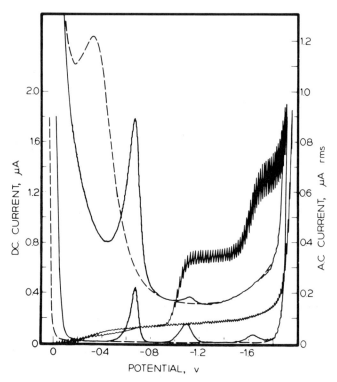

Figure 16. Direct- and alternating-current polarograms of NAD$^+$ (0.32mM) in Et$_4$NCl-carbonate buffers (pH 9.63). For ac polarograms: controlled drop time of 3.0 sec; $f = 50$ Hz; $\Delta E = 3.54$ mV rms; lower curves are for in-phase component and upper curves for quadrature components; dashed line is background electrolyte base current.

and the positively charged pyridinium ion exists in solutions of low ionic strength; such complex formation could increase the electron density in the nicotinamide nucleus (168). With increasing ionic strength, the complex dissociates, causing wave I to become more positive and wave II to appear and grow. Wave II of 1-methylnicotinamide, which has no phosphate grouping, is always well defined, even at low ionic strength (168). In addition, wave I of 1-methylnicotinamide shifts relatively little with ionic strength; however, it occurs at a more negative potential than wave I of the dissociated form of NMN$^+$, a difference which may be due to the lack of electron-withdrawing ability of the methyl substituent relative to the ribose substituent.

Although the exact nature of these effects is presently unclear, the discrepancies in Table III regarding $E_{1/2}$ values and the presence or absence of wave II are due in large part to the experimental conditions employed. In the case of NAD$^+$, for example, most workers who have used ionic strengths on the order of $0.1M$ in the absence of tetraalkylammonium salts (168,179, 195,196) report $E_{1/2}$ of wave I to be in the range of -0.91 to -0.94 V. With increasing ionic strength, $E_{1/2}$ becomes more positive (168); addition of tetraalkylammonium salts, on the other hand, makes the wave more negative (168,195,196) (Triton X-100 is also effective in making the wave slightly more negative). Wave II is not readily observed unless either solutions of high ionic strength are employed (168) or tetraalkylammonium salts are present (168,195,196).

B. ALTERNATING-CURRENT POLAROGRAPHY

The ac polarographic behavior of NMN$^+$ in KCl-carbonate buffer (pH 9.30) is relatively simple (168). Alternating-current polarograms exhibit faradaic peaks at -1.02 and -1.64 V, which correspond to the two dc waves. A slight depression in the differential DL capacity before the first reduction step indicates weak surface activity of the oxidized species. The capacity after the first reduction step is nearly the same as that of the supporting electrolyte alone, proving that neither the 1e nor the 2e reduction products are adsorbed at the electrode-solution interface under the experimental conditions employed.

The behavior of NAD$^+$ under the same conditions is somewhat more complicated (Figure 14). Alternating-current faradaic peaks at -0.97 and -1.65 V correspond to the two dc waves. In the potential region before the wave I process, adsorption of NAD$^+$ causes a depression in the differential DL capacity. After passing through a minimum at -0.65 V, the capacity begins to increase with rising applied potential. The occurrence of the capacity increase before initiation of the wave I process suggests that a tensammetric process due to desorption of NAD$^+$ is about to begin. Immediately thereafter, reduction begins, preventing the desorption process from being realized.

Results in the potential region of the first ac faradaic process are difficult to interpret. Little can be said concerning the differential capacity because of the concurrent faradaic ac current. In this region, the capacity appears, however, to be higher than that for the supporting electrolyte alone. At more negative potential, an ac peak at -1.29 V with only a small in-phase component is, therefore, almost entirely capacitive in nature and is not related to a faradaic process. These results can be accounted for by assuming that the wave I product is adsorbed at the interface and is desorbed at sufficiently negative potential, giving rise to the tensammetric peak at -1.29 V. However, the same tensammetric process appears to be in its early stages even in the potential region of the first ac faradaic process and accounts in part for the increased DL capacity at this potential, rather than the depression which would be expected from an adsorbed product. The increased capacity is also related, no doubt, to a tensammetric-type process involving a change from adsorbed NAD^+ to adsorbed dimer as a result of the reduction step.

Experimental variables, which would be expected to increase the surface activity of the wave I reduction product of NAD^+, did shift the ac desorption peak to more negative potential; for example, shifts of 200, 100, and 50 mV, respectively, were observed with increased ionic strength (0.1 to 2.0M), decreased temperature (37.0 to 1.4°C), and increased NAD^+ concentration (0.04 to 1.26mM) (168). In the potential region more negative than the tensammetric peak (Figure 14), the differential DL capacity is identical to that of the supporting electrolyte alone and indicates that the 2e reduction product is not adsorbed over the potential region where it is formed, that is, in the potential region of wave II. Finally, desorption phenomena involving the NAD^+ wave I product have considerable effect on the dc polarographic result (Figure 14). An anomalous dc wave which generally appears immediately after the ac tensammetric peak may be due to the desorption process, for example, by the latter causing stirring, which would increase the limiting portion of the wave I current.

a. Effect of Tetraalkylammonium Salts. Substitution of Et_4NCl for KCl causes little change in the ac polarographic behavior of NMN^+ (Figure 15); the behavior of NAD^+, however, is significantly altered (compare Figures 14 and 16) (168). In the case of NAD^+, a tensammetric peak at -0.67 V corresponds to a small-capacity wave seen under dc conditions at the same potential. At potentials before the peak, the differential DL capacity is depressed with respect to the supporting electrolyte alone, suggesting adsorption of NAD^+ at the electrode- solution interface. In the potential region of the tensammetric peak, which is more negative than the ecm under the experimental conditions employed, desorption of NAD^+ due to preferential adsorption of the tetraethylammonium ion increases the surface charge density to values normally observed for the supporting electrolyte alone. A plot of surface charge density versus potential would be expected to show a relatively rapid increase in charge density in this potential region, with an inflection point at -0.67 V; since the ac method measures the slope of such a plot (i.e., the differential DL capacity), a peak results. The presence of a resistive component in the tensammetric peak suggests that the electrode impedance is not entirely capacitive in this potential region (203,204). This may imply that the adsorption-desorption process under ac conditions is partially diffusion controlled (203).

With increasing NAD^+ concentration (0.13 to 1.26mM), the tensammetric peak shifts from -0.64 to -0.73 V and the current increases by a factor of 2.5. In the potential region following the tensammetric peak but preceding the first faradaic process, the capacity is nearly the same as that of the supporting electrolyte alone, indicating that NAD^+ is not adsorbed. Similarly, no change is observed in the capacity after the two faradaic processes, indicating that the reduction products are not adsorbed.

In Et_4NCl-carbonate buffers, the faradaic ac polarographic behaviors of NMN^+ and NAD^+ are similar (168). Peaks at -1.12 and -1.64 V corresponded to the two dc polarographic waves. Peak currents for the faradaic processes represent less than 3% of the total alternating current expected for a reversible 1e process; thus both steps are irreversible under ac conditions. In the case of NMN^+, this contrasts with the fact that wave I is reversible under dc conditions. An explanation for the discrepancy is essentially the same as that given for nicotinamide and 1-methylnicotinamide—that is, the ac method tests the reversibility of the overall electrochemical process, including the irreversible dimerization. The period of the alternating voltage used (20 msec at 50 Hz) would be expected to be much greater than the half-life of the NMN^+ radical (168). The dimerization rate constant of the NAD radical produced by pulse radiolysis (205), for example, is 5.6 $\times$ 10^7 l/mole-sec. The half-life of the NAD radical is estimated by current reversal chronopotentiometry to be less than 1 msec (194). Thus half-lives of a few tenths of a millisecond would be expected for both the NAD and NMN radicals.

The involvement of a follow-up chemical reaction in the wave I process of NMN^+ in Et_4NCl-carbonate buffer (Figure 15) is also suggested by the dc potential dependence of cot ϕ (168,435) (ϕ is the phase angle of the total faradaic alternating current with respect to the applied alternating voltage; cot ϕ equals the ratio of the in-phase and quadrature faradaic current components). At potentials more negative than $E_{1/2}$, cot ϕ approaches a value of 1.0; at more positive potential, it increases to a value considerably greater than 1.0. At the latter potentials, the most important terms in the overall faradaic impedance are the mass transfer impedance terms of the reduced species and the charge-transfer resistance. A chemical reaction following charge transfer would give rise to an added mass transfer term for the reduced species, whose resistive and capacitive components would not necessarily have the same impedance values (435). Therefore cot ϕ would deviate from the value of 1.0, which is commonly observed for a reversible, diffusion-controlled process. The deviation is known to be positive for following first-order chemical reactions (435), but its sign is not known for second- or higher-order reactions.

C. CYCLIC VOLTAMMETRY

The three nucleotides have been studied under cyclic voltammetric conditions at the H.M.D.E. (168,188,189). In slightly alkaline solution, sweeps to more negative potential give rise to two cathodic peaks (E_p of peak Ic = about -1.0 V; E_p of peak IIC = -1.6 to -1.7 V), corresponding to the two cathodic processes observed at the D.M.E. Peak Ic is well defined, but peak IIc appears as a shoulder on background discharge. The height of Ic is much greater than that of IIc, suggesting that the overall wave I reduction product is not further reduced (168).

At slow scan rates (1 V/sec or less), reversal of the voltage sweep after either peak Ic or IIc yields a single anodic peak Ia ($E_p = -0.15$ to -0.30 V), corresponding to the oxidation of the peak Ic 1e reduction product. Oxidation of the peak IIc 2e reduction product occurs at a potential more positive than mercury discharge and therefore is not observed at the H.M.D.E. Repetitive cycling around peaks Ic and Ia shows no diminution of the cathodic peak current and suggests that the oxidation process regenerates the original electroactive species. The potential separation indicates irreversibility in the overall wave I process (168,188,189), in agreement with the ac polarographic results.

When the scan rate is sufficiently rapid (10 V/sec or greater), all three nucleotides exhibit a previously unobserved anodic peak in the vicinity of -1 V, as well as the anodic peak Ia previously noted (168,188,189). In the case of NAD$^+$ and NADP$^+$, the new peak was interpreted as representing the reversible reoxidation of a free radical which, at the faster scan rates, has not sufficient time to disappear via the chemical step following electron transfer (188). The same interpretation was given for the anodic peak observed on reoxidation of the NMN radical (168). Attempts to measure the half-lives of the NAD and NADP radicals under the conditions of cyclic voltammetry yielded values on the order of 1 to 2 msec (188,189). The use of current-reversal chronopotentiometry also disclosed the presence of the NAD radical (194). Thus, under the conditions used, the dimerization of the NAD radical is very rapid, having a half-life of no more than a msec.

a. Possible Involvement of Adenine Reduction. In acidic solution, one report describes only cathodic peak Ic and the corresponding anodic peak Ia for both NAD$^+$ and NADP$^+$ (188). Other workers; however, observed a second cathodic peak for these compounds at more negative potential; this was interpreted as being due not to a second 1e addition to the pyridine ring, as seen in slightly alkaline solution, but rather to a direct reduction of the adenine moiety (168)(the reduction of adenine and its nucleosides and nucleotides is discussed in Section V). A second cathodic peak was also found for α-NAD$^+$ in acidic solution, but not for compounds

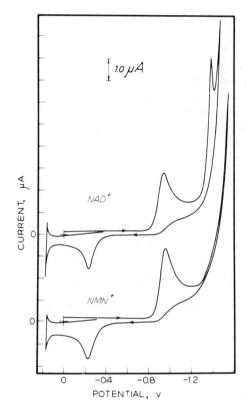

Figure 17. Cyclic voltammograms of NAD$^+$ and NMN$^+$ in acetate buffer (pH 5.50) at the H.M.D.E. Potential sweep rate = 100 mV/sec.

such as NMN$^+$ and 1-methylnicotinamide, which do not contain an adenine structure as part of their N-substituent (Figure 17). Nicotinamide-hypoxanthine dinucleotide and nicotinamide-hypoxanthine dinucleotide phosphate also did not exhibit the second cathodic peak (168); the hypoxanthine moiety in these compounds, if reduced, would be expected to involve only a 2e reduction similar to that of hypoxanthine itself (133) rather than a 4e reduction characteristic of adenine. In addition, the polarographic behavior of the parent hypoxanthine is less well defined than that of adenine, that is, it appears only as an ill-defined inflection on background discharge (133).

In the case of NAD$^+$, the second cathodic peak potential and current depend on pH in a manner similar to that found for adenine (133,168). With increasing pH, E_p becomes more negative:

$$E_p = -1.074 - 0.057 \text{ pH} \qquad [32]$$

and i_p decreases from a value five times the height of the first cathodic peak at pH 2 to values approaching zero around pH 6; above pH 6.5, the second peak is not seen. Similar results were found for NADP$^+$ and α-NAD$^+$, except that the second peak of NADP$^+$ disappeared at pH 5 and that of α-NAD$^+$ at pH 7 (168).

Other evidence supporting reduction of the adenine moiety was obtained by electrolytic reduction of NAD$^+$ at a large mercury pool cathode (an uncontrolled potential of 12 V versus an isolated platinum anode was employed) (206,207). In an acidic medium appreciable reduction of the adenine ring occurred, accompanied by an almost equivalent deamination. When NAD$^+$ was electrolyzed in a weakly alkline medium (pH 8), reduction of the adenine ring was sharply diminished and no deamination (as measured by ammonia evolution) was found (206,207).

Finally, it is of interest to compare the electrochemical reduction of the adenine moiety of NAD$^+$ with the results seen on pulse radiolysis. From experiments employing the latter technique, it was postulated that, when an electron reacts with the adenine portion of NAD$^+$, it transfers rapidly to the nicotinamide ring (205). Reductions by pulse radiolysis, however, generally involve only a single electron (NAD$^+$, e.g., forms a dimeric product rather than a dihydropyridine). Electrochemical reductions such as that of adenine and its nucleosides and nucleotides, on the other hand, may involve multiple electron transfers, in which electrons are transferred more or less simultaneously. When electrochemical experiments are carried out on NAD$^+$ in acidic solution at the H.M.D.E., virtually all of the NAD$^+$ in the vicinity of the electrode is converted to the dimeric product at a potential considerably more positive than the reduction potential of the adenine moiety; thus no transfer of electrons from the adenine moiety to the pyridine moiety would be expected under these conditions (168).

D. CONTROLLED-POTENTIAL ELECTROLYSIS

Most macroscale electrolysis studies of pyridine nucleotides have been concerned primarily with NAD$^+$. Attempts to prepare reduced nicotinamide-adenine dinucleotide (NADH) by controlled-potential electrolysis of NAD$^+$ at mercury electrodes (149,168,195,196,208–210) and at solid electrodes (209,211) have led to products of variable coenzyme activity, ranging from zero (208,211) to effective coenzyme concentrations of 30% (149,212), 50% (210), 54% (168), 56% (211), 59% (195), and 76% (209) of the value expected. On the other hand, it appears that some investigators did not recognize the existence of two reduction steps for NAD$^+$, and this may account, at least in part, for some of the variation in results.

a. NAD$^+$ (Wave I). Controlled-potential electrolysis at pH 9 to 10 on the limiting portion of wave I conclusively demonstrated that reduction of

NAD$^+$ under these conditions involves one electron (168,195). A polarogram of the electrolyzed solution revealed the loss of both NAD$^+$ cathodic waves and the presence of a diffusion-controlled anodic wave about 0.6 to 0.8 V more positive than $E_{1/2}$ for the original cathodic wave I (168,179,195, 196,202). The electroactive species responsible for the anodic wave had a diffusion coefficient considerably lower than that of NAD$^+$ [51% (213), 66% (168)]. The wave I electrolysis product could be reoxidized to NAD$^+$ at a mercury pool electrode at a potential on the limiting portion of the anodic wave (168,195), as well as by an enzyme obtained from mung bean seedlings (212–214).

The wave I product has been isolated as the salt Tris$_4$ dimerate·6H$_2$O (213). Molecular weight determinations by vapor pressure osmometry and ultracentrifuge sedimentation equilibrium confirmed the dimeric nature of the material. The ultraviolet spectrum of the isolated product was similar to that of NADH with maxima at 259 and 340 nm (213,215,216); however, solutions of the product do not exhibit NADH activity (168,195,208) and do not fluoresce as does NADH (208,213). The product is relatively stable in the presence of O$_2$ at pH 9 to 10 but exhibits slight O$_2$ dependence below pH 8 (168,213). By far the most important factor in the stability of the electrolysis product is pH; the rate of decomposition of the dimer increases with decreasing pH (168,213). At pH 7, the 340-nm band is lost with a concomitant increase in absorption at 280 nm (168,208). At pH 1 to 2, formation of the 280-nm band is nearly instantaneous, but its intensity then slowly decreases with time (168).

The overall behavior of the dimer is similar to that observed during the formation of the primary and secondary acid modification products of NADH (168,171,172). It also forms a mercury adduct, as does NADH (168). With excess of HgCl$_2$, the 340-nm band of the dimer is lost with a simultaneous increase in absorption at 282 nm (168); similarly, addition of HgCl$_2$ to NADH solutions causes a decrease in the 340-nm band and formation of a new band at 268 nm (171).

In addition to the anodic wave, solutions of the wave I electrolysis product also exhibit a small cathodic wave close to the background discharge (168,195). This wave ($E_{1/2} = -1.7$ V) has been attributed (195) to the further reduction of the dimer to the 1,4-dihydro compound, that is, NADH, since NADH activity developed slowly when further electrolysis of the wave I electrolysis solution was carried out at -1.85 V. It was suggested that the slow rate of reduction of the dimer to NADH was due to mercury surface adsorption phenomena (195). Other workers (168) attributed the small cathodic wave to free nicotinamide, which was present initially as a slight impurity in the NAD$^+$ sample used and was also produced as a result of the slow alkaline decomposition of NAD^+ during electrolysis (a similar cathodic wave was found for wave I-electrolyzed solutions of NMN^+ but not for those of 1-methylnicotinamide; in the latter case, alkaline decomposition results in hydrolysis of the amide group rather than in cleavage of the N-substituent from the nicotinamide ring).

The small cathode wave ($E_{1/2} = -1.61$ at pH 9) was diffusion controlled; the pH dependence of its $E_{1/2}$ was identical to that of nicotinamide. After controlled potential electrolysis on the limiting portion of the cathodic wave, a polarogram of the electrolyzed solution revealed that the cathodic wave was absent and that the anodic wave of the NAD dimer had increased in height because of oxidation of the $2e$ reduction product of nicotinamide (1,6-dihydronicotinamide), as well as the dimer (168). In addition, the concentration of nicotinamide before electrolysis (as estimated from the height of the cathodic wave) agreed both with the nicotinamide concentration predicted from the number of coulombs passed during electrolysis and with the concentration predicted from the change in the ultraviolet spectrum after electrolysis (the latter again being due to 1,6-dihydronicotinamide).

b. NAD^+ (Wave II). Results reported for the macroscale electrolysis of NAD^+ at a potential on the limiting portion of wave II differ somewhat. Early workers (209,212) reported only partial conversion of NAD^+ to 1,4-NADH (under optimal conditions, the yield was 76%) and stated that the reduction led to a mixture of 1,2- 1,4-, and 1,6-dihydropyridines.

Others (195), using $(n\text{-}Bu_4N)_2CO_3$ buffers, electrolyzed at -1.85 V and found that both 1,4-NADH and the dimeric $1e$ product were formed; under optimal conditions, about 59% of the NAD^+ was converted to the 1,4-dihydro product. Values of n between 1 and 2 were obtained, which correlated, within experimental error, with the enzymatic NADH assays. During later stages in the electrolysis NADH formed sowly; this was explained as being due to the slow reduction of the dimer at the potential employed. Addition of a cationic surfactant and vigorous ultrasonic agitation of the mercury-solution interface considerably increased the rate of NADH formation. The low yields of enzymatically active NADH were ascribed to the concurrent formation of dimer (195). Since the relative rates of formation of the two species depended on the presence of a surfactant and on the method of stirring, it was suggested that adsorption of NAD^+ and its reduction products at the electrode-solution interface altered the ratio of products obtained. In one experiment, designed to determine the amount of NAD^+ converted to both dimer and 1,4-NADH, assays accounted for about 80% of the initial NAD^+; the failure of the assays to add up to 100% was attributed to experimental error and to decomposition of the coenzyme solution (195).

Other workers (168), using $Et_4NCl\text{-}K_2CO_3$ buffers, who electrolyzed at

-1.8 V, found at least three different electrolysis products and virtually no adsorption of either NAD$^+$ or its reduction products, as determined from capacitance measurements using ac polarography. About 11% of the NAD$^+$ was converted to dimer (determined from the height of the anodic polarographic wave), about 54% formed enzymatically active 1,4-NADH, and about 35% was unaccounted for. Since an n value close to 2 was obtained during electrolysis, a $2e$ reduction seemed probable. The formation of 1,2-NADH ($\lambda_{max} = 395$ nm) (153,154) was ruled out because the electrolysis solutions remained colorless. The only other reasonable $2e$ product which might have been formed was 1,6-NADH.

In pH 9.6 Et$_4$NCl/K$_2$CO$_3$ buffer, NADH exhibits an oxidation peak at 0.42 V at the PGE (168).

c. NMN$^+$. The same experimental conditions used for NAD$^+$ were employed in a study of NMN$^+$ (168). Adsorption phenomena involving either NMN$^+$ or its reduction products were nonexistent. Electrolysis on the wave I plateau gave an n value of 1; the pyridinium absorption band (265 nm) was lost, and the final electrolyzed solution exhibited two new bands at 263 and 339 nm. The absorption at 263 nm suggests that at least a portion of the dimeric product involved a 1,6-reduced pyridine ring. The characteristic original cathodic polarographic waves were absent in the final electrolyzed solution, which gave two new waves: an anodic one ($E_{1/2} = -0.22$ V) corresponding to oxidation of the dimer to NMN$^+$, and a small cathodic wave ($E_{1/2} = -1.61$ V) which was attributed to reduction of free nicotinamide.

Electrolysis on the limiting portion of wave II gave an n value close to 2. A polarogram of the electrolyzed solution revealed that the two NMN$^+$ cathodic waves had vanished and that about 17% of the NMN$^+$ had gone to the $1e$ rather than to the $2e$ product. The electrolyzed solution also had maxima at 339 and 259 nm, which suggested formation of a 1,6-dihydropyridine species.

In pH 9.6 Et$_4$NCl/K$_2$CO$_3$ buffer, the dihydropyridine species gave an oxidation peak at 0.38 V at the PGE (168).

d. NADP$^+$. Macroscale electrolysis studies of NADP$^+$ (197), under virtually the same experimental conditions as those used in a similar study of NAD$^+$ (195), showed the behavior of NADP$^+$ to be essentially the same as that of NAD$^+$ (195). A solution electrolyzed at a potential on wave I exhibited an anodic polarographic wave ($E_{1/2} = -0.20$ V) which corresponded to the oxidation of the dimer to NADP$^+$, and a small cathodic wave ($E_{1/2} = -1.75$) which was attributed to reduction of the dimer. Further reduction at a potential on the cathodic wave produced NADPH activity very slowly

(2 to 3% conversion to NADPH in 3 hr). The solution had an absorption maximum at 340 nm and was devoid of NADP+ and NADPH activity.

Electrolysis at a potential on wave II produced a mixture of NADPH and dimer. In one experiment, for example, conversion of NADP+ to NADPH was 20% complete, as shown by enzymatic assay, and all of the starting material was transformed into substances with the 340-nm chromophore. The coulometric n value obtained (1.2 electrons per initial molecule of NADP+) was in accord with a 20% conversion to NADPH and an 80% conversion to dimer (197). It was suggested that strong adsorption of NADP+ and its reduction products at the electrode surface altered the product ratio. Other workers, who electrolyzed at a potential on NADP+ wave II, obtained yields of active NADPH of 94% (210) and 95% (209). These higher yields may have resulted from chemical reduction by nascent hydrogen rather than direct reduction at the electrode surface (197).

E. REDUCTION MECHANISM

On the basis of the voltammetric and electrolysis data, the reduction scheme shown in Figure 18 is likely, but some doubt still exists concerning the isomeric forms of the reduction products.

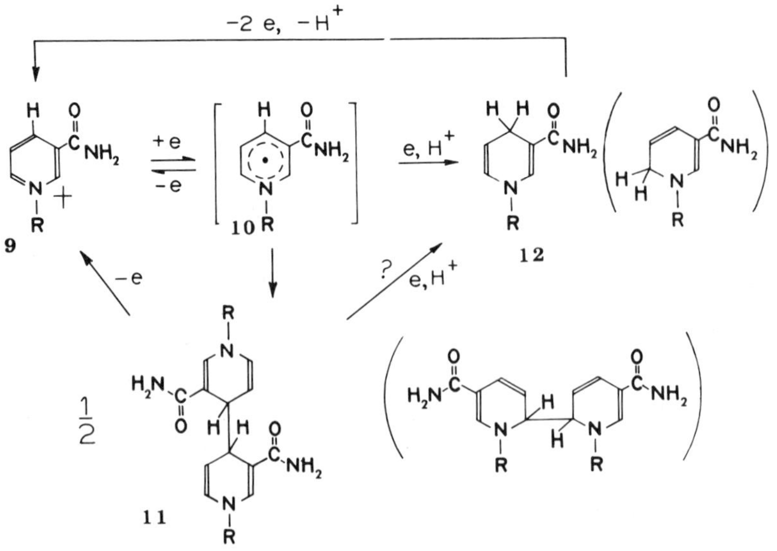

Figure 18. Interpretation of the electrochemical behavior observed for pyridine nucleotides and their reduction products.

As in the case of 1-methylnicotinamide, the first step generally involves a reversible 1e addition to the pyridinium ion (**9**) to produce a free radical (**10**) (source of wave I). This is followed by an irreversible dimerization, but the exact reaction site is presently unclear (cf. the subsequent discussion). At a potential corresponding to the second wave, **9** is reduced to a dihydro-pyridine species. In the case of NMN⁺, a major fraction (if not all) of the 2e product appears to be the 1,6-isomer. In the case of NAD⁺, a major fraction of the wave II product is generally enzymatically active 1,4-NADH; under some conditions, 1,6-NADH is also formed. At sufficiently positive potential, both the dimeric and 2e reduction products can be oxidized back to the original pyridine nucleotides. At potentials close to background discharge, the dimer is also slowly reduced to a dihydropyridine species; whether this reduction involves a direct electron transfer or occurs indirectly (e.g., via the formation of hydrogen radicals) remains to be determined.

a. Structure of the Dimer. The homogeneity of the wave I reduction products of the three pyridine nucleotides has not been determined. For each nucleotide, six different dimeric structures are possible, but because of the two asymmetric carbons at the point of dimerization as many as 21 stereo-isomers may exist (3 meso forms and 9 *dl* pairs). Only the NAD⁺-derived dimer has been isolated (213), but even in this case conclusive structural evidence is not available. It has been pointed out (213) that the ultraviolet absorption at 260 nm for the NAD dimer ($\varepsilon = 33,100$) is greater than that expected from its two adenine moieties ($2 \times 14,400 = 28,000$); this implies absorption at two wavelengths by the nicotinamide portion of the dimer (260 and 340 nm) (Table V), and in turn suggests dimerization at the 6-position (213). Nevertheless, the workers who first isolated the NAD⁺ dimeric product (213) regard the 4,4′-structure as the most likely and base their reasoning in part on NMR spectra.

The fact that the dimeric product undergoes acid decomposition reactions and forms a mercury compound similar to 1,4-NADH is significant (168). The postulated 4,4′-structure is also more acceptable than the 6,6′-structure with regard to steric interactions between the bulky N-substitutents of the nicotinamide rings. On the other hand, steric effects involving the 3-carb-amoyl group should be taken into consideration in addition to those of the N-substituent, since both groups protrude from the reduced pyridine ring. Molecular models of the *dl* form of the 4,4′-dimer, for example, suggest that at least one of the amide functions must be twisted out of the plane of the adjacent vinyl group in order to achieve dimerization; for the meso form, both amide groups must be twisted. For the 6,6′-dimer, however, very little deformation of the amide groups is necessary, although moderate interaction between N-substituents is seen. Amide deformation in the case of the 4,4′-dimer may be sufficient to hinder its formation. Thus on the basis of its

TABLE V

Summary of Spectral Data for Reduction Products of Various 1-Substituted Nicotinamide Salts[a]

Original species	Reduced species						
	Dimeric product			Dihydropyridine product			
	λ_{max}	ε	Method of reduction	λ_{max}	ε	Method of reduction	Ref.
Nicotinamide	266	12,080	Electrolytic	265	9,620	Electrolytic	168
	346	5,480		346	4,350		
1-Methylnicotinamide	277	12,480	Electrolytic	270	11,400	Electrolytic	168
	358	8,340		358	6,580		
	275	14,700	X-ray	266	8,140	Borohydride	205[b]
	355	9,600		358	5,890		347[b,c]
				360	7,000	Dithionite	150
				355	6,160	Dithionite	347[b,c]
1-Propylnicotinamide	255	3,800	Zinc	265	—	Borohydride	158[b,d]
	357	8,300		350	6,700		190[e]
				346	5,800	Dithionite	190[e]

Compound	λmax	ε	Method	λmax	ε	Method	
1-(2,6-Dichlorobenzyl)-nicotinamide	272 350	9,200 6,500	Zinc	265 355	9,400 7,500	Borohydride	158[b,d]
NMN+	263 339	7,820 7,540	Electrolytic	259 339	4,410 6,510	Electrolytic	168
NAD+	259 340	33,100 6,650	Electrolytic	259 340	14,400 6,220	Enzymatic or dithionite	213
	259 340	32,500 6,890	Electrolytic	259 345	— 6,220	Borohydride	168
	250 340	37,000 7,300	X-ray				205[b]
							215
							153

[a] Data taken from the literature for compounds where λmax and ε values could be found for both the dimer and dihydropyridine species; all data pertain to aqueous solutions unless otherwise stated.
[b] Data estimated from a published spectrum.
[c] Spectrum taken in ethanol.
[d] Spectrum taken in methanol.
[e] Spectrum taken in ether.

ultraviolet spectrum the NMN dimer appears to have a 6,6'-structure, even though its N-substituents are relatively bulky. It should be noted that the ultraviolet spectrum of the NMN dimer, unlike that of the NAD dimer, is due entirely to the nicotinamide portion of the molecule, that is, the 260-nm region is not marked by adenine absorption (Table V).

In connection with the structure of the dimers, a number of other questions have also arisen. A feature of many of the dimers shown in Table V is that their UV absorption spectra are qualitatively very similar to those of the corresponding dihydropyridines prepared by borohydride reduction. However, the long-wavelength band and, to a lesser extent, the short-wavelength band of a particular dimer have each a molar absorptivity which is only slightly greater than that found for the corresponding band of the fully reduced dihydro compound; this is true even though the dimer has twice as many chromophores per mole. The reason for such a finding is unclear. The two chromophores of the dimer are separated in all cases by two carbons and should therefore absorb independently of one another. The dimeric 1e products of pyrimidine and 2-aminopyrimidine show this behavior. In both cases the dimer chromophores are separated by two carbons, and in both cases the molar absorptivity of the dimer is about twice that of the corresponding dihydro compound. However, these dimers are not substituted with bulky side chains, as are the pyridine dimers.

Another question which arises concerns the inability of the dimers to fluoresce. Dimeric products prepared from 1-methylnicotinamide (217), 1-propyl-, 1-benzyl-, and 1-dichlorobenzylnicotinamide (158), and NAD$^+$ (213,218) have all been reported as nonfluorescent. This contrasts with the fact that the dihydropyridines such as 1-methyl-1,4-dihydronicotinamide (217) and 1,4-NADH (139) do fluoresce. An electronic substituent effect does not readily account for either the anomalous ultraviolet absorption behavior or the absence of fluorescence. The dihydroxyacetone adduct of NAD$^+$, for example, which contains a C—C single bond at the 4-position of the pyridine ring, has about the same absorption and fluorescence properties as NADH itself (139) and also undergoes an acid decomposition reaction, as does NADH. The lack of fluorescence from the dimeric structures may result from a molecular crowding which interferes very slightly with the coplanarity of the chromophoric structure (219).

The fact that a steric effect may be involved suggests that the dimeric structures are only moderately stable and that this may account for the ease with which the molecules are electrochemically oxidized or reduced. The electrochemical oxidation or reduction of C—C single bonds under polarographic conditions is virtually unknown, except in semiquinone-type dimers such as alloxantin, where the carbon atoms forming the single bond each carry a hydroxyl group (129). Oxidation of the NAD dimer in solutions of pH 9.6 at the P.G.E. occurs at a potential which is about 0.6 V less positive than that for the oxidation of NADH itself. Substitution of a methylene group for a hydrogen, as in the case of the dimer, would be expected to facilitate oxidation, since a methylene group relative to hydrogen is electron-donating. Because of the magnitude of the enhancement, however, the greater ease of oxidation of the dimer with respect to NADH cannot be accounted for in terms of simple inductive effects. A more reasonable explanation might be that, because of the molecular crowding brought about by interactions between the substituents of the two rings, the connecting bond is weakened. The same argument can also be applied to explain the relative ease with which the single bond can be reduced to form dihydro species, although in this case it is unclear as to whether the reduction involves a direct electron transfer from the electrode or occurs indirectly.

8. Biological Reduction Mechanism

In biological oxidation-reduction reactions involving the pyridine nucleotides an appropriate dehydrogenase catalyzes the stereo-specific interconversion of a pyridinium ring and a 1,4-dihydropyridine ring. The reduction of NAD^+ to 1,4-NADH in the presence of alcohol dehydrogenase, for example, involves what amounts to the transfer of a hydride ion from substrate (ethanol) to a coenzyme [27]. Although the overall mechanism may be viewed as a hydride transfer, this biological reaction can proceed a priori by four different reduction mechanisms (142): (a) transfer of a hydride ion from substrate to coenzyme, (b) simultaneous transfer of two electrons followed by transfer of one proton, (c) successive transfer of two single electrons and one proton, and (d) separate transfer of one electron and one hydrogen atom.

Differentiation among these four mechanisms is possible only if the rates of the different steps in the multistep mechanisms can be separated (142). Numerous enzymatic, chemical, and electrochemical studies have been conducted to elucidate the nature of this biological redox mechanism.

Three reviews (140,142,219a) summarize the various enzymatic and chemical approaches that have been used. In the absence of an enzyme, chemical approaches generally involve mechanistic studies of model systems in which either highly activated, abnormal substrates or coenzyme derivatives are used to demonstrate a net oxidation-reduction reaction. The model systems, however, do not provide sufficient information to determine conclusively whether the enzymatically catalyzed dehydrogenation proceeds by an ionic intermediate or a free radical intermediate. Both ionic and free-radical mechanisms have been observed (140,142,219a).

The electrochemical studies previously discussed were conducted mainly to elucidate the biological electron transfer mechanism. The electrochemical behavior of the NAD^+-NADH system, for example, is typical of that observed for the other pyridine nucleotide and model systems studied. The NAD^+-NADH system exhibits four distinct voltammetric steps at approximately the following potentials versus S.C.E. (potentials depend on experimental conditions; see references cited for details concerning the nature of the supporting electrolyte, solution pH, and electrode type):

1. NAD^+ reduced to a dihydropyridine species, that is, NADH: -1.6 to -1.8 V.
2. NAD^+ reduced to a free radical species, $NAD\cdot$; this reacts to form a dimer, DAN-NAD: -0.9 to -1.1 V.
3. DAN-NAD dimer oxidized to form NAD^+: -0.2 to -0.3 V.
4. NADH oxidized to form NAD^+: $+0.4$ to $+0.8$ V.

The two voltammetric reduction steps of NAD^+, as well as the single oxidation steps of the dimer and dihydropyridine species, differ significantly in potential relative to the formal potential E_0' of the NAD^+-NADH system. In this system the formal potential is -0.562 V versus S.C.E. $[-0.320$ V versus N.H.E. (219a)] at pH 7 and 25°C. The formal potential was calculated from thermal data and the equilibrium constants of alcohol dehydrogenase catalyzed ethanol-acetaldehyde and isopropanol-acetone reactions (219b).

The formal potential E_0' falls about midway between the voltammetric reduction and oxidation steps. Electrochemical reduction of NAD^+ to NADH, for example, occurs at least 1 V more negative than E_0', whereas oxidation of NADH to NAD^+ occurs at least 1 V more positive than E_0'. Both steps, which are complex in nature, occur irreversibly; considerable apparent overpotential is required before the reactions proceed at a measurable rate. Reduction is effected, essentially by forcing negative charge into the pyridinium ring system. Electrochemical reduction of NAD^+ is analogous to chemical reduction in the absence of enzyme. In the latter case reduction occurs only when highly reactive chemical substrates are used, for example, reasonably strong reducing agents such as dithionite or sodium borohydride (142).

At sufficiently negative potential, the electrochemical reduction of NAD^+ to NADH occurs in a stepwise sequence. First, a single electron is added to form a neutral free radical species and then a proton and a second electron are added to the initially formed radical to form a dihydropyridine species (the order of addition of the proton and second electron has not been determined).

Whether or not the biological reduction mechanism is similar to that of the electrochemical reduction mechanism has yet to be demonstrated. Under biological conditions energetic factors involving the conformation of substrate and/or coenzyme at the enzyme surface might be such that the three steps in the electrochemical mechanism occur in rapid succession, thus appearing overall as a hydride transfer.

On the other hand, the very fact that a substrate-coenzyme complex probably forms at the enzyme surface under biological conditions may mean that the biological reduction mechanism is entirely different from the electrochemical reduction mechanism.

It is interesting that in the absence of enzyme, ionic mechanisms seem to be favored in model chemical systems (142), whereas free radical mechanisms seem to be favored in electrochemical studies of the coenzymes themselves.

9. Analytical Application

The results of electrochemical investigations of 3-carbamoylpyridine and 1-substituted 3-carbamoylpyridinium ions can be applied to the development of analytical methods; for example, the polarographic determination of nicotinamide in various pharmaceutical or vitamin preparations has been described (220–232b). Polarography has also been used to study the formation of nicotinamide-ascorbic acid complexes (233) and the hydrolysis of nicotinamide in HCl solutions (234). Nicotinamide has been determined potentiometrically (235) and amperometrically (236) and has been identified oscillographically (237,238).

The polarographic determination of nicotinamide is generally carried out in slightly alkaline solution to avoid catalytic hydrogen evolution. Under these conditions, the $2e$ wave has a well-defined limiting portion, as previously discussed. In the potential region where reduction occurs, neither nicotinamide nor its reduction products show significant surface activity as determined by ac polarographic measurements (168); subtraction of the background current due to the supporting electrolyte from the total current will lead, therefore, to an accurate estimate of the nicotinamide diffusion current. Strongly alkaline solutions are to be avoided because of possible hydrolysis of the amide group; under these conditions the hydrolysis product, nicotinic acid, is not reduced (ref. 24, p. 814).

The determination of 1-substituted 3-carbamoylpyridinium ions can be carried out by measuring the diffusion current of the first $1e$ wave in neutral or slightly alkaline solution. With decreasing pH, this wave becomes less well defined because of the close proximity of background discharge; with increasing pH, decomposition of the electroactive species results. Since several of the first-wave reduction products are surface active, dc polarography is best carried out in a supporting electrolyte containing a tetraalkylammonium salt such as Et_4N^+, which tends to be preferentially adsorbed at the electrode-solution interface. The DL capacity of the test solution under these conditions is identical to that of the supporting electrolyte alone and allows background current to be used as base current.

Except for early determinations of NAD^+ and $NADP^+$ in various tissues and organs (180,239–242), little analytical use has been made of the results of electrochemical investigations of these compounds.

The *in situ* generation of NADH and NADPH by electrochemical techniques for synthetic purposes and diagnostic experiments has already been described (197a).

V. PYRIMIDINE AND PURINE DERIVATIVES

The electrochemical behavior of purines and pyrimidines, their nucleosides and nucleotides, and certain of their oligo and polymeric forms has been reviewed in recent years with varying degrees of completeness and for several reasons (6,13,14,124). This section is intended to summarize most of the electrochemical information available (generally through the middle of 1970) on the nonpolymeric states of purines and pyrimidines, with some emphasis on the electroanalytical determination of these compounds. Although the highly polymerized forms of nucleic acids, RNA and DNA, have also been extensively studied by means of polarographic techniques, a critical summary of this field is beyond the scope of this review. For a summary of the use of voltammetric techniques in the study of nucleic acid components, DNA, and synthetic polyribonucleotides, the reader is referred to the excellent review by Palecek (6).

1. Nomenclature

Purines and pyrimidines are found in fairly large amounts in most living cells. Occasionally they occur in the free state, but those which predominate in normal cells are usually present as glycosides, which are commonly referred to as nucleosides. These, in turn, occur primarily as their phosphoric esters or nucleotides, which play a vital role in all cells as group-transferring coenzymes and in highly polymerized condition as nucleic acids.

The numbering, nomenclature, and writing of formulas for purine and, more especially, for pyrimidine derivatives are not uniform (243). The trivial names common in biochemical literature and the numbering employed in *Chemical Abstracts* will be used in the present review; the abbreviations and symbols formulated by the IUPAC-IUB Commission on Biochemical Nomenclature have been used (244,245). The fundamental purine (a) and pyrimidine (b) structures may obviously be altered by tautomeric shifts. Hydroxy and amino derivatives have been named without regard to keto-enol and amino-imino tautomerism; possible acid-base equilibria may also have to be taken into account to reflect the actual state of a molecule.

(a)   (b)

The purine and pyrimidine nucleosides are, with very few exceptions, either β-D-ribofuranosides (ribonucleosides) or β-D-(2'-deoxy)ribofuranosides (deoxyribonucleosides), in which the sugar moiety is attached to N(9) of a purine or to N(1) of a pyrimidine. The commonly used nucleotide nomenclature is based on the trivial name of the nucleoside plus the position and number of the attached phosphate groups, for example, adenosine-5'-monophosphate or AMP (Figure 19).

2. Polarographic and Voltammetric Behavior

The electrochemical behavior of pyrimidines and purines will be treated separately for purposes of discussion, although in reality this division is purely artificial, since the reduction of purines generally occurs in the pyrimidine ring (13,14). Consideration of the polarographic behavior of purines and pyrimidines will then be further subdivided on the basis of important series of compounds in each class, beginning with the parent compound itself. Attempts will be made to summarize all pertinent electrochemical data available for the base and its nucleosides and nucleotides. Reaction path similarities and general behavioral characteristics will then be summarized in an attempt to provide (a) a framework for understanding the inherent electrochemical behavior of pyrimidines and purines, and (b) a basis on which predictions can be made concerning the behavior of compounds as yet unstudied.

Whereas the polarographic reduction of the parent purine and pyrimidine bases has been relatively well studied, little work has been done on their nucleosides and nucleotides (Tables VI, VII, and VIII). Although the reduction of nucleosides and nucleotides occurs in the purine or pyrimidine moiety, the sugar or sugar-phosphate group can influence the adsorbability, diffusion coefficient, electron density, and other characteristics, which, in turn, influence the ease of reduction and other electrochemical features.

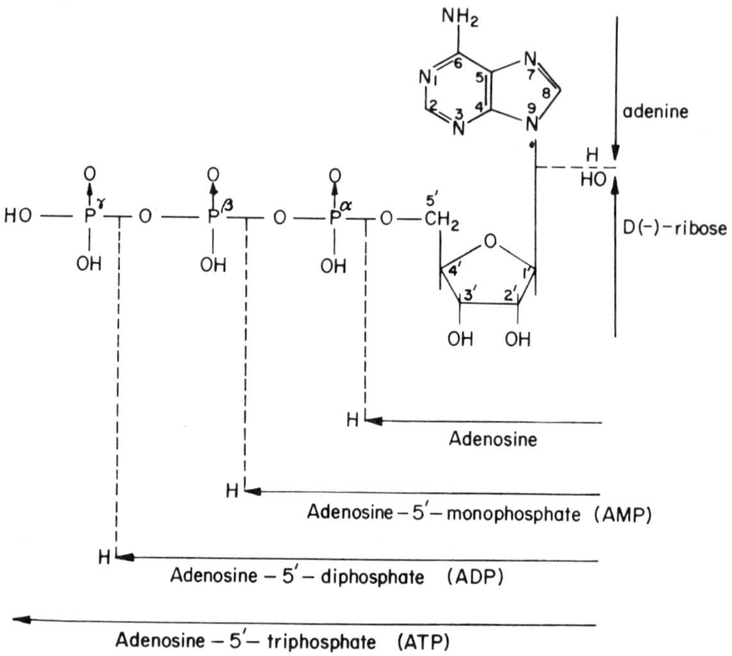

Figure 19. Formulas for adenine and derived nucleoside and nucleotide species. Configuration, dissociation equilibria, and tautomeric shifts are not shown. Taken with slight modification from Phillips (199).

There appear to be only two references to the polarographic reducibility of the sugars involved in nucleosides and nucleotides; these describe the reduction of ribose in neutral and basic media (246,247). At pH exceeding 6, ribose produces an unsymmetrical wave, which has an anomalous dependence of wave height on mercury column height, and whose temperature effect is very large and varies with mercury height (246). These unusual properties and the small height [0.92 μA at pH 11.0 for 20mM concentration (246); $E_{1/2} = -1.81$ V versus N.C.E., 25 μA at pH 7.0 for 0.10M ribose (247)] indicate a kinetically controlled wave, that is, reduction of ribose only in its aldehydic form (247). Thus the bound ribose and deoxyribose in nucleosides and nucleotides would not be expected to produce any additional polarographic waves.

3. Pyrimidines

The electrochemical behavior of pyrimidines is primarily concerned with their reduction, since none of the simpler pyrimidines are oxidizable under the experimental conditions normally employed. A wide variety of substituted pyrimidines have been studied polarographically; particular attention has been paid to the biologically important cytosine and uracil series. Table VI summarizes most of the polarographic data available on the reduction of pyrimidines; because of the large number of investigators of these important compounds and the wide range of literature sources involved, no claim is made for the completeness of these data. Furthermore, because of the wide disparity in the experimental conditions employed in regard to pH, buffer nature, ionic strength, reference electrode, temperature, and other factors that can affect polarographic characteristics and are often not specified, it is suggested that the original references be consulted for precise details.

A. PYRIMIDINE

Pyrimidine itself exhibits a complex reduction pattern, in which five polarographic waves appear over the pH range in aqueous solution (67) (Figures 4 and 20). In acid solution, only wave I appears, corresponding to the $1e$ reduction of the 3,4 N—C double bond with the acquisition of a proton

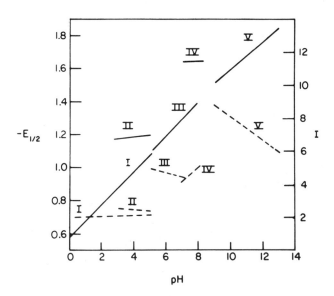

Figure 20. Variation with pH of half-wave potentials ($E_{1/2}$; solid lines) and diffusion current constants (I; dashed lines) for pyrimidine.

TABLE VI

Polarographic Reduction Data for Pyrimidines at the D. M. E.[a]

Compound	Medium[b]	pH	Wave[c]	$-E_{1/2}$ (V)[d]	i/C relation[e] Concentration (mM)	I	Refs.
Pyrimidine	2, 3	0.5–5	I	0.576 + 0.105 pH	0.3–4	2.0–2.5	67,113
	3	3–5	II	1.142 + 0.011 pH	0.3–0.5	2.2–2.7	67,113
	3	5–8	III	0.680 + 0.089 pH	0.3–5	4.3–4.7	67,113
	3	7–8	IV	1.600 + 0.005 pH	0.3	4.8	67,113
	6, 7	9–13	V	0.805 + 0.079 pH	0.4–1.4	8–9	67,113
	2	1.2		0.68	0.97	1.7	258
	8	6.8		1.30	0.97	3.5	258
2-Aminopyrimidine	2, 3	2–3	I	0.685 + 0.049 pH	0.5	2.0	67,113
	3, 4	4–7	I	0.425 + 0.121 pH		2.1–2.4	
	3, 4	4–7	II	1.360 + 0.004 pH		2.0–2.4	
	3, 6	7–9	III	0.680 + 0.090 pH	0.3–1.0	4.0–4.3	
	2	1.2		0.82	2.08	1.8	258
	8	6.8		1.40		3.5	
	8	9.2	I	1.54		0.9	
			II	1.91		2.4	
	2	1.1		1.45	0.6	2.4	277
	8[l]	7.4				2.74	302
4-Aminopyrimidine	2	1.2	I	1.13	0.66	5.7	258
			II	1.23		3.0	
	8	6.8		1.48		2.8	
2-Amino-4-methylpyrimidine	2, 3	2–4	I	0.770 + 0.063 pH	0.5	2.0	67
	3	4–7	I	0.550 + 0.113 pH		2.1	
	3	5–7	II	1.424 + 0.008 pH		2.0	
	3, 6	7–9	III	0.745 + 0.094 pH		4.0	
	8[l]	7.4		1.53	0.6	1.80	302

Compound		pH	Equation			Ref.
4-Amino-2,6-dimethylpyrimidine	3,4	2–8	$1.130 + 0.073$ pH			67
	3	2–5			6.8	113
4-Amino-2,5-dimethylpyrimidine	3	3–6	$1.060 + 0.076$ pH[f]		4.8	301
2,4-Diaminopyrimidine	2	1.2	1.19	0.82	4.8	258
	8	6.8	1.66		6.8	
4,6-Diaminopyrimidine	2	1.2	1.16	0.30	2.7	258
	8	6.8	1.52		15.4	
4,5,6-Triaminopyrimidine	2	1.2	1.17	0.40	7.4	258
	8	6.8	1.57		14.3	
2-Amino-4-methoxypyrimidine	2	1.2	1.21	1.16	4.3	258
	8	6.8	1.61		4.6	
2-Hydroxypyrimidine	3	2–8	$0.53 + 0.078$ pH	0.15–1.7	2.1	67, 113
		1.1			7.7	277
	8	7.0	1.19	10		251
4-Hydroxypyrimidine	8	6.8	1.60	1.01	3.8	258
	8	1–6	$1.05 + 0.088$ pH			
4-Amino-2-hydroxypyrimidine	2, 8, 18	4–6	$1.125 + 0.075$ pH	0.6–2.1	4–4.8	67, 113
(cytosine)	4	5.5			4.9	113
	4	5			6.4	277
4-Hydroxy-6-aminopyrimidine	2, 18	2.5–7	$1.070 + 0.084$ pH	0.99	3.4	258
	8	1–5	$1.06 + 0.099$ pH[a]		5–8	
		5–7	$1.33 + 0.042$ pH			
5-Methylcytosine	3, 11	5.6–6.5	1.65–1.73[f,h]			251
5-Hydroxymethylcytosine	3, 11	5.6–6.5	1.61–1.70[f,h]			251
6-Azauracil		8.3	1.336			269
	11	1–5.7	$0.51 + 0.127$ pH[f,g]	0.5		267
	11	5.7–12	$0.88 + 0.060$ pH			
5-Azauracil	8	7	1.30[f]		i	267
1,3-Dimethyl-6-azauracil	8	7	1.42[f]		i	267
5-Methyl-6-azauracil	8	2–5.8	1.35[f]		i	267
(6-azathymine)	11	2–5.8	$0.45 + 0.153$ pH[f,g]	0.5		267
	11	5.8–11	$0.99 + 0.57$ pH			
	11	8.3	1.408		i	269

385

(Continued)

TABLE VI (continued)

Compound	Medium[b]	pH	Wave[c]	$-E_{1/2}$ (V)[d]	i/C relation[e]		Refs.
					Concentration (mM)	I	
5-Ethyl-6-azauracil	11	8.3		1.419		i	269
5-Isopropyl-6-azauracil	11	8.3		1.426		i	269
5-t-Butyl-6-azauracil	11	8.3		1.475		i	269
5-Amyl-6-azauracil	8	7		1.39	0.5	i	267
5-t-Amyl-6-azauracil	11	8.3		1.417		i	269
5-Phenyl-6-azauracil	11	8.3		1.319		i	269
5-Chloro-6-azauracil	8	7		1.34ᶠ		j	269
5-Bromo-6-azauracil	8	7		1.28ᶠ	0.5	j	267
5-Iodo-6-azauracil	8	7	I	0.84	0.5	k	267
			II	1.30			
5-Amino-6-azauracil	8	7		1.55ᶠ	0.5		267
5-Azauracil-6-carboxylic acid (5-azaorotic acid)	8	7		1.30ᶠ	0.5	j	267
2-Thio-6-azauracil	8	7	I	1.09ᶠ	0.5	k	267
			II	1.30			
5-Azacytosine	8	7	I	1.40ᶠ	0.5	k	267
			II	1.80			
6-Azacytosine	8	7	I	1.17ᶠ	0.5	k	267
			II	1.40			
	11	3–8.5	I	$0.34 + 0.115$ pHᶠ,ᵍ	0.5		267
		8.5–11	I	$1.14 + 0.026$ pH			
		2–4.6	II	$1.17 + 0.009$ pH			
		4.6–9.7	II	$0.81 + 0.085$ pH			
		9.7–11	II	$1.34 + 0.030$ pH			
Pyrimidine-5-carboxamide	19			1.60			350
2-Amino-4-chloropyrimidine	8ˡ	7.4	I	1.35	0.6	1.23	302
			II	1.48		2.48	

386

Compound	Medium[b]	pH	Wave[c]	$E_{1/2}$[d]		I[e]	Ref.
2-Amino-4-chloro-6-methylpyrimidine	8	7.4	I	1.36	0.6	1.04	302
			II	1.56		2.64	
Alloxan	3, 4	3.6–5.6	I	0.060 + 0.031 pH	10	0.26	129
		4.6–5.6	II	1.44 + 0.015 pH	1	3.71	
Uracil-6-carboxylic acid (orotic acid)	11	2.1	II	0.68			312
3-Methyl-6-methoxythymine	2	1.2		1.18	0.88	6.0	258
	8	6.8		1.65		1.2	
5-Iodouracil	2, 4, 8	1.0–8.4		1.02	0.5	4.3	265a
	6, 7	8.4–13		0.18 + 0.1 pH			
5-Bromouracil	8	6.1–8.6		1.50 + 0.02 pH	0.5	4.7	265a
	6	8.6–11		0.98 + 0.08 pH		3.8	265a
1-Methyl-5-bromouracil	8	6.2–8.4		1.48 + 0.02 pH	0.5	2.9	265a
	6	8.4–11		1.06 + 0.07 pH		4.3	265a
1,3-Dimethyl-5-bromouracil	8, 6, 7	6.1–13		1.60	0.5		265a
5-Chlorouracil	8, 6	6.0–7.9		1.69 + 0.01 pH	0.5	4.3	265a
		7.9–10		1.37 + 0.05 pH			265a
1,3-Dimethyl-5-chlorouracil	8, 6	7.0–10		1.80	0.5	4.5	265a
5-Fluorouracil	8	7.4		1.845	0.5	8.9	265a

[a] $E_{1/2}$ versus S. C. E. unless otherwise stated; temperature is generally 25°C.

[b] Media are listed in Table IV.

[c] Roman numerals indicate successive polarographic waves.

[d] Equations given indicate the $E_{1/2}$ when extrapolated to zero pH and the variation of $E_{1/2}$ with pH.

[e] In some cases, the diffusion current constant, $I = i_d/Cm^{2/3}t^{1/6}$, was estimated by the present authors from polarographic curves and other data in the original reference.

[f] Potential versus N. C. E.

[g] Estimated by present authors.

[h] Ill-defined polarographic inflection on the background discharge reported as a wave.

[i] One 2e wave.

[j] One 4e wave.

[k] Two 2e waves.

[l] Three volume-percent ethanol in the background electrolyte.

TABLE VII

Polarographic Reduction of Purines at the D. M. E.[a]

Compound	Medium[b]	pH	Wave[c]	$-E_{1/2}$ (V)[d]	i/C relation[e] Concentration (mM)	I	Ref.
Purine	2, 4	2–6	I	0.697 + 0.083 pH	0.2–1	3.8	113,133
			II	0.902 + 0.080 pH		6.3	
6-Aminopurine (adenine)	3	2–6		0.975 + 0.90 pH	0.2	5.6	113,133
	2	1–3		f		10.2	277
	4	3.7–5.5		f		10.2	279
	9	1.0		1.1		7.0	351
	14	1		1.46[o]	0.1		278
	13	1.5		1.1	0.5	8.9	280
				0.95[p]	0.2		
		1.3		1.04			
6-Hydroxypurine (hypoxanthine)	2, 3	1.0–6.5		0.975 + 0.084 pH	0.125	9.8	70
	4	5.7		1.61		2.75	277
		1.1				2.6	
6-Methylpurine	2, 3	1.0–3.9	I	0.820 + 0.079 pH	0.125	8.5[g]	69
	3	3.9–6.0		0.745 + 0.091 pH			
	3	2.5–6.0	II	0.915 + 0.082 pH			
	4	3.9–5.9		0.765 + 0.162 pH			
	3	6.0–7.4	III	0.785 + 0.095 pH			
	3	7.4–7.8		-0.080 + 0.209 pH			
6-Methoxypurine		2.5–4.2		0.825 + 0.105 pH	0.125	8.5	69
		4.2–5.5		0.535 + 0.174 pH[h]			

388

Compound		pH range					Ref.
6-Methylaminopurine	2, 3	1.0–6.5		$0.995 + 0.081$ pH	0.125	9.8	69
2,6-Diaminopurine	13	1.5		1.14	0.5		351
	14	1		1.4°	0.1		279
2,6-Diaminopurine-1-N-oxide	13	1.5		1.1	0.5		351
6-Acetamidopurine (acetyladenine)	14	1	I	1.22°	0.1		279
			II	1.43			
2,6-Diacetamidopurine	14	1	I	1.27°	0.1		279
			II	1.5			
Bis(6-purinyl)disulfide	12	2.3	I	0.05			299
			II	1.05			
Purine-6-sulfonamide	6	9	I	1.04			290
			II	1.47			
			III	1.58			
6-Thiopurine	2, 4, 12, 20	0–5	I	$0.79 + 0.116$ pH			289
	2, 12, 20	0–2.3	II	$1.00 + 0.048$ pH			
	3, 4	5–8	III	$1.29 + 0.027$ pH			
	6	9.1	IV	1.74			
Purine-6-sulfinic acid	3, 4, 6, 12	2–9.1	I	$0.37 + 0.095$ pH			289
	3, 6, 7, 21	8–13	II	$0.79 + 0.075$ pH			
	3, 4, 6, 21	3–9	III	$0.99 + 0.080$ pH			
Purine-6-sulfonic acid	2, 3, 4, 12	1–7	I	$0.45 + 0.078$ pH			289
	3, 4, 6, 7, 21	3.6–12.5	II	$0.65 + 0.079$ pH			
	2, 3, 4, 6, 12, 21	1–9	III	$0.98 + 0.064$ pH			
	6	9.1	IV	1.450			

(Continued)

389

TABLE VII (continued)

Compound	Medium[b]	pH	Wave[c]	$-E_{1/2}$ (V)[d]	i/C relation[e] Concentration (mM)	I	Ref.
6-n-Hexylaminopurine	2	1.0–2.5		0.995 + 0.076 pH	0.125	8.3	69
	3	2.5–3.7		1.105 + 0.047 pH			
	3	3.7–6.5		0.995 + 0.076 pH			
6-Benzylaminopurine	2, 3	2.0–4.8		0.995 + 0.067 pH	0.125	9.8	69
	3	4.8–6.4		0.805 + 0.106 pH			
6-Phenylaminopurine	3	2.5–4.7		0.915 + 0.072 pH[i]	0.125	8.3	69
	3	4.7–7.9		0.640 + 0.131 pH			
6-Dimethylaminopurine	2, 3	2.0–4.5		1.025 + 0.068 pH[j]	0.125	8.9	69
	3	4.5–6.4		0.930 + 0.089 pH			
2-Hydroxy-6-aminopurine (isoguanine)	2, 3	1.0–4.6		0.990 + 0.072 pH	0.25	4.6	69
	3	4.6–7.2		0.820 + 0.190 pH			
	3, 10	7.2–9.6		1.210 + 0.055 pH			
	6	8.5–9.1		0.705 + 0.104 pH			
2-Azaadenine	13	1.5	I	0.45	0.5		351
			II	0.99			
	13	1–7	I	0.29 + 0.11 pH[k]	0.5	3.0[l]	351
			II	0.96 + 0.064 pH		5.6	
8-Azaadenine	8	7		1.40[m]	0.5	n	267
	11	2–7		0.82 + 0.086 pH[k,m]			

Adenine 1-N-oxide	13	1.5	I	0.87	0.5	2.9	351
			II	1.1		6.8	
2-Azaadenine 1-N-oxide	13	1.5	I	0.54	0.5		351
			II	1.02			
	17	1–6	I	$0.36 + 0.114$ pH[k]	0.5	5.1^l	351
		2–5	II	$0.96 + 0.064$ pH		2.8	

[a] $E_{1/2}$ versus S. C. E. unless otherwise indicated; temperature generally at 25°C.

[b] Media are listed in Table IV.

[c] Roman numerals indicate successive polarographic waves.

[d] Equations given indicate the $E_{1/2}$ when extrapolated to zero pH and the variation of $E_{1/2}$ with pH.

[e] In some cases, the diffusion current constant, I, was estimated by the present authors from polarographic curves and other data in the original reference.

[f] Dependency of $E_{1/2}$ on pH is similar to that given for McIlvaine buffers.

[g] Calculated from the total current of two reduction waves.

[h] Very ill-defined wave which emerges with background above pH 4.5.

[i] Maximum on the rising portion of the wave in the pH range indicated. The $E_{1/2}$ is not changed after the maximum is suppressed by Triton X–100 addition.

[j] Maximum on the crest of the wave in the pH range indicated. The $E_{1/2}$ is not changed after the maximum is suppressed by Triton X–100 addition.

[k] Estimated by the present authors.

[l] At pH 2.3.

[m] Potential versus N. C. E.

[n] One 4e wave.

[o] Potential versus the mercury pool anode.

[p] $E_{1/2}$ at pH = 0.

TABLE VIII

Polarographic Reduction of Nucleosides and Nucleotides at the D. M. E.[a]

Compound	Medium[b]	pH	Wave[c]	$-E_{1/2}$ (V)[d]	i/C relation[e] Concentration (mM)	i/C relation[e] I	Refs.
r-Ado	2, 3, 4	2.0–4.5		$1.040 + 0.070$ pH[f]	0.0–1.0	$7.5 - 2.5C$	14,70
		4.5–6.0		$1.180 + 0.041$ pH			
	13			1.17	0.5		279
	14			1.42[g]	0.1		278
	i			0.99[h]	0.2		
d-Ado	2, 3, 4	2.5–4.6		$1.060 + 0.069$ pH[f]	0.0–1.0	$7.9 - 2.4C$	14,70
		4.6–6.5		$1.205 + 0.037$ pH			
AMP	2, 3, 4	1.0–4.3		$1.015 + 0.083$ pH[f]	0.1–1.0	$7.9 - 2.1C$	14,70
		4.3–5.5		$1.115 + 0.060$ pH			
	14			1.34[g]	0.1		279
	i			0.97[h]	0.2		278
Poly-AMP	15	5.7–6.7		$1.005 + 0.058$ OH	0.1		262
d-AMP	2, 3, 4	2.0–6.5		$0.985 + 0.080$ pH[f]	0.0–1.0	$8.0 - 2.6C$	14,70
	i			0.98[h]	0.2		278
ATP	2, 3, 4	2.5–4.5		$1.035 + 0.083$ pH[f]	0.1–1.0	$6.4 - 1.8C$	14,70
		4.5–5.5		$1.175 + 0.052$ pH			
	i			0.98[h]	0.2		278
Ado 1-N-oxide	13		I	0.84	0.2		351
			II	1.14	0.5		
2',3',5'-Triacetyl-Ado				0.99[h]	0.2		278

392

Compound		pH					Ref.
2',3',5',N₆'-Tetraacetyl-Ado			I	0.78^h	0.2		278
			II	0.89			
Cyd	3, 11	2.5–7		1.105 + 0.072 pH			251,252
	3	5.0		1.38	0.1	4.1^k	252a
	4	4.1		1.38	1.0	6.3	252a
d-Cyd	3, 11	2.5–7		1.154 + 0.068 pH			251,252
CMP	3, 11	6.5		1.68^i	0.2		251,252
	3	5.0		1.46	0.1	3.2^k	252a
	4	4.1		1.37	0.99	4.6	252a
				1.37	0.49	3.0^k	252a
Poly-CMP	16	6.1–9.0		0.945 + 0.059 pH	0.1		262,356
d-CMP	3, 11	3.5–7.3		0.908 + 0.110 pH			251,252
		7.3–8.7		1.350 + 0.050 pH			
5-Methyl-d-Cyd	3, 11	4–7.3		0.775 + 0.118 pH			251,252
		7.3–10		1.325 + 0.042 pH			
5-Aza-Cyd	8	7	I	1.35^j	0.5		267
			II	1.68			
6-Aza-Urd	8	7		1.36^i	0.5		267
CpC	3	5.0	I	1.30	0.02–0.9	$5.5–0.4^k$	252a
			II	1.39 to 1.49	0.05–0.9	$1.6–5.9^k$	
			III	1.61	0.4–0.7	$1.3–1.7^k$	
	3	2.5–5.5	I	1.040 + 0.041 pH	0.05	10^l	252a
	3	2.5–5.9	II	1.085 + 0.056 pH			
	3, 21	5.9–8.1	II	0.800 + 0.104 pH			
	21	8.1–10.6	II	1.225 + 0.051 pH			
CpU	3	4.5		1.32	0.09	2.8^k	252a

(Continued)

393

TABLE VIII (continued)

Compound	Medium[b]	pH	Wave[c]	$-E_{1/2}$ (V)[d]	i/C relation[e] Concentration (mM)	I	Refs.
5-Iodo-Urd	2, 4, 8	1.0–8.9		0.92			265a
	6, 7	8.9–13		0.03 + 0.1 pH	0.5	3.3	
5-Bromo-d-Urd	8	6.2–8.3		1.23 + 0.04 pH	0.5	3.1	265a
	6, 7	8.3–11		0.90 + 0.08 pH			
5-Chloro-Urd	8	6.0–8.6		1.53 + 0.02 pH	0.5	3.3	265a
	6, 7	8.6–11		1.10 + 0.07 pH			

[a] $E_{1/2}$ versus S. C. E. unless otherwise indicated; temperature is usually 25°C. For $E_{1/2}$ values of the parent compounds, see Tables VI and VII.

[b] Media are listed in Table IV.

[c] Roman numerals indicate successive polarographic waves.

[d] Equations given indicate the $E_{1/2}$ when extrapolated to zero pH and the variation of $E_{1/2}$ with pH.

[e] In the diffusion current constant expressions, C stands for the concentration (mM) of the electroactive species.

[f] Average data for 0.125 and 0.50mM adenine derivative in chloride, acetate, and McIlvaine buffers of 0.5M ionic strength.

[g] Potential versus mercury pool anode.

[h] $E_{1/2}$ extrapolated to zero pH; in the pH range 0.65–5.3, $d(E_{1/2})/d(\text{pH}) = 0.080$ for both tetraacetyladenosine waves and −0.110 for the triacetyladenosine wave.

[i] Unspecified media.

[j] Potential versus the N. C. E.

[k] Temperature = 0°C.

[l] Sum of waves I and II in pH 2.5 McIlvaine buffer.

to form a free radical (Figure 21). The resulting free radical may then dimerize to 4,4'-bipyrimidine or be further reduced in a 1e process (wave II) to 3,4-dihydropyrimidine. At pyrimidine concentrations greater than about 3mM, wave II has been observed to split into two waves with unusual properties (248); the reasons for this split and the processes involved are not known. Since wave I is pH dependent and wave II shows little pH dependence, with increasing pH the two waves merge at about pH 5 to form pH-dependent 2e wave III. At pH 7, pH-independent wave IV appears at more negative potential, corresponding to the 2e reduction of 3,4-dihydropyrimidine to a tetrahydropyrimidine species. The differing pH dependences of waves III and IV cause them to merge at about pH 9 to form pH-dependent 4e wave V.

Cyclic voltammetry of pyrimidine at the H.M.D.E. and the P.G.E. shows cathodic processes corresponding to each of the five pyrimidine waves (114). No evidence for the reversibility of any of the processes is seen, even on fast-scan (36 V/sec) cyclic voltammetry, that is, no anodic peaks are observed. Alternating-current polarography of pyrimidine in aqueous media (68,114) also shows five peaks over the pH range (Figure 4), all moderately to highly irreversible in nature, corresponding to the five pyrimidine reduction waves. These results indicate that fast follow-up reactions, probably dimerization and protonation, occur after electron transfer, thereby making reoxidation very difficult. Macroscale controlled-potential electrolysis of pyrimidine at a mercury pool electrode has confirmed the n values for all five pyrimidine waves (67).

Studies (249) on the reduction of pyrimidine in acetonitrile media indicate that pyrimidine undergoes a 1e reduction at a very negative potential to give a highly reactive radical anion, which can then dimerize very quickly or abstract protons from residual water or other sources. Indications are that protons and water solvent have a determinative effect on the pyrimidine reduction mechanism in aqueous solution. Wave I is observed in aqueous media only at relatively low pH, where there is a sufficient concentration of labile protons to effect a preprotonation. At higher pH and more negative potential, a second reduction pathway becomes operative, in which the first step is electron addition, followed by proton *abstraction* by the C(4), leading to formation of the dihydro species. Proton abstraction probably occurs at the same carbon at which dimerization takes place.

All the data collected on the pyrimidine reduction to date tend to support the proposed pyrimidine reduction mechanism outlined in Figure 21.

a. General Pyrimidine-Purine Reduction Mechanism. The mechanistic processes undergone by pyrimidine itself are essentially followed by pyrimidine and purine derivatives and their nucleosides and nucleotides,

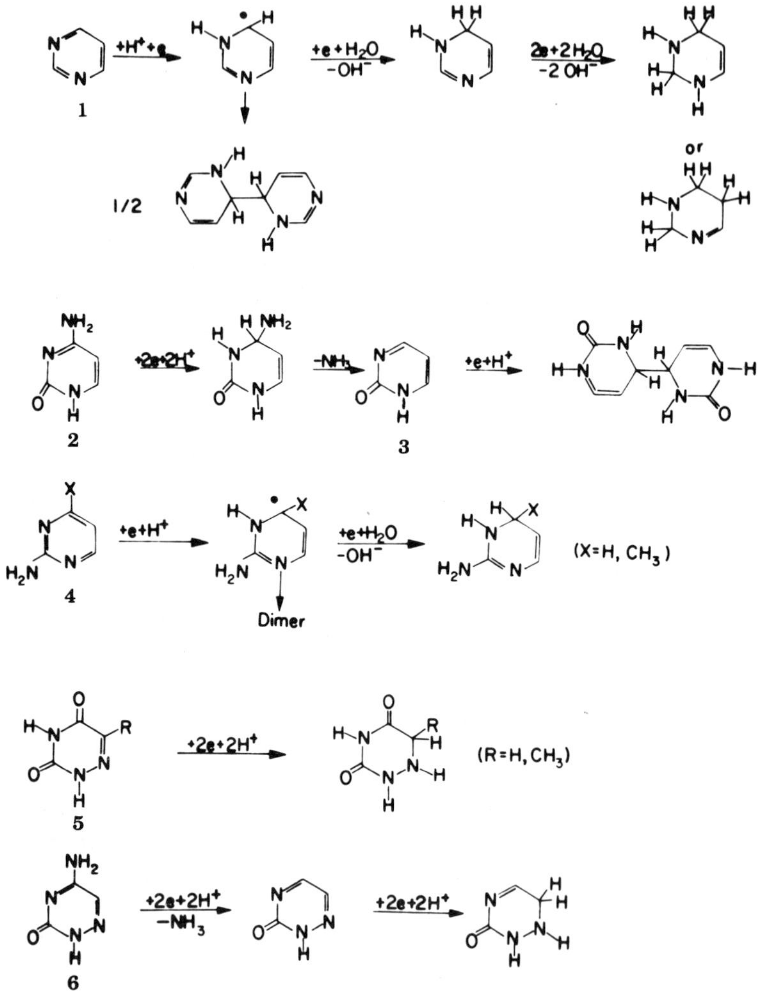

Figure 21. Interpretation of the electrochemical and chemical behavior observed for pyrimidine (**1**), cytosine (**2**), 2-hydroxypyrimidine (**3**), 2-aminopyrimidine or 2-amino-4-methylpyrimidine (**4**), 6-azauracil or 6-azathymine (**5**), and 6-azacytosine (**6**)

396

with the 3,4 N—C double bond in pyrimidines generally being the one most easily reduced (the 1,6 bond in purines). Several important factors may modify or alter the general pathway.

1. The addition of substituents may result in the fusing of two consecutive electron-transfer steps; for example, the free radical produced in an initial 1e process may be unstable at the potential of its formation and may be immediately reduced as formed to produce a 2e wave. Obviously, the splitting of a 2e wave into two 1e waves is possible, as is the merger of several separate faradaic processes into a single wave because of the instability of the various products at the potential required to introduce the first electron.

2. Tautomeric shifts due to substituents having divalent sulfur or oxygen bonded to carbon may remove reduction sites from the pyrimidine ring, thus either limiting reduction or making it much more difficult. For example, ultraviolet absorption spectra (250) indicate that all 2- and 4-hydroxy-pyrimidines should be represented as being in the keto form, thus resulting in the removal of a reducible C—N double bond from the ring.

3. Chemical reactions such as deamination or hydrolysis may accompany or follow electron-transfer processes and generate electroactive species.

B. CYTOSINE SERIES

a. Reduction of Cytosine. Detailed studies have been reported for cytosine and its nucleosides and nucleotides (67,251,252). Smith and Elving (67) postulated that cytosine undergoes an overall 3e reduction in pH 3.7 to 5.7 acetate buffer that is due to three successive processes (Figure 21): (a) a 2e reduction of the 3,4 N—C double bond (a combination of the first two 1e steps in pyrimidine), (b) a more or less rapid deamination of the product to 2-hydroxypyrimidine, and (c) a 1e reduction of the generated 2-hydroxypyrimidine to a free radical, which dimerizes under the experimental conditions before it can be further reduced. The diffusion current constant for the reduction of cytosine at the D.M.E. (4.0 to 4.8) indicates only a 2e to 2.5e reduction, but macroscale electrolysis at a mercury pool electrode definitely indicated a 3e process; hence the intervening chemical step, deamination, is somewhat slow on the time scale involved in classical polarography (drop-time) but occurs completely on prolonged electrolysis.

Recent results by Webb et al. (252a) confirmed this basic reduction mechanism for cytosine and several of its nucleosides and provided rate constants for several intervening chemical steps. For example, the experimental rate constant for the deamination process is 10 sec^{-1} for cytosine and 3 sec^{-1} for CMP. CpC does not deaminate under the experimental conditions. The sequence of steps in the cytosine reduction is rapid protonation at N(3), successive addition of two electrons to the 3,4 N—C double bond, protonation of the resulting carbanion, deamination to produce 2-hydroxy-pyrimidine, one-electron reduction of the 3,4 N—C double bond in the latter, and dimerization of the resulting free radical.

Palecek and Janik (251,252), however, reported that cytosine and its nucleosides and nucleotides undergo a $4e$ reduction in a variety of buffers: a $2e$ reduction of cytosine, fast deamination of the product to 2-hydroxy-pyrimidine, a $2e$ reduction of this compound, and, finally, hydrolysis in a pseudo-first-order process to form 3-uridoallyl alcohol:

$$+2e + 2H^+ \quad -NH_3 \qquad +2e + 2H^+ \qquad H_2O \tag{33}$$

R = H, ribosyl or ribosyl phosphate

where R = hydrogen, ribosyl, or ribosyl phosphate. The diffusion current obtained at the D.M.E. was estimated to be due to a $4e$ process by comparison with the currents obtained for two compounds whose reductions were said to be known to involve $4e$ processes: 5-acetamidomethyl-4-amino-2-methylpyrimidine (253) and 6-amino-1,3-dimethyl-5-nitrosouracil (254, 255). Moreover, macroscale electrolysis in pH 7.0 McIlvaine buffer and pH 6.78 phosphate-acetate buffer was said to indicate a $4e$ process. Unfortunately, no explicit diffusion current or electrolysis data were provided. The pseudo-first-order rate constants for the decomposition of the electrolysis products of both cytosine and 2-hydroxypyrimidine were identical within experimental error and agreed fairly well with those determined by Smith and Elving (67), considering the small differences in temperature, pH, and buffer composition.

Cytosine gives a single pH-dependent peak on cyclic voltammetry at the H.M.D.E., but no peaks at the P.G.E. (114). Analysis of various peak currents for the series of pyrimidine, purine, cytosine, and adenine could indicate anywhere between two and four electrons involved in the cytosine reduction, depending on experimental conditions and on the choice of the other compounds for comparison. Cytosine gives a fairly small peak current, Δi_s, on ac polarography, which at pH 4.7 is about 25% more than that for the known $1e$ initial reduction of pyrimidine and about 40% less than that for the known $2e$ initial reduction of purine. Ordinarily cyclic voltammetry and ac polarography are not good techniques for determining the number of electrons involved in an electrochemical process because of the influence of different mechanisms, reversibilities, and other parameters on the magnitude of the peak currents obtained. The adsorbabilities of cytosine and its nucleosides and nucleotides have been studied by ac polarography and bridge impedance methods (18,19,252a,256,257).

b. Isocytosine. The fact that isocytosine does not give a polarographic reduction wave at the D.M.E. (258) is an indication that the reduction of cytosine occurs at the 3,4 double bond. However, this compound does give an indentation on oscillographic polarography (264,326).

c. Nucleosides and Nucleotides. Regardless of the exact nature of the cytosine reduction, the reduction of its nucleosides and nucleotides is quite similar to that of the parent base. The products of macroscale electro-reduction of cytosine and cytidine at a mercury cathode were identical except for the sugar substituent. The $E_{1/2}$ for the reduction wave of the cytosine series becomes more positive in the order base < nucleotide < nucleoside and is linearly dependent on pH (Table VIII); the $E_{1/2}$-pH dependences for d-CMP and 5-methyl-d-Cyd show two linear segments intersecting at about pH 7. The diffusion-controlled character of the polarographic wave at low pH for all cytosine derivatives, its kinetic character at high pH, and the good agreement of the experimental S-shaped limiting current-pH relation with the theoretical polarographic dissociation curve calculated from Koutecky's equation (353), assuming recombination of the species with protons, led to the conclusion that the protonated form of the cytosine nucleus is probably the electroactive species.

C. HYDROXYPYRIMIDINES

a. 2-Hydroxypyrimidine. 2-Hydroxypyrimidine exhibits a single 1e reduction wave over the pH range from 2 to 12; the diffusion current constant is remarkably constant, $I = 2.1$, from pH 2 to 8 (67). This wave corresponds to the reduction to a free radical which dimerizes before further reduction occurs (Figure 21). Macroscale electrolysis (67) has confirmed the 1e nature of the 2-hydroxypyrimidine reduction, which is in accord with the cytosine reduction scheme proposed by Smith and Elving (67); the fact that $E_{1/2}$ for 2-hydroxypyrimidine is about 0.6 V more positive than that for cytosine is in accord with the postulation that the former compound is reduced immediately on formation during cytosine reduction. Although Janik and Palecek (251) present no explicit data for the diffusion current or electrolysis of 2-hydroxypyrimidine, it must be assumed that their values are very close to those for a 2e reduction, in accordance with their postulated reduction mechanism.

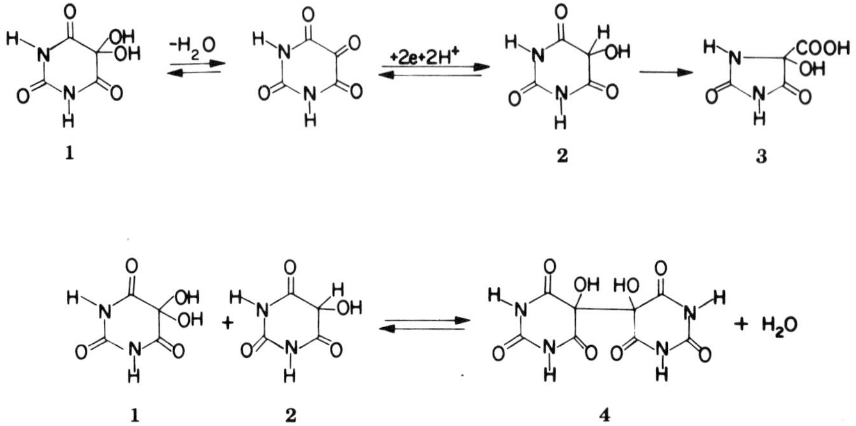

Figure 22. Interpretation of the electrochemical and chemical behavior observed for alloxan (1), dialuric acid (2), alloxanic acid (3), and alloxantin (4).

b. 4-Hydroxypyrimidine. 4-Hydroxypyrimidine produces a single reduction wave between pH 1 and 7, whose diffusion current constant steadily decreases from 9.5 in pH 1.2 chloride buffer to 3.8 in pH 6.8 phosphate buffer (258). Despite the rather high I values, it seems unlikely that 4-hydroxypyrimidine undergoes any greater than a direct $2e$ reduction; the added current is probably due to catalytic hydrogen evolution or, possibly, to the reduction of a species chemically generated after the initial reduction.

c. Alloxan System. Perhaps the most extensively studied pyrimidine redox system, which is also important from a biological viewpoint, is alloxan (1)-alloxantin (4)-dialuric acid (2), whose behavior is similar to that of the quinone-quinhydrone-hydroquinone system (Figure 22) (129, 259–261). Alloxan is reduced to dialuric acid; mixing the two compounds gives alloxantin, which is the dimer of the free radical formed in the $1e$ reduction of alloxan and in the $1e$ oxidation of dialuric acid.

Alloxan exhibits two pH-dependent reduction waves at the D.M.E. (129). Wave I involves the kinetically controlled $2e$ reduction to dialuric acid (the rate-controlling step is probably the dehydration of the *gem*-glycol at C(5) to produce the electroactive species). As a further complication, above about pH 6, alloxan rapidly rearranges to nonreducible alloxanic acid (3). Wave II, observed only between pH 4.6 and 6.0, involves further reduction; the latter is probably a $2e$ process, but little more is known about it.

Alloxantin gives a combined anodic-cathodic wave at the same potential as wave I of alloxan; the diffusion-controlled anodic portion is many times larger than the kinetically controlled cathodic portion. Dialuric acid is oxidized to alloxan at the same potential as cathodic wave I of alloxan and the combined wave of alloxantin; thus this system fulfills the classic requirements of electrode reversibility, that is, the reduced form is oxidized and the oxidized form is reduced at the same potential.

Volke (16, p. 285) and Clark (15, p. 439) have summarized potentiometric standard potential data on a number of alloxan derivatives.

d. Polyhydroxypyrimidines. With the exception of alloxan, pyrimidines with more than one hydroxy substituent appear to be polarographically nonreducible; uracil (2,4-dihydroxypyrimidine), 1,3-dimethyluracil, thymine (2,4-dihydroxy-5-methylpyrimidine), 1-methylthymine, and barbituric acid (2,4,6-trihydroxypyrimidine) are not reduced at the D.M.E. (258).

D. URACIL AND THYMINE SERIES

Uracil, thymine, and their nucleosides and nucleotides are not reduced at the D.M.E. (14,251,258,262). Some species, however, produce cathodic incisions on oscillopolarography in ammonium formate, sodium hydroxide, and potassium chloride backgrounds (263,264).

Uracil has been reported (265) to produce an anodic adsorption wave ($E_{1/2} = 0.020$ V) at the D.M.E. in borate buffer, which is due to mercury salt formation.

The mechanism for the electrochemical reduction of 5-halogeno derivatives of uracil, 1-methyluracil, 1,3-dimethyluracil, uridine, and oligonucleotides was investigated by classical polarography and macroscale electrolysis (265a). The reduction of the Cl, Br, and I compounds is a single, irreversible, $2e$ reduction of the carbon-halogen bond to produce the parent compound and halide ion. The reduction of 5-fluorouracil is an irreversible $4e$ process, corresponding to the reduction of the C—F bond and the 5,6 double bond to produce fluoride ion and 5,6-dihydrouracil. This reduction is partially kinetic in nature and is observed only in pH 7.4 phosphate buffer. The ease of reduction of a carbon-halogen bond follows the usual order: F < Cl < Br < I.

E. AZAPYRIMIDINES

A number of polarographic studies of the aza analogs of nucleic acid components and related compounds have been reported (266–269). Kittler and Berg (267) studied the electrochemical reduction of various 5- or 6-azapyrimidines (uracils and cytosines) and 8-azapurines. The azapyrimidines were generally reduced at the 5,6 double bond in a $2e$ process, but a $4e$ process occurred if a reducible group was substituted in the 4- or 5-position (halogen or amino). The aza compounds were more easily reduced than the parent uracil or cytosine compounds, with the 6-aza compound being more easily reduced than the 5-aza analog (the symmetrical species).

6-Azauracil and 6-azathymine are converted in a $2e$ reduction of the 5,6 double bond to the corresponding dihydro species. The similar polarographic behaviors of 6-azauracil and various N-methyl-substituted species indicate that, under the experimental conditions employed, 6-azauracil exists in the diketo form (269). This fact and experimental confirmation from the results of photohydration experiments (267) indicate that the reaction site is indeed the 5,6 C—N double bond. At higher pH (>9), the limiting current becomes kinetically controlled and decreases substantially, indicating that prior protonation is necessary for reduction.

6-Azacytosine exhibits two $2e$ waves at the D.M.E. Wave I corresponds to the $2e$ hydrogenation of the 3,4 double bond, followed by fast deamination at the 4-position, leading to regeneration of the double bond. The elimination of ammonia was confirmed by macroscale electrolysis; the amount of ammonia liberated could be correlated to the decrease in height of wave I on electrolysis. Wave II corresponds to the $2e$ hydrogenation of the wave I reduction product, 2-hydroxy-6-azapyrimidine, at the 5,6 C—N double bond to yield the corresponding dihydro species. At higher pH (>9), wave II displays kinetic control and decreasing height, indicating that the compound reduced is protonated. The similar nature of wave II to that of 6-azauracil, as well as ultraviolet absorbance data, indicates that the 5,6 bond, and not the 3,4 bond, is being reduced. This differs from the postulated reduction of cytosine itself at the regenerated 3,4 double bond.

F. THIOPYRIMIDINES

2-Thiocytosine and 2,4-dithiopyrimidine (thiouracil) give catalytic hydrogen reduction waves in a variety of buffers at the D.M.E. (270). The catalytic nature of these waves was determined by the effects of pH, buffer capacity, depolarizer concentration, and mercury column height on the shape and height of the reduction waves. The catalytic wave of 2,4-dithiopyrimidine, which is at a relatively positive potential ($E_{1/2} = -1.4$ V versus the Hg/Hg_2SO_4 reference electrode), has the shape of an ordinary diffusion wave at pH 7 to 9. At pH 5.2 (Britton-Robinson buffer), the height of the two catalytic waves is linear with concentration in the range of 0.1 to 0.5mM 2,4-dithiopyrimidine and thus can be used for analytical determination.

G. CHLOROPYRIMIDINES

In acidic medium, both 2,4-dichloro- and 2,4,6-trichloropyrimidine undergo a two-step reduction, the first product of which is 2-chloropyrimidine (271). In alkaline medium, 2,4,6-trichloropyrimidine undergoes a three-step reduction; the product of the first reduction wave is 2,4-dichloropyrimidine, and of the second 2-chloropyrimidine. At all pH values, the latter compound gives only one polarographic wave corresponding to simultaneous cleavage of the C—Cl bond and formation of dihydropyrimidine.

In the pH region of 7 to 9, 2-amino-4-chloropyrimidine and its 6-methyl derivative each exhibit two polarographic waves (302). In each case the first (more positive) wave results from the reductive dehalogenation process and the second wave from the reduction of the respective 2-aminopyrimidine nucleus, as determined by the coincidence of the second $E_{1/2}$ value with that of the single wave produced by the nonhalogenated analogs.

H. OXIDATION OF PYRIMIDINES

As previously mentioned, none of the simpler pyrimidines are oxidizable at graphite or mercury electrodes under polarographic conditions; there are, however, several reports of the oxidation of polyamino and polyhydroxy-pyrimidines (272–274). Table IX presents the known data for the voltammetric oxidation of pyrimidines.

All the oxidations of pyrimidines reported seem to involve (a) oxidation of substituted amino or hydroxy groups, or (b) mercury salt formation at the D.M.E. No evidence could be found in the literature for the actual oxidation of the pyrimidine nucleus.

Chiang and Chang (273), who reported the polarographic oxidation of some 4,5-dihydroxypyrimidines at a platinum microelectrode, concluded that pyrimidines without a 5-hydroxy group could not be oxidized and that pyrimidines with an enediol structure could be oxidized electrochemically to a diketo structure.

Glicksman (274) used thirteen amino- and hydroxypyrimidines as reducing agents in 1.44M NaOH in anodic half-cell discharge experiments. The higher potential of 5-amino- as compared to 6-aminouracil was attributed to the fact that the 5-aminouracil on oxidation can form a stable p-quinoidal structure, whereas a corresponding m-quinoidal structure is unlikely. The nature of the technique used, however, provides little other information pertinent to the voltammetric oxidation of these compounds, such as half-wave potentials, number of electrons transferred, or further insights into the mechanism of the oxidation process.

Cohen et al. (272), who studied some fourteen polyamino- and amino-hydroxypyrimidines at a rotating platinum electrode, concluded that the overall reaction is an irreversible 2e process to give an unstable product with a quinonimine structure. Limiting current-concentration curves were linear only at low concentration and had the shape of a Langmuir isotherm, which was attributed to slow desorption of the oxidation product or some condensation species resulting from this product.

TABLE IX

Half-Wave Potentials for Oxidation of Pyrimidines

Pyrimidine	pH	$E_{1/2}$ (V)[a]
4,5-Diamino-	3–6	$1.28 - 0.08$ pH[b]
	6–10	$0.96 - 0.03$ pH
2,5-Diamino-	1–6	$1.09 - 0.077$ pH[b]
4,5-Diamino-6-methyl-	3–6.6	$1.33 - 0.10$ pH[b]
	6.6–9	$0.78 - 0.017$ pH
2,4,5-Triamino-	3–6.2	$0.96 - 0.082$ pH[b]
	6.2–9	$0.64 - 0.041$ pH
4,5-Diamino-6-hydroxy-	3–9	$0.66 - 0.046$ pH[b]
4,5,6-Triamino-	3–6	$0.91 - 0.091$ pH[b]
2,4-Dihydroxy-5-methylamino-	3–6	$0.84 - 0.075$ pH[b]
(5-methylaminouracil)		
5,6-Diamino-2,4-dihydroxy-	4–9	$0.45 - 0.052$ pH[b]
(5,6-diaminouracil)		
5-Amino-2,3-dihydroxy-	3–10	$0.83 - 0.053$ pH[b]
(5-aminouracil)		
5-Amino-2,4-dihydroxy-6-methyl-	3–6.5	$0.77 - 0.051$ pH[b]
(5-aminothymine)		
2,4-Diamino-5-methylamino-6-hydroxy-	3–5.3	$0.54 - 0.078$ pH[b]
(5-methylamino-6-aminoisocytosine)	5.3–10	$0.30 - 0.032$ pH
2,4,5-Triamino-6-hydroxy-	3–5.5	$0.55 - 0.040$ pH[b]
(5,6-diaminoisocytosine)	5.5–9	$0.29 - 0.032$ pH
2,4,5-Triamino-6-methoxy-	3–5	$0.69 - 0.092$ pH[b]
	5–9	$0.36 - 0.027$ pH
4,5,6-Triamino-2-hydroxy-	2.8–9	$0.61 - 0.068$ pH[b]
2-Methyl-4,5,6-trihydroxy-		0.4[c]
2-Amino-4,5-dihydroxy-		0.43[c]
2,4,5-Trihydroxy-		0.47[c]
(5-hydroxyuracil)		
2-Methyl-4,5-dihydroxy-		0.79[c]
4,5-Dihydroxy-		0.8[c]

[a] Equations given indicate the $E_{1/2}$ when extrapolated to zero pH and the variation of $E_{1/2}$ with pH.

[b] Potentials are given versus the silver/silver chloride reference electrode for a R. P. E. anode. Concentration of electroactive species: usually 0.06mM. Buffers: pH 3–6, acetate; pH 6–7.5, phosphate; pH 7.5–10, borate. Except for two compounds, $E_{1/2}$ versus pH relationships were estimated by the present authors from data in the original reference (272).

[c] Potentials versus S. C. E. in acetic acid solution, 1M ammonium nitrate; anode is platinum microelectrode (273).

4. Purines: Reduction

The electrochemical reduction of purine and related compounds usually involves hydrogenation of the pyrimidine ring, which is finally followed by hydrolytic fission of the pyrimidine ring, leaving the imidazole ring intact. This general path is supported by the facts that imidazole has a benzene-like inertness (275,276) and is not reduced polarographically (258).

A. PURINE

Purine itself exhibits two 2e diffusion-controlled reduction waves in acidic media (133). Wave I corresponds to the 2e reduction of the 1,6 double bond to form 1,6-dihydropurine (Figure 23). The latter is slowly oxidized in the presence of oxygen to regenerate purine; the regeneration process probably involves the formation of 1-hydroxy-6-hydropurine, which subsequently dehydrates to regenerate the stable purine ring. In the absence of oxygen, the dihydropurine is converted to a yellowish, polarographically inactive compound. Wave II results from a further 2e reduction of 1,6-dihydropurine to 1,2,3,6-tetrahydropurine, which on subsequent hydrolysis forms the diazotizable amine, [(5-aminoimidazol-4-ylmethyl)amino] methanol (Figure 23). Unlike most other imidazoles, the 4- (or 5-)aminosubstituted compounds are highly unstable and have been isolated only as salts. The relatively large magnitude of wave II ($I = 6.0$) compared to wave I ($I = 3.7$) is associated with the concomitant hydrogen-ion reduction, which is catalyzed by the fully reduced form of purine (113,277).

On cyclic voltammetry, purine shows two well-formed cathodic peaks of constant and very reproducible current between pH 1 and 8, whose peak potentials vary linearly with pH (114). No evidence for an anodic process was observed on the return scan, except for a very small anodic peak at fast scan rate (36 V/sec). On starting at zero volt and scanning initially toward negative potential at the P.G.E., a single reduction peak appears at more negative potential than peak I at the H.M.D.E.; above pH 9, the peak disappears behind background discharge. On the return scan (toward positive potential), a well-formed oxidation peak appears at much more positive potential. This peak is due to the oxidation of the purine reduction product or some species derived from it, and not to purine itself, since the peak does not appear if the initial scan is toward positive potential.

Both electrocapillary and ac polarographic curves indicate adsorption of purine at mercury electrodes. A recent report discusses the mechanism and kinetics of the purine reduction in some detail (277a).

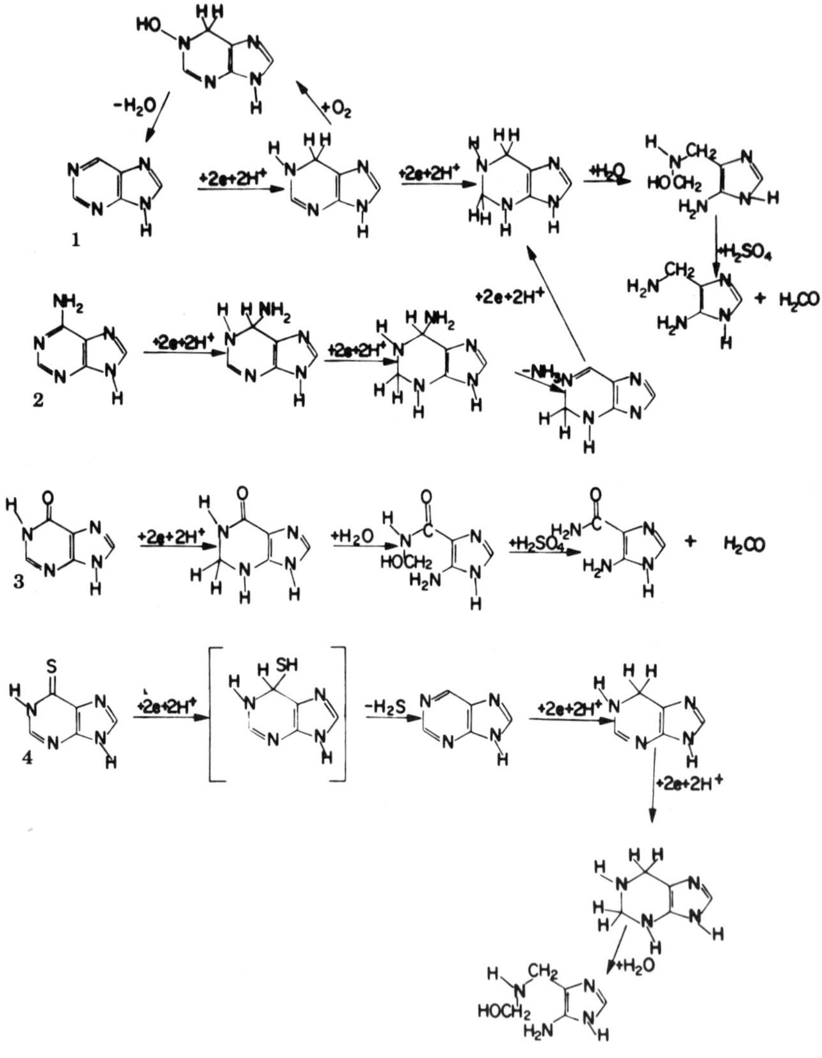

Figure 23. Interpretation of the electrochemical and chemical behavior observed for purine (1), adenine (2), hypoxanthine (3), and 6-thiopurine (4).

406

B. ADENINE SERIES

a. Adenine. On controlled-potential electrolysis, adenine undergoes a 6e reduction and gives the same product as purine does in its overall 4e reduction (Figure 23) (133). The adenine reduction probably involves (*a*) 2e reduction of the 1,6 double bond, (*b*) immediate 2e reduction of this product to give 1,2,3,6-tetrahydroadenine, (*c*) slow deamination at the 6-position to form 2,3-dihydropurine, (*d*) further 2e reduction of the regenerated 1,6 double bond to form 1,2,3,6-tetrahydropurine, and, finally, (*e*) hydrolytic cleavage at the 2,3-position to form the same diazotizable amine as in the purine reduction.

Under polarographic conditions, adenine exhibits only one reduction wave of approximately the same total current as purine itself, which is of the magnitude expected for a 4e wave; hence it is doubtful that deamination proceeds appreciably on the time scale involved in classical D.M.E. polarography. The polarographic patterns and data for purine and adenine have to be evaluated with some caution, since adenine and the completely reduced forms of adenine and purine lower the overpotential for hydrogen-ion reduction, resulting in some increase in current due to hydrogen evolution.

On cyclic voltammetry at the H.M.D.E., adenine shows a single pH-dependent cathodic peak, with no anodic peaks observed on the return scan even at very fast scan rate (114). The peak current, i_p, is constant to pH 5.5 and then decreases very rapidly to vanish above pH 6. At the P.G.E., adenine shows no cathodic activity over the available potential range from pH 1 to 12. However, a large pH-dependent anodic peak appears at 0.9 to 1.2 V between pH 3.6 and 10; this is probably due to oxidation of adsorbed adenine.

b. Adenine Nucleosides and Nucleotides. The electrochemical behavior over the available pH range of the fundamental sequence of adenine nucleosides and nucleotides at mercury electrodes is similar to that of the parent adenine (70,278). All exhibit a diffusion-controlled 4e wave due to reduction of the 1,6 and 3,2 N—C double bonds in the protonated species up to pH 4 or 5, where the wave begins to decrease sharply with increasing pH and disappears by pH 6 or 7 (70). The calculated diffusion current constants for the reduction wave of the adenine nucleosides and nucleotides agree within 10% with that for the adenine wave, implying a similar 4e process for all of the adenine species. The current, when measured at pH below that of the decreasing region, is linearly proportional to concentration, but the apparent diffusion current constant decreases with increasing concentration, similarly to the case for adenine (70,74,133).

Generally, the attachment of a sugar or sugar-phosphate group decreases the ease of reducibility relative to adenine, although the relative $E_{1/2}$ values within the adenine series are somewhat dependent on the depolarizer concentration and pH. The ease of reducibility at pH 4.1 and 1mM concentration is adenine > r-Ado > d-Ado > d-AMP, AMP > ADP > ATP. The difference in $E_{1/2}$, however, is rather small, for example, less than 50 mV for the series of base, nucleoside, and nucleotide monophosphate, compared to about 120 mV for the similar cytosine series. This is probably due to two factors. (a) The closer proximity of the sugar substituent to the reduction site in the cytosine series enhances the electron-withdrawing effect of the ribose ring, resulting in the ease of reducibility being nucleoside > nucleotide > base. (b) In the adenine series, the greater relative tendency to adsorbability (256) and intermolecular association (357), both of which factors tend to decrease the ease of reducibility, serves to counterbalance the effect of sugar attachment to a greater extent in the adenine series than in the cytosine series.

There is some evidence that adenine nucleosides and nucleotides may become more easily reducible than adenine itself at higher pH; that is, as the extent of protonation decreases, the electron-withdrawing effect of ribose and 2'-deoxyribose becomes operative.

Alternating-current polarography indicates that, with increasing substitution at the adenine N(9) position (adenine, adenosine, AMP, ATP), the faradaic peak current decreases appreciably and E_s becomes slightly more negative. Under identical experimental conditions, the peak current, Δi_s, for ATP is only 40% of that for adenine; with a similar $4e$ process for all the species, this fact indicates a substantial decrease in the reversibility of reduction.

Some of the uncertainties in the polarographic behavior of adenine nucleosides and nucleotides may be due to the appearance under certain conditions of an "abnormal" wave at more negative potential than the usual diffusion-controlled $4e$ wave (74). This wave, also observed with some 6-amino- and 6-alkylaminopurines, has some of the characteristics of a maximum of the second kind, primarily because of an enhanced supply of purine as a result of the onset of streaming of the solution over the electrode surface; the wave may also involve reduction of the 1,6 N—C double bond regenerated by deamination of the initial reduction product. The complicating presence of the abnormal wave can be avoided by the use of low-ionic-strength solutions and low temperature (e.g., 0.1M or lower ionic strength and temperature below 5°C); alternatively, 0.5M ionic strength solutions can be used at 25°C or the equivalent if the concentration of the electroactive species does not exceed 0.1mM.

The polarographic behavior of polyadenylicacid has been reported (262,262a).

A considerable quantity of useful data on the ionization, conformation, and metal complex formation of adenosine and adenine nucleotides has been summarized by Phillips (199); however no references to the use of polarographic techniques are cited.

c. Compounds Related to Adenine.

Luthy and Lamb (279) observed that adenine, 2,6-diamino-, 6-acetamido-, and 2,6-diacetamidopurine are reducible at the D.M.E. The last two compounds gave two waves; the first wave was attributed to reduction of the acetyl group, and the second to reduction of adenine and 2,6-diaminopurine, respectively, on the basis of $E_{1/2}$ coincidence. However, from inspection of the figures in the original paper, it would seem that the total current produced for adenine is about the same as that produced by an approximately equal concentration of acetyladenine or 2,6-diacetamidopurine. Moreover, similarly to the case of guanine itself, no reduction wave was observed for acetylguanine. Consequently, it would appear that the conclusion that the acetyl group itself is reduced is in error. It is more likely that the presence of the acetyl group promotes the reduction (lowers the $E_{1/2}$) for the first adenine reduction step, in effect, splitting the $4e$ reduction of the adenine moiety. Thus the reduction of acetyladenines resembles more that of purine or 6-methylpurine than that of adenine.

A variety of 6-substituted purines, primarily 6-alkylaminopurines, were studied, using dc and ac polarography to evaluate the effects of substitution on the electrochemical behavior of these compounds (69). When the 1,6 and 3,2 N—C double bonds are available in the pyrimidine moiety, a $4e$ polarographic wave due to hydrogenation of these bonds is generally observed.

C. GUANINE SERIES

Guanine, guanosine, and guanosine monophosphate give no indication of a polarographic reduction wave at the D.M.E. (133,277,279,280) but do show activity on oscillopolarography (264,281–283). The original explanation (282) of the anodic indentation on the oscillopolarographic (dE/dt versus E) curves of guanine and its nucleosides and nucleotides is that a product is formed at potentials very close to that of background discharge. The position of the reduction sites is still not clear, but it has been suggested that the 7,8 N—C double bond of the imidazole ring may be hydrogenated on reduction. At equal concentrations, the less soluble guanosine gives a deeper incision than guanylic acid; this, plus the decrease of the incision depth with increasing temperature, indicates that adsorption processes are probably involved.

Guanine and some of its derivatives have been studied by cyclic voltammetry (284). Guanine appears to be reduced in the region of background discharge to yield a product capable of undergoing a subsequent anodic reaction at about -0.2 V. The formation of this product seems to be conditioned upon the following: (a) both the C(2) and C(6) positions must be substituted by oxo, amino, or substituted amino groups, (b) positions N(7) and C(8) must be unsubstituted, and (c) substituents at positions N(1), N(3) and N(9) do not interfere with the formation of this product. Of the purines investigated, formation of a reduction product at a highly negative potential occurs only with those that do not produce a normal polarographic wave, with the exception of 2,6-diaminopurine. It was concluded that the reduction site is probably the 7,8 N—C double bond, as was previously assumed on the basis of oscillopolarographic measurements.

In contrast to the $4e$ wave usually observed with 6-aminopurines, isoguanine (2-hydroxy-6-aminopurine) gives only one $2e$ wave at the D.M.E., which corresponds to hydrogenation of the 1,6 N—C double bond, since the 2,3 C—N double bond has been removed from the pyrimidine ring because of the stability of the keto form. The isoguanine wave markedly deviates from the linear $E_{1/2}$ versus pH relations usually observed for purines and pyrimidines. Isoguanine also gives a second wave at more negative potential that has been identified as a catalytic hydrogen wave on the basis of its being several times higher than the normal wave, decreasing sharply with increasing pH, and being directly proportional to the mercury column height (74).

D. AZAPURINES

Only four azapurines appear to have been examined polarographically (267), three of which are irreducible: 8-azaguanine, 8-azaxanthine, and 8-azahypoxanthine. From pH 1 to 7, 8-azaadenine exhibits one $4e$ kinetically controlled polarographic wave, which was proposed to be due to $2e$ reduction of the 1,6 N—C double bond, fast deamination at the 6-position, and then additional $2e$ reduction to give a dihydro-8-azapurine species. One possibility that seems to have been overlooked is that the reduction involves the $4e$ hydrogenation of the 1,6 and 3,2 double bonds, and that the deamination step is slow on the time scale of D.M.E. polarography, in analogy to the behavior of adenine itself. Macroscale electrolysis might reveal that 8-azaadenine eventually undergoes a $6e$ reduction, the additional two electrons being consumed in reduction of the regenerated 3,4 double bond.

Reduction potentials in the 8-azapurine series were reported to change only slightly in relation to the parent purines, as opposed to the large differences for azapyrimidines. However, at pH 7, adenine and 8-azaadenine would have half-wave potentials of -1.59 and -1.36 V (ignoring various liquid junction potentials), a not inconsequential difference. The differences in the cytosine series are somewhat larger; $E_{1/2}$ values for cytosine, 5-aza-cytosine, and 6-azacytosine are -1.65, -1.38, and -1.13 V, respectively. 8-Azahypoxanthine was reported to be irreducible, although hypoxanthine itself produces a wave ($E_{1/2} = -1.61$ V in pH 5.7 acetate buffer) (113,133, 277). In this case, introduction of an additional ring nitrogen seems to have made reduction more difficult, in opposition to the usual trend.

8-Azaguanine produces an oscillopolarographic indentation in $1M$ sulfuric acid and ammonium formate backgrounds (266).

E. HYDROXYPURINES

Reduction of hypoxanthine (6-hydroxypurine) probably involves $2e$ hydrogenation of the 2,3 double bond, followed by hydrolysis to 4-amino-5-N-(hydroxymethyl)imidazole carboxamide; the 1,6 N—C double bond is unavailable for reduction because of the stability of the keto form. Further hydrolysis in strongly acid solution would result in formaldehyde and also in 5-amino-4-imidazolecarboxamide, which has been identified (284a) as one of the products in the reduction of hypoxanthine with zinc and sulfuric acid.

2-Hydroxypurine, which does not appear to have been studied polaro-graphically, can be predicted to have an $E_{1/2}$ close to that of purine on the basis that substituents in the 2-position have a minor effect on the ease of reduction. This prediction is supported by the small difference in $E_{1/2}$ produced by the addition of a 2-hydroxy group in pyrimidine (67) and by the fact that both purine and 2-hydroxypurine absorb 1 mole of hydrogen per mole of compound on room-temperature hydrogenation over a palladium catalyst, whereas adenine and hypoxanthine are unaffected (285).

Although there do not appear to be any reports to this effect, it seems unlikely, by analogy to the general behavior of pyrimidines, that any purine with more than one hydroxy substituent would be polarographically reducible within the available potential range.

F. THIOPURINES

a. 6-Thiopurines. Contrary to previous reports (286–288) of 6-thio-purine being irreducible under polarographic conditions, it exhibits several polarographic reduction waves in a variety of buffers (289). At low pH, wave I corresponds to the $4e$ reduction of 6-thiopurine to 1,6-dihydropurine, with concomitant loss of hydrogen sulfide. Wave II, which is probably identical to the second reduction wave of purine, corresponds to the $2e$ hydrogenation of the first product to give 1,2,3,6-tetrahydropurine, which then hydrolyzes to form a 4-aminoimidazole. Above pH 3, the two waves merge to produce a composite $6e$ wave, which has the same potential dependence as wave I up to pH 5.5. Above pH 5.5, wave III occurs, which corresponds to the same 6e composite process of waves I and II with a different rate-controlling process involving a different number of protons in the initial step; this accounts for the change in pH dependence of the composite process. One further wave (IV) of unknown nature is observed only in pH 9.1 ammonia buffer. Over the entire pH range, 6-thiopurine and/or its reduction products seem to catalyze hydrogen evolution, resulting in high diffusion and electrolysis currents and subsequent difficulty in interpretation of the reduction mechanism of this compound.

A variety of 6-substituted S-methyl and S-carboxyalkyl purines are polarographically reducible (287), generally producing either one 4e wave or two 2e waves, except when there is a 2-hydroxy substituent (thereby removing the 2,3 C—N double bond) and one 2e or two 1e waves are produced. Unfortunately, no $E_{1/2}$ values or current data were presented; only the number of waves and the number of electrons involved were given. Although the polarographic processes involved in the reduction of these compounds can only be guessed at with the scanty information available, it seems highly likely that these compounds are reduced in an overall 4e hydrogenation of the 1,6 and 3,2 N—C double bonds to produce 1,2,3,6-tetrahydro derivatives. Naturally, further chemical reactions such as hydrolysis could occur to produce ultimately a variety of products. The rather rapid splitting off of hydrogen sulfide which occurs with 6-thiopurine, however, is probably too slow a reaction to occur to a measurable degree with the corresponding S-methyl and S-carboxyalkyl purines, and the polarographic reduction reactions stops at 4e instead of involving the composite 6e reduction for 6-thiopurine.

Although 2-hydroxy-6-thio- and 2-methyl-6-thiopurine were reported to be polarographically inactive in a variety of buffers (287), a recent paper (284) has stated that these two compounds produce a normal polarographic reduction wave in acetate medium.

The very unstable bis(6-purinyl)disulfide produces two reduction waves in pH 2.3 acetate buffer. The mechanism of reduction is not known (289).

b. Purine-6-Sulfinic Acid. Because of the instability of purine-6-sulfinic acid in aqueous solution, particularly when acidic, and the number of electroactive decomposition products derived from it, the observed polarographic behavior of this compound is extremely complex (289,290). Between pH 2 and 9.1, wave I is attributed to a 2e reduction of the monoanionic species to 1,6-dihydropurine-6-sulfinic acid, which dehydrates fairly slowly to purine-6-sulfenic acid; this is immediately reduced in a further 2e process to 6-thiopurine. Wave II is attributed to a 2e reduction of the dianionic form of purine-6-sulfinic acid to the corresponding 1,6-dihydro derivative, which decomposes to yield purine and the sulfoxylate anion. Wave III is attributed to the further overall $4e$ reduction of the 1,6-dihydro derivative to produce eventually 1,2,3,6-tetrahydropurine, which subsequently hydrolyzes. As with 6-thiopurine, catalytic hydrogen evolution also occurs with purine-6-sulfinic acid, producing, at times, extremely high coulometric n values.

c. Purine-6-Sulfonic Acid. Purine-6-sulfonic acid exhibits four reduction waves over the normal pH range which are considered to result from a rather complex mechanism (289). Wave I, which is seen from pH 1 to 7, is attributed to the $2e$ reduction of the monoanionic form to the 1,6-dihydro species, which can then decompose by at least two pathways to form purine or 6-thiopurine. Wave II, which occurs from pH 3.6 to 12, supposedly arises from a similar reduction of the dianionic species to give a product which slowly decomposes to the reducible purine. Wave III of purine-6-sulfonic acid probably corresponds to the further reduction of the initial wave I and wave II reduction product, 1,6-dihydropurine-6-sulfonate, at the 2,3-position to give the 1,2,3,6-tetrahydro derivative. Presumably, subsequent breakdown of this product to yield sulfurous acid followed by further reduction, analogous to the adenine reduction mechanism, would eventually result in the formation of a 4-aminoimidazole.

In view of the experimental value (291) of pK_2 of 8.56 for purine-6-sulfonic acid, it seems highly unlikely that sufficient amounts of the dianion would exist below pH 7 to enable any polarographic wave to be seen. It must be concluded, therefore, that either the pK_2 value is too high by some 3 pK units or the interpretation of the electrode mechanism for wave II is in error.

Purine-6-sulfonamide produces three diffusion-controlled reduction waves in pH 9 ammonia buffer of unknown mechanism (289).

5. Purines: Oxidation

Before the work in this laboratory (4,5,113), investigation of the electrolytic oxidation of purines appears to be limited to a single report by Fichter and Kern (291a), who electrolyzed uric acid, theobromine, 6-desoxytheobromine, caffein, and 6-desoxycaffein in aqueous lithium carbonate, sulfuric acid, and acetic acid solutions. Since this work was done at a time when there was little appreciation of the importance of controlled-potential electrolysis, it is quite possible that considerable oxygen generation and chemical oxidation occurred concomitantly with anodic oxidation.

In general, except in the case of reversible redox couples, the electrolytic oxidation of organic compounds in aqueous media, even under the controlled-potential conditions which prevail in many polarographic techniques, is not as straightforward or simple a reaction as the electrolytic reduction of organic compounds so frequently is (13). Electrochemical reduction often involves simple hydrogenation of one or more double bonds; purines and pyrimidines, as has been shown, can commonly accept up to four electrons in a reduction process before chemical reactions such as hydrolysis and ring cleavage begin to play important roles. With oxidations, various chemical reactions and multiplicity of electrochemical pathways are much more prevalent and often begin at a much earlier stage in the oxidation process. One of the reasons for this is that, at the solid electrodes employed in oxidation (platinum, graphite), the problem of adsorption and specific interaction with the electrode surface complicates mechanistic pathways more than at the mercury electrodes usually used for reduction.

Although purine itself is not oxidizable under normal voltammetric conditions, many amino- and hydroxy-substituted purines exhibit voltammetric oxidation peaks at graphite electrodes. Table X presents some of the available electrochemical data for the oxidation of purines at graphite; peak current data, i_p or i_p/C, have not been given because of the variation in these parameters with such variables as scan rate and surface area of the electrode.

A. ADENINE

Adenine gives a single well-defined voltammetric oxidation wave at the P.G.E. (4,114), which appears to follow initially the same path as the enzymatic oxidation, but further oxidation and fragmentation of the purine ring system occur. Thus adenine is oxidized in a process involving a total of six electrons to give as the primary product an unstable dicarbonium-ion intermediate:

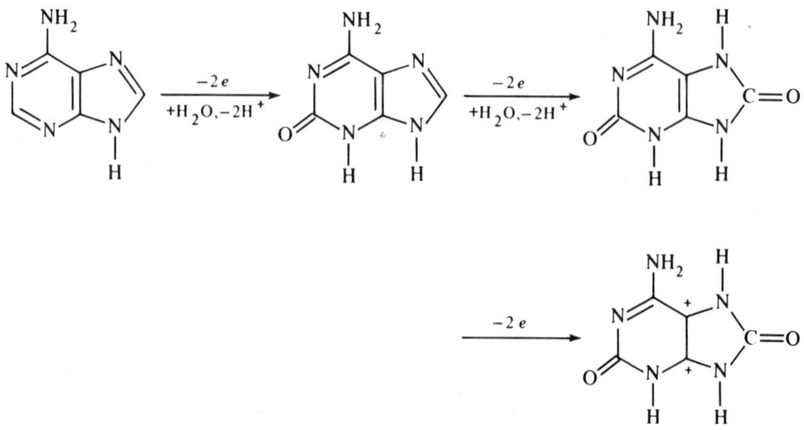

(34)

TABLE X

Half-Peak Potentials for Oxidation of Purines

Purine	Medium[a]	pH	Anode[b]	Concentration (mM)	$E_{p/2}$ (V)	Ref.
2-Amino-(adenine)	4	5.5	S. G. E.	0.4–1.04	1.03 – 1.07	352
	1		S. G. E.	1.0	1.34	113
	12		P. G. E.	1.0	1.30	4
	4	4.7	P. G. E.	1.0	1.14	4
2-Amino-6-hydroxy-(guanine)	1		S. G. E.	0.2–1.0	1.02	113
		0–12.5	P. G. E.		1.115 − 0.065 pH[c]	294
6-Amino-2-hydroxy-(isoguanine)	1		S. G. E.	0.5–1.0	1.06	113
6-Hydroxy-(hypoxanthine)	3, 4	2.3–5.7	S. G. E.	0.5	1.26 − 1.04	113
2,6-Dihydroxy-(xanthine)	1		S. G. E.	0.5	1.01	113
	3, 4	2.3–5.7	S. G. E.		0.84 − 0.71	113
2,6,8-Trihydroxy-(uric acid)	1		S. G. E.		0.62	113
	3, 4	3.7		0.15	0.45	113
	3, 4	5.7		0.15	0.33	113
	12			0.3	0.58	5
6-Thio-	4, 6, 12	1–12	P. G. E.	0.08–0.43	0.83 − 0.071 pH[c]	294
	4, 6, 7, 12	2–8	P. G. E.		I 0.51 − 0.047 pH[c]	299
	4, 6, 12	0–12			II 0.805 − 0.052 pH[c]	
		2–9			III 1.88 − 0.136 pH[c]	
Purine-6-sulfinic acid	6	9.1	P. G. E.	0.2–1	0.70	290

[a] Media are listed in Table IV.
[b] S. G. E. = stationary wax-impregnated graphite electrode; P. G. E. = stationary pyrolytic graphite electrode.
[c] E_p.

Continuation of oxidation to the dicarbonium-ion stage, as soon as the electrolytic oxidation of adenine is initiated, is due to the fact that the ease of oxidation of purines generally increases with the number of hydroxyl groups on the molecule (13).

After the primary electrochemical oxidation to the dicarbonium ion, the nature and ratio of the products ultimately produced indicate that at least three further distinct chemical and electrolytic reactions occur: (a) electrochemical oxidation to parabanic acid (some of which is further hydrolyzed to oxaluric acid), urea, carbon dioxide, and ammonia; (b) electrochemical reduction to give ultimately 4-aminopurpuric acid, carbon dioxide, and ammonia; and (c) simple hydrolysis to allantoin, carbon dioxide, and ammonia.

B. HYDROXYPURINES

a. Uric Acid. The study of Struck and Elving (5) of the electrolytic oxidation of uric acid (2,6,8-trihydroxypurine) at graphite electrodes proposed a single pathway with a common intermediate for the formation of both alloxan and allantoin, the two major enzymatic oxidation products. This mechanism agreed with the suggestion of Canellakis and Cohen (292) that the uricase oxidation of uric acid proceeded via a common primary oxidation intermediate to produce both alloxan and allantoin, depending on experimental conditions. Before this report, it has been implied that two separate and distinct pathways occurred in the enzymatic oxidation of uric acid, producing alloxan as the major product under strongly acid conditions and allantoin under weakly acid, neutral, and alkaline conditions (293).

In $1M$ acetic acid, uric acid gives a well-defined $2e$ anodic wave; this is due to formation of a primary short-lived dicarbonium-ion intermediate which undergoes three simultaneous transformations (Figure 24): (a) hydrolysis to the allantoin precursor, (b) hydrolysis to alloxan and urea, and (c) further oxidation and hydrolysis leading to parabanic acid and urea; the secondary oxidation accounts for the nonintegral number of electrons obtained on coulometry at controlled potential, that is, 2.2 electrons per molecule of uric acid. The primary oxidation intermediate ultimately produces both allantoin and alloxan, indicating that this intermediate may be common to all uric acid oxidations and that the ultimate product is probably controlled by experimental conditions.

Dryhurst (293a) has presented evidence to indicate that the shortlived intermediate may, in fact, be a 4,5-diol instead of a dicarbonium ion.

b. Guanine. Guanine (2-amino-6-hydroxypurine) is electrochemically oxidized by a mechanism very similar to that observed for other naturally occurring purines such as uric acid and adenine (Figure 24). An initial $2e$

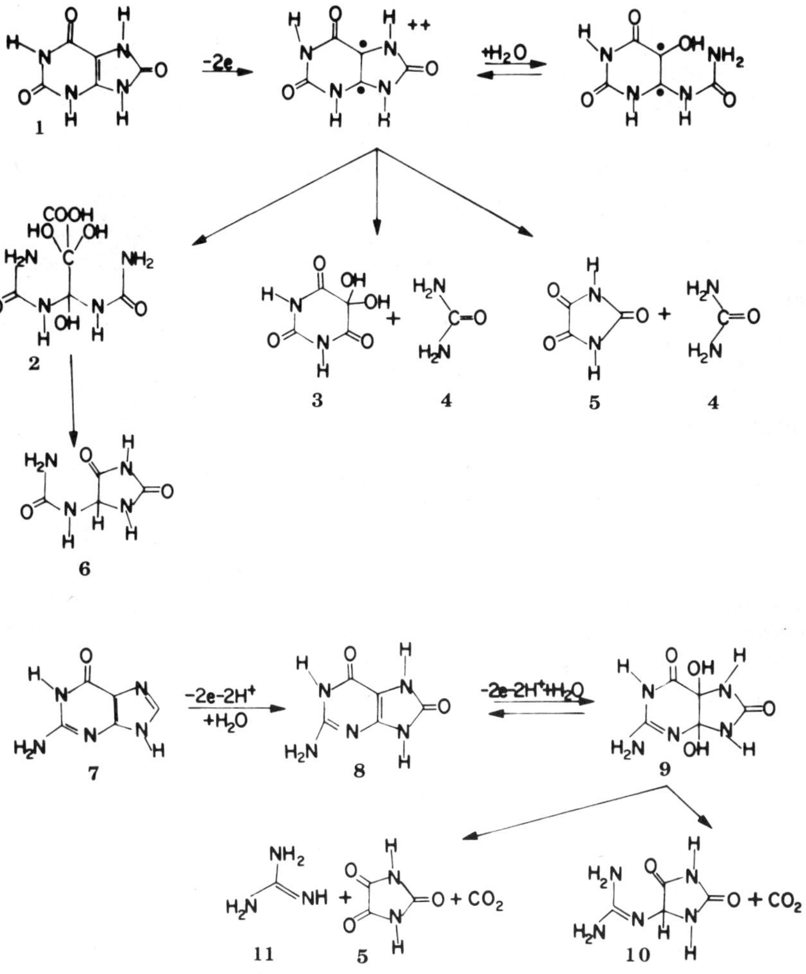

Figure 24. Proposed pathways for the electrolytic oxidation of uric acid (**1**) in acetic acid solution and of guanine (**7**) in moderately acidic media: "Behrend compound" (**2**), alloxan (**3**), urea (**4**), parabanic acid (**5**), allantoin (**6**), 2-amino-6,8-dioxypurine (**8**), 2-amino-6,8-dioxypurine-4,5-diol (**9**), oxalyl guanidine (**10**), and guanidine (**11**).

417

attack at the 7,8 N—C double bond is followed by a further $2e$ oxidation of the 4,5 C—C double bond to give 2-amino-6,8-dioxypurine-4,5-diol, which, being unstable, undergoes further reaction by two routes: (a) further electrochemical oxidation to parabanic acid, guanidine, and carbon dioxide, and (b) hydrolysis to oxalylguanidine and carbon dioxide (294). The site of initial electron removal is the same for both enzymatic and electrochemical oxidations and is in accord with MO predictions of being the most readily oxidized site. Unlike most enzymatic reactions, however, the deamination reaction at the C(2) does not occur, and further electrochemical reaction is centered at the 4,5 C—C double bond. Strikingly, the photosensitized degradation (295–297) of guanine and DNA appears to be identical to the electrochemical oxidation of guanine.

Guanine and isoguanine both give well-formed anodic peaks in $2M$ sulfuric acid (113).

c. Miscellaneous. It has been suggested (298) that the oxidation of xanthine (2,6-dihydroxypurine) involves about 4 electrons; this may indicate a two-step oxidation to uric acid and then to a dicarbonium ion. The oxidation of a series of hydroxypurines has been investigated, primarily from the analytical viewpoint (113). At constant pH, the $E_{p/2}$ values for the series of hypoxanthine (6-hydroxypurine), xanthine, and uric acid (2,6,8-trihydroxypurine) become less positive, and the i_p/C values decrease with the increasing number of hydroxy substituents. Relative current magnitudes from linear scan voltammetry indicate the participation of 3 to 4 electrons for xanthine and 4 to 5 electrons for hypoxanthine and isoguanine.

A preliminary report (298) on the cyclic voltammetric oxidation of several purines at the P.G.E. at moderately fast voltage scan rates (3 to 6 V/sec) indicates that, with uric acid, xanthine, hypoxanthine, 2,8-dihydroxy-6-aminopurine, adenine, and guanine, the primary electrochemical oxidation product is a highly reactive dicarbonium ion or a 4,5-diol intermediate (although there may be some preliminary oxidation processes such as those exhibited by adenine), as indicated by the appearance of a nearly reversible $2e$ redox couple centered at about 0.44 V. The pH dependence of this peak for uric acid and guanine (294) indicates that it is the 4,5-diol that is formed. Significantly, the two nucleosides studied, adenosine and guanosine, did not show this same reversible redox couple (298) but did exhibit an oxidation peak of unknown origin at 0.50 V on second and subsequent scans.

C. THIOPURINES

In general, purines containing an exocyclic sulfur atom appear to be oxidized, not in the purine ring, but rather at the sulfur substituent (299). Several thiopurines—2-thio-, 2-thio-6-hydroxy-, 2-hydroxy-6-thio-, and

2,6-dithiopurine—produce anodic adsorption waves at the D.M.E. in borate and acetate buffers, whose $E_{1/2}$ values are generally between 0 and -0.3 V (265). Although no mention was made of the observation of any reduction waves at the D.M.E., it would seem probable that 2-thiopurine would be reducible under polarographic conditions.

a. 6-Thiopurine. On electrochemical oxidation at the P.G.E., 6-thiopurine exhibits three pH-dependent waves in a complicated reaction scheme that appears to be quite different from the major enzymatic oxidation route (299). Wave I, which corresponds to the oxidation of adsorbed 6-thiopurine to bis(6-purinyl)disulfide on the electrode surface, as indicated by the very low current observed for the wave and other characteristics, may be similar to the oxidation wave with adsorption character observed at the D.M.E. (286,288). Wave II is a true $1e$ faradaic oxidation process to give bis(6-purinyl)disulfide:

$$2 \quad \text{(structure)} \longrightarrow \text{(structure)} + 2\text{H}^+ + 2e \qquad (35)$$

The disulfide decomposes in acetate and ammonia buffers to yield a variety of products. Wave III in pH 9 ammonia buffer represents a multielectron oxidation of 6-thiopurine to purine-6-sulfinic acid, purine-6-sulfonamide, and a small amount of purine-6-sulfonic acid, as well as the oxidation of the decomposition products of bis(6-purinyl)disulfide. In the absence of oxygen and ammonia in pH 9 carbonate buffer, wave III corresponds to the quantitative $6e$ oxidation of 6-thiopurine to purine-6-sulfonic acid;

$$\text{(structure)} + 3\text{H}_2\text{O} \longrightarrow \text{(structure)} + 6\text{H}^+ + 6e \qquad (36)$$

b. Purine-6-Sulfinic acid. At the P.G.E., purine-6-sulfinic acid is oxidized in ammonia buffer in a $2e$ process to purine-6-sulfonamide (a trace of purine-6-sulfonic acid is also formed) (299):

$$+ \; NH_3 \longrightarrow \qquad\qquad + \; 2H^+ + 2e \qquad (37)$$

The exact mechanism of oxidation is not known.

6. Correlation of Electrode Reaction Paths and Substitution

Certain generalizations regarding the polarographic reduction of purines and pyrimidines should serve to systematize and correlate the various electrochemical mechanisms that have been described in some detail. Although the subsequent discussion is centered on the behavior of pyrimidines, the effect of substituents is usually the same for purine derivatives as it is for pyrimidine itself, since electrochemical reduction occurs in the pyrimidine ring.

Electrochemical reduction occurs first at the 3,4 N—C double bond in pyrimidines. Although MO calculations (300) predict the order of reactivities of the pyrimidine carbon atoms to nucleophilic attack to be 2 > 4(6) > 5, there is ample evidence to indicate that the 3,4 bond is the one initially reduced (67,133,258). When this bond is not available for reduction, as, for example, when a thio or hydroxy group is substituted in this position, thereby removing the double bond by means of a tautomeric shift, reduction is then generally much more difficult and occurs at the 1,2 bond, which is usually the site for the second series of reduction steps.

In general, the ease of electrochemical reduction of a pyrimidine decreases with the number of added amino and hydroxy substituents; part of this effect seems to involve saturation of the ring by means of tautomeric shifts, thereby removing reduction sites in the ring. For example, the extrapolated $E_{1/2}$ values at pH 0 for pyrimidine, 2-aminopyrimidine, and 4-amino-2-hydroxypyrimidine are -0.58, -0.69, and -1.13 V, respectively. Pyrimidines polysubstituted with tautomeric groups are likely to be either nonreducible within the available potential range or very difficult to reduce.

Alkyl-substituted pyrimidines are generally slightly more difficult to reduce than the corresponding nonsubstituted compounds; this difficulty increases the closer the substituent is to the reaction site. For example, the $E_{1/2}$ values for reduction of 6-azauracil and the 5-methyl- and 1,3-dimethyl-substituted compounds at pH 7 are -1.30, -1.39, and -1.35 V; the reaction site is the 5,6 C—N double bond. The $E_{1/2}$ values for purine and 6-methylpurine at pH 2 are -0.86 and -0.98 V.

Pyrimidines having a 4-amino or 4-hydroxy substituent are more difficult to reduce than the corresponding 2-substituted compounds. For example, the $E_{1/2}$ values for 2-hydroxy- and 2-aminopyrimidine at pH 2.3 are -0.71 and -0.80 V (67), while the values for the corresponding 4-substituted compounds are -1.25 and -1.23 V (258). The greater stability of the 4-hydroxy-substituted compounds has also been observed in the purine series (133) and probably accounts for the difference in ease of reduction of cytosine and isocytosine.

In many cases, reduction of a compound with a 4-amino substituent results in deamination and further reduction of the regenerated 3,4 double bond. Reductions of cytosine, 4-amino-2,6-dimethylpyrimidine, and 4-amino-2,5-dimethylpyrimidine appear to be $3e$ processes, involving $2e$ reduction of the 3,4 double bond, deamination, and $1e$ reduction of the resulting compound (67,301).

Quite often, the first reduction wave for substituted pyrimidines and purines exhibits a change in slope in the linear $E_{1/2}$-pH relation at about pH 5 to 7 (67,113,251,258,267). Generally, this can be attributed to the fact that at lower pH a protonated form of the compound is undergoing reduction, whereas at higher pH the nonprotonated form is being reduced. When the nonprotonated form is nonreducible, as is the case with adenine, the reduction current decreases fairly rapidly as the wave becomes kinetically controlled and reaches zero within about 2 pH units.

Substitution of a halogen on the pyrimidine ring generally seems to result in cleavage of the C—X bond in a $2e$ process, giving rise either to a new $2e$ wave at more positive potential or to an increase in current for a wave present for the nonhalogenated compound (267,302). For example, in the series consisting of 6-azauracil and the 5-chloro- and 5-iodo-substituted compounds (267), the following waves are observed: one $2e$ wave, one $4e$ wave, and two $2e$ waves. In the reduction of 2-chloropyrimidine in acetonitrile media (249), a $2e$ wave corresponding to the cleavage of the C—Cl bond is observed before the $1e$ wave at very negative potential corresponding to reduction of the pyrimidine moiety.

7. Adsorption on Electrodes

A. MECHANISM OF ADSORPTION

Most pyrimidines are only very slightly adsorbed on mercury electrodes, at least at the low concentrations at which many purine compounds show significant adsorption effects. Alternating-current polarographic data indicate that all purine-based compounds are adsorbed on mercury and, with the exception of phosphate derivatives, are adsorbed near the potential of zero charge on the electrode, that is, in the region of the ecm. The phosphate derivatives appear to be more strongly adsorbed on the positive potential side of the ecm; this suggests adsorption of a negatively charged species formed by acidic dissociation of the phosphate grouping as, for example, with adenosine phosphates:

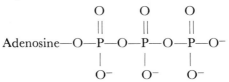

For example, purine exhibits a significant base current depression between about -0.2 and -0.8 V, whereas pyrimidine shows almost no tendency to either raise or lower the base current up to concentrations of about $10 mM$ (68,114). These and other relevant facts indicate that this strong adsorption involves an uncharged portion of the purine molecule, which would be the imidazole ring (70,114), and, specifically, since position 9 is occupied in the adenosines, either position 7 or 8. Since position 7 contains a heterocyclic nitrogen, adsorption may be more favorable at this position. It is more likely, however, that adsorption would involve π-bonding to the aromatic ring.

From the maximum surface excess of adsorbed guanine molecules (Γ_m), it was calculated (256) that the area occupied by an adsorbed guanine molecule is approximately 35 $\mathring{A}^2$. Since this area corresponds approximately to the area of the purine moiety, it was concluded that guanine is adsorbed planar to the electrode, similarly to the behavior of other heterocyclic bases (303,304).

At more negative potential than the ecm region, gradual desorption of adsorbed compounds occurs, as shown by a broad peak on ac polarographic and differential capacity curves (256). Adenine derivatives (70) and other 6-substituted purines seem to be readsorbed just before the main faradaic reduction peak; this may involve reorientation of adsorbed molecules on the electrode surface as the potential becomes more negative, so that the positively charged site of the molecule is attached to the mercury surface. The latter site is generally identified with the protonated N(1) (69,70).

Alternating-current polarograms of guanine in $0.2M$ H_2SO_4, where the guanine species is protonated, exhibit a desorption peak which shifts with increasing concentration from -0.6 to -0.9 V. The fact that this type of desorption maximum is not observed in neutral $1M$ NaCl media, where the guanine species is nonprotonated, may be taken as an indication that the desorption maximum involves reorientation of the adsorbed molecule. At potentials near the ecm (-0.6 V), 2,6-dichloro-9-methylpurine exhibits a small peak which may be due to reorientation of the compound from flat adsorption of the ring to adsorption of the more hydrophobic end of the molecule near or at the methyl substituent.

B. PARENT BASES

Several studies of the adsorption of purine and pyrimidine bases on the electrode surface, employing bridge impedance and ac polarographic techniques, have been published (68,70,114,256,257,277a,305,306). Differential capacity data (256) for $1M$ NaCl solutions of the common DNA bases indicate that the surface activity increases in the order of cytosine $<$ thymine $<$ adenine $<$ guanine. Generally, the order of increasing surface activity of purine and pyrimidine bases can be qualitatively correlated with the decreasing dipole moment of the compounds, which is also related to the increasing hydrophobic nature of the species.

Significantly, it was observed (256,257,305) that the greatest tendency toward intermolecular association on the electrode surface, as indicated by formation of a sharp pit in the capacity curve (Figure 6), occurred with derivatives that are common components of nucleic acids. Derivatives that occur only in certain nucleic acids (e.g., 5-hydroxymethylcytosine) and those that are only occasionally incorporated into nucleic acids, such as 8-azaguanine, exhibit much less tendency to intermolecular association; purines and pyrimidines that do not occur in nucleic acids do not associate on the electrode surface.

For guanine in $0.2M$ H_2SO_4, the rate of attainment of intermolecular association equilibrium is approximately one hundredth of the value for initial adsorption equilibrium, although the rates are dependent on both the potential and the guanine concentration (306).

C. NUCLEOSIDES AND NUCLEOTIDES

The shapes of the ac polarograms of nucleosides and nucleotides are generally much more complex than those of the corresponding bases, and the causes behind some of the observed effects are not well understood. All of the nucleosides and nucleotides that have been examined appear to be surface active and to be adsorbed at potentials more positive than about -1.6 V (18,19,70,114,115a).

Alternating-current polarograms of uridine, thymidine, and cytidine show a simple shape with a minimum at about -0.4 to -0.6 V and a cathodic desorption maximum between -0.9 and -1.5 V. The pits characteristic of the association of adsorbed molecules are not observed even at concentrations approaching the saturation level (55, 42, and 17mM, respectively), whereas uracil, thymine, and cytosine themselves begin to show association pits at about 6, 13, and 39mM, respectively. Alternating-current polarograms of deoxycytidine, however, do show association pits in the potential range -0.2 to -0.6 V; deoxyribonucleosides, for some reason, seem to exhibit more complicated adsorption behavior than ribonucleosides.

Other nucleosides studied—Ado, d-Ado, Guo, and d-Guo—also exhibit intermolecular association pits and other more complicated behavior.

The surface activity of DNA and RNA components, as measured by the change in differential capacity, increases in the order of base $<$ nucleotide $<$ nucleoside, with the exception of the guanine series, which exhibits the reverse order. The surface activity of nucleosides and nucleotides increases in the order of d-Cyd $<$ Thd $<$ d-Ado $<$ d-Guo and d-CMP $<$ TMP $<$ d-AMP $<$ d-GMP. Among the factors that may have to be considered in explaining the surface activity of nucleosides and nucleotides, in addition to the specific adsorption and intermolecular association already mentioned, are multilayer association films and reorientation of the molecules on the electrode surface.

Typical of what can be investigated is the time-dependence of adsorption of RNA and denatured DNA, using ac polarography (306a). Since native DNA behaves differently from denatured DNA, photochemical or thermal denaturation can be followed.

8. Applications: Analysis

Electroanalytical procedures based on various polarographic techniques can be used to follow the course of chemical reactions involving purines and pyrimidines, to test the effectiveness of fractionation or isolation procedures involving these compounds, and to identify and/or estimate quantitatively with speed and precision several purines and pyrimidines. In some cases, analyses of mixtures difficult to accomplish by other methods are readily performed.

Emphasis will be placed in this section on classical polarography and simple linear scan voltammetry, both of which can be performed with a simple commercial polarograph. Although other methods which have been used, particularly oscillopolarography, may be quite useful, unfamiliarity with the technique or unavailability of equipment may limit these methods of analysis for many groups. The ready availability of modern, highly versatile electroanalytical instrumentation assemblies at a moderate cost, however, can place many voltammetric techniques at the disposal of even small laboratories.

A. CONTROL OF TEST SOLUTION COMPOSITION

The importance of careful control of experimental conditions in analytical procedures is indicated by the fact that only the protonated form of adenine and its nucleosides and nucleotides is reducible (70); at pH exceeding about 5, the limiting current due to these compounds decreases rapidly and becomes kinetically controlled. Often the choice of the specific buffer used may be critical; for example, cytosine, which gives in pH 4 to 8 McIlvaine buffer a very poorly defined wave very close to background discharge and probably unsuitable for analysis, shows a fairly well-defined wave in acetate buffer of pH 3.7 to 5.7. It has been suggested (267) that borate buffers may form complexes with purines and pyrimidines and thus complicate electrochemical behavior.

Generally, as long as the buffer concentration is sufficiently high to provide precise pH control (each component is present at 50 to 100 times the concentration of the electroactive species), little further thought is necessary. However, ionic strength effects should be investigated. For example, adenine derivatives sometimes show a somewhat distorted reduction wave at higher ionic strength ($0.5M$) at 25°C, because of the appearance of an "abnormal" wave which is not present at low ionic strength and temperature ($0.1M$ and 1.5°C) (70,74).

B. DETERMINATION OF PYRIMIDINES: REDUCTION

Table VI summarizes analytically orientative data with respect to half-wave potentials and diffusion current constants for reduction of a number of pyrimidines at the D.M.E. From these data and the preceding discussion on the general characteristics of pyrimidine reductions, the analytical potentialities and limitations are generally deducible, both for compounds that have been studied and for those that have not yet been examined.

The selectivity possible in analysis can be illustrated by a consideration of the polarographic determination of pyrimidine itself, which gives five distinguishable polarographic waves over the normal pH range. Pyrimidine itself is the only pyrimidine or purine thus far investigated which yields an additional reduction wave in alkaline solution; the appearance of a new wave at about pH 7.2 positively identifies pyrimidine in the presence of a host of other reducible compounds, including even the 2-aminopyrimidines and 2-hydroxypyrimidine, which are also reduced in alkaline solution. The heights of all five of the pyrimidine waves are directly proportional to concentration over a wide range (at least 0.1 to 5mM), when pH, buffer component, and ionic strength are properly controlled (67, 133).

The polarographic determination of several pyrimidinyl-substituted sulfanilamides present in medicinal tablets, syrups, and injections was accomplished by the method of standard addition with results comparable to the usual analytical procedure of diazotization (307,308).

The polarographic behavior of 2-methyl-4-aminopyrimidine and its 5-bromomethyl analog was investigated and the best conditions for polarographic analysis were determined (308a). The quantitative determination of some pyrimidine derivatives—etimidine (2,4-bis[ethylenimino]-5-chloropyrimidine) and allacil (1-allyl-3-ethyl-6-amino-1,2,3,4-tetrahydropyrimidinedione-2,4)—was reported (308b), using potentiometric titration with perchloric acid in acetic acid-acetic anhydride mixtures.

a. Hydroxypyrimidines. 2-Hydroxypyrimidine gives a single diffusion-controlled $1e$ wave over the pH range from 2 to 8; the general behavior of this wave corresponds to that of pyrimidine wave I in acidic solution. Except for a significant decrease in pH 3.7 acetate buffer, I is remarkably constant (2.06 ± 0.03 at the 0.5mM concentration level); $I = 2.12 \pm 0.05$ between 0.15 and 1.67mM in pH 2.9 McIlvaine buffer (113).

Cytosine gives a fairly well-defined wave, useful for analysis, which lies close to background discharge in pH 3.7 to 5.7 acetate buffer; in 0.5M acetate buffer of pH 5.5, $I = 4.9 \pm 0.3$ between 0.6 and 2.1mM cytosine (113). The fact that a wave is not observed for cytosine between pH 1.2 and 2.9 in chloride solution is the basis for the determination of adenine in adenine-cytosine mixtures, since adenine does give a wave in this pH range. Cytosine and its nucleosides and nucleotides have half-wave potentials that are too close together to permit accurate determination of the separate components in a mixture (251); total cytosine content, however, can be determined.

Cyclic voltammetry at the H.M.D.E. of pyrimidine and cytosine indicates an accuracy of 2 to 3% relative (114). The data on cytosine at a scan rate of 0.26 V/sec show that the peaks are much better formed and easier to measure than at lower scan rates; this indicates that, for compounds which give slow-scan voltammetric peaks close to background, fast-scan cyclic voltammetry could occasionally provide a rapid and convenient analytical method.

b. Uracil Derivatives. Although uracil and its nucleosides and nucleotides are polarographically inactive, 6-azauracil shows a well-defined wave (309) from pH 2 to 10. Specific procedures have been published on the polarographic determination of 6-azauracil and 6-azauradine in blood and urine (268,309–311). A procedure for analyzing mixtures of these two compounds (268) is based on a differential technique. The composite polarographic wave is measured (a) in pH 6.9 phosphate buffer, where the limiting current for both compounds reaches a maximum, and (b) in pH 9.4 borate buffer, where the wave height of 6-azauracil is the same but the height of the

6-azauridine wave has decreased to 50% of its original value. By means of a simple algebraic calculation, the concentrations of both components can be determinated. The accuracy of this method is said to be ±4%; uracil-5- and uracil-6-carboxylic acids interfere.

Specific polarographic procedures have been published for other compounds related to uracil: 5'-O-stearyl-6-azauridine in hydrolyzates of 2',3'-O-isopropylidene-5'-O-stearyl-6-azauridine (311); 1,3-dimethyl-6-amino-5-nitrosouracil in production samples (254,255); 2-methyl-4-amino-5-acetaminomethylpyrimidine and 2-methyl-4-amino-5-aminomethylpyrimidine sulfate in pilot samples (253); and orotic acid (uracil-6-carboxylic acid) (312).

C. DETERMINATION OF PYRIMIDINES: OXIDATION

No procedures have been reported for the analytical determination of pyrimidines using oxidation at platinum or graphite electrodes, although such determinations should be possible for the types of polyamino and polyhydroxy compounds that are oxidizable (Table IX). Apparently because of the adsorption of the oxidation products and the resulting non-linear current-concentration plots, the upper concentration limit for determination of these compounds at platinum electrodes appears to be quite low, on the order of 0.2mM (272).

Several thiopyrimidines—5-thiouracil, 2-thiouracil, 2-thiocytosine, 2-thiothymine, and 2,4-dithiopyrimidine—exhibit anodic waves at the D.M.E. in borate buffer because of mercury salt formation (265). The heights of these waves are proportional to concentration at higher pH and could thus be used for analysis. The anodic wave of uracil in pH 9.3 borate buffer has been suggested for its determination in concentrations up to 0.1mM, and that of 6-methyl-2-thiouracil in 1M NaOH for its determination up to 1mM concentration (313).

D. DETERMINATION OF PURINES: REDUCTION

Purine, adenine, various N-substituted adenines, hypoxanthine, isoguanine, azaadenines, and various 6-thiopurines are reducible in aqueous solution, whereas guanine, xanthine, and uric acid are not reduced within the potential range normally available.

a. Purine. Either of the two purine waves can be used for its determination; when a choice is possible, wave I is recommended. When wave I is employed, purine can be determined in the presence of all other purines, except adenine 1-N-oxide, 2-azaadenine, and 2-azaadenine-1-N-oxide. The concomitant reduction of hydrogen ion during the purine wave II accounts for its diffusion current constant being larger than expected for a $2e$ reduction (133). A maximum appears on wave II at concentrations greater than about 0.1mM in HCl-KCl buffer; addition of the nonionic surfactant Triton X-100 eliminates this maximum and shifts both purine waves to more negative potential by about 20 mV per 0.002% Triton added. The height of wave I is not appreciably affected by increasing the Triton X-100 concentration, but wave II decreases until it approximately equals wave I. Wave II is less well defined at higher purine concentration because of the catalytic effect of the reduction product on the evolution of hydrogen; the greater the product concentration at the mercury drop, the more positive is the potential of hydrogen evolution.

At the H.M.D.E., purine shows two well-formed cathodic peaks suitable for analysis. Both peak I and the sum of peaks I and II show linear i_p—C relationships from 0.04 to 1mM. At the P.G.E., purine gives one cathodic peak and one anodic peak, both suitable for analysis up to about 0.2mM concentration (114).

b. Adenine. Reduction of adenine at the D.M.E. has been successfully applied to its determination in biological samples, for example, in blood (314–317), in RNA hydrolyzates (318), and in some tissues (319–321). These methods are generally based on Heath's original observation (280) that adenine gives a wave of $E_{1/2} = -1.1$ V in 0.1M HClO$_4$-KClO$_4$ solution, which permits its determination in the presence of cytosine, guanine, guanosine, and a few other substances whose presence might be expected in hydrolyzates of desoxyribo- and ribonucleic acids. However, in most samples, prior separations are usually required, for example, precipitation of the adenine with silver oxide from the hydrolyzates of animal tissues (320) and deproteination with trichloroacetic acid, followed by silver oxide precipitation in the case of blood samples (315). Adenine, adenosine, 3'-AMP, 5'-AMP, and ATP all have approximately the same $E_{1/2}$ in 0.1N perchloric acid, and therefore individual components of adenine mixtures could probably not be determined. Addition of equimolar amounts of GMP, CMP, UMP, or IMP to AMP solutions had no effect on the wave height for AMP (318).

Experimental factors such as buffer system employed, pH selected, and use of a maximum suppressor may markedly affect the diffusion current and consequently are important in the determination of adenine. The adenine diffusion current constant (5.6 ± 0.3) is about 40% lower in pH 2 to 5 McIlvaine buffer than in chloride and acetate buffers, where the average I value (pH 1.2 to 5.5) is 10.2 ± 0.7 at the 0.2mM adenine level. The reason for this variation in current is the presence of the adenine abnormal wave (70,74). Although data for the normal wave obtained in the presence and absence of the abnormal wave are essentially identical, the proximity of the two waves makes data for the normal wave obtained in the absence of the abnormal wave more accurate. Consequently to avoid the complicating presence of the abnormal wave, it is recommended that analysis be carried out in solutions of low ionic strength and at low temperatures ($0.1M$, $<5°C$). Alternatively, at $0.5M$ ionic strength and $25°C$, the concentration of the electroactive species should not exceed $0.1mM$ (74). A pH 2 chloride buffer containing 0.002% Triton X-100 has been recommended (113).

At the H.M.D.E., adenine shows a single cathodic peak and no anodic peak, even at very fast scan rate. Adenine shows no cathodic activity at the P.G.E. from pH 1 to 12 but exhibits a poorly formed anodic peak (114). Both purine and adenine, however, show poor reproducibility of current at the P.G.E., probably because of adsorption phenomena, and nonlinear i_p—C curves; consequently, the concentration range in which they can be accurately determined may have to be severely limited.

c. Hypoxanthine. The polarographic waves of hypoxanthine are generally ill-defined inflections on the background discharge, little suited for analysis. However, hypoxanthine yields a fairly well-defined wave in pH 5.7 acetate buffer after point-by-point subtraction of the residual current (113, 133) and also at pH 1.1 (277). The anodic wave of hypoxanthine at the stationary graphite electrode (S.G.E.) is recommended for quantitative analysis (see below).

d. Thiopurines. Because of the catalytic hydrogen evolution at low pH and the proximity of waves I and II, pH 3.5 to 5.7 acetate buffer has been recommended for the determination of 6-thiopurine (290). In view of the nonlinear current-concentration relation, the 6-thiopurine concentration should be kept below $0.1mM$ or a calibration curve should be employed. The i_d/C ratio for 6-thiopurine in pH 3.5 acetate buffer decreased from 41.4 $\mu A/mM$ at $0.03mM$ to 30.9 at the $0.4mM$ concentration level.

Procedures have been published for the determination of the antimetabolite 6-(4-carboxybutyl)thiopurine, which gives two $2e$ waves in acidic and neutral solution (286,288). These procedures could probably also be adapted for other 6-(carboxyalkyl)thiopurines; 6-thiopurine is said not to interfere under the conditions used. Up to 2% 6-thiopurine present as an impurity in batches of the carboxybutylthiopurine could be determined by recording its anodic adsorption wave at the D.M.E.

The optimum background for determination of purine-6-sulfinic acid was determined to be pH 9.1 ammonia buffer (290). In this medium, reduction waves I and III have fairly constant i_d/C ratios of 5.9 $\pm$ 0.1 and 6.4 $\pm$ 0.5 $\mu A/mM$ between 0.2 and $1mM$ concentration. Because the wave III i_d/C

ratios decrease slightly with increasing concentration, wave I is recommended for quantitative analysis.

Purine-6-sulfonic acid could probably be determined accurately at various pH values, utilizing any one or several of its reduction waves. The interferences present would probably determine the best conditions to be used. If the sulfinic acid is present, as is often likely, a pH 9.1 ammonia buffer has been recommended (290). In this medium, the i_d/C ratios for all three reduction waves are constant in the concentration of 0.2 to 1mM: 6.36 ± 0.13 for wave II, 6.14 ± 0.19 for wave III, and 8.3 ± 0.6 μA/mM for wave IV. Any combination of these waves has a relatively constant i_d/C ratio that affords an accuracy of ±2 to 3%.

At pH 9, purine-6-sulfonamide produces three diffusion-controlled waves (290). The most positive wave ($E_{1/2} = -1.04$ V) could be used for determination of this compound in the presence of all of the other thiopurines discussed in this section. Because of the difficulty of obtaining sufficient pure purine-6-sulfonamide to prepare calibration curves, the method of standard additions was employed for its determination (299).

E. DETERMINATION OF PURINES: OXIDATION

Of the purines investigated, only purine itself appears to be nonoxidizable within the potential range available at graphite electrodes. The compounds in order of decreasing ease of oxidation (increasingly more positive half-peak potentials, $E_{p/2}$, at constant pH) are uric acid > 6-thiopurine > xanthine > guanine > isoguanine > adenine > hypoxanthine. With the exception of xanthine at pH 3.7 and 6-thiopurine, all of the oxidizable purines show one well-formed anodic peak. The $E_{p/2}$ becomes more positive (oxidation becomes more difficult) with decreasing pH for all of the compounds and is also affected by the nature of the background electrolyte and concentration of the electroactive species. Because of the generally observed nonlinear i_p/C plots and the many different types and sizes of graphite and platinum electrodes employed, the use of calibration curves obtained under the experimental conditions is suggested. Except for adenine (4) uric acid (5), 6-thiopurine (299), and guanine (294), the actual anodic processes undergone at graphite electrodes are unknown.

Several references to analytically usable oxidation waves have been made in the preceding section on the determination of purines by their electrochemical reduction.

a. **Hydroxypurines.** Difficulty soluble guanine and isoguanine give well-defined anodic waves in 2M H$_2$SO$_4$. As with most oxidizable purines, i_p/C ratios increase with decreasing concentration. In 2M H$_2$SO$_4$, either guanine or isoguanine can be determined in the presence of adenine, which gives a peak about 0.3 V more positive, but not in the presence of each other. Xanthine would interfere in the determination of either guanine or iso-guanine, but hypoxanthine would not.

The $E_{p/2}$ values for hypoxanthine, xanthine and uric acid become less positive and the i_p/C values decrease with the increasing number of hydroxy substituents (113). Although the $E_{p/2}$ values of these three compounds differ sufficiently to allow detection of one in the presence of the others, simultaneous quantitative determination of the three hydroxypurines is not practical, since simple current-concentration relationships do not exist for consecutive peak currents in linear scan voltammetry; it may be possible to develop empirical methods. More likely, the use of rotating platinum or graphite electrodes, where convection-limited currents are obtained, may permit simultaneous determination of the three, when present, within certain concentration limits.

Troy and Purdy (321a) reported a method for the coulometric determination of uric acid in serum and urine, using coulometrically generated iodine, which has a relative standard deviation of ±2 to 3%. The major advantages of this method over the usual uricase-spectrophotometric technique are that there is no need to prepare standards or standard solutions and that the occasional interference with the spectrophotometric assay of high concentrations of ultraviolet adsorbing materials in some uricase preparations and serum samples is avoided.

b. Thiopurines. In pH 7.1 McIlvaine buffer, 6-thiopurine exhibits a well-developed anodic wave at the D.M.E. at $E_{1/2} = -0.26$ V, which can be used for analysis (288). Hypoxanthine, which interferes in the iodometric and bromometric determination of 6-thiopurine, does not interfere in this procedure. 6-Thiopurine impurity in batch samples of 6-(4-carboxybutyl)-thiopurine could be determined by use of this wave.

For analytical purposes, wave II in $1M$ acetic acid or wave III in pH 9.1 ammonia buffer has been recommended for the determination of 6-thio-purine at the P.G.E. (290). In acetic acid medium, waves I and II are well separated, a fact which is important, since only wave II is proportional to concentration because of the adsorption character of wave I. Although the average deviation of replicate measurements on one solution is on the order of ±6%, probably because of the variation in surface area of the graphite electrode, the i_p/C ratios for these two peaks are constant to within about 1 to 2% within the concentration range of 0.08 to 0.4mM. Thus, if repeated measurements are made on the same solution, the accuracy of determination could be within 3%, although the average deviation of all the values may be above 6%.

Increased sensitivity can be obtained by rotating the graphite electrode (at 600 rpm, i/C ratios are increased by a factor of about 4) with some loss in the precision and accuracy of determination.

In pH 9.1 ammonia buffer, purine-6-sulfinic acid exhibits a single anodic peak at the stationary P.G.E. that is suitable for analysis. The i/C ratios are approximately constant in the concentration range of 0.2 to 1mM, with a precision and accuracy of determination on the order of 10%.

Other thiopurines—purine-6-sulfonic acid, purine-6-sulfonamide and bis(6-purinyl)disulfide—do not exhibit analytically useful oxidation peaks at the graphite electrode.

F. OSCILLOPOLAROGRAPHY

The oscillopolarographic activity of nucleic acid bases and nucleosides has been utilized for the analysis of nucleic acids and their constituents (264,266,281,282,322,323); for the determination of adenine in RNA hydrolyzates (324); for following changes induced in purine and pyrimidine bases by ultraviolet radiation (325–329); for the determination of the latter compounds in growing cultures of Escherichia coli (330–333); and for the identification of chromatographic spots (263) and fractions (334).

Characteristic electrochemical processes such as oxidation-reduction and adsorption-desorption are manifested by incisions or indentations on the roughly circular dE/dt versus E plots obtained. In addition to incisions which are due to normal faradaic processes and have a corresponding dc polarographic wave (adenine, cytosine, etc.), solutions of pyrimidine and purine derivatives also exhibit capacity incisions and incisions due to electrolysis of chemically or electrochemically altered species. Various criteria proposed for distinguishing the nature of oscillopolarographic incisions (335) were applied to explain the incisions observed (263,323). Cathodic incisions for adenine, cytosine, and their derivatives in sulfuric acid and ammonium formate media are due to faradaic reduction; incisions in sodium hydroxide medium are caused by mercury salt formation. The anodic incision for guanine and its derivatives is due to oxidation of some product formed at potentials near cathodic background discharge; indications are that some form of adsorption process is involved.

A table (14) has been compiled of the analytically useful incisions observed for the common nucleic acid components. This will not be duplicated here; readers are urged to consult the table and the original literature.

Most nucleic acid components and related compounds can be detected or determined alone or in known simple mixtures down to 10^{-5} to $10^{-6}M$ concentration levels (322,323). Mixtures of nucleotides separated by gradient elution from ion-exchange columns can be analyzed by this method (322). However, because of the variety of processes undergone and the resulting curved plots of incision depth versus concentration, concentration levels must often be within a rather restricted range for even moderately accurate quantitative analysis. The precision of determination of several nucleic acid components was found to be about $\pm 5\%$ (266). Although ultraviolet spectroscopy generally affords better accuracy, most of these compounds absorb in about the same wavelength range, thus presenting problems when dealing with mixtures. In some cases, several different backgrounds can be used for oscillopolarography, permitting detection and quantitative analysis of mixtures.

In the determination of adenine in RNA hydrolyzates, an accuracy of about $\pm 7\%$ was reported (324) for a spiking method; an excess of cytidine or CMP does not interfere with the adenine determination (263). Generally, however, a greater degree of accuracy in the determination of adenine is obtained with normal D.M.E. polarography.

G. ANALYSIS OF MIXTURES

Only a few examples will be considered of the polarographic and voltammetric analysis of mixtures of purines and pyrimidines. Other possible situations in which polarographic techniques might be effective will be apparent on considering the data and behavior patterns already presented in reference to the individual components of the mixture in question; the analyst should be able to tailor experimental conditions and voltammetric techniques in order to obtain optimal experimental results with a minimal amount of interference.

An explicit procedure was developed for the analysis of mixtures of adenine, cytosine, and guanine, three of the most important nucleic acid components (113). The determination of adenine and cytosine is based on a differential procedure; at low pH (about 2) only adenine gives a wave at the D.M.E., whereas at higher pH (about 5) both adenine and cytosine contribute to the cathodic current observed, since their $E_{1/2}$ values differ by less than 80 mV. With care, accuracy should be better than $\pm 2\%$ for adenine and $\pm 5\%$ for cytosine. Guanine is then determined independently in $2M$ H_2SO_4 by anodic voltammetry at a graphite electrode, where adenine and cytosine do not interfere. Guanine (0.2 to $1mM$) can be determined with about $\pm 4\%$ accuracy. Uracil and thymine do not interfere in any of the procedures.

It may be possible to determine thymine and uracil by applications of the analytically useful oscillopolarographic incision (263,264,266) for uracil in $1N$ NaOH ($Q_c = 0.12$; adenine, cytosine, and guanine give reversible incisions very close to anodic mercury dissolution, $Q = 0$, but might not interfere seriously) and for thymine in $2N$ ammonium formate ($Q_c = 0.11$;

however, adenine, guanine, and uracil may interfere). Alternatively, one could use the anodic adsorption waves of thymine and uracil at the D.M.E. (265,313) to determine the sum of the two, although it is not known whether adenine, cytosine, and guanine would interfere.

Thus, on the basis of the two preceding paragraphs, it may be possible to establish a completely voltammetric procedure for analyzing mixtures of all five of the important nucleic acid bases simply by changes in background electrolyte and voltammetric technique, without the need for prior separation or fractionation procedures.

As previously discussed, a procedure for the determination of 6-azauracil and 6-azauridine in mixtures of the two is based on differences in their diffusion current constants at two pH values (268).

Generally, the electrochemical characteristics of a nucleic acid base and its nucleosides and nucleotides are so similar that some type of differential procedure is necessary for their simultaneous determination in mixtures. Such procedures could probably be devised for the analysis of adenine-adenosine and cytosine-cytidine mixtures; for example, at pH 1.2 and 5.6, the diffusion current constants for adenine are approximately equal, but the diffusion current constant for adenosine at pH 5.6 has decreased to approximately 50% of its value at pH 1.2.

A 2-amino- or 2-hydroxy-substituted purine or pyrimidine, or pyrimidine itself, can be easily determined in the presence of 4-amino- or 4-hydroxy-substituted compounds, since the latter are much more difficult to reduce.

Purine can be determined in the presence of the other purines, except adenosine 1-N-oxide, 2-azaadenine, and 2-azaadenine 1-N-oxide. Since purine does not give a wave in alkaline solution, purine and pyrimidine can be determined in the presence of each other.

The voltammetric analysis of various mixtures of 6-thiopurines (6-thiopurine, purine-6-sulfinic and sulfonic acids, and purine-6-sulfonamide) has been described in detail (290). Analysis of mixtures containing between 0.2 and 0.8 mM of each component could be expected to achieve a relative accuracy on the order of 4 to 10% for each component.

A method has been reported (335a) for the determination of guanine in the presence of guanosine via oxidation at the pyrolytic graphite electrode.

9. Applications: Correlations Involving $E_{1/2}$

Correlations of $E_{1/2}$ with various molecular parameters—pK_a, MO energy levels, and Hammett σ values—for purines and pyrimidines have not been extensive, probably because of the great variety of these compounds and the various complicating factors, often unknown, which influence electrochemical behavior. Successful correlations and/or discussion of such correlations, however, have been presented for the reduction of various 6-substituted purines (69), of adenine nucleosides and nucleotides (70), and of 5-alkyl- and aryl-substituted 6-azauracils (269), as well as for the oxidation of a series of amino- and hydroxy-substituted pyrimidines (272).

A. CORRELATIONS WITH pK_a

The correlation of $E_{1/2}$ with acid dissociation constants rests on the twin premises that the ease of reducibility of a compound (within a fairly closely related series of compounds) is approximately inversely proportional to the electron density within the immediate vicinity of the reaction site and that the basicity of a compound, as measured by its pK_a, should as a first approximation be linearly proportional to the electron density at the reaction site. Thus, if reduction and protonation sites are identical (or at least closely related in some manner), the ease of reducibility as a first approximation should be inversely proportional to the acid dissociation constant, pK_a.

Since apparently only the protonated form of many purine derivatives is polarographically reducible (69,70), correlation of $E_{1/2}$ with acidic dissociation constants might be expected. Adenine (and presumably other 6-substituted purines) is first protonated at N(1), that is, at the most basic nitrogen (3,101,336), and a linear correlation would be expected on the basis that the initial potential-determining step is associated with reduction of the 1,6 N—C double bond. Six of the seven 6-substituted purines for which pK_a data are available show a reasonable linear correlation of $E_{1/2}$ and pK_a (69).

With the adenine nucleosides and nucleotides, the generally expected behavior of decreasing ease of reducibility with increasing pK_a is observed for Ado, d-Ado, AMP, and ATP, whereas adenine and d-AMP strongly deviate in being more easily reduced than expected for an inverse pK_a-$E_{1/2}$ correlation for the series (70). This is probably due in large part, if not entirely, to the more complex structure of the nucleosides and nucleotides, resulting in increased adsorption and greater possibility for intermolecular association.

The oxidation potentials for several amino- and hydroxy-substituted pyrimidines (272) appear to show no evident correlation with pK_a values, probably because of multiple protonation, adsorption phenomena, and possible multiplicity of (potential-determining) oxidation pathways among the different compounds.

B. CORRELATIONS WITH LINEAR FREE-ENERGY PARAMETERS

A fairly linear correlation (69) was obtained for a plot of $E_{1/2}$ versus the total polar substituent constant, σ_p, which is dependent on the kind and position of the substituent as well as the nature of the aromatic ring, for eight

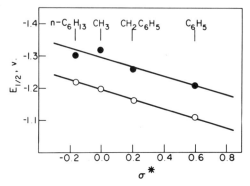

Figure 25. Variation of $E_{1/2}$ of N-substituted 6-aminopurines with the polar substituent constant, σ^* (substituent indicated). $E_{1/2}$ determined in McIlvaine buffer (open circles, pH 2.5; solid circles, pH 4.0)   Slope, ρ, is 0.14 V for pH 4.0 and 0.15 V for pH 2.5.

6-substituted purines; the slope, ρ, was 0.46 V. Plots of $E_{1/2}$ versus other constants, which are frequently used in modifications of the Hammett equation, did not result in reasonably linear correlations. A linear correlation was obtained, however, between $E_{1/2}$ and the polar substituent constant, σ^*, for four N'-substituted 6-amino purines (hexyl, methyl, benzyl, and phenyl); ρ^* is 0.14 V at pH 4.0 and 0.15 V at pH 2.5 (Figure 25).

The $E_{1/2}$ values for reduction at pH 8.3 for 6-azauracil and for six 5-alkyl- and 5-aryl-substituted homologs show a good linear correlation with Taft σ^* constants except for the t-butyl compound, where steric influences would be expected (269). Since the nucleophilic attack is at the 5,6 C—N double bond, such correlation would be expected. The effect of substitution is relatively small, that is, $\rho^* = 0.14$ V.

The positive ρ values for the reduction of azauracils and 6-substituted purines indicate that the mechanisms of the potential-determining step is a nucleophilic one (337), with an electron being the most probable nucleophilic agent. Thus any primary electrophilic attack, for example, protonation before electron transfer, has no significant role in the potential-determining step (338). Therefore any protonation before the electron-transfer step must be very rapid.

The $E_{1/2}$ values for the anodic adsorption waves of uracil and 6-methyl- and 6-carboxyuracil correlated with polar substituent constants, with $\rho = +0.40$ V (313). The $E_{1/2}$ for the similar waves of four 5,5'-dialkyl-substituted barbituric acids correlated linearly with the sum of the polar substituent constants of the substituents on the C(5), $\Sigma\sigma^*$; the effect of substitution was quite small, that is, $\rho = +0.07$ V (355).

C. CORRELATIONS WITH MOLECULAR ORBITAL PARAMETERS

A plot of $E_{1/2}$ values extrapolated to pH 0, for the oxidation of eight of the eleven amino- and hydroxypyrimidines for which the energy of the HOMO could be calculated, versus k_i gave a fairly good linear correlation (272), which fitted the relation

$$E_{1/2} = 0.467 + 1.708k_i \qquad [38]$$

Three of the compounds deviated markedly from this relation.

Janik and Elving (69) obtained a fair linear correlation between $E_{1/2}$ (reduction) and k_i for the LEMO of five 6-substituted purines; purine, 6-methylpurine, 6-methoxypurine, adenine, and isoguanine (Figure 26). At pH 2.5,

$$E_{1/2} = 0.34 + 1.72k_i \qquad [39]$$

At pH 4.0, the slope was about the same but the intercept was different. ATP ($E_{1/2} = -1.260$; $k_i = -0.913$) also falls on the line of [39], whereas ADP (-1.260, -0.985) and, especially, AMP (-1.235, -1.144) deviate markedly (70). Correlations with bond orders at selected pairs of atoms and net electronic charges on atoms were also investigated (Figure 27) (69).

It may be that the similarity in slopes for the two $E_{1/2}$ versus k_i relations for the oxidation of aminopyrimidines and for the reduction of 6-substituted purines is fortuitous, considering possible differences in reversibility, adsorption properties, and solvation and known differences in mechanisms for the two series of compounds and the two processes.

The $E_{1/2}$ values for the cytosine series (base, nucleosides, and nucleotides) are more negative than the $E_{1/2}$ values for the corresponding adenine deriva-

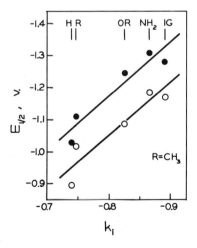

Figure 26. Variation of $E_{1/2}$ of 6-substituted purines (substituent indicated) with LEMO energies (k_i). IG = isoguanine. $E_{1/2}$ determined in McIlvaine buffer (open circles, pH 2.5; solid circles, pH 4.0).

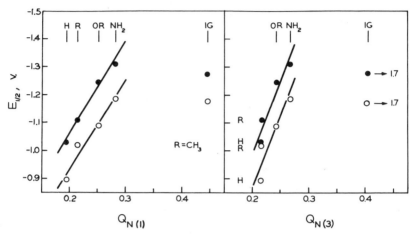

Figure 27. Variation of $E_{1/2}$ of 6-substituted purines (substituent indicated) with net negative charges on N(1) and N(3). IG = isoguanine. $E_{1/2}$ determined in McIlvaine buffer (open circles, pH 2.5; solid circles, pH 4.0).

tives, while LEMO calculations predict the opposite. Two of the assumptions implicit in correlating half-wave potentials with MO parameters is that the electrode reactions are reversible or nearly so and that solvation energies for the series of compounds studied are constant or vary in a regular manner. Alternating-current polarographic data (114) indicate that neither adenine nor cytosine is reduced reversibly and that, if the reductions are assumed to involve an identical number of electrons, adenine is reduced far more reversibly than cytosine. Hence the departure of $E_{1/2}$ from the order predicted by LEMO values would be less for adenine than for cytosine. Furthermore, it appears to be unlikely that the assumption involving the solvation energies would be true for the different purines and pyrimidines in aqueous solution because of the variation in nature, number, and strength of the hydrogen-bonding donor and acceptor groups in the different compounds and in the reduction products derived from them. The nice linear correlations between k_i and $E_{1/2}^{ox}$ and $E_{1/2}^{red}$ obtained for a large number of aromatic hydrocarbons (140–142) will probably never be so extensive in number for purines and pyrimidines for reasons of the complicating factors noted above.

10. Applications: Miscellaneous

A. CHEMICAL KINETICS

Polarographic techniques have long been used as an analytical tool for following the kinetics of chemical reactions. One advantage of polarography is that it can serve to monitor reactions in small volumes of solutions at low concentrations, thus allowing minute amounts of very expensive chemicals to be studied. For example, a 10-ml solution of $1mM$ concentration requires only 10^{-5} mole of compound; a typical voltammetric procedure consumes only 0.1 to 0.01% of the compound present, thus permitting many *in situ* analyses during the course of a reaction without essentially disturbing the system under study.

Oscillopolarography was used to determine the rates of ultraviolet-induced photodimerization of thymine and 5-hydroxymethyluracil (328). The radiolytic (γ-ray) splitting of cytosine was followed by means of D.M.E. polarography (339).

Polarographic techniques can be applied to the study of rates of attainment of adsorption and association equilibria of purine and pyrimidine species (306).

B. SYNTHESIS: MACROSCALE ELECTROLYSIS

In general, electrolytic methods do not appear to have been utilized to any great extent for the synthesis of purine and pyrimidine derivatives.

At zinc or cadmium cathodes, macroscale reduction of 2-amino-4-chloro-pyrimidine and 2-amino-4-chloro-6-methylpyrimidine produces the corresponding nonchlorinated derivative in almost quantitative yields (302). Theophylline can be prepared by the reduction of 8-chlorotheophylline in sodium hydroxide solution (340). The general procedure of dehalogenation reduction could be employed in some cases to prepare purines and pyrimidines difficult to obtain by other methods. The removal of a substitutent halogen atom in both saturated and unsaturated hydrocarbons by controlled-potential electrolysis generally proceeds smoothly without side reactions (341,342).

The macroscale oxidation of 6-thiopurine at graphite electrodes could be a useful synthetic route to purine-6-sulfinic and sulfonic acids, and particularly to the sulfonamide (299). The oxidation of 1 mole of 6-thiopurine in pH 9 ammonia buffer produces 0.5 mole of the sulfonamide, which can be prepared only with difficulty by other methods (343).

Since the $1e$ reduction of several pyrimidines leads to the formation of dimers (67), various bipyrimidines could be synthesized if a gentle oxidation method could be used on the dimers to effect a rearomatization without destroying the newly formed C—C' bond. Both 2,2'- and 4,4'-bipyrimidine are rather difficult to prepare by conventional methods (344,345).

C. ASSOCIATION, CONFORMATION, AND ORIENTATION

Recent studies (70,115) have indicated that polarography can be a sensitive method for the detection and evaluation of intermolecular association and orientation phenomena of biologically significant purine and pyrimidine compounds. Since this application of electrochemical techniques has been covered in Sections III.5 and V.7, the discussion will not be duplicated here.

VI. FLAVIN NUCLEOTIDES

1. Nomenclature

Flavin is the trivial name of the redox-active prosthetic group of a class of respiratory enzymes, that is, the flavin enzymes or flavoproteins, which occur widely in animals and plants. A prosthetic group is generally considered to be a cofactor firmly bound to the enzyme molecule; here the general term "cofactor" refers to factors which, in combination with an inactive protein (apoprotein), form an active enzyme. Cofactors may be divided rather loosely into three types: (a) prosthetic groups, (b) coenzymes, and (c) metal activators.

Many of the flavoproteins work directly on substrates, while others act as intermediates in the respiratory chain by bridging the electron-transport gap between the pyridine nucleotides, which they dehydrogenate, and the cytochromes, to which they transfer the hydrogen. In the latter case, the flavin enzymes constitute only part of a complex enzyme system that effects the cascade of hydrogen from a substrate to molecular oxygen, forming water. Two main groups of flavoproteins can be distinguished; one contains

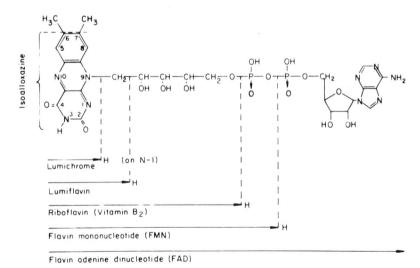

Figure 28. Formulas for the flavin nucleotide sequence. Configuration, dissociation equilibria, and tautomeric shifts are not shown.

iron (and sometimes also molybdenum) as an essential cofactor, whereas the other is free of metals.

The flavin group constitutes a part of the hydrogen-transferring coenzymes, that is, flavocoenzymes, which are linked with the apoproteins. Two flavoco-enzymes [flavin mononucleotide (FMN) and flavin-adenine dinucleotide (FAD)] are called flavin nucleotides, although this term is inappropriate, since these flavins are not N-glycosides of ribose, but derivatives of ribitol, that is, 6,7-dimethyl-9-D-ribitylisoalloxazine (Figure 28). Their most important structural component is riboflavin (vitamin B_2, 6,7-dimethyl-9-ribitylisoalloxazine), which occurs in nature almost exclusively as a constituent of FMN and FAD. The essential role of riboflavin undoubtedly stems from its importance in the production of the two nucleotides under physiological conditions.

2. Redox Activity

The isoalloxazine moiety constitutes a reversible redox system:

$$\tag{40}$$

which has been studied polarographically (357a). Recent reviews (358,359) cover rather thoroughly the chemistry and molecular biology of flavins and flavocoenzymes.

Flavins form thermodynamically reversible redox systems, irrespective of whether one or two electrons are being transferred per flavin molecule, thus fulfilling the criteria of a quinoid redox system (Figure 29). Depending on pH, one cationic, one neutral, and one anionic species can be distinguished for each redox state. The flavohydroquinone to flavoquinone conversion is readily produced by oxygen.

Flavins are very sensitive to light. On absorption of light, hydrogen is transferred from the hydroxylic side chain to the isoalloxazine ring with subsequent elimination of the side chain. Irradiation of riboflavin in alkaline solution gives the isoalloxazine lumiflavin (360); in acid solution the alloxazine, lumichrome, is obtained (361). Such irreversible photolysis can be prevented by the addition of external electron donors that are themselves more easily dehydrogenated by the light-activated flavin in its triplet state, so that only a photoreduction occurs, which is reversible on addition of oxygen.

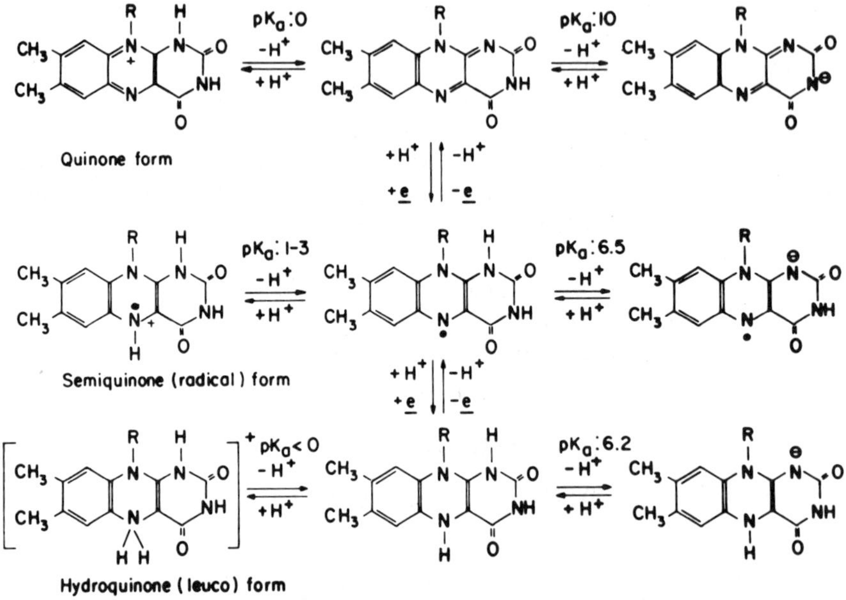

Figure 29. The flavin redox system: flavin species are shown as a function of redox state and pH [adapted from Hemmerich et al. [358]]. The pK_a values of the reduced, semiquinone, and totally oxidized forms of FMN have been reported as 6.72, 8.55, and 10.35, respectively; pK_a for the semiquinone form of riboflavin has been reported to be 8.27, rather than 6.5 as shown (372).

3. Polarographic and Voltammetric Behavior

All compounds derived from flavin, for example, lumiflavin, riboflavin, and the flavin nucleotides, which are in general N(9)-substituted isoalloxazines, form reversible redox systems, whose half-wave potentials are identical with the potentiometrically determined formal potentials (cf. ref. 15).

The literature up to about 1954 on the polarographic behavior of flavin nucleotides and the analytical application of such behavior has been largely reviewed by Brezina and Zuman (20). The polarography of isoalloxazine derivatives, as well as riboflavin and flavin nucleotides, has also been reviewed more recently (14,362).

A. RIBOFLAVIN AND RELATED COMPOUNDS

Riboflavin (vitamin B₂, lactoflavin), which is an important growth factor and an essential component of a number of flavin-containing enzymes, has been extensively studied potentiometrically (363–372) and polarographically, for example, by dc polarography at the D.M.E. (363,364,366–368,370,372), ac polarography (373), and chronopotentiometry (193,374).

Potentiometric studies (363,364,366–368,370,372) have shown that riboflavin is reduced in two overlapping $1e$ steps. Formation constants for the intermediate semiquinone were calculated from potentiometric (364,366–369,372) and polarographic data (375), using Michaelis' index potential method (368,376). Standard redox potentials for riboflavin and related compounds (364,365,367–370) and for a number of flavin derivatives, that is, 9-substituted isoalloxazines (364,368), have been determined at various pH values; critical summaries of these data are available (15,16).

Riboflavin (346,369,375,377–385) and compounds related to it (362,383, 385) exhibit a reversible $2e$ reduction wave over the whole pH range on dc polarography. Analysis of this wave has indicated, (346,375,377,382,386) as did the potentiometric studies, that polarographic reduction of these compounds involves two overlapping $1e$ steps with the intermediate formation of a semiquinone (Figure 29).

The pH dependence of $E_{1/2}$ for the riboflavin reduction wave (Figure 30) is linear with two changes in slope (293,312,316,348) at pH values which correspond to the acid dissociation constants of the reduced and oxidized forms (375,387) (the pK_a for riboflavin itself is 9.69 at 25°C). Equation [41] for $E_{1/2}$ versus N.C.E., which was

$$E_{1/2} = 0.188 + 0.029 \log \frac{5.01 \times 10^{-7} + [H^+]}{6.31 \times 10^{-10} + [H^+]} - 0.058 \text{ pH} \qquad [41]$$

calculated from dc polarographic data (375), fits fairly well the potentiometrically estimated formal potentials (363–370). The apparent semiquinone formation constant at 25° calculated from dc polarographic data (346), $K = 1.49$ at pH 7.4, also agrees well with potentiometric data. The diffusion coefficient, D, was calculated to be 7.41 and 5.64×10^{-6} cm^2/sec for 0.0475mM and 0.38mM solutions of riboflavin, respectively (377) (in the original paper the units are erroneously given as "mm/sec").

In interpreting Figure 30, it should be noted that the slope of each linear segment of the $E_{1/2}$-pH curve is related to the ratio of protons to electrons in the controlling electrode process in that pH region (cf. [1] and [2]). After an inflection, a change in the slope to a smaller value with increasing pH indicates that the pH at the inflection point pertains to the pK_a of the reduced form; if the slope changes to a higher value, the pK_a pertains to the oxidized form.

Although free riboflavin is readily reducible, protein-bound riboflavin is not reduced at the D.M.E. (388). From the curve for the titration of riboflavin with protein based on polarographic estimation of the riboflavin con-

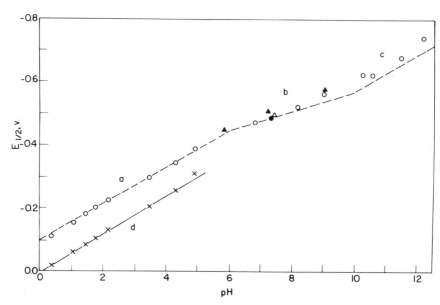

Figure 30. Variation with pH of $E_{1/2}$ for riboflavin. Potentials are versus N.C.E. (add 0.037 V to obtain potential versus S.C.E.). Reduction (main) wave: dashed line, concentration: 0.3mM (375,386); open circles, 0.475mM (387); solid circles (382); open triangles, 1mM (346); solid triangles (381). Adsorption prewave crosses, 0.475mM (387). Calculated (375,386) $d(E_{1/2})/d(pH)$: (a) 0.0582; (b) 0.0304; (c) 0.0590; (d) 0.0165. First inflection point is at pH 6.05, and second at pH 9.95.

centration, a molar ratio of 1:1 was calculated for the protein-riboflavin complex.

a. Adsorption of Riboflavin and Leucoriboflavin. In addition to the reduction wave mentioned, riboflavin and other flavin derivatives exhibit a reduction prewave in acid solution (346,358,375,377–380,382–384,387, 389–391). In very dilute riboflavin solutions (below 0.05mM), only one wave, whose height reaches a limiting value with increasing concentration (377) (limiting concentration not specified), is seen. At higher concentration, a second, more negative wave appears, whose height is directly proportional to concentration, while the height of the original wave remains unchanged.

To interpret the cause and behavior of the riboflavin prewave, as well as the prewave shown by methylene blue (392), an adsorption theory was developed by Brdička (377,386) (and subsequently generalized), in which it is assumed that the reduction product of the riboflavin (leucoriboflavin or semiquinone) is more strongly adsorbed than riboflavin on the mercury electrode; this results in the appearance of the prewave, since the reversible reduction of a compound to an adsorbed species requires less energy than its reduction to the same species in solution. Since pK_a for leucoriboflavin is 6.3, the anion of leucoriboflavin is the form mainly present in solutions of higher pH (Figure 29); furthermore, since the anion is more soluble and hence less strongly adsorbed than the uncharged molecule, no dc polarographic prewave is found under conditions in which no uncharged species is present (382), that is, above pH 7.3.

The kinetics of the adsorption were investigated by means of i-t curves (377), which exhibit two maxima in the potential region of the prewave; one was ascribed to the adsorption of the semiquinone and the other to that of the leuco form. From the shapes of the i-t curves, delayed establishment of adsorption equilibrium was assumed for the leuco form and was explained as being due to the conversion of an adsorption-active form of the reduction product into an adsorbable form (the difference between the two forms was not made clear); this change was considered to be probably autocatalytic. The adsorption prewave is not suppressed by the addition of urea but disappears on the addition of pyridine (389) (presumably because of the stronger adsorption of pyridine on the electrode).

Adsorption of leucoriboflavin has been confirmed by the use of ac polarography (373,393). A depression of the base current on one or both sides of the ac reduction wave of riboflavin (E_s = about -0.05 and -0.25 V in $0.1M$ $HClO_4$ and pH 4.7 acetate buffer, respectively) is observed up to pH 9.5. These results were interpreted to indicate that slight adsorption of leucoriboflavin on the D.M.E. persists even at a pH as high as 9.5 and, moreover, that both the uncharged and anionic forms of riboflavin are appreciably adsorbed. Adsorption of riboflavin is accompanied by the appearance of an ac polarographic wave at pH 3 and above, whose E_s shifts from 0.28 V at pH 3 to -0.1 V in $0.1M$ NaOH and whose height increases with increasing riboflavin concentration to a limiting value at about $0.06mM$.

No dc polarographic wave corresponding to the ac wave was found, although the residual current in the presence of riboflavin, at potentials more positive than that of the riboflavin reduction wave, is slightly below the value observed for supporting electrolyte alone. Moreover, in the potential region corresponding to this ac wave, a slight kink could often be observed on the dc polarograms. The behavior of the ac wave was interpreted as being due to the formation of an insoluble covalent compound with mercury by the adsorbed riboflavin molecules.

Adsorption of riboflavin and of leucoriboflavin at the D.M.E. was also studied by chronopotentiometry (193,374,390,391,394) and by measurement of the DL capacity (395) and electrocapillary curves (193). The adsorption of both compounds was confirmed (193,374), with the leuco form being the more strongly adsorbed. The surface excess values (Γ) are of the order of 0.2×10^{-9} mole/cm^2 for both (373,395), which agrees well with the value of 0.14×10^{-9} mole/cm^2 obtained from dc polarographic data (377). However, only limited physical significance can be attributed to these values, since the assumption that all of the product remains on the electrode surface is obviously not correct (395); in addition, the method of calculating (193) the extent of adsorption of electroactive riboflavin onto the mercury electrode from chronopotentiometric data has been questioned (396).

A chronopotentiometric study (390) indicated that, in the potential region of the riboflavin adsorption prewave, the D.M.E. surface is covered by riboflavin and that the adsorption (relative coverage of electrode surface?) of the leucoriboflavin produced on reduction is less than the calculated value. In connection with this finding and the preceding discussion, the point has been made that chronopotentiometry has only limited applicability to the study of adsorption (397).

b. Catalytic Wave. In addition to the waves mentioned, riboflavin (398,399), FMN (399), and FAD (398) exhibit up to pH 8 a catalytic hydrogen wave in the potential region between the normal reduction wave and the background discharge. The catalytic wave height at equimolar depolarizer concentration decreases in the following order (relative wave heights shown in parentheses): FAD (6) > riboflavin (2) > FMN (1). The catalytic wave height for isoalloxazine derivatives without the sugar moiety is about one tenth of that of riboflavin; lumichrome, which is the product of photolytic degradation of riboflavin, does not give the catalytic wave (398).

A second catalytic wave, which appears (potential region not specified) in riboflavin solutions above 3mM, is independent of the first one (398).

It has been suggested that the protonated reduced forms of riboflavin, FAD, and FMN in their adsorbed states are the catalytically active species (399).

c. Behavior in Dimethyl Sulfoxide. The electrochemistry of riboflavin and related compounds such as lumiflavin and lumichrome has also been studied in the aprotic solvent dimethyl sulfoxide (DMSO) (400), in order to avoid the complications normally seen in aqueous solutions, for example, the existence of a number of protonated and deprotonated forms of electrode participants, as well as the strong adsorption of the various forms at the mercury-solution interface. Riboflavin is reduced in DMSO in a 1e reduction step to the riboflavin anion radical, which is not stable but decomposes by two parallel reactions. A definite assignment of the products and mechanisms of the following chemical reactions was not possible (400).

B. FLAVIN NUCLEOTIDES

The polarographic behavior of FMN and FAD in aqueous solutions is similar to that of riboflavin; a $2e$ reduction wave produced on dc polarography is also due to two overlapping $1e$ steps, indicating semiquinone formation (387,401–404). The $E_{1/2}$ values for FMN and FAD (402) correspond to the potentiometrically determined standard redox potentials of the thermodynamically reversible systems (364,372). The $E_{1/2}$–pH plots are linear, with inflection points (Figure 31) corresponding to the acid dissociation constants of the reduced and oxidized forms (387,402). Semiquinone formation constants, calculated from the polarographic data by Michaelis' index potential method (402,404), increase with increasing concentration (402) and change with pH (404); for example, $\log K = -2$ in pH 2.0 citrate-phosphate buffer $+ 0.9N$ KNO_3, -0.2 in $0.1N$ $HNO_3 + 0.9N$ KNO_3, and 0.8 in $1N$ HNO_3.

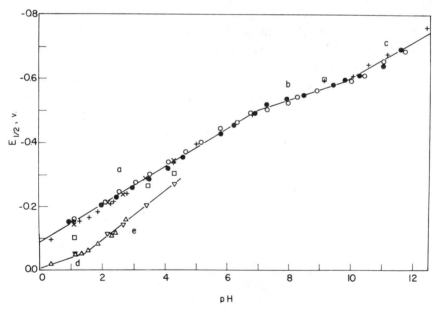

Figure 31. Variation of $E_{1/2}$ for FMN and FAD with pH. Potentials are versus N.C.E. (add 0.037 V to obtain potential versus S.C.E.). Reduction (main) wave of FAD: solid circles (402); pluses, concentration: 0.46mM (387); crosses, 0.092mM (387); squares, 0.0184mM (387). Adsorption prewave of FAD: triangles, 0.46 mM (387); reversed triangles, 0.092mM (387). Calculated $d(E_{1/2})/d(\mathrm{pH})$: (a) 0.0597; (b) 0.0310; (c) 0.0568; (d) 0.0346; (e) 0.0772. First inflection point in upper curve is at pH 7.0, and second at pH 9.75; inflection point in lower curve is at pH 1.5.

a. Adsorption Prewave. Like riboflavin, FMN and FAD show an adsorption prewave at the D.M.E. in acidic solution (362,387,401,404), which has been interpreted by Brdička's adsorption theory on the basis that the reduced form is strongly adsorbed (at least more strongly than the oxidized form) on the mercury. The sum of the heights of the main reduction wave and the prewave is independent of pH and is directly proportional to concentration, with the height of the prewave becoming constant at about 0.06mM (387,404).

The dc polarographic results have been unambiguously confirmed by ac polarography (404–406) and by cyclic voltammetry and chronopotentiometry (401,407).

Alternating-current polarograms of FMN and FAD in acidic solution exhibit two peaks, whose summit potentials correspond quite well to the $E_{1/2}$ values of the dc adsorption prewave and main reduction wave (404–406) (Figure 32); the potentials and heights of the two peaks remain constant for different ratios of oxidized to reduced forms of FMN in the bulk solution (404). The ac adsorption wave of FMN is several times greater than the main ac reduction wave; at very low FMN concentration, only the adsorption wave is observed, whose height is proportional to concentration even though the height is almost 10 times that expected for a simple inorganic redox system (404) (presumably for a reversible 1e faradaic process). An equation for an ac polarogram involving a reversible redox system with an intermediate semiquinone step and strong adsorption of the reduced form at the D.M.E. has been theoretically derived (404); on the basis of the degree of conformity of behavior to this equation, the flavin redox system does not seem to be completely reversible by ac polarography.

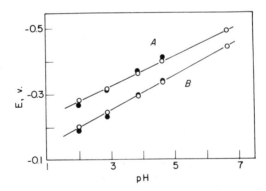

Figure 32. Comparison of dc and ac polarographic data for 0.2mM FMN (404). Potentials given are versus N.C.E. (add 0.037 V to obtain potential versus S.C.E.). (A) Reduction (main) wave. (B) Absorption prewave. Open circles, $E_{1/2}$; solid circles, E_s. Calculated $d(E_{1/2})/d(\text{pH})$: (a) 0.0458; (b) 0.0520.

b. Adsorption Postwave. In neutral and alkaline solutions of FMN and FAD, where the prewave is not observed, a dc polarographic wave appears at more negative potential than the main reduction wave (probably close to the background discharge, since this postwave merges with background discharge below pH 5.6) and is accompanied by abnormal current oscillations (402). The FAD postwave has a slightly more positive $E_{1/2}$ and a slightly smaller height than the corresponding FMN postwave at identical concentrations (402). The FMN postwave appearing in basic solution has also been described as ill defined; a corresponding ac postwave is seen at a potential 15 mV more negative than that for the main ac wave (404).

At very low concentration, only the postwave is observed. When both the main wave and the postwave appear, their total height is proportional to concentration; the height of the postwave is independent of concentration (apparently above a certain limit) and directly proportional to h, indicating its adsorption character. The diffusion current constants for the total wave heights of FMN and FAD (main wave plus postwave) are 1.22 and 1.60, respectively (402); the ratio of 1.31 for the square roots of the diffusion coefficients agrees only moderately well with the value of 1.10 for the similar ratio calculated on the basis of the Stokes-Einstein equation (24).

On raising the temperature from 25° to 45°, the postwave almost disappears, while the height of the main wave increases by about 50% (this is equivalent to a 2% increase per degree) and its $E_{1/2}$ becomes 25 mV more negative.

The appearance and behavior of the postwave were explained as being due to the strong adsorption of the oxidized form, that is, FMN or FAD, at the D.M.E. (402,404).

Alternating-current polarography of FMN in neutral and slightly basic solution (404) revealed, in agreement with the behavior of riboflavin (408), the adsorption of not only the reduced form but also FMN itself. An anodic wave observed at about 0.0 V is probably due to oxidation of the mercury, involving formation of an insoluble mercury-nucleotide compound (404).

In pH 7 phosphate buffer FMN exhibits a very high wave, accompanied at more negative potential by a small wave (about one tenth of the height of the former wave), on both the cathodic and anodic sweeps on oscillographic square-wave polarography (409). [In the latter technique, cathodic and anodic processes can be observed continuously as the applied dc voltage sweeps in triangular pattern between 0 and −2 V with a superimposed alternating square-wave voltage of small amplitude, while synchronized to the mercury drop (433).] The summit potentials of the two waves on the cathodic and anodic sweeps almost coincide, and they correspond, respectively, to the summit potentials of the adsorption and main reduction waves seen on ac polarography.

4. Electrochemically Calculated Molecular Areas

The areas occupied by the adsorbed molecules of riboflavin and FAD on mercury electrodes were calculated (method not specified) to be 117 and 280 $\mathring{A}^2$, respectively (387). Since these values are close to the maximum cross sections of these molecules calculated from molecular models, it was suggested (387) that the molecules form a monomolecular layer on the electrode with their planes parallel to the mercury surface.

Recent chronopotentiometric and cyclic voltammetric measurements (401,407) indicate that the adsorption of FMN is not a simple process; experiments on the hanging mercury drop, which allows greater time for the attainment of adsorption equilibrium, show the disappearance of the prewave (401). This behavior has been attributed (401) to slow adsorption of FMN, implying the reorientation of the originally adsorbed FMN into a stable film, in which the ring system is parallel to the electrode.

Equilibrium surface coverage calculated (401,407) from chronopotentiometric data, assuming the applicability of several possible adsorption mechanisms and using a method of graphical evaluation (192), is 0.7×10^{-10} mole/cm^2 for FMN and 1.22×10^{-10} mole/cm^2 for its reduced form; these values correspond to areas of 135 and 409 $\mathring{A}^2$, respectively. However, these data must be accepted very cautiously, because, as previously mentioned, use of chronopotentiometry for the study of adsorption involves uncertainties (397).

5. Complexation

Since riboflavin, FMN, and FAD form polarographically reversible redox systems, polarography can be used to study their complexation with cations and other chemical species. Thus investigation by dc polarography and chronopotentiometry (410) of mixtures of iron(III) and FMN in neutral oxalate medium indicated that two Fe(III) ions combine with one FMN molecule to form a stable complex, which produces a kinetically controlled Fe(III) reduction wave because of its dissociation.

6. Biological Fuel Cells

Enzymatic redox systems have been examined in biological fuel cells in conjunction with an oxygen cathode (411). The redox mechanism involves electron transfer in the case of flavoprotein enzymes and hydrogen transfer in the case of pyridine nucleotide enzymes, so that only the former system produces a voltage in the biological fuel cell. Elemental iron promotes the flavoprotein systems, that is, the glucose oxidase and the D-amino acid oxidase systems, causing an increase in voltage output from 175–350 mV to 635–750 mV. The promoting effect of iron was considered to be due to an increased rate of oxidation of FADH to FAD, coupled with the greater oxidation potential of this element.

7. Analytical Application

The monograph by Brezina and Zuman (20), as previously mentioned, contains considerable useful information on the analytical application of the polarographic behavior of riboflavin.

A number of procedures have been published for the polarographic determination of riboflavin, mostly in pharmaceutical preparations (223,228, 375,381,412–415). The fact that the reduction wave of riboflavin occurs at relatively positive potential in acidic and neutral media (i.e., in the region of -0.1 to -0.5 V) allows it to be used for the determination of riboflavin in mixtures with other polarographically reducible vitamins, such as thiamine, nicotinic acid, panthotenic acid and pyridoxine (381), folic acid (223,413, 416), and ascorbic acid (223,228,416). The polarographic determination of riboflavin is preferred (20,389,414,415) to other analytical methods, because it is quicker than the current fluorimetric, microbiological, and spectrophotometric methods and has approximately the same or better accuracy. Riboflavin has also been detected oscillopolarographically in drugs (417).

Polarography has been successfully used for the determination of riboflavin in food (418), in milk (419), and in tissues and biological fluids (420,421), as well as for following the formation of riboflavin in butylogenous *Clostridium* cultures (422) and the change in riboflavin concentration of fruit juices during storage (423).

Riboflavin can be determined in pharmaceutical preparations and biological material by means of an ac polarographic method in concentrations as low as $2 \times 10^{-7}M$ (408).

The catalytic wave observed on dc polarography of riboflavin and FAD solutions can be used to determine the latter compounds in solutions as dilute as $10^{-8}M$ (399). A similar determination of riboflavin down to $10^{-6}M$ has also been reported (398).

The possible use of thin-layer electrophoresis for the identification of vitamins, including riboflavin in polyvitamin mixtures, has been considered (424).

The potential application of polarographic and oscillopolarographic methods for the determination of intermediates in the synthesis of riboflavin has been discussed (425). In addition, the photolysis of riboflavin under various conditions has been followed polarographically (377–380,382,389, 426–431), and some of the products have been identified and studied by the same technique, for example, lumichrome (375,378,380–385), lumiflavin

TABLE XI
Electrochemical Symbols and Acronyms

A	Ampere
A	Area of an electrode.
ac	Alternating current.
α	Alpha; transfer coefficient involved in heterogeneous electrode kinetics.
AR, SR	An acronym referring to a type of electrochemical process in which an adsorbed species is first reduced, and then the species in the bulk of the solution is reduced.
C	Concentration of electroactive species; usually expressed in units of millimoles per liter.
D	Diffusion coefficient, usually of an electroactive species, in square centimeters per second.
dc	Direct current.
DL	The double layer, generally referring to that around an indicating electrode.
D. M. E.	The dropping mercury electrode.
E	Potential in volts.
E^0	Standard or normal potential, defined for unit activity.
$E_c{}^0$	Standard potential, defined for unit concentration; also indicated as $E^{0\prime}$ and E_0'.
$E_{1/2}$	Half-wave potential; the potential at which the current for a polarographic wave reaches half of its limiting value.
E_p	Peak potential; the potential at which the current obtained on linear scan or cyclic voltammetry at a constant-area stationary electrode reaches a maximum, thus producing a peak.
E_s	Summit potential; the potential at which the current obtained on ac polarography reaches a maximum, thus producing a peak.
ECE	An acronym referring to a type of electrochemical process in which there occur first an electrochemical step (electron transfer), then a chemical step, and finally a second electrochemical step.
ecm	Electrocapillary maximum; the potential at which the droptime of the D. M. E. is at a maximum.
η	Eta; overpotential or difference between the applied potential and the reversible potential for the half-reaction system involved.
F	Faraday; 1 mole of electrons; 96,454 coulombs.
H. M. D. E.	Hanging mercury drop electrode.
HOMO	Highest occupied molecular orbital.
I	Diffusion current constant; $I = i_d/Cm^{2/3}t^{1/6}$.
i	Current, usually expressed in microamperes.
i_d	The (average) diffusion current in microamperes obtained at the D. M. E.
i_l	Limiting current in microamperes measured at the D. M. E.; may or may not be corrected for the residual or background current.

(Continued)

TABLE XI (*continued*)

i_p	Peak current in microamperes obtained at a stationary electrode in linear scan voltammetry.
Δi_s	Peak current in microamperes at the summit potential on ac polarography.
LEMO	Lowest empty molecular orbital.
m	Mercury flow rate at the D. M. E. in milligrams per second.
μA	Microamperes, 10^{-6} ampere.
mA	Milliamperes, 10^{-3} ampere.
MO	Molecular orbital.
mV	Millivolts, 10^{-3} volt.
mM	Millimolar; the usual concentration unit for the electractive species.
n	The number of electrons transferred as expressed in the half-reaction equation for the faradaic process, which results in current flow.
n_a	The number of electrons involved in the potential-determining step of the faradaic process; is generally one and is, in any event, not greater than n for the faradaic process.
N. C. E.	Normal calomel reference electrode.
N. H. E.	Normal hydrogen reference electrode.
nm	Nanometers, 10^{-9} meter; the recommended unit for wavelength in the ultraviolet region, which has replaced millimicrons (mμ), the former unit.
ω	Omega; speed (angular frequency) or a rotating electrode in radians per second; equals $2\pi f$, where f is the number of revolutions per second.
P. G. E.	Pyrolytic graphite electrode.
Q	Represents in oscillopolarography the ratio (or quotient) of (distance of the incision from the potential of anodic mercury dissolution)/(distance between the latter potential and that of cathodic background discharge).
R	Resistance in ohms.
R. P. E.	Rotating platinum microelectrode.
rms	The rms (root-mean-square) alternating current or voltage is the square root of the instantaneous currents or voltages that have been squared and averaged over one cycle: $i_{rms} = \sqrt{2}/2 i_{peak} = 0.707 i_{peak}$
S. C. E.	Saturated calomel reference electrode.
S. G. E.	Stationary graphite electrode; usually refers to some form of wax-impregnated spectroscopic graphite rod.
t	Droptime at the D. M. E. in seconds.
TMAC	Tetramethylammonium chloride background electrolyte.
TBAI	Tetra-n-butylammonium iodide background electrolyte.
TEAC	Tetraethylammonium chloride background electrolyte.
V	Volts.
V	Amplitude of the applied alternating voltage in ac polarography.
v	Rate of potential variation; scan, sweep, or polarization rate in volts per seconds.

453

(389), rhodoflavin (427), deuterioriboflavin (428), and deuteriolumiflavin (428). Polarographic data were used to establish the reversibility of the photoreduction system involving riboflavin (382); the photoreduction can be inhibited by oxygen, certain metal ions, dinitrophenol, p-nitrobenzaldehyde, and sodium azide. Oxygen also protects riboflavin from photolysis (382). Photochemical oxidation of EDTA and methionine by riboflavin was also followed polarographically (432).

Acknowledgments

The authors thank the National Science Foundation, which is supporting the investigation of the electrochemical behavior of biologically significant compounds in connection with which the present review was prepared. One of the authors (J.E.O.) is grateful to the National Science Foundation for a Predoctoral Fellowship.

References

1. R. N. Adams, *Electrochemistry at Solid Electrodes*, Dekker, New York, 1969.
2. P. J. Elving, I. Fried, and W. R. Turner, in *Polarography 1964*, G. J. Hills, Ed., Interscience, New York, 1966, pp. 227–297.
3. B. Pullman and A. Pullman, *Quantum Biochemistry*, Interscience, New York, 1963.
4. G. Dryhurst and P. J. Elving, *J. Electrochem. Soc.*, *115*, 1014 (1968).
5. W. A. Struck and P. J. Elving, *Biochemistry*, *4*, 1343 (1965).
6. E. Palecek, in *Progress in Nucleic Acid Research and Molecular Biology*, Vol. 9, J. N. Davidson and W. E. Cohn, Eds., Academic Press, New York, 1969, pp. 31–73.
7. H. Berg and H. Bär, *Monatsber. Deut. Akad. Wiss. Berlin*, *7*, 210 (1965).
8. H. Berg, H. Bär, and F. A. Gollnick, *Biopolymers*, *5*, 61 (1967).
9. *Bibliography of Polarographic Literature 1922–1967*, Sargent-Welch Scientific Co., Skokie, Ill., 1969.
10. J. Heyrovsky, *Bibliography of Publications Dealing with the Polarographic Method*, Czechoslovak Academy of Science, Prague. (Latest issue available to the authors was issued in 1966.)
11. Centro de Studio per la Polarografia (Padua), *Bibliografia Polarografia*, Consiglio Nazionale delle Ricerche, Rome. (Latest issue available to the authors was issued in 1968.)
12. P. Zuman, *Organic Polarographic Analysis*, Pergamon Press, New York, 1964.
13. P. J. Elving, W. A. Struck, and D. L. Smith, *Mises Point Chim. Anal. Org. Pharm. Bromatol.*, *14*, 141 (1965).
14. B. Janik and P. J. Elving, *Chem. Rev.*, *68*, 295 (1968).
15. W. M. Clark, *Oxidation-Reduction Potentials of Organic Systems*, Williams & Wilkins, Baltimore, 1960.
16. J. Volke, in *Physical Methods in Heterocyclic Chemistry*, Vol. I, A. R. Katritzky, Ed., Academic Press, New York, 1963, pp. 217–323.
17. J. Volke, *Talanta*, *12*, 1081 (1965).
18. V. Vetterl, *Biophysik*, *5*, 255 (1968).
19. V. Vetterl, *J. Electroanal. Chem.*, *19*, 169 (1968).

20. M. Brezina and P. Zuman, *Polarography in Medicine, Biochemistry and Pharmacy*, Interscience, New York, 1958.
21. O. H. Müller, in *Methods of Biochemical Analysis*, Vol. 11, D. Glick, Ed., Interscience, New York, 1963, pp. 329–403.
22. J. Homolka, *Polarography of Proteins and Its Clinical Applications*, Czechoslovak Medical Press, Prague, 1964.
23. B. Pullman and A. Pullman, *Progress in Nucleic Acid Research and Molecular Biology*, Vol. 9, J. N. Davidson and W. E. Cohn, Eds., Academic Press, New York, 1969, pp. 327–402.
23a. A. L. Underwood and R. W. Burnett, in *Electroanalytical Chemistry*, Vol. 6, A. J. Bard, Ed., Dekker, New York, 1972, Chap. 1.
23b. R. M. Izatt, J. J. Christensen, and J. H. Rytting, *Chem. Rev.*, *71*, 439 (1971).
24. I. M. Kolthoff and J. J. Lingane, *Polarography*, 2nd ed., Interscience, New York, 1952.
25. L. Meites, *Polarographic Techniques*, 2nd ed., Interscience, New York, 1965.
26. L. Meites, in *Treatise on Analytical Chemistry*, I. M. Kolthoff and P. J. Elving, Eds., Part I, Vol. 4, Interscience, New York, 1963, pp. 2303–2379.
27. O. H. Müller, in *Physical Methods of Organic Chemistry*, A. Weissberger, Ed., Part IV, 3rd ed., Interscience, New York, 1960, pp. 3155–3379.
28. C. L. Rulfs and P. J. Elving, in *Encyclopedia of Electrochemistry*, C. A. Hampel, Ed., Reinhold, New York, 1964, pp. 944–963.
29. Symposium on Operational Amplifiers, *Anal. Chem.*, *35*, 1770 (1963).
30. R. Bezman and P. S. McKinney, *Anal. Chem.*, *41*, 1560 (1969).
31. R. S. Nicholson, *Anal. Chem.*, *37*, 1351 (1965).
32. R. S. Nicholson, *Anal. Chem.*, *38*, 1406 (1966).
33. R. S. Nicholson and I. Shain, *Anal. Chem.*, *36*, 706 (1964).
34. R. S. Nicholson and I. Shain, *Anal. Chem.*, *36*, 1212 (1964).
35. R. S. Nicholson and I. Shain, *Anal. Chem.*, *37*, 178 (1965).
36. R. S. Nicholson and I. Shain, *Anal. Chem.*, *37*, 190 (1965).
37. R. S. Nicholson, J. M. Wilson, and M. L. Olmstead, *Anal. Chem.*, *38*, 542 (1966).
38. M. L. Olmstead and R. S. Nicholson, *Anal. Chem.*, *38*, 150 (1966); *41*, 862 (1969).
39. G. L. Booman and D. T. Pence, *Anal. Chem.*, *37*, 1366 (1965); *41*, 746 (1969).
40. D. S. Polcyn and I. Shain, *Anal. Chem.*, *38*, 370 (1966).
41. R. L. Myers and I. Shain, *Anal. Chem.*, *41*, 980 (1969).
42. B. Kastening, *Anal. Chem.*, *41*, 1142 (1969).
43. M. H. Hulbert and I. Shain, *Anal. Chem.*, *42*, 162 (1970).
44. M. S. Shuman, *Anal. Chem.*, *42*, 521 (1970).
45. J. M. Saveant and E. Vianello, *Electrochim. Acta*, *12*, 1545 (1967).
46. M. L. Olmstead, R. G. Hamilton, and R. S. Nicholson, *Anal. Chem.*, *41*, 260 (1969).
47. M. Mastragostino, L. Nadjo, and J. M. Saveant, *Electrochim. Acta*, *13*, 721 (1968).
48. B. Breyer and H. H. Bauer, *Alternating Current Polarography and Tensammetry*, Interscience, New York, 1963.
49. D. E. Smith, in *Electroanalytical Chemistry*, Vol. 2, A. J. Bard, Ed., Dekker, New York, 1966, pp. 21–155.
50. B. Breyer, *Pure Appl. Chem.*, *15*, 313 (1967).
50a. D. E. Smith, *Crit. Rev. Anal. Chem.*, *2*, 247 (1971).
51. B. Timmer, M. Sluyters-Rehbach, and J. H. Sluyters, *J. Electroanal. Chem.*, *14*, 169 (1967).
52. B. Timmer, M. Sluyters-Rehbach, and J. H. Sluyters, *J. Electroanal. Chem.*, *14*, 181 (1967).

53. D. E. Smith and T. G. McCord, *Anal. Chem.*, *40*, 474 (1968).
54. R. Kalvoda, *Techniques of Oscillographic Polarography*, 2nd ed., Elsevier, Amsterdam, 1965.
55. J. Bohacek and E. Palecek, *Collect. Czech. Chem. Commun. 30*, 3456 (1965).
56. E. Palecek, *Biochim. Biophys. Acta*, *94*, 293 (1965).
57. E. Palecek, *J. Mol. Biol.*, *11*, 839 (1965).
58. E. Palecek, *Collect. Czech. Chem. Commun.*, *31*, 2360 (1966).
59. P. Zuman, *J. Polarog. Soc.*, *13*, 53 (1967).
60. L. Meites, *Rec. Chem. Progr.*, *22*, 81 (1961).
61. L. Meites, in *Physical Methods of Organic Chemistry*, 3rd ed., A. Weissberger, Ed., Interscience, New York, 1960, pp. 3281–3333.
62. C. L. Perrin, in *Progress in Physical Organic Chemistry*, Vol. III, S. G. Cohen et al., Eds., Interscience, New York, 1963, pp. 165–316.
63. P. J. Elving, *Pure Appl. Chem.*, *7*, 423 (1963).
64. W. R. Turner and P. J. Elving, *J. Electrochem. Soc.*, *112*, 1215 (1965).
65. H. Lund, *Acta Chem. Scand.*, *11*, 1323 (1957).
66. P. J. Elving and B. Pullman, in *Advances in Chemical Physics*, Vol. III, I. Prigogine, Ed., Interscience, New York, 1961, pp. 1–31.
67. D. L. Smith and P. J. Elving, *J. Am. Chem. Soc.*, *84*, 2471 (1962).
68. J. E. O'Reilly and P. J. Elving, *J. Electroanal. Chem.*, *21*, 169 (1969).
69. B. Janik and P. J. Elving, *J. Electrochem. Soc.*, *116*, 1087 (1969).
70. B. Janik and P. J. Elving, *J. Am. Chem. Soc.*, *92*, 235 (1970).
71. R. Brdička, V. Hanus, and J. Koutecky, in *Progress in Polarography*, P. Zuman, Ed., Interscience, New York, 1962, pp. 145–199.
72. A. Bewick, M. Fleischmann, and J. N. Hiddleston, *Polarography 1964*, G. J. Hills, Ed., Interscience, New York, 1966, pp. 57–78.
73. J. W. Ashley and C. N. Reilley, *J. Electroanal. Chem.*, *7*, 253 (1964).
74. B. Janik and P. J. Elving, *J. Electrochem. Soc.*, *117*, 457 (1970).
75. P. G. Grodzka and P. J. Elving, *J. Electrochem. Soc.*, *110*, 225 (1963).
76. P. G. Grodzka and P. J. Elving, *J. Electrochem. Soc.*, *110*, 231 (1963).
77. M. Suzuki and P. J. Elving, *J. Phys. Chem.*, *65*, 391 (1961).
78. M. Suzuki and P. J. Elving, *Collect. Czech. Chem. Commun.*, *25*, 3202 (1960).
79. W. H. Reinmuth, *Anal. Chem.*, *40*, 185R (1968).
80. G. C. Barker, in *Polarography 1964*, G. J. Hills, Ed., Interscience, New York, 1966, pp. 25–47.
81. P. Delahay, in *Advances in Electrochemistry and Electrochemical Engineering*, Vol. I, P. Delahay and C. Tobias, Eds., Interscience, New York, 1961, pp. 233–318.
82. K. J. Vetter, *Elektrochemische Kinetik*, Springer-Verlag, Berlin, 1961.
83. K. J. Vetter, *Electrochemical Kinetics: Theoretical and Practical Aspects*, Academic Press, New York, 1967.
84. K. J. Vetter, *Electrochemical Kinetics: Theoretical Aspects*, Academic Press, New York, 1967.
85. B. B. Damaskin, *Principles of Current Methods for the Study of Electrochemical Reactions*, McGraw-Hill, New York, 1967.
86. S. G. Mairanovskii, *Electrochim. Acta*, *9*, 803 (1964).
87. G. E. Cheney, H. Freiser, and Q. Fernando, *J. Am. Chem. Soc.*, *81*, 2611 (1959).
88. T. R. Hawkins and H. Freiser, *J. Am. Chem. Soc.*, *80*, 1132 (1958).
89. I. Bayer, E. Posgay, and P. Majlat, *Pharm. Zentralhalle*, *101*, 476 (1962).
90. W. Lorenz, *Z. Elektrochem.*, *62*, 192 (1958).
91. P. Delahay, *Double Layer and Electrode Kinetics*, Interscience, New York, 1965.

92. E. Gileadi, Ed., *Electrosorption*, Plenum, New York, 1967.
93. A. N. Frumkin and B. B. Damaskin, in *Modern Aspects of Electrochemistry*, Vol. 3, J. O'M. Bockris, Ed., Butterworths, London, 1964, pp. 149–223.
94. R. Parsons, *J. Electroanal. Chem.*, *8*, 93 (1964).
95. S. G. Mairanovskii, *J. Electroanal. Chem.*, *4*, 166 (1962).
96. D. M. Mohilner, in *Electroanalytical Chemistry*, Vol. 1, A. J. Bard, Ed., Dekker, New York, 1966, pp. 241–409.
97. C. N. Reilley and W. Stumm, in *Progress in Polarography*, P. Zuman, Ed., Interscience, New York, 1962, pp. 81–121.
98. C. A. Barlow and J. R. MacDonald, in *Advances in Electrochemistry and Electrochemical Engineering*, Vol. VI, P. Delahay and C. Tobias, Eds., Interscience, New York, 1967, pp. 1–199.
99. R. Parsons, in *Advances in Electrochemistry and Electrochemical Engineering*, Vol. I, P. Delahay and C. Tobias, Eds., Interscience, New York, 1961, pp. 1–64.
100. G. J. Gleicher and M. K. Gleicher, *J. Phys. Chem.*, *71*, 3693 (1967).
101. H. Berthod, C. Geissner-Prettre, and A. Pullman, *Theor. Chim. Acta*, *5*, 53 (1966).
102. E. S. Pysh and N. C. Yang, *J. Am. Chem. Soc.*, *85*, 2124 (1963).
103. J. W. Sease, F. G. Burton, and S. L. Nickol, *J. Am. Chem. Soc.*, *90*, 2595 (1968).
104. W. W. Hussey and A. J. Diefenderfer, *J. Am. Chem. Soc.*, *89*, 5359 (1967).
105. R. Zahradnik and C. Parkanyi, *Talanta*, *12*, 1289 (1965).
106. I. Rosenthal, C. H. Albright, and P. J. Elving, *J. Electrochem. Soc.*, *99*, 227 (1952).
107. P. Zuman, *Substituent Effects in Organic Polarography*, Plenum, New York, 1967.
108. A. Streitwieser, *Molecular Orbital Theory for Organic Chemists*, Wiley, New York, 1961.
109. P. J. Elving, in *Organic Analysis*, Vol. II, J. Mitchell, Ed., Interscience, New York, 1954, pp. 195–236.
110. P. J. Elving, in *Progress in Polarography*, P. Zuman, Ed., Interscience, New York, 1962, pp. 625–648.
111. P. J. Elving, in *Advances in Electroanalytical Chemistry*, Vol. II, H. H. Nürnberg, Ed., Interscience, New York, in press.
112. W. C. Purdy, *Electroanalytical Methods in Biochemistry*, McGraw-Hill, New York, 1965.
113. D. L. Smith and P. J. Elving, *Anal. Chem.*, *34*, 930 (1962).
114. G. Dryhurst and P. J. Elving, *Talanta*, *16*, 855 (1969).
115. J. W. Webb and P. J. Elving, work in progress.
115a. P. J. Elving and J. W. Webb, *The Purines—Theory and Experiment*, The Jerusalem *Symposia on Quantum Chemistry and Biochemistry*, *IV*, The Israel Academy of Sciences and Humanities, Jerusalem, 1972, pp. 371–391.
116. D. L. Smith, D. R. Jamieson, and P. J. Elving, *Anal. Chem.*, *32*, 1253 (1960).
117. E. Smith, L. F. Worrel, and J. E. Sinsheimer, *Anal. Chem.*, *35*, 58 (1963).
118. J. T. Stock, *Amperometric Titrations*, Interscience, New York, 1965.
119. M. R. F. Ashworth, *Titrimetric Organic Analysis*, Parts I and II, Interscience, New York, 1964–65.
120. D. D. DeFord and J. W. Miller, in *Treatise on Analytical Chemistry*, Part I, Vol. 4, I. M. Kolthoff and P. J. Elving, Eds., Interscience, New York, 1963, pp. 2475–2531.
121. G. Semerano, *Ric. Sci., Suppl.*, *22*, 197 (1952).
122. P. Zuman, *Chem. Eng. News*, *46*, No. 12, 94 (1968).
123. H. Strehlow, in *Investigations of Rates and Mechansims of Reactions*, Part II, S. L. Friess, et al. Eds., Interscience, New York, 1963, pp. 799–843.
124. P. J. Elving, *Ann. N. Y. Acad. Sci.*, *158*, 124 (1969).
125. E. A. Abrahamson, *J. Am. Chem. Soc.*, *81*, 2692 (1959).

458   P. J. ELVING, J. E. O'REILLY, AND C. O. SCHMAKEL

126. E. A. Abrahamson, *J. Am. Chem. Soc.*, *81*, 3919 (1959).
127. D. C. Olson, V. P. Mayweg, and G. N. Schrauzer, *J. Am. Chem. Soc.*, *88*, 4876 (1966).
128. D. C. Olson, *Anal. Chem.*, *39*, 1785 (1967).
129. W. A. Struck and P. J. Elving, *J. Am. Chem. Soc.*, *86*, 1229 (1964).
129a. H. Berg and H. Schütz, *Proc. 2nd Conf. Appl. Phys. Chem. (Veszprem)*, *1*, 479–489 (1971).
129b. B. Janik and R. G. Sommes, *Biochim. Biophys. Acta*, *269*, 15 (1972).
129c. B. Janik, R. G. Sommes, and A. M. Bobst, *Biochim. Biophys. Acta*, *281*, 152 (1972).
130. M. P. Schweizer, A. D. Broom, P. O. P. Ts'o, and D. P. Holis, *J. Am. Chem. Soc.*, *90*, 1042 (1968).
131. A. E. V. Haschemeyer and A. Rich, *J. Mol. Biol.*, *27*, 369 (1967).
132. S. S. Danyluk and F. E. Hruska, *Biochemistry*, *7*, 1038 (1968).
133. D. L. Smith and P. J. Elving, *J. Am. Chem. Soc.*, *84*, 1412 (1962).
134. S. Wawzonek, *Science*, *155*, 39 (1967).
135. S. Swann, in *Catalytic, Photochemical and Electrolytic Reactions*, 2nd ed., A. Weissberger, Ed., Interscience, New York, 1956, pp. 385–523.
136. F. D. Popp and H. P. Schultz, *Chem. Rev.*, *62*, 19 (1962).
137. P. Zuman, in *Progress in Physical Organic Chemistry*, Vol. 5, A. Streitwieser and R. W. Taft, Eds., Interscience, New York, 1967, pp. 81–206.
138. H. H. Bauer, P. J. Herman, and P. J. Elving, in *Modern Aspects of Electrochemistry*, Vol. 7, J. O'M. Bockris and B. E. Conway, Eds., Butterworths, London, 1972, pp. 143–197.
139. N. O. Kaplan, in *The Enzymes*, Vol. III, P. D. Boyer, H. Lordy, and K. Myrböck, Eds., Academic Press, New York, 1960, pp. 105–169.
140. E. M. Kosower, *Molecular Biochemistry*, McGraw-Hill, New York, 1962, p. 166.
141. H. Sund, H. Diekmann, and K. Wallenfels, in *Advances in Enzymology*, Vol. 26, F. F. Nord, Ed., Interscience, New York, 1964, p. 115.
142. H. Sund, in *Biological Oxidations*, T. P. Singer, Ed., Interscience, New York, 1967, p. 603.
143. S. Chaykin, *Ann. Rev. Biochem.*, *36*, 149 (1967).
144. E. E. Conn and P. K. Stumpf, *Outlines of Biochemistry*, Wiley, New York, 1963, p. 130.
145. H. F. Fisher, E. E. Conn, B. Vennesland, and F. H. Westheimer, *J. Biol. Chem.*, *202*, 687 (1953).
146. M. E. Pullman, A. San Pietro, and S. P. Colowick, *J. Biol. Chem.*, *206*, 129 (1954).
147. H. von Euler, E. Adler, and H. Hellström, *Z. Physiol. Chem.*, *241*, 239 (1936).
148. E. Adler, H. Hellström, and H. von Euler, *Z. Physiol. Chem.*, *242*, 225 (1936).
149. M. B. Yarmolinsky and S. P. Colowick, *Biochim. Biophys. Acta*, *20*, 177 (1956).
150. G. W. Rafter and S. P. Colowick, *J. Biol. Chem.*, *209*, 773 (1954).
151. H. E. Dubb, M. Saunders, and J. H. Wang, *J. Am. Chem. Soc.*, *80*, 1767 (1958).
152. R. F. Hutton and F. H. Westheimer, *Tetrahedron*, *3*, 73 (1958).
153. S. Chaykin and L. Meissner, *Biochem. Biophys. Res. Commun.*, *14*, 233 (1964).
154. S. Chaykin, L. King, and J. G. Watson, *Biochim. Biophys. Acta*, *124*, 13 (1966).
155. K. Wallenfels and H. Schüly, *Angew. Chem.*, *67*, 517 (1955); *69*, 505 (1957).
156. K. Wallenfels and H. Schüly, *Ann.*, *621*, 106 (1959).
157. K. Wallenfels, in G. E. W. Wolstenholme and C. M. O'Connor, Eds., *Ciba Foundation Study Group*, *2*, 10 (1959).
158. K. Wallenfels and M. Gellrich, *Chem. Ber.*, *92*, 1406 (1959).
159. H. Diekmann, G. Englert, and K. Wallenfels, *Tetrahedron*, *20*, 281 (1964).

160. W. L. Meyer, H. R. Mahler, and R. H. Baker, Jr., *Biochim. Biophys. Acta*, *64*, 353 (1962).
161. G. Maggiora, H. Johansen, and L. L. Ingraham, *Arch. Biochem. Biophys.*, *131*, 352 (1969).
162. E. Knobloch, *Chem. Listy*, *39*, 54 (1945); *Collect. Czech. Chem. Commun.*, *12*, 407 (1947).
163. J. Sancho, P. Salmeron, and J. G. Hurtado, *Anales Real Soc. Espan. Fis. Quim.*, *Sec. B*, *55*, 23 (1959).
164. P. C. Tompkins and C. L. Schmidt, *Univ. Calif. (Berkeley) Publ. Physiol.*, *8*, 247 (1943).
165. V. Moret, *Arch. Sci. Biol.*, *39*, 456 (1955).
166. E. Peingor and G. Farsang, *Acta Chim. Acad. Sci. Hung.*, *27*, 175 (1961).
167. G. A. Pozdeeva and E. G. Novikov, *Zh. Prikl. Khim.*, *40*, 213 (1967).
168. C. O. Schmakel and P. J. Elving, Unpublished work.
168a. D. D. Perrin, *Dissociation Constants of Organic Bases in Aqueous Solution*, Butterworths, London, 1965.
169. D. J. McClemens, A. K. Garrison, and A. L. Underwood, *J. Org. Chem.*, *34*, 1867 (1969).
170. R. M. Burton and N. O. Kaplan, *Arch. Biochem. Biophys.*, *101*, 150 (1963).
171. A. Stock, E. Sann, and G. Pfleiderer, *Ann. Chem.*, *647*, 188 (1961).
172. S. G. A. Alivisatos, F. Ungar, and G. J. Abraham, *Biochemistry*, *4*, 2616 (1965).
172a. S. Kato, J. Nakaya, and E. Imoto, *Rev. Polarogr.*, *17*, 46 (1971).
173. J. N. Burnett and A. L. Underwood, *J. Org. Chem.*, *30*, 1154 (1965).
174. W. Ciusa, P. M. Strocchi, and G. Adamo, *Gazz. Chim. Ital.*, *80*, 604 (1950).
175. S. J. Leach, J. H. Baxendale, and M. G. Evans, *Australian J. Chem.*, *6*, 395 (1953).
176. F. Sorm and Z. Sormova, *Chem. Listy*, *42*, 82 (1948).
177. V. Moret, *Arch. Sci. Biol.*, *40*, 635 (1956).
178. V. Moret, *Giorn. Biochim.*, *5*, 51 (1956).
178a. D. Thevenot and R. Buvet, *J. Electroanal. Chem.*, *39*, 447 (1972).
179. I. Bergmann, in *Polarography 1964*, G. J. Hills, Ed., Interscience, New York, 1966, p. 985.
180. C. Carruthers and V. Suntzeff, *Arch. Biochem. Biophys.*, *45*, 140 (1953).
180a. K. K. Lovecchio, *Diss. Abstr.*, *30B*, 5397 (1970).
180b. J. Kuthan, V. Simonek, and V. Volveo, *Z. Chem.*, *11*, 111 (1971).
180c. S. Kato, J. Nakaya, and E. Imoto, *Rev. Polarogr.*, *18*, 29 (1972).
181. J. Volke and V. Volkova, *Collect. Czech. Chem. Commun.*, *34*, 2037 (1969).
182. J. Volke, *Chem. Listy*, *61*, 429 (1967).
183. R. F. Homer, G. C. Mees, and T. E. Tomlinson, *J. Sci. Food Agr.*, *11*, 209 (1960).
184. R. F. Homer and T. E. Tomlinson, *Nature*, *184*, 2012 (1959).
185. J. Engelhardt and W. P. McKinley, *J. Agr. Food Chem.*, *14*, 377 (1966).
186. W. R. Boon, *Chem. Ind. (London)*, 782 (1965).
187. J. Heyrovsky and J. Kuta, *Principles of Polarography*, Academic Press, New York, 1966.
188. A. J. Cunningham and A. L. Underwood, *Biochemistry*, *6*, 266 (1967).
189. A. J. Cunningham, Ph.D. Thesis, Emory University, Atlanta, Ga., 1966.
190. Y. Paiss and G. Stein, *J. Chem. Soc.*, 2905 (1958).
191. W. J. Blaedel and R. G. Haas, *Anal. Chem.*, *42*, 918 (1970).
192. H. A. Laitinen and L. M. Chambers, *Anal. Chem.*, *36*, 5 (1964).
193. S. V. Tatwawadi and A. J. Bard, *Anal. Chem.*, *36*, 2 (1964).
194. A. M. Wilson and D. G. Epple, *Biochemistry*, *5*, 3170 (1966).

195. J. N. Burnett and A. L. Underwood, *Biochemistry, 4*, 2060 (1965).
196. B. Ke, *Biochim. Biophys. Acta, 20*, 547 (1956).
197. A. J. Cunningham and A. L. Underwood, *Arch. Biochem. Biophys., 117*, 88 (1966).
197a. R. J. Day, S. J. Kinsey, E. T. Seo, N. Weliky, and H. P. Silverman, *Trans. N. Y. Acad.*, 558 (1972).
197b. D. Thevenot and R. Buvet, *J. Electroanal. Chem., 39*, 429 (1972).
198. J. N. Burnett, Ph.D. Thesis, Emory University, Atlanta, Ga., 1965.
199. R. Phillips, *Chem. Rev., 66*, 501 (1966).
200. J. Nakaya, *Nippon Kagaku Zasshi, 81*, 1459 (1960).
201. V. Moret, *Giorn. Biochim., 4*, 192 (1955).
202. V. Moret, *Giorn. Biochim., 5*, 318 (1956).
203. A. N. Frumkin and V. I. Melik-Gaikazyan, *Dokl. Akad. Nauk SSSR, 77*, 855 (1951).
204. R. DeLevie and A. A. Husovsky, *J. Electroanal. Chem., 20*, 181 (1969).
205. E. J. Land and A. J. Swallow, *Biochim. Biophys. Acta, 162*, 327 (1968).
206. I. A. Gorskaya, A. V. Kotelnikova, Kh. F. Sholts, and S. Yu. Drizovskaya, *Biokhimiya, 30*, 315 (1965).
207. A. V. Koletnikova and V. V. Solomatina, *Biokhimiya, 30*, 816 (1965).
208. B. Ke, *Arch. Biochem. Biophys., 60*, 505 (1956).
209. T. Kono and S. Nakamura, *Bull. Agr. Chem. Soc. Japan, 22*, 399 (1958).
210. R. F. Powning and C. C. Kratzing, *Arch. Biochem. Biophys., 66*, 249 (1957).
211. B. Ke, *J. Am. Chem. Soc., 78*, 3649 (1956).
212. T. Kono, *Bull. Agr. Chem. Soc. Japan, 21*, 115 (1957).
213. R. W. Burnett and A. L. Underwood, *Biochemistry, 7*, 3328 (1968).
214. T. Kono and M. Suekane, *Bull. Agr. Chem. Soc. Japan, 22*, 404 (1958).
215. J. M. Siegel, G. A. Montgomery, and R. M. Bock, *Arch. Biochem. Biophys., 82*, 288 (1959).
216. A. D. Winer, *J. Biol. Chem., 239*, PC3598 (1964).
217. G. Stein and A. J. Swallow, *J. Chem. Soc.*, 306 (1958).
218. A. J. Swallow, *Biochem., J., 61*, 197 (1955).
219. H. H. Willard, L. L. Merritt, and J. A. Dean, *Instrumental Methods of Analysis*, Van Nostrand, Princeton, N. J., 1965, p. 373.
219a. T. Bruice and S. J. Benkovic, *Bioorganic Mechansims*, W. A. Benjamin, New York, 1966, p. 301.
219b. K. Burton and T. H. Wilson, *Biochem. J., 54*, 86 (1953).
220. W. A. Dewjatnin and L. A. Kusnetzowa, *Med. Prom. SSSR, 58*, 58 (1964).
221. W. Kemula and J. Chodkowski, *Rocz. Chem., 29*, 839 (1955).
222. G. Matta and E. S. Lopes, *Rev. Port. Farm., 14*, 336 (1964).
223. Y. Asahi, *Vitamins (Kyoto), 13*, 490 (1957).
224. R. Cernatescu, M. Poni, and R. Ralea, *Stud. Cercet. Stiin., 4*, 117 (1953).
225. Y. Asahi, *J. Vitaminol. (Osaka), 4*, 118 (1958).
226. A. Anastasi, E. Micarelli, and L. Novacic, *Ind. Chim. Belge, 20*, 468 (1955).
227. L. Murea, A. Iacob, and I. Calafeteanu, *Farmacia, 6*, 333 (1958).
228. G. P. Tikhomirova, S. L. Belenkaya, R. G. Madievskaya, and O. A. Kurochkina, *Vopr. Pitaniya, 24*, 32 (1965); *Chem. Abstr. 62*, 12978d (1965).
229. O. Enriquez and V. Kubac, *Rev. Fac. Farm., Univ. Centr. Venez., 3*, 249 (1962).
230. K. Wenig and M. Kopecky, *Čas. ces. likarnictva, 56*, 6 (1943).
231. K. Knobloch, in *Die Polarographie in der Chemotherapie, Biochemie und Biologie*, (Proceedings of the First Jena Symposium, 1962), Akademie-Verlag, Berlin, 1964, p. 176.
232. R. Pleticha, *Pharmazie, 13*, 622 (1958).
232a. J. M. Moore, *J. Pharm. Sci., 58*, 1117 (1969).

232b. M. E. Schertel and A. J. Sheppard, *J. Pharm. Sci.*, *60*, 1070 (1971).
233. Y. Okazaki, T. Otsuki, K. Miyasaka, and T. Nabikawa, *Japan Analyst*, *17*, 1228 (1968).
234. H. H. G. Jellinek and A. Gordon, *J. Phys. Colloid Chem.*, *53*, 996 (1964).
235. G. A. Miana and M. Ikram, *Pakistan J. Sci. Ind. Res.*, *5*, 201 (1962).
236. V. Vajgand and T. Pastor, *Glasnick Hem. Drustva Beograd*, *28*, 73 (1963).
237. G. Dusinsky and L. Faith, *Pharmazie*, *22*, 475 (1967).
238. J. Volke and V. Volkova, *Chem. Listy*, *48*, 1031 (1954).
239. C. Carruthers and J. Tech, *Arch. Biochem. Biophys.*, *56*, 441·(1955).
240. C. Carruthers and V. Suntzeff, *Cancer Res.*, *12*, 253 (1952).
241. C. Carruthers and V. Suntzeff, *Cancer Res.*, *14*, 29 (1954).
242. C. Carruthers, V. Suntzeff, and P. N. Harris, *Cancer Res.*, *14*, 845 (1954).
243. D. J. Brown, *The Pyrimidines*, Interscience, New York, 1962.
244. *J. Biol. Chem.*, *241*, 527 (1966).
245. *Handbook of Biochemistry*, H. A. Sober, Ed., Chemical Rubber Company, Cleveland, 1968, pp. A8–9 and G3–8.
246. T. Tsukamoto and T. Tono, *Rev. Polarog.* (*Kyoto*), *14*, 377 (1967).
247. S. M. Cantor and Q. P. Peniston, *J. Am. Chem. Soc.*, *62*, 2113 (1940).
248. D. Thevenot, G. Hammouya, and R. Buvet, *Compt. rend.*, *Ser. C*, *268*, 1488 (1969); *J. Chim. Phys. Physicochem, Biol.*, *66*, 1903 (1969).
249. J. E. O'Reilly and P. J. Elving, *J. Am. Chem. Soc.*, *93*, 1871 (1971).
250. J. R. Marshall and J. J. Walker, *J. Chem. Soc.*, *1004* (1951).
251. B. Janik and E. Palecek, *Arch. Biochem. Biophys.*, *105*, 225 (1964).
252. E. Palecek and B. Janik, *Arch. Biochem. Biophys.*, *98*, 527 (1962).
252a. J. W. Webb, B. Janik, and P. J. Elving, *J. Am. Chem. Soc.*, in press.
253. M. Konupcik and O. Manousek, *Cesk. Farm.*, *9*, 78 (1960).
254. O. Manousek, M. Konupcik, and J. Davidek, *Cesk. Farm.*, *6*, 593 (1957).
255. H. Marciszewski and A. Grzeszkiewicz, *Chem. Anal.* (*Warsaw*), *5*, 509 (1960).
256. V. Vetterl, *Collect. Czech. Chem. Commun.*, *31*, 2105 (1966).
257. V. Vetterl, *Abh. Deut. Akad. Wiss. Berlin, Kl. Med.*, 493 (1966).
258. L. F. Cavalieri and B. A. Lowy, *Arch. Biochem. Biophys.*, *35*, 83 (1952).
259. G. Sartori and A. Liberti, *Ric. Sci.*, *16*, 313 (1946).
260. S. Ono, M. Takazi, and T. Wasa, *Bull. Chem. Soc. Japan*, *31*, 364 (1958).
261. W. A. Hoffman, Jr., *J. Sci. Labs.*, *Denison Univ.*, *45*, 215 (1962).
262. E. Palecek, *J. Electroanal. Chem.*, *22*, 347 (1969).
262a. J. A. Reynaud and M. Leng, *Compt. rend.*, *Ser. C*, *271*, 854 (1970).
263. E. Palecek, *Collect. Czech. Chem. Commun.*, *25*, 2283 (1960).
264. E. Palecek and D. Kalab, *Chem. Listy*, *57*, 13 (1963).
265. G. Horn and P. Zuman, *Collect. Czech. Chem. Commun.*, *25*, 3401 (1960).
265a. M. Wrona and B. Czochralska, *Acta Biochim. Pol.*, *17*, 351 (1970).
266. A. Humlova, *Collect. Czech. Chem. Commun.*, *29*, 182 (1964).
267. L. Kittler and H. Berg, *J. Electroanal. Chem.*, *16*, 251 (1968).
268. J. Krupicka and J. Gut, *Collect. Czech. Chem. Commun.*, *25*, 592 (1960).
269. J. Krupicka and J. Gut, *Collect. Czech. Chem. Commun.*, *27*, 546 (1962).
270. P. Zuman and M. Kuik, *Collect. Czech. Chem. Commun.*, *24*, 3861 (1959).
271. B. Czochralska, *Rocz. Chem.*, *44*, 2207 (1970).
272. D. Cohen, M. Koenigsbuch, and M. Sprecher, *Israel J. Chem.*, *6*, 615 (1968).
273. K-C. Chiang and P. Chang, *Hua Hsueh Hsueh Pao*, *24*, 300 (1958); *Chem. Abstr.*, *53*, 20080f (1959).
274. J. Glicksman, *J. Electrochem. Soc.*, *109*, 352 (1962).

275. K. Hofmann, *Imidazole and Derivatives*, Part I, Interscience, New York, 1953.

276. E. S. Schipper and A. R. Day, in *Heterocyclic Compounds*, Vol. V, R. C. Elderfield, Ed., Wiley, New York, 1957.

277. D. Hamer, D. M. Waldron, and D. L. Woodhouse, *Arch. Biochem. Biophys.*, *47*, 272 (1953).

277a. P. J. Elving, S. J. Pace, and J. E. O'Reilly, *J. Am. Chem. Soc.*, *95*, 647 (1973).

278. V. P. Skulachev and L. I. Denisovich, *Biokhimiya*, *31*, 132 (1966); *Chem. Abstr.*, *64*, 14467f (1966).

279. N. G. Luthy and B. Lamb, *J. Pharm. Pharmacol.*, *8*, 410 (1956).

280. J. Heath, *Nature*, *158*, 23 (1946).

281. B. Janik and E. Palecek, *Abh. Deut. Akad. Wiss. Berlin, Kl. Med.*, 513 (1966).

282. E. Palecek and B. Janik, *Chem. Zvesti*, *16*, 406 (1962).

283. B. Janik and E. Palecek, *Z. Naturforsch.*, *21B*, 1117 (1966).

284. B. Janik, *Z. Naturforsch.*, *24B*, 539 (1969).

284a. S. Friedman and J. S. Gots, *Arch. Biochem. Biophys.*, *39*, 254 (1952).

285. A. Bendich, P. J. Russell, Jr., and J. J. Fox, *J. Am. Chem. Soc.*, *76*, 6073 (1954).

286. J. Vachek, *Cesk. Farm.*, *14*, 216 (1965).

287. J. Vachek, B. Kakac, A. Cerny, and M. Semonsky, *Pharmazie*, *23*, 444 (1968).

288. J. Vachek, *Cesk. Farm.*, *9*, 126 (1960).

289. G. Dryhurst, *J. Electrochem. Soc.*, *116*, 1357 (1969).

290. G. Dryhurst, *Anal. Chim. Acta*, *47*, 275 (1969).

291. I. L. Doerr, I. Wempen, D. A. Clarke, and J. J. Fox, *J. Org. Chem.*, *26*, 3401 (1961).

291a. F. Fichter and W. Kern, *Helv. Chim. Acta*, *9*, 429 (1926).

292. E. S. Canellakis and P. P. Cohen, *J. Biol. Chem.*, *213*, 385 (1955).

293. H. Blitz and H. Schauder, *J. Prakt. Chem.*, *106*, 114 (1923).

293a. G. Dryhurst, *J. Electrochem. Soc.*, *118*, 699 (1971).

294. G. Dryhurst and G. F. Pace, *J. Electrochem. Soc.*, *117*, 1259 (1970).

295. J. S. Sussenbach and W. Berends, *Biochim. Biophys. Acta.* *95*, 184 (1965).

296. L. A. Waskell, K. S. Sastry, and M. P. Gordon, *Biochem. Biophys. Acta*, *129*, 49 (1966).

297. J. S. Sussenbach and W. Berends, *Biochim. Biophys. Acta*, *76*, 154 (1963).

298. G. Dryhurst, *J. Electrochem. Soc.*, *116*, 1411 (1969).

299. G. Dryhurst, *J. Electrochem. Soc.*, *116*, 1097 (1969).

300. R. D. Brown and M. L. Heffernan, *Australian J. Chem.*, *9*, 83 (1956).

301. Y. Asahi, *Yakugaku Zasshi*, *80*, 1222 (1960).

302. K. Sugino, K. Shirai, T. Sekine, and K. Ado, *J. Electrochem. Soc.*, *104*, 667 (1957).

303. B. E. Conway and R. G. Barradas, *Electrochim. Acta*, *5*, 319 (1961).

304. R. G. Barradas and B. E. Conway, *Electrochim. Acta*, *5*, 349 (1961).

305. V. Vetterl, *Experientia*, *21*, 9 (1965).

306. V. Vetterl, *Collect. Czech. Chem. Commun.*, *34*, 673 (1969).

306a. J. Flemming and H. Berg, *Abh. Deut. Akad. Wiss., Kl. Med.*, 559 (1966).

307. Y. Okazaki, *Bunseki Kagaku*, *11*, 1142 (1962); *Chem. Abstr.*, *58*, 7786e (1963).

308. Y. Okazaki, *Bunseki Kagaku*, *11*, 1239 (1962); *Chem. Abstr.*, *58*, 8854g (1963).

308a. S. L. Belenkaya and G. P. Tikhomirova, *Zh. Anal. Khim.*, *24*, 227 (1969).

308b. L. I. Rapaport and H. V. Verzina, *Formatsevt. Zh.*, *22*, 62 (1967).

309. F. Icha, *Cesk. Farm.*, *8*, 384 (1959).

310. V. Bulant, M. Urks, and H. Parizkova, *Antibiotiki 9*, 545 (1964).

311. F. Sorm, J. Beranek, J. Smrt, J. Krupicka, and J. Skoda, *Collect. Czech. Chem. Commun.*, *27*, 575 (1962).

312. F. Icha, *Pharmazie*, *14*, 684 (1959).

313. O. Manousek and P. Zuman, *Collect. Czech. Chem. Commun.*, *20*, 1340 (1955).
314. S. Eguchi, *Nichidai Igaku Zasshi*, *20*, 2368 (1961); *Chem. Astr.*, *61*, 3496e (1964).
315. W. Leyko and H. Panusz, *Bull. Soc. Sci. Lettres Lodz*, Cl. III, *5*, 1 (1954).
316. S. Golewski and H. Panusz, *Polish Acad. Sci.*, *Vth Symposium of Biochemists*, Warsaw, 1954.
317. W. Leyko and H. Panusz, *Polski Tygod. Lekar.*, *9*, 1 (1954).
318. S. Matsushita, F. Ibuki, and A. Aoki, *J. Agr. Chem. Soc. Japan*, *37*, 67 (1963).
319. S. Fiala and H. E. Kasinski, *J. Natl. Cancer Inst.*, *26*, 1059 (1961).
320. B. Filipowicz and W. Leyko, *Bull. Soc. Sci. Lettres Lodz*, Cl. III, *4*, 1 (1953).
321. W. Leyko and S. Gross, *Polish Acad. Sci.*, *Vth Symposium of Biochemists*, Warsaw, 1954.
322. E. Palecek, *Naturwissenschaften*, *45*, 186 (1958).
323. E. Palecek, *Chem. Zvesti*, *14*, 798 (1960).
324. J. Bohacek, *Experientia*, *19*, 435 (1963).
325. D. Kalab, *Experientia*, *19*, 392 (1963).
326. D. Kalab, *Chem. Zvesti*, *18*, 435 (1964).
327. D. Kalab, *Abh. Deut. Akad. Wiss. Berlin, Kl. Med.*, 519 (1966).
328. D. Kalab, *Experientia*, *22*, 23 (1966).
329. L. Kittler and H. Berg, *Abh. Deut. Akad. Wiss. Berlin, Kl. Med.*, 547 (1966).
330. D. Kalab, *Sb. Ved. Praci. Biovet.*, 56, (1960–61).
331. D. Kalab, *Chem. Zvesti*, *14*, 823 (1960).
332. D. Kalab, *Experientia*, *17*, 275 (1961).
333. D. Kalab, *Abh. Deut. Akad. Wiss. Berlin, Kl. Med.*, 333 (1964).
334. V. Habermann, E. Maidlova, and R. Cerny, *Collect. Czech. Chem. Commun.*, *31*, 139 (1966).
335. R. Kalvoda, in *Progress in Polarography*, P. Zuman, Ed., Interscience, New York, 1962, p. 449.
336. C. A. Dekker, *Ann. Rev. Biochem.*, *29*, 463 (1963).
337. P. Zuman, *Ric. Sci.*, *Suppl. Contrib. Teor. Sper. Polarogr.*, *5*, 229 (1960).
338. P. Zuman, *Collect. Czech. Chem. Commun.*, *25*, 3225 (1960).
339. R. Pleticha-Lansky and J. J. Weiss, *Anal. Biochem.*, *16*, 510 (1966).
340. N. Urabe and K. Yasukochi, *J. Electrochem. Soc. Japan*, *22*, 525 (1954).
341. M. von Stackelberg and W. Stracke, *Z. Elektrochem.*, *53*, 118 (1949).
342. P. J. Elving and J. T. Leone, *J. Am. Chem. Soc.*, *82*, 5076 (1960).
343. A. G. Beaman and R. K. Robins, *J. Am. Chem. Soc.*, *83*, 4038 (1961).
344. D. D. Bly and M. G. Mellon, *J. Org. Chem.*, *27*, 2945 (1962).
345. F. Effenberger, *Chem. Ber.*, *98*, 2260 (1965).
346. R. C. Kaye and H. I. Stonehill, *J. Chem. Soc.*, 3244 (1952).
347. W. Thaber and W. Karrer, *Helv. Chim. Acta*, *14*, 2066 (1958).
348. H. T. S. Britton and R. A. Robinson, *J. Chem. Soc.*, 458 (1931).
349. H. T. S. Britton, *Hydrogen Ions*, 4th ed., Vol. I, Van Nostrand, Princeton, N. J., 1956.
350. E. F. Rogers, W. J. Leanza, H. J. Becker, A. R. Matzuk, R. C. O'Neill, A. J. Basso, G. A. Stein, M. Solotorovsky, F. J. Gregory, and K. Pfister, *Science*, *116*, 253 (1952).
351. F. A. McGinn and G. B. Brown, *J. Am. Chem. Soc.*, *82*, 3193 (1960).
352. P. J. Elving and D. L. Smith, *Anal. Chem.*, *32*, 1849 (1960).
353. J. Koutecky, *Collect. Czech. Chem. Commun.*, *18*, 597 (1953).
354. P. J. Elving, J. M. Markowitz, and I. Rosenthal, *Anal. Chem.*, *28*, 1179 (1956).
355. K. R. Voronova and A. G. Stromberg, *Zh. Obshch. Khim.*, *31*, 2786 (1961).

464   P. J. ELVING, J. E. O'REILLY, AND C. O. SCHMAKEL

356. V. Brabec and E. Palecek, *J. Electroanal. Chem.*, *27*, 145 (1970); *Biophysik*, *6*, 290 (1970).
357. M. P. Schweizer, S. I. Chan, and P. O. P. Ts'o, *J. Am. Chem. Soc.*, *87*, 5241 (1965).
357a. Y. Mori and K. Murata, *Vitamins (Kyoto)*, *2*, 24 (1949).
358. P. Hemmerich, C. Veeger, and H. C. S. Wood, *Angew. Chem., Intern. Ed. Engl.*, *4*, 671 (1965).
359. K. Yagi, *Flavins and Flavoproteins*, University Park Press, Baltimore, 1968.
360. R. Kuhn, H. Rudy, and T. Wagner-Jauregg, *Ber.*, *66*, 1950 (1933).
361. P. Karrer, H. Salomon, K. Schöpp, E. Schlittler, and H. Fritzsche, *Helv. Chim. Acta*, *17*, 1010 (1934).
362. Y. Asahi, *Rev. Polarog. (Kyoto)*, *8*, 1 (1960).
363. E. S. G. Baron and A. B. Hastings, *J. Biol. Chem.*, *105*, Proceedings vii (1934).
364. R. Kuhn and P. Boulanger, *Chem. Ber.*, *69B*, 1557 (1936).
365. R. Kuhn and G. Moruzzi, *Chem. Ber.*, *67*, 120 (1934).
366. L. Michaelis, *Chem. Rev.*, *16*, 243 (1935).
367. L. Michaelis and G. Schwarzenbach, *J. Biol. Chem.*, *123*, 527 (1938).
368. L. Michaelis, M. D. Schubert, and S. V. Smythe, *J. Biol. Chem.*, *116*, 587 (1936).
369. F. J. Stare, *J. Biol. Chem.*, *112*, 223 (1935–1936).
370. K. G. Stern, *Biochem. J.*, *28*, 949 (1934).
371. M. Valentinuzzi, L. E. Cotino, and M. Portnoy, *Anales Soc. Cienti. Arg.*, *147*, 45 (1949).
372. R. D. Draper and L. L. Ingraham, *Arch. Biochem. Biophys.*, *125*, 802 (1968).
373. B. Breyer and T. Biegler, *Collect. Czech. Chem. Commun.*, *25*, 3348 (1960).
374. H. Herman, S. V. Tatwawadi, and A. J. Bard, *Anal. Chem.*, *35*, 2210 (1963).
375. R. Brdička and E. Knobloch, *Z. Elektrochem.*, *47*, 721 (1941).
376. R. Brdička, *Z. Elektrochem.*, *47*, 314 (1941).
377. R. Brdička, *Z. Elektrochem.*, *48*, 686 (1942).
378. R. Brdička, *Chem. Listy*, *36*, 286 (1943).
379. R. Brdička, *Chem. Listy*, *36*, 299 (1943).
380. R. Brdička, *Collect. Czech. Chem. Commun.*, *14*, 130 (1949).
381. J. J. Lingane and O. L. Davis, *J. Biol. Chem.*, *137*, 567 (1941).
382. J. R. Merkel and W. J. Nickerson, *Biochim. Biophys. Acta*, *14*, 303 (1954).
383. Y. Mori and K. Murata, *Vitamins (Kyoto)*, *2*, 14 (1949).
384. K. Murata and Y. Mori, *Vitamins (Kyoto)*, *2*, 68 (1949).
385. K. Schwabe, *Polarographie and Chemische Konstitution Organischer Verbindungen*, Akademie-Verlag, Berlin, 1957.
386. R. Brdička, *Collect. Czech. Chem. Commun.*, *12*, 522 (1947).
387. Y. Asahi, *J. Pharm. Soc. Japan*, *76*, 378 (1956).
388. Z. Ostrowski and A. Krawczyk, *Acta Chem. Scand.*, *17* (*Suppl.*), 241 (1963).
389. S. Fowler and R. C. Kaye, *J. Pharm. Pharmacol.*, *4*, 748 (1952).
390. G. A. Tedoraze, E. Yu. Khmelnitskaya, and Ya. M. Zolotovitskii, *Elektrokhimiya*, *3*, 200 (1967); *Chem. Abstr.*, *66*, 10351 (1967).
391. E. Yu. Khmelnitskaya, G. A. Tedoradze, and Ya. M. Zolotovitskii, *Elektrokhimiya*, *4*, 886 (1968).
392. R. Brdička, *Z. Elektrochem.*, *48*, 278 (1942).
393. J. Sancho, J. G. Hurtado, and P. Salmeron, *Anales Real Soc. Espan. Fis. Quim.*, *58B*, 7, 511 (1962).
394. H. B. Herman and H. N. Blount, *J. Electroanal. Chem.*, *25*, 165 (1970).
395. T. Biegler and H. A. Laitinen, *J. Phys. Chem.*, *68*, 2374 (1964).
396. P. J. Lingane, *Anal. Chem.*, *39*, 541 (1967).

397. P. J. Lingane, *Anal. Chem.*, *39*, 485 (1967).
398. B. Jambor, *Magy. Kem. Folyoirat*, *73*, 178 (1967); *Chem. Abstr.*, *67*, 3706 (1967).
399. E. Knobloch, *Collect. Czech. Chem. Commun.*, *31*, 4503 (1966).
400. S. V. Tatwawadi, K. S. V. Santhanam, and A. J. Bard, *J. Electroanal. Chem.*, *17*, 411 (1968).
401. A. M. Hartly and G. S. Wilson, *Anal. Chem.*, *38*, 681 (1966).
402. B. Ke, *Arch. Biochem. Biophys.*, *68*, 330 (1957).
403. V. Moret and P. Pinamonti, *Giorn. Biochim.*, *9*, 223 (1960).
404. M. Senda, M. Senda, and I. Tachi, *Rev. Polarog.* *(Kyoto)*, *10*, 142 (1962).
405. T. Tachi and Y. Takemori, *Vitamins* *(Kyoto)*, *9*, 441 (1955).
406. Y. Takemori, M. Senda, and M. Senda, *Vitamins* *(Kyoto)*, *16*, 492 (1959).
407. Y. Takemori, *Rev. Polarog.* *(Kyoto)*, *12*, 63 (1964).
408. B. Breyer and T. Biegler, *J. Electroanal. Chem.*, *1*, 453, (1959/60).
409. K. Okamoto, *Rev. Polarog.* *(Kyoto)*, *11*, 225 (1964).
410. Y. Takemori, *Rev. Polarog.* *(Kyoto)*, *12*, 176 (1965).
411. A. T. Yahiro, S. M. Lee, and D. O. Kimble, *Biochim. Biophys. Acta*, *88*, 375 (1964).
412. A. Maguinay and N. Brouhon, *J. Pharm. Belg.*, *12*, 350 (1957).
413. R. Portillo and G. Varela, *Anales Bromatol.* *(Madrid)*, *2*, 251 (1950).
414. W. J. Seagers, *J. Am. Pharm. Assoc., Sci. Ed.*, *42*, 317 (1953).
415. A. J. Zimmer and C. L. Huyck, *J. Am. Pharm. Assoc., Sci. Ed.*, *44*, 344 (1955).
416. I. E. Kruze, *Farmatsiya*, *18*, 59 (1969).
417. G. Dusinsky and L. Faith, *Pharmazie*, *22*, 475 (1967).
418. E. Kevei, M. Kizzel, and M. Simek, *Acta Chim. Acad. Sci. Hung.*, *6*, 345 (1955).
419. A. L. Markman and S. I. Vityaeva, *Izv. Vysshikh Uchebn. Zavedenii, Pishchevaya, Tekhnol.*, *2*, 153 (1969).
420. R. Portillo and G. Varela, *Anales Real Acad. Farm.*, *15*, 787 (1949); *Chem. Abstr.*, *44*, 10782 (1950).
421. M. A. Hoffman, *Intern. Z. Vitaminforsch.*, *20*, 238 (1948); *Chem. Abstr.*, *43*, 1076 (1949).
422. O. Sebek, *Chem. Listy*, *41*, 238 (1947).
423. D. E. Pratt, C. M. Balkcom, J. J. Powers, and L. W. Mills, *J. Agr. Food Chem.*, *2*, 367 (1954).
424. E. N. Aksenova, *Tr. 1 (Pervogo) Mosk. Med. Inst.*, *61*, 241 (1968).
425. E. Knobloch and K. Mnovcek, *Cesk. Farm.*, *17*, 196 (1968).
426. K. Enns and W. H. Burgess, *J. Am. Chem. Soc.*, *87*, 1822 (1965).
427. A. Kocent, *Chem. Listy*, *47*, 195 (1953).
428. A. Kocent, *Chem. Listy*, *47*, 652 (1953).
429. L. I. Korshunov, Ya. M. Zolotovitskii, V. A. Benderskii, and L. A. Blyumenfel'd, *Dokl. Akad. Nauk SSSR*, *176*, 1381 (1968).
430. W. M. Moore and C. R. Baylor, Jr., *J. Am. Chem. Soc.*, *91*, 7170 (1969).
431. L. I. Korshunov, Ya. M. Zolotovitskii, and V. A. Benderskii, *Polyarogr. Opred. Kisloroda Biol. Obektakh*, 54 (1968).
432. K. Enns and W. H. Burgess, *J. Am. Chem. Soc.*, *87*, 5766 (1965).
433. Y. Saito and K. Okamoto, *Rev. Polarog.* *(Kyoto)*, *10*, 227 (1962).
434. A. J. Swallow, *Biochem. J.*, *54*, 253 (1953).
435. D. E. Smith, *Anal. Chem.*, *35*, 602 (1963).

Integrated Ion-Current (IIC) Technique of Quantitative Mass Spectrometric Analysis: Chemical and Biological Applications

JOHN R. MAJER, *Department of Chemistry, University of Birmingham, Birmingham, England,*

AND

ALAN A. BOULTON, *Psychiatric Research Unit, University Hospital, Saskatoon, Saskatchewan, Canada*

I.	Introduction	468
II.	Methods and Materials	471
III.	Factors Influencing Instrumental Sensitivity	471
	1. Sample Factors	471
	2. Instrumental Factors	472
IV.	Analysis of Trace Metals	473
	1. Mercury	473
	2. Nickel	473
	3. Various Oxinates	474
	4. Thermal Stability of Certain Oxinates	475
	5. Some Diketones	476
	6. Thermal Reactions of Metal Chelates	480
	7. Applications	481
V.	Pollution Studies	481
	1. Detection and Analysis of Polycyclic Hydrocarbons in the Atmosphere	481
	2. Detection and Analysis of Polycyclic Hydrocarbons in Extracts from Solid Samples	485
	3. Estimation of Lead in the Atmosphere and in Petroleum	487
	4. Other Applications	487
VI.	Surface Decontamination and Detergency	487
	1. Applications	490
VII.	Analysis of Complex Biological Extracts	490
	1. Urine Extracts	491
	2. *p*-Tyramine in the Rat Brain	494
	3. β-Phenylethylamine in Some Tissues of the Rat	495
	4. Applications	498

VIII. Detection of Isomerism... 498
    1. Dioximes.. 499
    2. Structural Isomers... 499
    3. Cis-trans Isomers.. 500
    4. Isomeric Polycyclic Hydrocarbons............................ 501
    5. Conformational Isomers...................................... 502
    6. Metal Chelates... 503
    7. Discussion of Detection of Isomerism....................... 504
  IX. Summary.. 505
    Acknowledgments.. 514
    References.. 514

I. INTRODUCTION

It has been common practice, when carrying out quantitative analysis with a mass spectrometer, to provide a reservoir of sample vapor, the pressure of which remains substantially constant while a mass spectrum is recorded. The partial pressure of a component in the sample is then proportional to the ion current produced and hence to the height of peaks on the recorder trace which are due to that component only. The proportionality constant or sensitivity for each component is determined by recording the mass spectrum of the pure component at a known pressure. This method is inapplicable, however, when only very small quantities of relatively involatile materials are available. Indeed, the only satisfactory method of recording the mass spectra of such components is by their direct evaporation into the ion source from a probe which can be inserted into the vacuum system of the mass spectrometer. When such a device is used, two problems arise. First, the partial pressure of the sample does not remain constant during the time required to record a mass spectrum, so that it is no longer possible to apply previously determined sensitivities; second, at very low sample concentrations, evaporation may be complete before the mass spectrum is recorded. There is a further instrumental complication: the sensitivity of the instrument itself may change because of deliberate or unintentional variations in the gain of the electron multiplier detector. For these reasons it is necessary to provide an internal standard of constant and known partial pressure. The mass spectrum of this standard is recorded at the same time as that of the sample. This is most simply achieved by providing a reservoir of heptacosafluorotri-*n*-butylamine (a compound commonly used as a mass marker in mass spectrometry) vapor at a known pressure and by allowing it to leak into the ion source at a constant rate at the same time as the sample is evaporating.

As the complete mass spectrum of the sample at low concentration cannot be recorded in the evaporation time, and because even if this were possible the mass peak sensitivities would not remain constant, it is necessary to use a single mass peak characteristic of the sample and to record the rise and fall of the ion current at this particular mass value during the course of the evaporation. The integrated ion current is then directly proportional to the amount of sample evaporated, and hence to the concentration of this compound in the sample, provided that no irreproducible thermal breakdown occurs during evaporation. The most suitable ion for this purpose is the one most characteristic of the sample, that is, the molecule ion itself. Because some molecule ions are unstable, however, not every molecule exhibits a mass spectrum in which the molecule ion is represented by a mass peak. In such cases an intense peak at a high m/e value is selected.

There is a difficulty associated with recording the rise and fall of such peaks, in that no magnetic sector mass spectrometer can provide a means of tuning-in the ion current at a selected m/e value before the ion current itself has been established. It was necessary, therefore, to devise a method in which a magnetic sector instrument could be made to record the ion current at the selected m/e value before, during, and after the evaporation of the sample. The method which was developed involves the use of a second substance, usually heptacosafluorotri-n-butylamine, as a means of calibrating the mass scale. A constant partial pressure of this substance is maintained within the ion source, using the conventional gas-handling facilities of the instrument. It is then possible to select a peak from the mass spectrum of this compound which is close to the characteristic peak. The peak selected is displayed on the oscilloscope facilities of the instrument by the normal tuning-in procedure. Using the peak comparison technique, it is possible to display alternately on the oscilloscope the reference peak and the characteristic peak. This is achieved on the AEI MS-9 mass spectrometer by adjusting the decadic resistance to a value given by the ratio of the precise masses of the ions responsible for the two selected peaks. A recorder trace taken at this stage shows only the reference peak repeated at intervals (see Figure 1a). When the sample is lowered into the heating region of the ion source, and evaporation begins, a second peak appears between the repeated reference peaks. In sharp distinction from the reference peak, which has a constant height, this second peak increases in height, reaches a maximum, decays, and finally disappears. It is the envelope of these secondary peaks (see Figure 1b) which provides a record of the rise and fall of the ion current, and therefore it is the area under this envelope which is proportional to the integrated ion current and hence to the concentration of substance being evaporated.

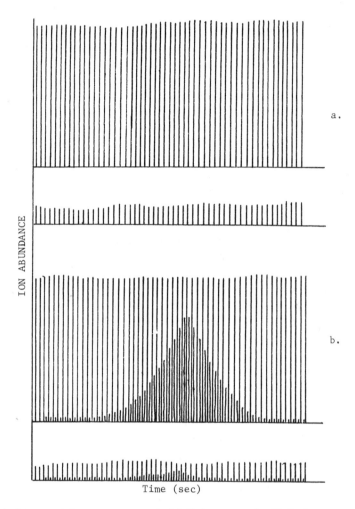

Figure 1. Integrated ion current record. (*a*) Reference peak. (*b*) Reference peak and sample peak.

It can now be seen that the reference compound serves a two-fold purpose. It enables the mass scale to be calibrated, so that the rise and fall of the selected sample peak can be completely recorded, and it provides an internal standard against which any variation of instrument sensitivity can be measured.

The peak comparison technique, although providing an interrupted rather than a continuous trace, ensures that the full peak height is always recorded.

II. METHODS AND MATERIALS

A measured volume (1 to 5 μl) of a dilute solution of the compound or mixture to be analyzed in a volatile solvent is introduced into the probe tip with the aid of a calibrated microsyringe. The probe tip consists of a Pyrex tube 27 mm long and 1 mm in internal diameter. The same tube serves for any series of measurements and is discarded after use. The probe is then inserted through the vacuum lock and into the vacuum system of the mass spectrometer so that the tip and the sample are located in the cool part of the ion source region. During this stage the volatile solvent evaporates, but the compound under investigation remains in the solid state within the probe tip. A measured pressure of heptacosafluorotri-n-butylamine is admitted to the gas-inlet system, and the reference peak is selected and displayed on the oscilloscope. The decadic resistance is set to a value corresponding to the ratio of the ions responsible for the reference and the characteristic mass peaks, and the peak-switching system so adjusted that the oscilloscope gives alternate displays of the reference and characteristic peaks. At this stage, of course, the characteristic peak does not appear. The galvanometer recorder is then set in operation, so as to record only the peak pair displayed on the oscilloscope, and the sample tube is lowered into the heated region of the source to allow evaporation of the sample. When evaporation begins, the characteristic peak appears on the oscilloscope and is recorded. Evaporation is usually complete within 10 to 20 sec, depending on sample size; the characteristic peak then disappears from the oscilloscope and the recorder trace.

III. FACTORS INFLUENCING INSTRUMENTAL SENSITIVITY

The sensitivity of the integrated ion current method is dependent on two groups of variables (1), the first group being associated with the sample and the second with the instrument.

1. Sample Factors

The most important sample characteristic determining sensitivity is the sublimation temperature. The method relies on the identification of a moderately sharp rise and fall in the ion current at a selected m/e value when the sample is lowered into the heating region of the source. If the rate of evaporation is too slow, the individual peak heights will be small and separation from the background difficult, even though the total integrated ion current, as represented by the total area under the peaks, will be significant. The measurement of broad, low peaks in gas chromatography presents a similar problem. Also, the sample may decompose on evaporation, either dissociating or polymerizing, so that the intensity at any selected m/e value will be lower. Sometimes, even if the sample does not decompose, it may not be possible to use the most intense mass peak because of the superposition of reference compound peaks or the unavailability of a suitable adjacent reference compound peak. It may therefore be necessary on occasions to use a sample peak of low intensity with corresponding loss in sensitivity.

There are two factors which may alter sensitivity and therefore reproducibility. It is possible that the evaporation of a minor component from a mixture may be affected by the nature or state of the matrix. In the absence of such a matrix it is still possible that the disposition of the sample within the evaporation tube may affect the transient change in source pressure and hence sensitivity. Neither effect has yet been identified. Finally, in the case of trace metal analysis the metal may have a number of isotopes of equal abundance so that the ion current for any specific ionization process is distributed over a number of m/e values with corresponding loss of sensitivity, since only one of the isotopes will be involved. A specific example is the analysis of mercuric chloride, in which the peak at m/e 268 carries only 6% of the ion current carried by all the molecule-ion species.

2. Instrumental Factors

The AEI MS-9 readout is on a Honeywell-Brown ultraviolet recorder, which provides a number of simultaneous records at different degrees of attenuation. All sensitivities quoted are based on the least sensitive record, an increase in sensitivity by at least two orders of magnitude being obtained by using the most sensitive record. Because of the problems of calibrating the instrument at the higher sensitivity, measurements are first made at low sensitivity. A second factor controlling instrumental sensitivity is the electron multiplier gain, which is determined by the voltage applied to the dynodes. The gain is normally low because of calibration difficulties when the instrument is operated at high gain, but a further hundred-fold increase in sensitivity can be obtained by increasing the electron multiplier gain. The limitation here is the random noise in the instrument. A further increase in sensitivity (by a factor of 3) is possible if the electron beam current is increased from the conventional 100 μA to 500 μA. Finally, an apparent gain in sensitivity is achieved if the recorder paper speed is increased. This, however, is not a real gain permitting lower levels of sample size to be measured, since the peak height is not increased. At low levels the discrimination against noise is given by the peak height and not the peak area.

IV. ANALYSIS OF TRACE METALS

Very small quantities of impurities in metals are commonly estimated with the aid of a spark source mass spectrometer. With this instrument a sample of the metal is vaporized before the ionization, and the ion currents corresponding to the components of the metal mixture are recorded on a photographic plate. The limit of detection of impurities in a sample is set at 1 part in 10^9. The chelates of many metals vaporize at temperatures considerably lower than those at which the parent metals have detectable vapor pressures. As a result it is possible to detect such metals as their chelates in the source of a conventional mass spectrometer such as the AEI MS-9 at temperatures between 200° and 500°. Using the present technique, it is possible to detect subpicogram amounts of metal in the form of a volatile chelate in the probe tip. The solution of a metal chelate, therefore, need only contain a few nanograms of the metal chelate per milliliter. Trace metals in biological material can be estimated either directly or by ashing and converting to the appropriate volatile metal chelate.

1. Mercury

Some metal salts are sufficiently thermally stable and volatile at elevated temperatures to permit their direct determination without the necessity for prior conversion to a chelate. Mercuric chloride ($HgCl_2$) has a boiling point of 302°C and does not decompose upon evaporation, nor does it evaporate at an appreciable rate at room temperature. It has proved possible (2) to measure very small quantities of mercuric chloride by integrating the ion current at an m/e value of 268, corresponding to the ionic species $^{198}Hg^{35}Cl_2{}^+$, during evaporation. The analysis is complicated by the fact that at the temperature where no mercuric chloride can be detected in the vapor phase by the mass spectrometer a complex mass spectrum is obtained. This has been shown to be caused by impurities of mercuric iodide and mercuric bromide existing in the mercuric chloride sample. There is also a peak corresponding to the ion $HgBrI^+$, which may exist uncharged in the sample or may be formed on evaporation. Because of their higher volatility the concentrations of these impurities cannot be estimated by the present technique.

2. Nickel

The familiar complexes of nickel with dimethylglyoxime and benzildioxime are sufficiently volatile and stable to be evaporated at a temperature of 250°C into the ion source of a mass spectrometer (3). The amount of nickel dimethylglyoximate evaporated may be determined by integrating the ion current at an m/e value of 288, corresponding to the molecule ion. The response of the instrument was linear over five orders of magnitude, and at the highest electron multiplier gain and instrumental sensitivity the limit of detection with a signal-to-noise ratio of 3:1 corresponded to 10^{-14} g of nickel dimethylglyoximate being evaporated.

The extreme sensitivity and specificity of this technique permitted the detection of the gradual solvolysis of the metal chelate which occurred on standing at room temperature. When determinations were repeated with stock solutions which had been allowed to stand for a day, it was found that the sensitivity had apparently dropped. This effect was traced to the instability of nickel dimethylglyoximate in chloroform, by the observation of the appearance of a small subsidiary peak at $m/e = 288$, due to an ion with a mass sufficiently different from that of $C_8H_{14}N_4O_4{}^{58}Ni^+$ to be resolved by the instrument at low resolution. The determination of the precise mass revealed that the ion contained no nickel. A complete mass spectrum was therefore recorded for the stock solution at various times after its preparation. The spectrum revealed the gradual decay of the nickel dimethylglyoximate molecule-ion peak, and the rise of peaks at m/e values above 288. Precise mass measurements of these peaks revealed that the corresponding ions contained no nickel atoms. Most probably, the compounds responsible for these spectra are formed by the polymerization of dimethylglyoxime molecules or their breakdown products. It can be demonstrated that several compounds are involved, by recording a mass spectrum very rapidly (8 sec) as soon as the sample is lowered into the heating region of the source, and before evaporation of the chelate has begun. Such a spectrum contains no peaks due to ions containing nickel atoms, and must be derived from a compound more volatile than the metal chelate. However, the spectrum obtained from the chloroform solution on prolonged standing does not correspond to a simple superposition of this spectrum or that of the metal chelate.

3. Various Oxinates

8-Hydroxyquinoline is the most widely used chelating agent and forms complexes with most metal ions. A preliminary study of the mass spectra of metal oxinates has shown them to be very simple with intense peaks due to the molecule ions (4). The volatility and stability of metal oxinates vary considerably, but many satisfy the requirements of the integrated ion current technique. It is possible, therefore, to consider analyzing a complex mixture of metal ions by converting them to their oxinates, followed by evaporation into a mass spectrometer. Table I summarizes the results obtained with a number of metal oxinates which could be evaporated without apparent decomposition. The sensitivities are virtually expressed in arbitrary units as square millimeters of integrated-ion-current-record area per

TABLE I

Quantitative Analysis of Some Metal Oxinates

Oxinate	m/e value used	Ion formula	Sensitivity $(mm^2/\mu g)$	Source (°C)
Aluminum	459	$Al(Ox)_3^+$	1350	400
Beryllium	296	$Be(Ox)_2^+$	400	380
Cobalt	347	$Co(Ox)_2^+$	200	370
Copper	351	$Cu(Ox)_2^+$	6480	255
Gallium	357	$Ga(Ox)_2^+$	2250	320
Indium	403	$In(Ox)_2^+$	32	395
Iron	344	$Fe(Ox)_2^+$	200	405
Manganese	343	$Mn(Ox)_2^+$	700	360
Nickel	346	$Ni(Ox)_2^+$	300	370
Zinc	352	$Zn(Ox)_2^+$	350	380

microgram of metal oxinate evaporated. The significance of this sensitivity is discussed later.

4. Thermal Stability of Certain Oxinates

Although many of the metal oxinates exist in the solid phase in the form of hydrates, no peak in their mass spectra has been shown to correspond to a hydrated form. Even if the hydrate were to survive evaporation at an elevated temperature, it is unlikely that the comparatively weak forces would survive electron impact. In many of the cases studied, for example, copper oxinate, dehydration is a smooth process which precedes evaporation. In other oxinates a more general decomposition occurs, evaporation is incomplete, and the resulting sensitivities are low. Not all metal oxinates are soluble in organic solvents, so that when a quantitative analysis is required it may be necessary in certain cases to form the metal oxinates directly in the probe tip by mixing appropriate quantities of the dilute metal salt and oxine solutions. One further complication is that the mass spectra of some metal oxinates contain peaks with a higher m/e value than that of the molecule ion. It is not possible to determine whether these are formed as a result of polymerization processes taking place at the elevated temperatures needed for evaporation or whether such polymers exist in the solid phase and are merely being evaporated. The degree of polymerization sometimes varies very widely, beryllium produces only a single small peak corresponding to the ion $M_2Ox_3^+$, whereas the rare-earth oxinates (dysprosium and neodynium, e.g.) produce a large number of polymer peaks extending to high m/e values that correspond to ions of the formula $M_4Ox_4^+$. These complications limit the general usefulness of oxine as a chelating agent in the IIC technique.

5. Some Diketonates

The volatility of the metal derivatives of 1,3-diketones and ketoesters has been known for 40 years, and recently this property has been exploited in a number of elegant gas chromatographic studies by Moshier and Sievers (5). The volatility of the metal derivatives is increased markedly by the substitution of fluorine atoms in the diketone molecule, and some transition metal diketonates, such as the chromium derivative of 1,1,1-trifluoro-2,4-pentanedione, are too volatile at room temperature to be determined by the integrated ion current technique. The rare-earth metal derivatives of 1,3-diketones are in general less volatile and more suitable for the application of IIC method. However, not all of the rare-earth chelates are sufficiently thermally stable, and the integrated ion current curve gives an excellent illustration of stability. If the ion responsible for the mass peak being integrated is formed from a molecule in the gas phase arising by a simple evaporation process, then the IIC curve is approximately symmetrical and Gaussian in shape. If, however, the ion is formed from a species produced from a thermal decomposition, the resulting IIC curve is asymmetric with a long tail, as illustrated in Figure 2. Metal chelates exhibiting this type of curve are not suitable for examination by the IIC technique.

Table II summarizes the results of studies on a number of the holmium derivatives of substituted 1,3-diketones. As the salts of holmium can be quantitatively converted into volatile chelates using any of the 1,3-diketones listed in Table II, as little as 10^{-10} g of this element may be determined (6). The best results were obtained using the diketone heptafluoro-7,7-dimethyoctan-4,6-dione $(C_3F_7COCH_2—COC(CH_3)_3]$.

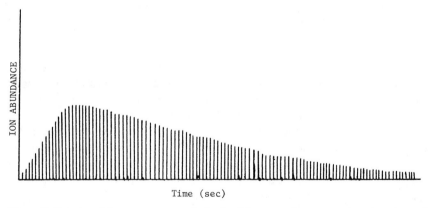

Figure 2. Integrated ion current record obtained from a substance undergoing thermal decomposition.

TABLE II

Holmium Chelates of the Form HoL_3 (L = 1,3-diketone)

1,3-Diketone (L)	Metal chelate	Molecular Weight	m/e Value used	Ion formula	Sensitivity ($mm^2/$ (μg)
Heptafluorobutanoylpivaloylmethane	$Ho(HFM)_3$	1050	755	$Ho(HFM)_2^+$	700
Pentafluoroproprionoylpivaloylmethane	$Ho(PFM)_3$	900	655	$Ho(PFM)_2^+$	580
Trifluoroacetylpivaloylmethane	$Ho(TPM)_3$	750	555	$Ho(TPM)_2^+$	560
Dipivaloylmethane	$Ho(DPM)_3$	714	531	$Ho(DPM)_2^+$	488
Trifluoroacetylacetone	$Ho(TFA)_3$	624	471	$Ho(TFA)_2^+$	616
Thenoyltrifluoroacetone	$Ho(TTA)_3$	828	607	$Ho(TTA)_2^+$	444
Benzoyltrifluoroacetone	$Ho(BTA)_3$	810	595	$Ho(BTA)_2^+$	200
Dibenzoylmethane	$Ho(DBM)_3$	834	611	$Ho(DBM)_2^+$	404
Acetylacetone	$Ho(A)_3$	462	363	$Ho(A)_2^+$	480

A similar procedure using this diketone has been applied to the detection of very small quantities of lead (7). The lead salts were converted to the chelate, and the integrated ion current determined at an m/e value of 503. Figure 3 shows a plot of peak area against sample size in the microgram range. The slope of this graph gives a sensitivity of 410, and the limit of detection at highest instrumental gain corresponds to 10^{-14} g of lead in the original sample.

Because of the closely similar size of the ionic and atomic radii of zirconium and hafnium, the properties of compounds of these two metals are very similar. As a result the separation of the metals is difficult, and analytical procedures for the determination of one metal in the presence of the other have been difficult to devise. Methods involving differences in molecular weight have been suggested in which a mixture of the pure metal salt is weighed and converted quantitatively to a mixture of a second type of compound (e.g., oxide). The second mixture is then weighed, and the relative amounts of the two metals are calculated from a knowledge of their atomic weights and the appropriate conversion factors. A variation of this technique uses the molar extinction coefficients of complexes of the metals with dyes such as alizarin red S and xylenol orange, rather than their molecular weights. An alternative approach is to utilize differences in the stability of metal complexes in solution by measuring absorbances at different pH values. The IIC technique, however, offers the advantage of high sensitivity and

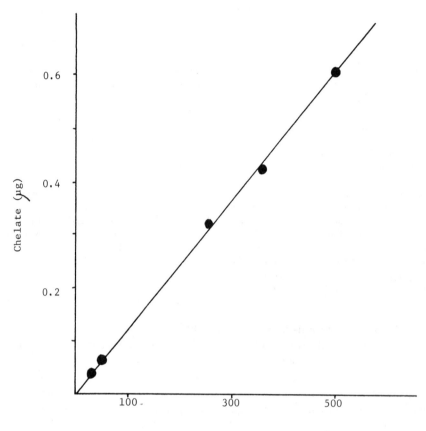

Figure 3. Calibration curve for Pb(C₃F₇COCHCOC₄H₉)₂.

selectivity and the ability to carry out the determination of mixtures of zirconium and hafnium in the presence of another metal such as titanium. Table III summarizes the results of a study in which benzoyltrifluoroacetone ($C_6H_5COCH_2COCF_3$) was used to form chelates with the salts of hafnium, zirconium, and titanium (8).

The success of the method in the analysis of heavy metal ions prompted an investigation into the more difficult problem of the production of volatile derivatives of the alkali metals. Unexpectedly, the alkali metal derivatives of certain fluorinated 1,3-diketones and keto esters have been found to possess vapor pressure curves which permit their use in the ICC technique. In the mass spectra of the alkali metal chelates, the alkali metals provided

TABLE III

Analysis of the 1,3-Diketones of Hafnium, Zirconium, and Titanium

Metal chelate	m/e Value used	Ion formula	Sensitivity (μg/ mm^2)	Limit of detection (g)	Analysis of mixture	
					Amount introduced (μg)	Amount found (μg)
Ti(BTFA)$_3$	497	Ti(BTFA)$_2$F$^+$	108	1×10^{-11}	2.0	1.7
Zr(BTFA)$_4$	539	Zr(BTFA)$_2$F$^+$	876	5×10^{-14}	2.0	1.8
Hf(BTFA)$_4$	629	Hf(BTFA)$_2$F$^+$	35	5×10^{-12}	3.0	3.1

the most abundant ions; these peaks were chosen, therefore, for the application of the technique to the lithium, sodium, and potassium derivatives of pentafluoro-6,6-dimethylheptan-3,5-dione and heptafluoro-7,7-dimethyloctan-4,6-dione. The reference compound, heptacosafluorotri-n-butylamine, used in other studies of metal chelates proved unsuitable for the present work and was replaced in each case by an inert gas of appropriate mass (see Table IV). Quantitative measurements were made by taking a solution of the appropriate alkali metal hydroxide, adding the diketone, and extracting the chelate into ether. Aliquots of the ethereal solution were then introduced directly into the mass spectrometer, and the integrated ion current curve at the appropriate alkali metal ion peak was recorded and measured. Straight-

TABLE IV

Analysis of the Alkali-Metal 1,3-Diketone Chelates

Metal chelate	m/e Value used	Ion formula	Reference gas	Sensitivity (mm^2/μg)
Lithium pentafluoropropanoyl-pivaloylmethane	6	^{6}Li$^+$	^{4}He$^+$	49
Sodium pentafluoropropanoyl-pivaloylmethane	23	Na$^+$	^{20}Ne$^+$	206
Lithium heptafluorobutanoyl-pivaloylmethane	6	^{6}Li$^+$	^{4}He$^+$	48
Sodium heptafluorobutanoyl-pivaloylmethane	23	Na$^+$	^{20}Ne$^+$	220
Potassium heptafluorobutanoyl-pivaloylmethane	39	^{39}K$^+$	^{40}Ar$^+$	823

line calibration curves relating the peak area to the initial quantity of alkali metal hydroxide were obtained. Amounts of alkali metal between 10^{-6} and 10^{-9} g were measured in this way at low instrumental gain; the sensitivities are shown in Table IV.

6. Thermal Reactions of Metal Chelates

Although it has been demonstrated that a variety of metal chelates can be analyzed using a mass spectrometer, there are complicating reactions which must be taken into account in any proposed scheme of analysis. Metal chelates undergo thermal reactions of two types, exchange reactions and association reactions.

In the first of these, a metal ion or a ligand may be exchanged between a pair of chelates. A simple example of this type of reaction occurs during the simultaneous evaporation of sodium heptafluoro-7,7-dimethyloctan-4,6-dione and potassium pentafluoro-6,6-dimethylheptan-3,5-dione. The combined mass spectrum (see Figure 4) reveals the existence of ions formed as a result of the interchange of the alkali metal atoms (9).

Association reactions of metal chelates occurring within the ion source may be exemplified by the formation of tetrakis chelates of the general formula $MM^1L^1L_3$ from rare-earth tris chelates (ML_3) and alkali

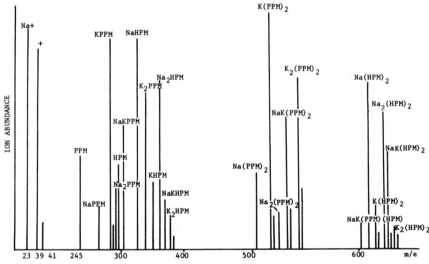

Figure 4. Mass spectrum (simplified) of a mixture of $Na(C_3F_7COCHCOC_4H_9)$ and $K(C_2F_5COCHCOC_4H_9)$. $PPM = C_2F_5COCHCOC_4H_9$; $HPM = C_3F_7COCHCOC_4H_9$.

metal chelates (M^1L^1). Complex chelates of this form, such as NaHo-($CF_3COCHCOCH_3$)$_4$, have been isolated by normal chemical procedures (10), and their mass spectra have been shown to be identical with those of mixtures of the two separate chelates. Similar association reactions have been shown to occur between the alkali metal (M^1L^1) and alkaline earth metal (ML_2) derivatives of keto esters. Finally, many chelates exhibit association reactions between two or three molecules of the same compound. Their mass spectra then possess polymeric ions containing several metal atoms. This polymerization process has been shown to be temperature dependent by integrating the ion current at monomer and polymer mass peaks at a variety of temperatures. By this means it has been shown that the degree of polymerization in $Ca(CF_3COCHCOC_4H_9)_2$ increases with rising temperature.

7. Applications

Despite the complicating reactions that can occur during the IIC analysis of metal chelates, the specificity and sensitivity are such that applications in some areas are likely to be profitable. In medicine the analysis of certain trace metals in the various body fluids from healthy and diseased individuals will lead to an increase in our knowledge of the distribution and significance of these trace metals. In physiology it is possible to envisage the use of the IIC technique in studies of such phenomena as nervous transmission and membrane transport. In biochemistry and, more especially, neurochemistry a qualitative and quantitative analysis of the metals associated with certain enzymes and a regional analysis within the brain substance would be most desirable. In pollution, studies of the lead complexes formed during the combustion of petroleum products would obviously shed light on some of the toxic substances released into the atmosphere in our cities. Of particular interest these days would be an assessment of the various mercury-containing substances and, of course, other pesticide and herbicide residues in fish and game.

V. POLLUTION STUDIES

1. Detection and Analysis of Polycyclic Hydrocarbons in the Atmosphere

The identification and estimation of polycyclic aromatic hydrocarbons in the atmosphere have become increaingly important because of the demonstration of the carcinogenic properties of some of these compounds. The presence of benzo(a)pyrene and benzo(e)fluoranthene, in particular, in the atmosphere has received great attention in recent years, although it should be remembered that dibenzanthracene is just as carcinogenic, and many of the other polycyclic compounds are also moderately active. It is highly desirable, therefore, that methods be developed for the rapid analysis of individual hydrocarbons.

Several of the existing analytical methods have been reviewed by Sawicki et al. (11). In the main these methods consist of the isolation of the polycyclic group of compounds, followed by estimation of the individual species. Column, paper, and thin-layer chromatography have all been used in the initial separation, and this has been followed by ultraviolet absorption or fluorimetric examination. The limits of polycyclic estimation by these methods range from the microgram region for the column chromatography/ultraviolet method to the nanogram region for the thin-layer/fluorescence method.

The application of gas chromatography to the analysis of polycyclic compounds has also been developed over the last few years. To some extent this technique has been criticized for its lack of sensitivity. The detection limit for benzo(a)pyrene, for instance, using a flame ionization detector, is only approximately 5 μg; this sensitivity is much lower than that obtainable with any of the fluorimetric procedures. Furthermore, peaks arising from benzo(a)- and benzo(e)pyrene are hardly resolved at all using conventional analytical columns. Gas chromatographic methods, however, possess the advantage that a simultaneous analysis can be carried out for many of the more volatile polycyclic compounds. More recently, the sensitivity of detection has improved, particularly as a consequence of the use of capillary columns and electron capture detectors.

In a recent study on the mechanism of formation of polycyclic compounds during the incomplete combustion of ethylene, the advantages of using high-resolution mass spectrometry in the identification and estimation of polycyclic compounds became apparent. Use of the integrated ion current technique allows the assay of polycyclic compounds in the subpicogram range (12).

A series of calibration curves for typical polycyclic compounds, using quantities in the nanogram range, are shown in Figure 5. The calibration curves were linear in all the ranges used. The differences in the slopes of the various curves are accounted for by the variation in sensitivity of the instrument to the different polycyclic aromatic hydrocarbons. By decreasing the concentrations of the solutions used, the limits of detection were found for several polycyclic aromatic hydrocarbons; these limits are shown in Table V. These results do not represent the ultimate sensitivity of the IIC method, since further amplification of the ion current is possible. Such amplification would lower the limit of detection by at least another two orders of magnitude.

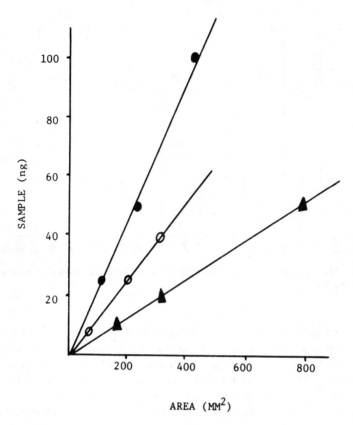

Figure 5. Calibration curves for various polycyclic hydrocarbons. ● Anthracene; ○ benzanthracene; ▲ benzopyrene.

TABLE V

Detection Levels of Some Polycyclic Hydrocarbons

Polycyclic hydrocarbon	m/e Value used (M$^+$)	Sensitivity (mm^2/μg)	Limit of detection (g)
Anthracene	178	4,260	1×10^{-11}
Benzanthracene	228	6,720	5×10^{-13}
Benzo(a)pyrene	252	15,800	5×10^{-14}
Fluorene	166	6,200	1×10^{-12}
Pyrene	202	5,500	1×10^{-11}

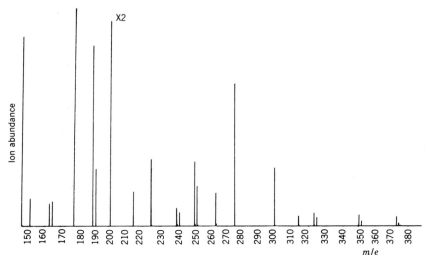

Figure 6. Mass spectrum of a complex hydrocarbon mixture at low electron-accelerating voltage (10 V).

By using successive aliquots of a solution at appropriate m/e values, the various components of a mixture can be determined by this technique, as shown in Table VI. Polycyclic aromatic hydrocarbons are particularly suited to the IIC method because their molecular ion carries a large proportion of the total ion current. Furthermore, if the ionizing electron voltage is reduced to 10 V, the mass spectra of the polycyclic hydrocarbons reduce to a series of single peaks, each characteristic of an individual compound. Such a spectrum is shown in Figure 6; it was obtained from the soot deposited by a low-pressure acetylene flame. It is possible to integrate the ion current obtained at each of the intense mass peaks and to relate it to the amount of hydrocarbon evaporated. There is the possibility, however, that two isomeric hydrocarbons may provide the same mass peak. This possibility may be obviated by a prior chromatographic separation, and a quantitative study has been carried out in which thin-layer chromatography was combined with the IIC technique (13). Provided that the thin-layer plate is calibrated using standard samples, so that the position of any hydrocarbon on the final plate is known, it is possible to transfer samples in the picogram range from the chromatogram to the mass spectrometer (12). Extraction of the polycyclic compound from the cellulose absorbent can be achieved in a micro centrifuge tube with as little as 20 μl of solvent, thus enabling a large part of the eluted sample to be used in the mass spectral assay. A decrease in the thickness of the absorbent layer, with consequent reduction in the volume of the eluting solvent required, will increase the sensitivity of the method still further.

TABLE VI

Determination of Mixtures of Polycyclic Hydrocarbons

Polycyclic hydrocarbon	Amount (ng)	Calculated area from graph (mm²)	Measured area (mm²)
Benzo(a)pyrene	20	310	333
Benzanthracene	20	132	150
Anthracene	20	82	86.3

The chromatographic properties of some polycyclic compounds separated as a mixture by thin-layer chromatography are shown in Table VII. Satisfactory separation was obtained, as well as a ready identification of each substance, using the IIC technique. In some cases, however, the isomeric constitution may be determined directly as described in Section VIII.

TABLE VII

Analysis of Some Polycyclic Compounds First Separated by
Thin-Layer Chromatography

Polycyclic compound	R_f Value[a]	Quantity introduced into mixture (ng)	Quantity detected (ng)
Fluoroanthene	0.73	10	9.8
Benzo(a)anthracene	0.68	10	10.0
Benzo(a)pyrene	0.62	10	11.6
Benzo(g,h,i)perylene	0.56	10	12.1
Coronene	0.45	10	9.4

[a] Separated on silica gel in the solvent system dimethylformamide-water (70–30). See ref. 13 for further details.

2. Detection and Analysis of Polycyclic Hydrocarbons in Extracts from Solid Samples

A mass spectrum of a methylene chloride extract of the soot produced by an acetylene flame showed peaks on most mass numbers up to m/e 378. Many of these peaks were composite in nature, the ions being derived both directly from the molecular ionization process and also as a result of the dissociation of heavier molecule ions. In order to eliminate the uncertainties introduced by integrating the current at such m/e values, the process was carried out at

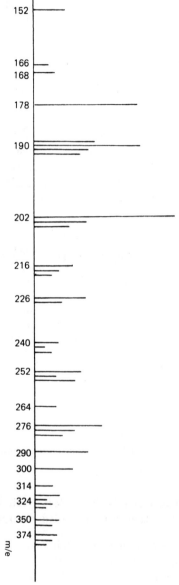

Figure 7. Mass spectrum of a methylene chloride extract of soot at low electron-accelerating voltage (10 V).

reduced electron energy. At an electron-accelerating energy of 10 V virtually all the ion current for any molecule is carried by the molecule ion. It is possible, therefore, to obtain a mass spectrum of a complicated mixture of polycyclic hydrocarbons at an electron-accelerating energy of 10 V and be sure that every peak in the spectrum corresponds to a hydrocarbon molecule ion. A typical example is shown in Figure 7. It is then possible to carry out a series of integrated ion current determinations at all these m/e values and to estimate the concentrations of the polycyclic hydrocarbons in the initial extracts. All the polycyclic hydrocarbons responsible for peaks in the spectrum indicated in Figure 7 were estimated in this way.

3. Estimation of Lead in the Atmosphere and in Petroleum

A preliminary study has been made on the estimation of lead alkyls in petroleum and in the atmosphere. Tetramethyl- and tetraethyllead are too volatile to be estimated by the simple IIC technique. It was possible to eliminate the loss of the lead alkyl from the probe during the initial pumping-down period, however, by using a different introduction method. The glass probe tip was filled with a conventional chromatographic column material, in this case SE 52 silicone gum coated on 80–100 mesh Chromasorb (10% w/w). The sample was injected into this material and then examined in the usual way. Satisfactory IIC curves were obtained for tetraethyllead measured at m/e 293. A linerar calibration graph was obtained in the range 1–100 ng, and the ultimate limit of detection was 1.7×10^{-12} g.

For an analysis of lead alkyl in the atmosphere, air was passed through a probe filled with the silicone gum absorbent. This was then transferred to the mass spectrometer and analyzed in the conventional manner.

4. Other Applications

Perhaps the most fruitful future application of the technique now being studied involves the area of water pollution. The methods used for the assay of polycyclic hydrocarbons in flame gases, air, and soots can be applied very simply to the assay of these compounds in river and in drinking water. In addition, many of the herbicides and pesticides have sufficiently characteristic mass spectra, with substantial molecule-ion peaks, to permit their determination in a similar way.

VI. SURFACE DECONTAMINATION AND DETERGENCY

The interrelated problems of the accurate assessment of the efficiency of detergent solutions and the extent of decontamination of surfaces by such solutions are difficult to solve. An early approach to this problem was to measure the optical density of clean quartz disks in the near-ultraviolet region of the spectrum. The disks were then contaminated with films of substances absorbing in this region. After washing the disks in the detergent solution, the optical densities were again measured and the residual additional density was taken as a measure of the degree of contamination. Such a method is ambiguous, however, since the detergent solution itself may deposit some absorbing material; in addition, the sensitivity of the method is low unless materials of very high molar absorptivity are used as contaminants.

An improvement in the precision and sensitivity of the method is obtained if the disks are contaminated with a fluorescent substance or if the contaminant itself is mixed with a small quantity of a fluorescent substance. A suitable combination would be anthracene and a hydrocarbon grease. The residual fluorescence of the quartz disk after treatment with a detergent solution may be related to the weight of material still adhering to the disk. The disadvantages of this method are that a special apparatus must be constructed in order to measure the fluorescence intensity, and the surface itself must not fluoresce when irradiated with the ultraviolet light (some lower-grade fused silica and quartz surfaces do, in fact, fluoresce).

Currently the most sensitive method for the detection of residual contamination on a surface is radioactive counting. This procedure, however, possesses the disadvantages of any radioactive method, namely, the need for careful handling of all radioactive materials, the use of expensive tracers, and the requirement for counting apparatus.

The integrated ion current technique is ideally suited to a study of the decontamination of surfaces by detergent solutions. The technique has been somewhat adapted by using a small cylindrical rod as the specimen surface. The material to be used as contaminant should be capable of evaporating at temperatures below 500° but should not exhibit an appreciable rate of evaporation at room temperature. It should have a mass spectrum with an easily identified intense peak in an accessible region of the spectrum. The compound chosen for the present study was benzo(a)pyrene, a carcinogenic substance present in the exhaust gases from internal combustion engines. Known weights of this material were evaporated from the probe, and a calibration curve was plotted (see Figure 8). Decon 75 (Decon Laboratories, Ellen Street, Portslade, Brighton, BN4, U.K.) was chosen as the decontaminating agent because of its high efficiency in cleaning contaminated surfaces. A Pyrex glass rod 28 mm long and 1.7 mm in diameter was immersed in a solution of benzo(a)pyrene in benzene to a depth of 10 mm, removed, and dried. The amount of benzo(a)pyrene remaining on the surface was estimated by the above method. The rod was reimmersed, dried, and then

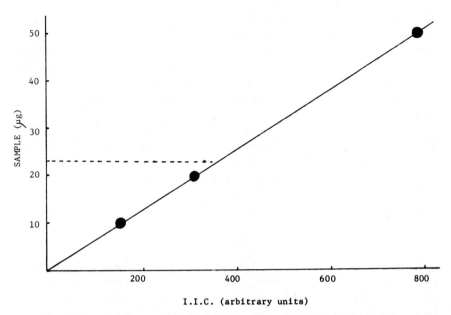

Figure 8. Calibration curve for benzo(a)pyrene. The dotted line indicates the weight required to form a single monolayer on the glass rod (see text for further details).

washed in various concentrations of Decon 75 solution for a number of different times. After drying, the amount of benzo(a)pyrene remaining on the glass rod was again determined by the same technique. The results showed that washing at room temperature removed all of the contaminant down to a certain fixed concentration, whatever strength of Decon 75 was used. A similar result was obtained, although the removal of benzo(a)pyrene was slower, when other detergent mixtures were employed.

The cross-sectional area of a benzo(a)pyrene molecule is approximately 1.1 nm², and the surface area of the glass rod is 0.53 cm² or 53 × 10¹² nm. The molecular weight of benzo(a)pyrene is 252, and the amount of this material required to produce a molecular monolayer on the glass rod can be calculated to be approximately 22.7 ng, assuming that the molecules lie flat on the surface. The amount of benzo(a)pyrene remaining on the glass surface after cold washing was found to be approximately 37 ng. It is seen, therefore, that the treatment of a contaminated surface with detergent solutions at room temperature leads to a stripping of the contaminant down to the state where the surface is covered by a single layer of molecules. This finding was confirmed using a different hydrocarbon (*cis*-stilbene) as the contaminant. When the temperature of Decon 75 solution is raised during the washing process, the amount of contaminant remaining on the glass surface is reduced below that required for a monolayer. A single wash in a boiling detergent solution leaves less than one fiftieth of this amount. Washing with distilled water after the initial detergent wash has little effect on the amount of contaminant removed; in fact, the only factor which influenced the degree of contamination was the length of exposure to the hot solution.

This technique provides a unique method for studying the efficiency of detergent mixtures and the degree of decontamination of surfaces. A further variation of the method permits the determination of the amount of detergent itself which becomes absorbed on the surface during the decontamination process. In the case of Decon 75, no evidence was found for such an absorption process.

1. Applications

A possible further application of this technique could be the identification of various lipid and other moieties in extracts of cell membranes and their quantitative estimation. If the surface area of the membrane were calculated [and this would seem to be a possibility, at least in the case of red cell membranes, bacterial membranes, and/or yeast protoplast membranes (14)] the existence of possible monolayers or bimolecular layers of lipid material could be assessed. Additionally it might be possible to assess the frequency of occurrence or density of rarer molecules (perhaps those possessing interesting functional groups or connected in some way with the transport process) by this method.

VII. ANALYSIS OF COMPLEX BIOLOGICAL EXTRACTS

Non-mass-spectrometric analysis of substances present in biological fluids nearly always includes extraction of the substance, or group of substances, of interest, followed by reaction of the extract with some reagent to produce a colored or fluorescent derivative. The derivative so formed can sometimes be quantitated. The specificity of such analytical procedures is clearly limited unless extensive fractionation and separation precede the quantification stage. Most frequently the methods adopted involve organic solvent extraction, ion exchange, or absorption chromatography, or various combinations of all three. The group of substances obtained in this way is then usually separated further by chromatography on various thin media. After chromatographic separation the substances of interest are visualized after conversion to chromaphores or fluorophores (sometimes it is the derivatives themselves that are chromatographically separated). An enhancement of specificity occurs if the chromogenic or fluorogenic reaction applied to a particular metabolite is specific for that metabolite. Because of the large number of substances present in biological extracts, however, it is difficult to be sure that the analytical procedure used results in an unambiguous analysis.

The lower limit of detection with these classical procedures varies widely but in the main is in the 10^{-7} to 10^{-5} g range (15,16). In favorable cases where fluorescent derivatives are used it is sometimes possible to detect as little as 10^{-8} g. For quantitative purposes, however, even in these cases the lower level is seldom better than about 25×10^{-9} g. Gas chromatography or sophisticated ion-exchange separations do not offer improved sensitivities; in fact, these methods approach the above sensitivity levels only for special groups of substances. In all cases the procedures are time consuming and involve frequent transfers and separative procedures.

There is nearly always some degree of ambiguity in any identification. To cite a specific example we recently developed a thin-layer chromatographic procedure for the separation and quantitative analysis of some noncatecholic primary aromatic amines in rat brain. The overall process involved organic solvent or ion-exchange isolation of an amine fraction from a brain homogenate, removal of catecholic substances by adsorption on alumina, paper chromatographic separation, elution of the appropriate amine zone, dansylation of the substances present in the zone eluate, and two further chromatographic separations on thin layers of silica gel, followed by a final elution of the amine of interest (delineated by running standards in parallel) and measurement (365 mμ, 520 mμ) in the spectrophotofluorimeter. Although this analysis requires 3 full working days and the lower level of detection is about 5×10^{-9} g (12.5×10^{-9} g in the initial extract, since the overall recovery is about 40%), we felt confident of the identification and quantification only after confirmation by precise mass analysis and use of the IIC technique (see refs. 17–21 for further details). Moreover, the whole process of extraction, identification, and quantification using the MS-9 involves less than 2 hr of laboratory time.

1. Urine Extracts

p-Tyramine in a phenolic amine extract (22) of human urine was analyzed after paper chromatographic separation and formation of the 1-nitroso-2-naphthol fluorophore (15). Quantification of urinary p-tyramine by this procedure is time consuming (18 hr, of which 14 is overnight chromatographic separation) but fairly accurate and reproducible (±5.2% fiducial limits) in the range 5×10^{-7} to 10^{-5} g.

p-Tyramine was identified in a similar phenolic amine extract by precise

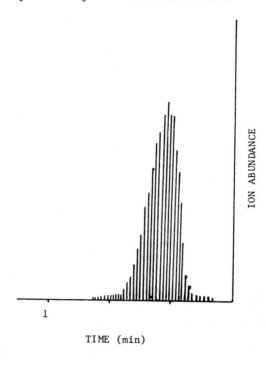

ION ABUNDANCE

1

TIME (min)

Figure 9. Integrated ion current curves AT^m/e 108 during evaporation at 220°C of p-tyramine (1 ng).

mass analysis using the MS-9, giving at m/e 137 a precise mass of 137.0841 $C_8H_{11}ON^+$. A record of the ion current for p-tyramine (1 ng of free base) dissolved in ethanol-water, 70:30 (v/v), at m/e 108 during evaporation at 220°C from the probe is reproduced in Figure 9. A typical calibration curve (overall range for this method of analysis is 10^{-12} to 10^{-5} g with an error ±5.0% fiducial limits) is shown in Figure 12.

When an aliquot (5 μl) of the urinary phenolic amine extract dissolved in ethanol-water, 70:30 (v/v), was evaporated from the probe at 220°C, the ion current curve was elongated as shown in Figure 10. Quantitative values could only be obtained by applying arbitrary limits to the curve width. This shape of curve suggested that the p-tyramine was complexed in a way that limited its ability to evaporate from the probe. Because a sodium acetate treatment of the ion-exchange resin was included in the preparation of the urinary phenolic amine extract (22), we suspected that the amine existed as its acetate salt. By dissolving the extract in ethanol–ammonia, 70:30 (v/v),

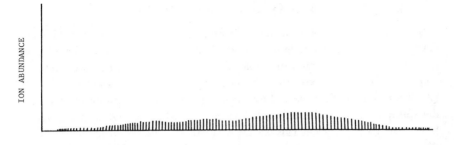

Figure 10. Integrated ion current curves at m/e 108 during evaporation at 220°C of a urinary phenolic amine extract dissolved in 70% ethanol.

the amines were converted to the free base and an evaporation from the probe produced the usual record (see Figure 11). Values obtained for the p-tyramine content of several urinary phenolic amine extracts, some supplemented by the addition of p-tyramine, agreed with the expected values and with those obtained using the nitrosonaphthol chromatographic procedure.

It is clear that most of the constituents present in crude urinary extracts could be unambiguously identified and quantitated by using the mass spectrometric techniques described in this review. Kryptopyrrole (2,4-dimethyl-3-ethylpyrrole) (23), and p-hydroxyphenylacetic acid have already been positively identified in urine, the former for the first time.

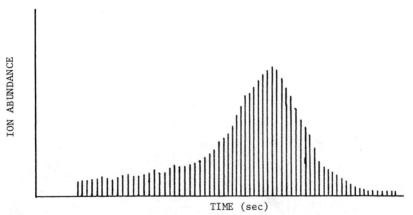

Figure 11. Integrated ion current curves at m/e 108 during evaporation at 220°C of a urinary phenolic amine extract dissolved in ethanol-ammonia, 70:30 (v/v).

2. *p*-Tyramine in Rat Brain

Several noncatecholic primary aromatic amines [*p*-tyramine (17–21), octopamine (24), phenylethylamine and tryptamine (25)] have been identified and quantitated in whole rat brain and certain rat brain regions. The dissection and extraction procedures have been described in detail (21,26). Briefly, in the case of the analysis of *p*-tyramine, rats were stunned and decapitated and the heads chilled in an ice-salt freezing mixture. Whole brains or dissected areas (dissection was carried out on an ice-chilled Petri dish) were washed in ice-cold saline solution, blotted, weighed, and homogenized in $0.4N$ $HClO_4$ solution. At this stage an aliquot of *p*-tyramine (1 mg) was added in those cases where recovery values for the procedure were to be determined. The homogenate maintained at 4°C was then centrifuged at $10,000g$ for 20 min, the supernatants adjusted to pH 5.0 to 6.0 and percolated through a column (1 × 2.5 cm) of Biorad AG5OW–X2 resin (H^+ form), and the phenolic amine fraction eluted in normal NH_4OH (5 ml), after washing with $0.1N$ sodium acetate (10 ml), and water (5 ml). The extract was rotary evaporated to dryness under reduced pressure, transferred in 500 µl 70% ethanol solution containing 3 drops concentrated HCl to a 5-ml conical flask and reevaporated to dryness. This final evaporate was dissolved in 50 µl ethanol-ammonia, 70:30 (v/v), and 5 µl placed on the probe of the MS-9 mass spectrometer. A typical calibration curve and concentration of

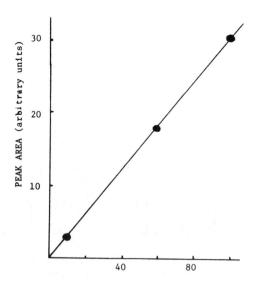

Figure 12. Typical calibration curve for *p*-tyramine.

p-tyramine in whole rat brain and five regions (hypothalamus, caudate nucleus, cerebellum, stem, and "rest") are shown in Figure 12 and Table VIII, respectively.

TABLE VIII

Concentration (ng/g Tissue) of p-Tyramine in the Rat Brain

Brain region	Weight[a] (g)	p-Tyramine (ng/g)
Hypothalamus	0.044 ± 0.007	1940 ± 224[b]
Caudate	0.76 ± 0.010	892 ± 256[b]
Cerebellum	0.269 ± 0.010	316 ± 49[b]
Stem	0.228 ± 0.011	198 ± 21[b]
"Rest"	1.307 ± 0.035	53 ± 3[b]
Whole brain	1.959 ± 0.101[c]	189 ± 33[c]

[a] Mean ± standard deviation (n = 6).
[b] Mean ± mean deviation (n = 5).
[c] Mean ± standard deviation (n = 9).

A complication arising during the direct analysis of p-tyramine in the cerebral phenolicamine fraction—but adequately resolved by the IIC procedure (see Figure 13)—was the presence of contaminant ions at both m/e 108 and m/e 137. Precise mass analysis allocated 137.0841 $C_8H_{11}ON^+$ to p-tyramine, 137.1330 $C_{10}H_{17}^+$ to a hydrocarbon, and 108.0575 $C_7H_8O^+$ to p-tyramine, 108.0939 $C_8H_{12}^+$ to a hydrocarbon, respectively. The origin or nature of the hydrocarbon moiety is not at present understood, but it seems that lipid-like materials affect both the recovery and analysis of β-phenylethylamine and tryptamine in rat brain (27). Consequently, a somewhat modified procedure has been developed for the analysis of β-phenylethylamine in several tissues of the rat. This recent work has been performed using the DNS derivative of phenylethylamine, deuterated phenylethylamine as internal standard, an MS-902S mass spectrometer, and the active collaboration of Philips, Durden, and Davis in Saskatoon.

3. β-Phenylethylamine in Some Tissues of the Rat

Rats were stunned; the brain, liver, lung, heart, and kidneys were removed, chilled on solid CO_2, weighed, and homogenized in 20 ml ice-cold 0.4N $HClO_4$ and Triton X-100 (Canadian Laboratories, Toronto) added to a final concentration of 0.5% (w/v). Phenylethylamine 1,1-didentero-2-phenylethylamine (25 or 100 ng) was then added. After centrifugation at

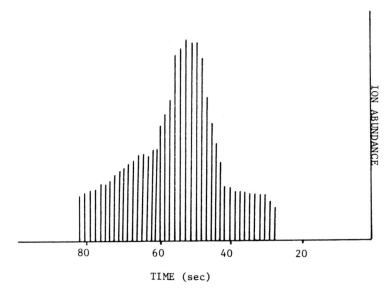

Figure 13. Integrated ion current curves at m/e 108 during evaporation at 220°C of a rat brain phenolic amine extract dissolved in ethanol-ammonia, 70:30 (v/v).

12,000g for 10 min the supernatant was neutralized and percolated through a Biorad AG50W-X2 (H⁺) resin column, and after washing with water, sodium acetate and water eluted with 10 ml methanol-HCl, 73:27 (v/v). After drying, this eluate was dissolved in 1 ml sodium carbonate (10% w/v) mixed with 1.5 ml DNS reagent (Calbiochem, Los Angeles) (1 mg/ml in acetone) and allowed to stand at 37°C for 2 hr. After precipitation of the sodium carbonate with excess acetone (9 ml) the resultant solution (after drying) was transferred in benzene-acetic acid, 99:1 (v/v) (2 × 100 µl) to the origin of a thin layer of silica gel (Mondray Ltd., Montreal) and developed with chloroform-butyl acetate, 5:2 (v/v). After separation the damp plate was sprayed with isopropanol-triethanolamine, 4:1 (v/v) to stabilize the fluorescence. The DNS-phenylethylamine zone was then eluted and separated on Mondray plates in benzene-triethanolamine, 8:1 (v/v), then on Kodak 6061 plates in the organic phase of a mixture of light petroleum (b.p. 100 to 120°C)-toluene-acetic acid-water, 66:33:50:50 (v/v). After this final separation the DNS-phenylethylamine zone was quickly eluted without fluorescence stabilization and dried. The residue was dissolved in 500 µl ethanol and 5 µl ethanol was introduced in the probe. A suitable reference peak (m/e 337.9839) of heptacosafluoro-tri-n-butylamine was selected to locate the DNS-phenylethylamine molecule ion (m/e 354.1402) and the deuterated DNS-phenylethylamine (m/e 356.1528). The

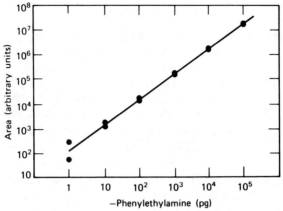

Figure 14. Typical calibration curve of β-phenylethylamine (as its DNS derivative).

machine was then arranged to record the ion currents (IIC procedure) switching between the two amines 354.1402 and 356.1528. The amount of phenylethylamine in any and all samples was then calculated from the following formula:

$$\frac{\text{area } m/e \ 354.1402}{\text{area } m/e \ 356.1528 \ - \ 7\% \ \text{area } m/e \ 354.1402} \times y \text{ ng}$$

where y is the amount of deuterated phenylethylamine originally introduced as internal standard. The value 7% represents the contributions at m/e 356.1528 ± 0.03 of the ^{13}C, ^{34}S and ^{18}O isotopes of DNS-phenylethylamine. A typical calibration curve (crystalline DNS derivate kindly provided by Nikolaus Seiler) and the endogenous concentration of β-phenylethylamine in several tissues of the rat are illustrated in Figure 14 and Table IX, respectively.

TABLE IX

Concentration of β-Phenylethylamine in Several Tissues of the Rat

Tissue	Weight[a] (g)	β-Phenylethylamine[b] (ng/g)
Brain	1.830 ± 0.060	1.7 ± 0.4
Heart	0.571 ± 0.066	6.6 ± 2.6
Kidney	1.625 ± 0.308	22.8 ± 2.7
Liver[c]	—	1.9 ± 0.7
Lung	1.180 ± 0.181	4.2 ± 1.1
Spleen	0.618 ± 0.099	6.1 ± 1.6

[a] Mean ± Standard deviation ($n = 10$).
[b] Mean ± standard deviation ($n = 8$).
[c] In the case of liver approximately 3.3 g of tissue was used.

4. Applications

It is clear that precise mass analysis and the IIC technique of quantification will find extensive application in the identification and analysis of minute amounts of metabolites present in the various body fluids (urine, blood, cerebrospinal fluid, synovial fluids, etc.), especially substances that are controversial, such as the various chromatographically separated "spots" supposed to refer to specific metabolites in urine collected from psychotic individuals. As has been described here, the IIC technique can be used to show that any procedure developed is indeed as specific for the metabolite in question as it is supposed to be. In addition to extending the method to include analysis of numerous different metabolites in body fluids and tissue extracts, it is possible to extend the studies of cerebral regionalization to even smaller anatomical substructures, as well as to the various subcellular fractions obtained on differential or density gradient separation of cells.

Probably it is now feasible to determine the chemical constitution of small biopsy samples, and in this laboratory we hope to extend our studies to an analysis of cerebral biopsy specimens obtained in various psychiatric and neurological conditions. Obviously it will be possible to analyze biopsy specimens taken from the brains of certain primates. The analysis of large single cells or homogenous cell populations (neurones, glia, bacteria, etc.) obtained either by dissection, fractionation, synchronous culture, or tissue culture techniques may now be feasible. In neuropharmacology it will clearly be possible to investigate the distribution within the body or the brain of a variety of drugs and their respective metabolites, as well as the effects of these drugs on the distribution of normal metabolites both regionally and subcellularly. It ought to be possible, in view of the subnanogram sensitivity of the IIC procedure, to test directly whether certain of the assumed transmitter substances actually are released on stimulation. Furthermore, in this regard, it will be possible to postulate and to look for new transmitter substances and cerebral constituents and metabolites.

VIII. DETECTION OF ISOMERISM

When a sample consists of two substances each of which exhibits a peak in the mass spectrum at a specific m/e value, the integrated ion current curve at this m/e value may show fine structure because the rates of evaporation of the two components may be different. This effect is most marked when recording the molecule ion current of a substance which can exist in two isomeric forms. In a recent study (28) the IIC curves were recorded for samples prepared by mixing known proportions of pure isomers and also for samples known to be mixtures of naturally occurring isomers.

In all cases where the sample contained more than one isomeric species, the IIC curve exhibited fine structure, although the temperature at which this was revealed was sometimes rather critical. In no case did the curve for a single pure substance show any fine structure. Using an instrument equipped with a total ionization detector, it is possible to compare the integrated total ion current with the curve obtained at a specific m/e value. It is then possible to deduce how much of the total sample evaporated is due to a species the spectrum of which can contribute at the selected m/e value.

1. Dioximes

Theoretically dioximes may exist in three isomeric forms. It has been suggested that furil dioxime may exist only in two of these forms, namely, the α and γ. The integrated ion current curve obtained with a sample of furil dioxime is shown in Figure 15a; only two peaks can be detected. A similar result is obtained with phenylglyoxime (Figure 15b). The difference in the size of the two peaks must reflect the relative concentrations of the two isomeric species in both compounds, and it is interesting that for phenyl-glyoxime one form is predominant. Unfortunately, in the absence of pure samples of the separate isomers it is not possible to identify the peaks.

2. Structural Isomers

The ready availability of samples of pairs of isomers of disubstituted benzenes made it possible to investigate this form of isomerism and to make a positive identification of the separate peaks. Figure 16 shows the curves obtained with mixtures of o- and p-nitrophenol and o- and p-nitroaniline. It can be seen that almost complete separation has been achieved and that the first peak to be recorded is much sharper and narrower than the second. In order to avoid evaporation of the samples before the probe was lowered, they were initially cooled in liquid air. The ion source temperature required for resolution was critical and was found to be 215° for the nitrophenols and 220° for the nitroanilines.

Figure 17 shows the integrated ion current curves obtained with mixtures of o-, m-, and p-tyramine, recorded at m/e 137 and a source temperature of 220°C. The presence of isomers is clearly revealed by the fine structure in the curves. The IIC curves obtained on evaporation under identical conditions of the phenolic amine extract obtained from a rat brain (as described in Section VII and [26]) revealed no such structure (Figure 18a), nor did the

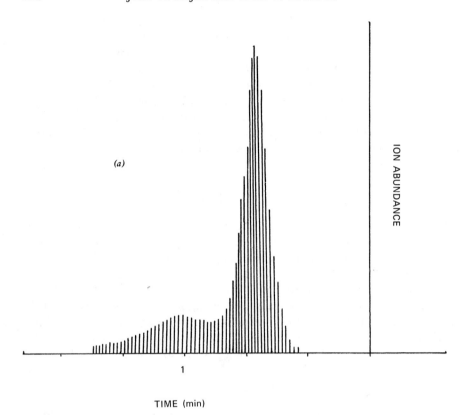

(a)

ION ABUNDANCE

1

TIME (min)

Figure 15. Integrated ion current record of (a) furildioxime and (b) phenylglyoxime.

same extract after supplementation with pure *p*-tyramine (Figure 18*b*). This result was confirmed by measuring the ratio of the IIC curve areas at *m/e* 108 and *m/e* 137 (value 6.43) and showing that they matched almost exactly the ratio of the two corresponding peak heights (value 6.50) attained in a normal mass spectrum of *p*-tyramine.

3. Cis-trans Isomers

Figure 19 shows the integrated ion current curve obtained by evaporating a synthetic mixture of *cis*- and *trans*-stilbene at a temperature of 280°. Unexpectedly, the resolution of these two isomers was poor despite the difference in their melting points and the demonstration that they can be successfully separated by gas chromatography with Apiezon M as the

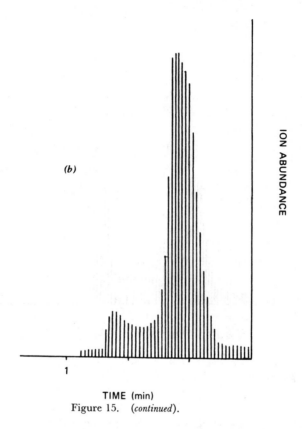

TIME (min)

Figure 15. (*continued*).

stationary phase. Nevertheless, the curve showed sufficient structure to reveal the presence of two isomers.

4. Isomeric Polycyclic Hydrocarbons

Figure 20 illustrates the integrated ion current curves obtained by evaporating a mixture of phenanthrene and anthracene at a temperature of 232° and a mixture of chrysene, naphthacene, and benzanthracene at a temperature of 320°. In order to prevent premature evaporation the samples were initially cooled in liquid air. Whereas the former pair of isomers was clearly separated, the resolution of the triplet was only partial, although there was sufficient structure and peak width to reveal the presence of three separate species.

Benzo(a)pyrene is a carcinogenic substance which is usually found in products of incomplete combustion, while benzo(e)pyrene and perylene are two accompanying comparatively noncarcinogenic isomers. Because of the

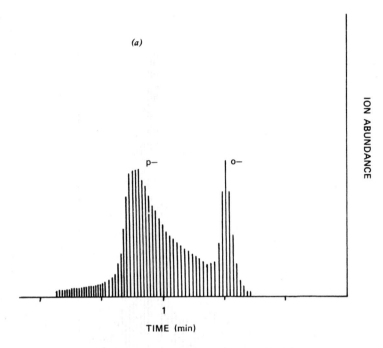

Figure 16. Integrated ion current curves for mixtures of (*a*) *o*- and *p*-nitrophenol and (*b*) *o*- and *p*-nitroaniline.

great sensitivity of the integrated ion current method, it is possible to envisage the identification of the carcinogenic benzo(a)pyrene in submicrogram samples of smokes and soots. It has already been shown that such a separation is very difficult to achieve by routine gas chromatography. Figure 21 illustrates the IIC curves obtained on evaporation at a temperature of 270° of mixtures of benzo(a)pyrene and benzo(e)pyrene (Figure 21*a*), benzo(a)-pyrene and perylene (Figure 21*b*), and benzo(s)pyrene, benzo(e)pyrene, and perylene (Figure 21*c*), respectively. It can be seen that benzo(1)pyrene can be separated from perylene or from benzo(e)pyrene, although benzo(e)-pyrene cannot be separated from perylene.

5. Conformational Isomers

Cloran (see structure in Figure 22) can exist in six isomeric forms. When a sample was evaporated at a temperature of 200°, the integrated ion current curve obtained had the form illustrated in Figure 22. Although no complete resolution of the isomers was obtained, five distinct peaks may be identified.

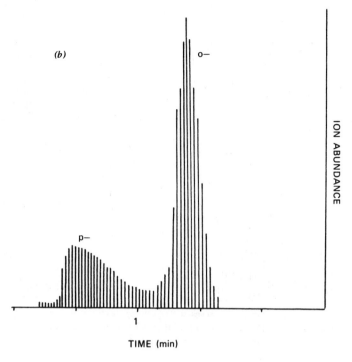

(b)

o—

p—

ION ABUNDANCE

1

TIME (min)

Figure 16.   (*continued*).

6. Metal Chelates

The rare-earth metal derivatives of unsymmetrical 1,3-diketones, such as the holmium derivative of trifluoro-5,5-dimethoxyhexan-2,4-dione $[Ho(CF_3COCHCOC_4H_9)_3]$, can exist in two isomeric forms. On evaporation at a temperature of 190° the resulting integrated ion current curve. exhibits two peaks of equal height (Figure 23). Similar results were obtained for $Ho(C_2F_5COCHCOC_4H_9)_3$ at 225° and $Ho(C_3F_7COCHCOC_4H_9)_3$ at 220°.

No fine structure was observed in records obtained from tris chelates of symmetrical 1,3-diketones. 8-Hydroxyquinoline is an unsymmetrical ligand; metal derivatives may exhibit isomerism so long as the structure is planar (i.e., in the case of divalent metals) (29). Figure 24 illustrates the record obtained for zinc oxinate and reveals the existence of the two planar isomers. A single peak was obtained for the tetrahedral copper oxinate. The trivalent metal oxinates can also exist in two forms, and this isomerism is demonstrated in the records obtained from iron and aluminum oxinates (Figure 25).

a.   b.

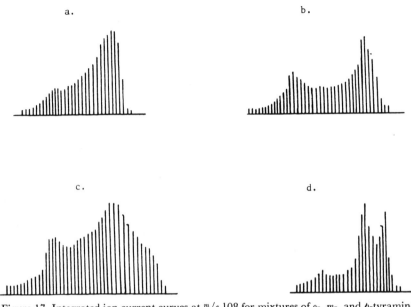

c.   d.

Figure 17. Integrated ion current curves at $^{m}/e$ 108 for mixtures of o-, m-, and p-tyramine: (a) o- and p-tyramine; (b) m- and p-tyramine; (c) o- and m-tyramine; (d) o-, m-, and p-tyramine. In these figures the ordinate represents ion abundance, and the abscissa time (in seconds).

7. Discussion of Detection of Isomerism

The fine structure in the ionization curves is due to differences in the rates of evaporation of the isomers. In all cases where isomers were known to be present in the sample, the curves either exhibited fine structure or else were broader than those obtained with samples of single isomers. The evaporation is not an equilibrium process, and it occurs under rather unusual conditions. The sample is present in the form of an extremely thin film and is flash-heated in a high-vacuum system. The factors influencing the differences in the rates of evaporation are, therefore, difficult to identify; the only clue is the critical nature of the flash temperature for a successful resolution. The fact that isomerism can be demonstrated with sample sizes below 10^{-9} g makes this technique unique.

When an integrated ion current curve is recorded together with a total ionization current curve, further information about the purity of the sample and the rates of evaporation of the individual components becomes available. It is possible to study the structure on the total ionization current curve and to identify it by reference to the IIC curve; if the instrument possesses

a.

b.

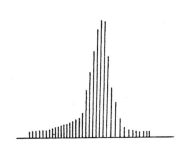

Figure 18. Integrated ion current curves at $^m/e$ 108 for (*a*) rat brain phenolic amine extract and (*b*) rat brain phenolic amine extract supplemented with *p*-tyramine. In these figures the ordinate represents ion abundance, and the abscissa time (in seconds).

sufficiently fast recording facilities, it is then possible to observe the rise and fall of the total ionization current curve and to record mass spectra at times when different isomers are being evaporated from the sample probe. The mass spectra of single isomers can thus be recorded from a sample mixture, and differences observed. If the recording facilities are not sufficiently fast, the same information can be obtained by comparison of the IIC curves for the molecule ion and for the fragment ions. This is a slower but more sensitive way of revealing slight differences in the mass spectra of isomers. The technique may be particularly useful when it is necessary to study changes in the concentration of isomers during the course of a reaction, as, for example, in photoinitiated cis-trans interconversion or in substitution reactions of aromatic compounds.

IX. SUMMARY

Unless extensive measures are taken, it is difficult (and on occasion not feasible) with most of the frequently used physicochemical techniques

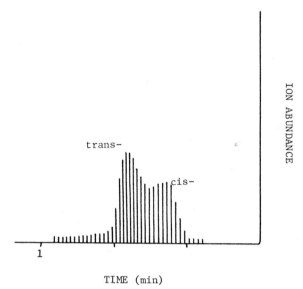

TIME (min)

Figure 19. Integrated ion current curve for a mixture of *cis*- and *trans*-stilbene.

(ultraviolet, infrared, and visible spectrometry and paper, thin-layer or ion-exchange chromatography) to be certain that the identity of an unknown has been established beyond doubt, especially for metabolites present in micro- and submicroamounts in complex biological fluid extracts. Lately gas chromotagraphic-mass spectrometer, fast-scan, single-focusing facilities are being increasingly used. Although such systems provide unambiguous identification, their application is somewhat limited since the substances to be analyzed have first to be separated on the gas column; furthermore, the lower limit of detection sensitivity is restricted to that obtainable by the method of detection used to detect solutes in the column effluent.

The enhanced certainty of identification provided by precise mass analysis using large, double-focusing mass spectrometers makes this the method of choice when the facility is available. The resolution now offered (2 ppm with the new-generation machines) makes positive identification possible even in the presence of gross contamination at any particular m/e value. The integrated ion current technique permits quantification in the 10^{-12} to 10^{-5} range even in the presence of contaminant ions. Most substances, whether of chemical or biological interst, are amenable to analysis in this way so long as they are volatile in high vacuum in the temperature range $-50°C$ to $+500°C$. Of course, nonvolatile substances can frequently be made volatile by conversion to some suitable derivative. In the rare cases where the

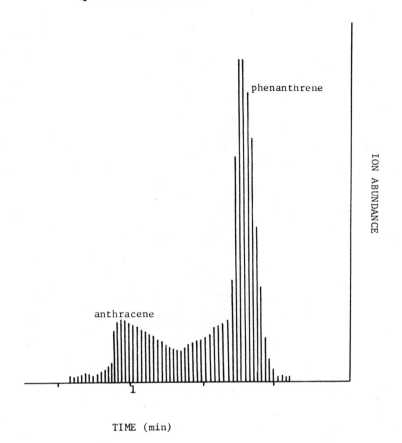

Figure 20. Integrated ion current curve for a mixture of phenanthrene and anthracene.

overlapping contamination for a particular ion is too serious to allow identification or quantification, a different ion can be selected or else the substance can be converted to some suitable derivative.

The applications discussed in this review—trace metal analysis, studies of pollutants, surface decontaminants, and detergents, separation of isomers, and the analysis of various constituents of complex biological fluids—are only a small proportion of all possibilities. Because of the unambiguity of identification, sensitivity of detection, and quantification in the submicro-region, the IIC technique is the method of choice in biochemical applications, especially in neurochemistry and neuropharmacology. The only snag would seem to be the expense of the installation (approximately $100 to $150K) and the need for skilled and knowledgeable personnel.

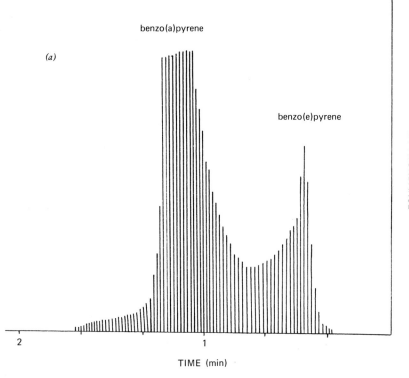

ION ABUNDANCE

benzo(a)pyrene

(a)

benzo(e)pyrene

2   1

TIME (min)

Figure 21. Integrated ion current curves for mixtures of (a) benzo(a)pyrene and benzo(e) pyrene; (b) perylene and benzo(a)pyrene; (c) benzo(e)pyrene, benzo(a)pyrene, and perylene.

508

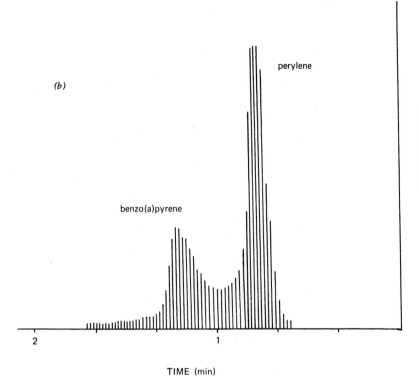

Figure 21. (*continued*).

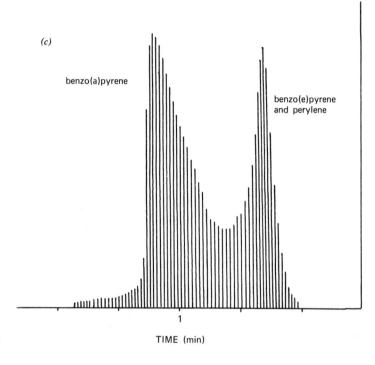

Figure 21. (*continued*).

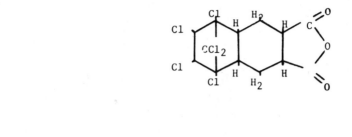
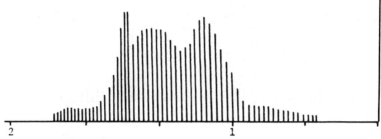

TIME (min)

Figure 22. Integrated ion current record for cloran.

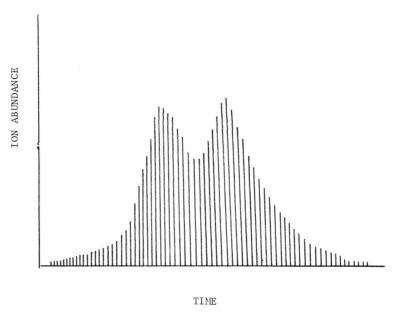

Figure 23. Integrated ion current curve for $Ho(CF_3COCHCOC_4H_9)_3$.

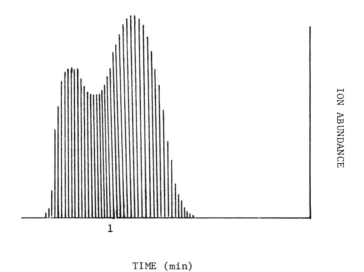

Figure 24. Integrated ion current record for zinc oxinate.

512

a.

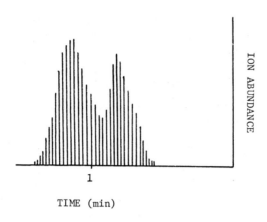

ION ABUNDANCE

1

TIME (min)

b.

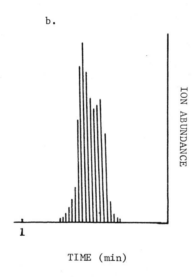

ION ABUNDANCE

1

TIME (min)

Figure 25. Integrated ion current record for (*a*) iron oxinate and (*b*) aluminum oxinate.

Acknowledgments

One of us (A.A.B.) thanks the Psychiatric Services Branch, Province of Saskatchewan, and the Medical Research Council of Canada for continuing financial support.

References

1. J. R. Majer, M. J. A. Reade, and W. I. Stephen, *Talanta*, *15*, 373 (1968).
2. J. R. Majer, *Talanta*, *16*, 420 (1969).
3. A. E. Jenkins and J. R. Majer, *Talanta*, *14*, 777 (1967).
4. A. E. Jenkins, J. R. Majer, and M. J. A. Reade, *Talanta*, *14*, 1213 (1967).
5. R. W. Moshier and R. E. Sievers, *Gas Chromatography of Metal Chelates*, Pergamon Press, Oxford, 1965.
6. R. Belcher, J. R. Majer, R. Perry, and W. S. Stephen, *Anal. Chim. Acta*, *43*, 451 (1968).
7. R. Belcher, J. R. Majer, W. I. Stephen, I. J. Thomson, and P. C. Uden, *Anal. Chim. Acta*, *50*, 423 (1970).
8. M. G. Allcock, R. Belcher, J. R. Majer, and R. Perry, *Anal. Chem.*, *42*, 776 (1970).
9. J. R. Majer and R. Perry, *Chem. Commun.*, *454* (1969).
10. R. Belcher, J. R. Majer, R. Perry, and W. I. Stephen, *J. Inorg. Nucl. Chem.*, *31*, 471 (1969).
11. E. Sawicki, T. W. Stanley, W. C. Elbert, J. Meeker, and S. McPherson, *Atmospheric Environment*, *1*, 131 (1967).
12. J. R. Majer and R. Perry, I. V. P. A. C. Meeting on Air Pollution, Cortina, Italy, June, 1969.
13. J. R. Majer, R. Perry, and M. J. A. Reade, *J. Chromatog.*, *48*, 328 (1970).
14. A. A. Boulton, *Exptl. Cell Res.*, *37*, 343 (1962).
15. A. A. Boulton, *Chromatographic and Electrophoretic Techniques*, Vol. I, I. Smith, Ed., William Heinemann, London, 1969.
16. A. A. Boulton, *Methods Biochem. Analy.*, *16*, 328 (1968).
17. A. A. Boulton and Lillian Quan, 2nd International Meeting, Intern. Soc. Neurochemistry, Milan, 1–5, 1969.
18. A. A. Boulton and J. R. Majer, *J. Chromatog.*, *48*, 322 (1970).
19. A. A. Boulton and J. R. Majer, *Nature*, *225*, 658 (1970).
20. A. A. Boulton and Lillian Quan, *Can. J. Biochem.*, *48*, 1287 (1970).
21. A. A. Boulton and J. R. Majer, *Methods Neurochem.*, *1*, 341 (1972).
22. Y. Kakimoto and M. C. Armstrong, *J. Biol. Chem.*, *237*, 208 (1962).
23. D. Irvine, W. Bayne, H. Miyahsita, and J. R. Majer, *Nature*, *224*, 811 (1969).
24. P. B. Molinoff and J. Axelrod, *J. Neurochem.*, *19*, 157 (1972).
25. J. M. Saavedra and J. Axelrod, *J. Pharmacol. Exp. Ther.*, *182*, 363 (1972).
26. A. A. Boulton and P. H. Wu, *Can. J. Biochem.*, in press.
27. A. A. Boulton, P. H. Wu, and S. Philips, *Can. J. Biochem.*, *50*, 1210 (1972).
28. J. R. Majer and R. Perry, *J. Chem. Soc.*, A, 822 (1970).
29. J. R. Majer and M. J. A. Reade, *Chem. Commun.*, *58* (1970).

Author Index

Numbers in parentheses are reference numbers and show that an author's work is referred to although his name is not mentioned in the text. Numbers in *italics* indicate the pages on which the full references appear.

Abou-Issa, H. M., 2(6), *35*
Abraham, G. J., 344(172), 369(172), *459*
Abrahamson, E. A., 329, *457, 458*
Abramowitz, A., 222(202), 224(202), *279*
Ackerman, C. J., 252(453), 255(453), *285*
Ackermant, E., 169(56), *187*
Adams, J. B., 243(366), 259(473), *283, 285*
Adams, R. N., 292(1), *454*
Adamo, G., 344(174), 345(174), 346(174), 347(174), 348(174), *459*
Adler, E., 335(147,148), *458*
Adler, J., 200(30), 238(30), *275*
Ado, K., 384(302), 386(302), 387(302), 403(302), 421(302), 439(302), *462*
Agarwal, S. S., 239(335), *282*
Agranoff, B. W., 253(456), *285*
Agren, G., 80(4), *152*
Airas, R. K., 253(458), 254(458), *285*
Airi, L. T., 253(458), 254(458), *285*
Ajl, S. J., 252(447,448), *285*
Aksenova, E. N., 451(424), *465*
Albers, A., 250(427), *284*
Albersheim, P., 230(280), *281*
Albright, C. H., 319(106), *457*
Algranati, I. D., 213(131), 230(131), *277*
Alivisatos, S. G. A., 344(172), 369(172), *459*
Al-Khalidi, U. A. S., 220(180), *278*
Allan, J. E., 54, *76*
Allaway, W. H., 40(8), 41, 43, 48(11), 53, 70, *74*
Allcock, M. G., 478(8), *514*
Allende, J. E., 254(465), *285*
Almens, A., 225(235), *280*
Al-Mudhaffer, S., 252(453), 255(453), *285*
Alpert, A., 218(165), 226(165), *278*
Alverez, R., 57(116), *76*
American Public Health Association, 49(61), *75*, 163(31), *187*
Amrhein, N., 223(212), *279*
Anastasi, A., 379(226), *460*
Anders, M., 224(223), *279*

Anders, O. U., 65(172), 66, *78*
Anderson, J. S., 255(463), *285*
Anderson, L. E., 251(440), *285*
Anderson, N. G., 80(5–7), 89, 95(6,7), 116(5,6), 134(5), 136(6), *152*
Anderson, R. A., 88(21), 94(21), *152*
Anderson, R. L., 185(121), *189*, 220(182), *279*
Aoki, A., 428(318), *463*
Applemann, M. M., 203(57), 242(356), *275, 283*
Aposhian, H. V., 200(22), 239(22), 249(420), *274, 284*
Arand, S. R., 239(336), *282*
Argrett, L., 58(128), 65(128), 66(128), *77*
Argyrakis-Vomvoyannis, M. P., 235(322), 236(322), *282*
Arion, W., 335(317), *282*
Arion, W. J., 2(9), 16(20,21), 19(21), 31(56), *35–37*
Ariyoshi, H., 51(76), *75*
Armstrong, M. C., 491(22), 492(22), *514*
Arribas-Jimeno, S., 56(104), *76*
Arroyo, A. A., 63(161), *78*
Asato, R., 170(68), *188*
Asahi, Y., 379(223,225), 385(301), 421(301), 442(362), 443(362,387), 444(387), 447(387), 448(362,387), 450(387), 451 (223), *460, 462, 464*
Ashley, J. W., 312, *456*
Ashworth, M. R. F., 388(119), *457*
Association of Official Agricultural Chemists, 43, 67, *74, 78*
Asher, C. J., 40, *74*
Ashworth, J. M., 240(347), *282*
Asrow, G., 169(55), *187*
Atkins, I. C., 211(126), 225(126,246), *277, 280*
Atkinson, M. R., 204(79), 231(79,298), *276, 281*
Attril, J. E., 80(7), 95(7), *152*
Attwood, D., 28(40), *36*

Axelos, M., 241(350), *283*
Axelrod, J., 216(151–153), 221(186), 224(152,153,217), 225(229,230,232, 233), 251(442), *278–280, 285,* 494(24, 25), *514*
Aures, D., 221(192), 251(192,426), *279, 284*
Aw, S. E., 169(63), *187*

Babad, H., 213(132), 230(132,272), *277, 281*
Babson, A. L., 182(118), *189*
Bachynsky, N., 215(147), *278*
Baeyer, A., 3(11), *35*
Baeyer, H. V., 178(103), *188*
Baguley, B. C., 224(224), *279*
Bailey, J. M., 156(5), 158, 159, *186*
Baird, R. B., 54(93c), *76*
Baker, N., 35(67), *37*
Baker, R. H., Jr., 336(160), *459*
Balasubramanian, A. S., 218(168), 244(168), 247(168), *278*
Balis, M. E., 204(80), 248(80), *276*
Balkcom, C. M., 451(423), *465*
Ballard, F. J., 210(119), 219(172), 251(119, 172,439), 255(119), *277, 278, 285*
Ballou, C. E., 230(278), *281*
Bapat, M. G., 56(107), *76*
Bär, H., 295(7,8), *454*
Bär, H.–P., 212, 243(359), 246(391), *277, 283*
Barber, G. A., 219(175), 230(277), *278, 281*
Barchas, J. D., 225(234), *280*
Barchas, R. E., 225(234), *280*
Bard, A. J., 351(193), 443(193,374), 445 (193,374), 446(193,400), *459, 464, 465*
Bardos, T. J., 234(309), *282*
Barger, M. L., 80(5), 89(5), 116(5), 134(5), *152*
Barker, G. C., 314(80), *456*
Barlow, C. A., 316, *457*
Barnes, J. K., 31(55), *37*
Baron, E. S. G., 442(363), 443(363), *464*
Barradas, R. G., 422(303,304), *462*
Barritt, G. J., 209(115), 210(115), 255, *277*
Barthelmai, W., 178, *188*
Baserga, R., 204(74), 206(74), *276*
Basso, A. J., 386(350), *463*

Batt, R. D., 257(472), *285*
Bauer, H. H., 299(48), 316, *455, 458*
Baugnet-Mahieu, L., 203(55), *275*
Baumann, C. A., 44(48), 48(48), 53(48), *75*
Baumann, K., 180(113), *189*
Bayer, I., 315(89), *456*
Baylor, C. R., Jr., 451(430), *465*
Bayly, R. J., 263(475), *285*
Bayne, W., 493(23), *514*
Baxendale, J. H., 344(175), 345(175), 347(175), 348(175), *459*
Beach, E. F., 178(100), *188*
Beaman, A. G., 439(343), *463*
Beath, O. A., 40, 41(10), 47, *74*
Beaufay, H., 2(3), 17(22), *35, 36*
Beck, W. S., 220(178), *278*
Becker, A., 246(388), *283*
Becker, F. F., 243(360), *283*
Becker, H. J., 386(350), *463*
Beckman, E. D., 66, *78*
Behrens, N., 213(131), 230(131), *277*
Belcher, R., 476(6), 477(7), 478(8), 481(10), *514*
Belenkaya, S. L., 379(228), 426(308a), 451(228), *460, 462*
Belfield, A., 245(382), *283*
Belleau, B., 264(477), *285*
Belpaire, F., 215(146), 221(146), 224(146), 251(441), *278, 285*
Benderskii, V. A., 451(429,431), *465*
Bendet, I., 217(163), *278*
Bendich, A., 411(285), *462*
Benedict, C. R., 199(15), *274*
Benkovic, S. J., 377(219a), 278(219a), *460*
Bennett, L. L., 231(297), *281*
Benson, A. M., 87(16), 116(16), *152*
Bentley, R., 157(10), 166(10), *186*
Bera, B. C., 51(80), *75*
Beranek, J., 426(311), 427(311), *462*
Berends, W., 418(295,297), *462*
Berg, H., 295, 330, 385(267), 386(267), 390(267), 393(267), 401, 402(267), 410(267), 421(267), 424(306a), 425(267), 432(329), *454, 458, 461–463*
Berg, P., 202(50), *275*
Berger, A., 224(219), *279*
Bergmann, I., 344(179), 345(179), 346(179), 354(179), 355(179), 355(179), 357(179), 363(179), 369(179), *459*
Bergren, W. R., 204(77), 205(77), 234(306), 240(77), *276, 281*

Bergkvist, R., 116(58), *153*
Bergmeyer, H. U., 184, *189*
Bergmeyer, H-U., 195(5), 197(10), *274*
Bergstrand, A., 30(49), *36*
Berlinguet, L., 208(108), 209(108), 250 (108), *277*
Berthet, J., 17(22), *36*
Berthod, H., 317(101), 435(101), *457*
Bessman, M. J., 213(137,138), 214(137, 138), 236(137,138), 237(137,138), 239 (336), 249(418,419), *277, 282, 284*
Betteridge, D., 58(130), 77
Bewick, A., 312, *456*
Bezman, R., 296(30), *455*
Beyersmann, D., 239(337), 246(337), *282*
Bibliography of Polarographic Literature (1922-1967), 295(9), *454*
Biegler, T., 443(373), 445(373,395), 446 (373,395), 449(408), 451(408), *464,465*
Birnbaumer, L., 242(357), 243(357), *283*
Birnie, G. D., 200(24), 249(424), *275, 284*
Biswas, S. D., 49, 75
Bitensky, M. W., 204(84), 205(84), 243(360, 361), *276, 283*
Blaedel, W. J., 169(41), 176, 187, 188, 350 (191), *459*
Blair, D. G. R., 231(286), *281*
Blakley, R. L., 221(194), *279*
Blanco, M., 243(361), *283*
Blatt, L. M., 228(263), *280*
Blattner, F. R., 116(67), *153*
Blincoe, C., 42(20), 53(20), *74*
Blitz, H., 416(293), 443(293), *462*
Blount, H. N., 445(394), *464*
Blum, H. E. C., 237(330), *282*
Bly, D. D., 439(344), *463*
Blyumenfel'd, L. A., 451(429), *465*
Boardman, N. K., 94(29,30), *152*
Bobst, A. M., 330(129c), *458*
Bock, R. M., 202(51,52), *275*, 369(215), 375(215), *460*
Bogue, D. C., 88(21), 94(21), *152*
Bohacek, J., 300(55), 432(324), 433(324), *456, 463*
Böhme, E., 243(362,363), *283*
Bollum, F. J., 200(25), 201, 202(43), 211, 238(43,334), 246(25,43), 253(43), *275, 282*
Bonhorst, C. W., 47, 51, *75*
Bonnelycke, B. E., 116(62), *153*

Booman, G. L., 299(39), *455*
Boon, W. R., 345(186), *459*
Borek, E., 224(228), *280*
van den Bosch, H., 244(373), *283*
Bosmann, H. B., 230(276), *281*
Bosselmann, R., 234(308), *282*
Botsford, J. L., 225(239), *280*
Boulanger, P., 115, *153*, 442(364), 443(364), 447(364), *464*
Boulton, A. A., 490(14), 491(15-21), 494(17-21,26), 495(27), *514*
Bourke, J. B., 208(106), *277*
Bowen, H. J. M., 58(119,125), 59(125), 62, 65(119,125), *77*
Boyer, J., 244(370), *283*
Boyer, P. D., 249(421), *284*
Brabec, V., 393(356), *464*
Bradley, A., 246(389), *283*
Bradley, R. M., 247(399), *284*
Brady, R. O., 247(399), *284*
Brandt, R., 165, 172, *187*
Brante, G., 244(371), 248(371), *283*
Braun, T., 246(391), *283*
Bray, G. A., 204, *276*
Brdička, R., 312, 443(375-380,386), 444, 445(377), 446(377), 451(375,377-380), *456, 464*
Brech, W., 255(469), *285*
Breitman, T. R., 204(67), 206(67), 231(288), 234(67), *276, 281*
Bremer, J., 227(253), *280*
Bremerskov, V., 234(308), *282*
Brennan, P., 230(278), *281*
Bresnick, E., 204(69), *276*
Bretthauer, E., 42(20), 53(20), *74*
Bretthauer, R. K., 230(279), *281*
Breuer, H., 229(270), *281*
Breyer, B., 299(48,50), 316, 443(373), 445(373), 446(373), 449(408), 451(408), *455, 464, 465*
Brezina, M., 295, 325(20), 337(20), 344(20), 355(20), 442, 451, *455*
Bridger, W. A., 249(421), *284*
Bridges, W. F., 246(385), *283*
Brignac, P. J., Jr., 173(88), 174(89), *188*
Brin, M., 226(247), *280*
Britton, H. T. S., 357(348,349), 443(348), *463*
Brodie, B. B., 242(358), *283*
Brooker, G., 137, *153*, 203(57), 242(356), *275, 283*

Broom, A. D., 331(130), *458*
Broome, J. D., 248(413), *284*
Brouhon, N., 451(412), *465*
Brown, D. D., 225(231), *280*
Brown, D. H., 178(97), *188*
Brown, D. J., 380(243), *461*
Brown, D. M., 247(400), *284*
Brown, G. B., 388(351), 389(351), 390 (351), 391(351), 392(351), *463*
Brown, J. M. M., 58(124), *77*
Brown, P. R., 94, 126, 134, 137, 139–141, *152*
Brown, R., 57, *76*
Brown, R. D., 420(300), *462*
Bruice, T., 377(219a), 378(219a), *460*
Brumm, A. F., *153*
Buccino, R. J., 236(329), *282*
Buckley, G., 218(166), 226(166), 259(166), 266(166), *278*
Buhler, D. R., 208(102), *276*
Bulant, V., 426(310), *462*
Burger, M. M., 228(254), *280*
Burgess, W. H., 451(426), 454(432), *465*
Burk, R. F., Jr., 53(87), *76*
Bürk, R. R., 204(83), 205(83), 243(83), *276*
Burke, G., 204(85), 205(85), *276*
Burnett, F. R., 17(24), 18(24), *36*
Burnett, J. N., 344(173), 345(173), 347(173), 348(173), 349(173), 350(173), 354(195), 355(195), 357(195), 358(195,198), 359 (195,198), 360(195,198), 363(195), 368(195), 369(195), 370(195), 371(195), *459, 460*
Burnett, R. W., 296, 369(213), 373(213), 375(213), 376(213), *455, 460*
Burnstine, R. C., 13(17), *36*
Burriel-Marti, F., 56(104), *76*
Burris, R., 49(62), *75*
Burstein, S., 200(21), 220(21), 221(21), 247(21), *274*
Burtis, C., 111, 144, 145, *153*
Burtis, C. A., 95(38), 100, 116(38), 134, 140, 141, 142, 143(74), 146, *153*
Burton, F. G., 319(103), 322(103), *457*
Burton, K., 378(219b), *460*
Burton, R. M., 343(170), *459*
Busch, E. W., 116(61), *153*
Buvet, R., 344(178a), 354(197b), 395(248), *459–461*

Byers, G. S., 225(234), *280*
Byers, H. G., 42(15), 43(15), 57(15), 72(15), *74*
Byrne, W. L., 25(35a), *36*

Calib, E., 213(131), 228(259), 230(131), *277, 280*
Calafeteanu, I., 379(227), *460*
Caldwell, I. C., 116(64), *153*
Camargo, E. P., 228(255), 229(255), *280*
Camiener, G. W., 249(416), *284*
Campbell, J., 248(411), *284*
Canellos, G. P., 215(148), *278*
Canellakis, E. S., 416, *462*
Cannon, M., 254(464), *285*
Cannon, P. L., Jr., 175(94), *188*
Cantoni, G. L., 147(75), *153*
Cantor, S. M., 382(247), *461*
Carminatti, H., 213(131), 230(131), *277*
Carr, H. S., 270(484), *286*
Carroll, J. J., 182, *189*
Carruthers, C., 344(180), 346(180), 355 (180,239), 380(180,239–242), *459, 461*
Carter, C. E., 80(3), 116(3), *152*
Cartwright, G. E., 208(104), *277*
Carulli, N., 208(97), 209(109), 249(97), 250(109), *276, 277*
Cary, E. E., 40(8), 41, 43, 48(11), 53, 70, *74*
Casida, J. E., 210(124), 244(124), *277*
Cater, B. R., 25, *36*
Cavalieri, L. F., 384(258), 385(258), 387 (258), 399(258), 400(258), 401(258), 405(258), 420(258), 421(258), *461*
Cawley, L .P., 169(46), 171(46), *187*
Cawse, P. A., 58(125), 59(125), 62, 65(125), *77*
Cedar, H., 204(81), 248(81), *276*
Centro de Studio per la Polarographia (Padua), 295(11), *454*
Cernatescu, R., 379(224), *460*
Cerny, A., 411(287), 412(287), *462*
Cerny, R., 432(334), *463*
Chakrabartty, M. M., 51(80), *75*
Chakrabati, C. L., 54, *76*
Chambers, L. M., 351(192), 450(192), *459*
Chan, F. L., 51(79), *75*
Chan, S. I., 408(357), *464*
Chang, F. N., 223(209), *279*
Chang, P., 403, 404(273), *461*

Chase, A. M., 159, *186*
Chauveau, J., 29(45), *36*
Chaykin, S., 333(143), 334(143), 335(153, 154), 336(153,154), 344(154), 371(153, 154), 375(153), *458*
Cheetham, R. D., 30(50), *37*
Chem. Eng. News, 222(199,200), *279*
Cheney, G. E., 315(87), *456*
Cheng, K. L., 50, 51, *75*
Cherayil, J. D., 202, *275*
Chester Beaty Research Institute, 171(86), *188*
Chester, M. A., 213(134), 230(134), *277*
Chiang, K-C., 403, 404(273), *461*
Chick, W. L., 183, *189*
Chino, H., 244(369), *283*
Chiriboga, J., 208(99), *276*
Chmielewicz, Z. F., 234(309), *282*
Chodkowski, J., 379(221), *460*
Chrambach, A., 249(420), *284*
Christella, R., 43(30), 53(30), 58(30), *74*
Christensen, J. J., 296, *455*
Christian, G. D., 44(38,41), 49(38), 55, 57(38), *74, 76*
Christianson, D. D., 116(56), *153*
Ciaranello, R. D., 225(234), *280*
Cihak, A., 251(443), *285*
Cimerman, C., 49, *75*
Ciusa, W., 344(174), 345(174), 346(174), 347(174), 348(174), *459*
Clark, A., 170(70), 171(70), *188*
Clark, R. E. D., 47, 48(56), *75*
Clark, W. M., 295, 401, 442(15), 443(15), *454*
Clarke, D. A., 413(291), *462*
Claude, A., 2(1), *35*
Cleland, W. W., 2(6), 8(14c), 34(65), 35(65), *35–37,* 204(70), 205(70), 234(307), 235(70), *276, 281*
Clemena, G. G., 57(114,115), *76*
Cocking, E. C., 200(34), 254(34), *275*
Cohen, D., 403, 404(272), 427(272), 435(272), 437(272), *461*
Cohen, P. P., 416, *462*
Cohen, S. S., 225(241), *280*
Cohn, R., 217(157), *278*
Cohn, W. E., 80, 94(8), 95(8), 115, 116 (1–3,8), 130(8), 131(52), 134(52), 138, 142(8), 143(8), *152, 153*
Coleman, W. C., 56, *76*

Collins, C. G. S., 221(187), *279*
Collins, D. C., 230(282), *281*
Colombo, J. P., 169(49), *187*
Colowick, S. P., 335(146,149,150), 368 (149), 374(150), *458*
Comar, D., 58(127), 62(127), 63(127), 65(127), 66(127), *77*
Comb, D. G., 116(57), *153*
Conn, E. E., 223(211), 252(452), *279, 285,* 333(144), 335(145), *458*
Conrad, F. J., 63, 65(158), *78*
Consolo, S., 218(166), 226(166), 259(166), 266(166), *278*
Conway, B. E., 422(303,304), *462*
Conway, T. W., 200(37), 254(37), *275*
Corcoran, J. W., 200(33), 254(33,466), *275, 285*
Cori, C. F., 178(98), *188*
Cori, G. T., 178, *188*
Cooper, J. A., 58(123,137), 61(137), 62 (137), 63(123,137), 65(123,137,170), 66(123,137), *77, 78*
Cooper, R. A., 246(395), *284*
Coryell, M. E., 252(451), *285*
Costa, E., 218(167), 243(167), *278*
Cotino, L. E., 442(371), *464*
Cousen, A., 49, *75*
Cousins, F. B., 44(36), 48, 53, *74*
Crambes, M. R., 66(176), *78*
Crampton, C. F., 87, 116(16), *152*
Cremona, T., 20(28), *36*
Cresser, M. S., 51(81), 52, *75*
Creveling, C. R., 223(206), *279*
Cukor, P., 43, 48, 51(21,59), 53(21,59), 57(59), *74, 75*
Cummins, L. M., 44(46,47), 51, *75*
Cunningham, A. J., 347(188,189), 350(188, 189), 354(197), 355(197), 357(197), 358(197), 359(189,197), 366(188,189), 371(197), 372(197), *459, 460*
Cunningham, W. P., 30(50), *37*
Curl, A. L., 43(26), 56, *74*
Czochralska, B., 387(265a), 394(265a), 401(265a), 402(271), *461*
Czok, R., 178, *188*

Daikuhara, Y., 217(159), 226(159), 251 (159), *278*
Dallner, G., 29, 30(46,48,49), *36*

Daly, J., 212(128), 216(151–154), 222(201, 203,204), 223(203,204), 224(152–154, 213), 225(230,232), *277–280*

Damaskin, B. B., 314, 316, *456, 457*

Dams, R., 63(159), 65(159), 66(159), *78*

Daneo-Moore, L., 240(343), *282*

Daniels, A., 202(48), 246(48), *275*

Danyluk, S. S., 331(132), *458*

Danzuka, T., 51(71), *75*

D'Ari, L., 239(281), *281*

Davidek, J., 398(254), 427(254), *461*

Davidson, J. D., 265(478), *286*

Davidson, W. D., 209(110), *277*

Davies, D. R., 147(75), *153*

Davies, J. W., 200(34), 254(34), *275*

Davis, O. L., 443(381), 444(381), 451(381), *464*

Dawson, D. M., 255(467), *285*

Day, A. R., 405(276), *462*

Day, R. J., 354(197a), 380(197a), *460*

Dean, B. M., 231(293), *281*

Dean, J. A., 376(219), *460*

De Bersaques, J., 240(406), *284*

Debrun, Jean-Luc, 61(155a), *78*

Dedonder, R., 228(257), *280*

De Duve, C., 2(3), 17(22), *35, 36*

DeFord, D. D., 328, *457*

Dekker, C. A., 435(336), *463*

Delahay, P., 314(81), 316, *456*

DeLand, F. H., 209(111), *277*

DeLange, R. J., 246(395), *284*

Delellis, R., 33(58,59), *37*

DeLevie, R., 364(204), *460*

Demeyere, D., 51(77), *75*

Denisovich, L. I., 355(278), 388(278), 392(278), 393(278), 407(278), *462*

Deroo, J., 249(415), *284*

Deshmukh, G. S., 56(107), *76*

Deus, B., 237(330), *282*

Deutsch, A., 116(58), *153*

DeVoe, J. R., 66(176), *78*

DeWet, P. J., 58(124), *77*

Dewjatnin, W. A., 379(220), *460*

De Wulf, H., 200(38), 228(38,258), 229, *275, 280*

Dey, A. K., 49, *75*

Diamond, I., 227(249), *280*

Dickson, R. C., 58(126), 61(126), *77*

Diefenderfer, A. J., 319(104), 322(104), *457*

Diekmann, H., 333(141), 336(159), 438 (141), *458*

Dietrich, C. P., 228(255), 229(255), *280*

Dietz, F. W., 239(340), *282*

Dietzler, D. N., 255(463), *285*

DiPetro, D. L., 208(94,95), *276*

Disney, R. W., 210(123), 244(123), 245 (380), *277, 283*

Dixon, M., 196, 197(8), 211(8), *274*

Doerr, I. L., 413(291), *462*

Doing, A. R., Jr., 176(96), *188*

Domenjoz, R., 218(169), 244(169), *278*

Donley, J., 2(7), 26(7), *35*

Donnell, G. N., 204(77), 205(77), 234(306), 240(77), *276, 281*

Doppelmayr, H. A., 52, *76*

Dorfman, R. I., 200(21), 220(21), 221(21), 247(21), *274*

Down, J. L., 44, *74*

Draper, R. D., 442(372), 443(372), 447 (372), *464*

Drizovskaya, S. Yu., 368(206), *460*

Dryhurst, G., 294(4), 301(4), 327(114), 389(289,290,299), 395(114), 398(114), 405(114), 407(114), 411(289), 412(289, 290), 413(4,289), 414(4,114), 415(4,290, 294,299), 416, 418(294,298,299), 419 (299), 422(114), 423(114), 426(114), 428(114), 429(114,290), 430(4,290,294,299), 431(290), 434(290), 438(114), 439(299), *454, 457, 462*

Dubb, H. E., 335(151), 336(151), *458*

Dubois, F., 244(373), *283*

Duckworth, D. H., 213(137), 214(137), 236(137), 237(137), *277*

Dudley, H. C., 42(15), 43(15), 57(15), 72 (15), *74*

Durham, J. P., 204(67b), 206(67b), 207 (67b), 234(67b,311), 236(67b), 245(67b), 246(67b), *276, 282*

Dus, K., 116(62), *153*

Dusinsky, G., 379(237), 451(417), *461, 465*

Duttera, S. M., 25, *36*

Dutton, G. J., 4(12), *36*

Dye, W. B., 42, 53(20), *74*

Dyer, F. F., 64(166), 65, *78*

Ebner, E., 238(332), *282*

Edwards, A. M., 243(366), 259(473), *283, 285*

Eeckhaut, Z., 59(148), *77*
Effenberger, F., 439(345), *463*
Eggerer, H., 253(456), *285*
Eggermont, E., 247(404), *284*
Eguchi, S., 428(314), *463*
Ehlig, C. F., 40, *74*
Ehmann, W. E., 58(135), 59(147), 61(135, 147), 63, 65(135,147), 66, *77*
Eigner, J., 225(240), *280*
Eisenberg, F., 218(170), *278*
Eisenman, R. A., 218(168), 244(168), 247 (168), *278*
Elbein, A. D., 229(268,269), 230(277), *280, 281*
Elbert, W. C., 482(11), *514*
Ellenbogen, L., 224(218), 251(435), *279, 284*
Ellingson, J. S., 2(8), *35*
Elsasser, S., 238(332), *282*
El-Sheikh, M., 34(62), *37*
Elving, P. J., 292(2), 294(4,5), 295, 296 (28), 298, 301(4,5,63), 302(63), 304(2), 307(64,66), 308(67,68), 309(76,133), 311(69,70), 313(74–78), 315(69,70), 316(138), 318(63), 319(106), 320(69,70), 323(69,70), 324(69), 325, 326(63), 327, 328(116), 329(5,124), 330(70,115,129), 333(14,133), 337(168), 338(168), 339 (168), 340(168), 341(168), 343(168), 344(168), 345(168), 347(168), 348(168), 349(168), 350(168), 351(168), 354(5, 168), 355(168), 356(70), 357(168,354), 358(168), 359(168), 360(168), 362(168), 363(168), 364(168), 365(168), 366(168), 367(133,168), 368(168), 369(168), 370 (168), 371(168), 373(168), 374(168), 375(168), 376(168), 377(129), 379(168), 380(13,14,124), 381(13,14), 383(67), 384(67,113), 385(67,113), 387(129), 388(69,70,113,133), 389(63,64,66–70), 390(69), 392(14,70), 393(252a), 395(67, 68,114,249), 397–399, 400(129), 401 (14), 405(113,114,133,277a), 407(70, 74,114,133), 408(74), 409(69,133), 410 (74), 411(67,113,133), 413(4,5,113), 414(4,13,14), 415(4,5,113,352), 416, 418(113), 420(67,133), 421(67,113,133, 249), 422(68–70,114), 423(68,70,114, 277a), 425(67,70,74,133), 426(113,114), 428(70,74,114,133), 429(74,113,114,

133), 430(4,5), 431(113), 432(14), 433 (109–111,113–115,115a,116), 435(69, 70), 437, 438(114), 439(67,70,115,342), 442(14), *454–459, 461–463*
Engel, J. L., 210(124),244(124), *277*
Engelhardt, J., 345(185), *459*
Englert, G., 336(159), *458*
Enns, K., 451(426), 454(432), *465*
Enriquez, O., 379(229), *460*
Epple, D. G., 351(194), 365(194), 366(194), *459*
Eppson, H. F., 40(5,6), 43(33), *74*
Epstein, S. E., 246(393), *283*
Erickson, H. P., 116(67), *153*
Ernster, L., 29, 30(46), *36*
Ettre, L. S., 81(10), 94(23), *152*
vonEuler, H., 335(147,148), *458*
Evans, C. A., Jr., 57, *76*
Evans, C. S., 40(7), *74*
Evans, E. A., 263(476), 267, 268(482), *285, 286*
Evans, M. G., 344(175), 345(175), 347(175), 348(175), *459*
Ewan, R. C., 44(48), 48(48), 53(48), *75*
Eylar, E. H., 230(276), *281*

Fain, J. N., 169(48), 171(48), *187*
Faith, L., 379(237), 451(417), *461, 465*
Falaschi, A., 200(23,30), 238(30), 239(23), 246(23), *274, 275*
Fales, F. W., 169(48), 171(48,83), *187, 188*
Falkers, K., 46(50), *75*
Fangman, W. L., 213(141), 253(141), *277*
Fardy, M. A., 208(100), *276*
Farese, G., 169(51), 179, *187, 188*
Farsang, G., 337(166), *459*
Farstad, M., 215(149), 254(149), *278*
Faulkner, A. G., 41, 55, *74*
Faust, M., 208(106), *277*
Fausto, N., 231(295), 235(323), 237(323), 251(295), *281, 282*
Fay, H., 56, *76*
Feingold, D. S., 200(39), 251(444), 252 (454), *275, 285*
Feitelson, J., 94(28), *152*
Fellman, J. H., 221(191), *279*
Fernandez, F. J., 54(93a,93b), *76*
Fernando, Q., 315(87), *456*
Fiala, S., 428(319), *463*
Fichter, F., 413, *462*

Fiegelson, P., 208(98), 220(98), *276*
Fiers, W., 240(344), *282*
Filby, R., 65(170), *78*
Filby, R. H., 58(131), 63(131), *77*
Filipowicz, B., 428(320), *463*
Finch, L. R., 254(459), *285*
Finch, P. R., 179, *189*
Findley, J. K., 252(451), *285*
Finkelstein, J. D., 225(237), 252(450), *280, 285*
Fischer, E. H., 250(430), *284*
Fisher, H. F., 335(145), *458*
Fishman, P. H., 156(5), 159(17), *186*
Fishman, W. H., 33(58,59), *37*
Fitt, P. S., 239(340), *282*
Flaks, J. G., 225(241), *280*
Flavin, M., 227(251), *280*
Fleischer, B., 29(52), 30(51,52), *37*
Fleischer, S., 26(38), 29(52), 30(51,52), *36, 37*
Fleischmann, M., 312(72), *456*
Fleming, R., 221(192), 251(192), *279*
Fleming, W. H., 249(418), *284*
Flemming, J., 424(306a), *462*
Flickinger, R. A., *284*
Foa, P. P., 169(49), *187*
Fogg, D. N., 43(34), 49, 57(34), *74*
Foltz, C. N., 58(134,140,142), *77*
Fonnum, F., 199(14), 216(155), 218(155), 226, *274, 278*
Fould, W. T., 215(150), 229(150), *278*
Fouts, J. R., 28, 29(43), *36*
Fowler, S., 444(389), 445(389), 451(389), 454(389), *464*
Fox, A. L., 246(384), *283*
Fox, C. F., 216(156), 227(156), *278*
Fox, J. J., 411(285), 413(291), *462*
Fox, S. M., 200(24), 249(249), *275, 284*
Fraenkel-Conrat, H., 246(389), *283*
Franke, K. W., 40(2), 49, *74, 75*
Frankel, F. R., 87(16), 116(16), *152*
Fraser, T. R., 254(460), *285*
Free, A. H., 156, *186*
Freedland, R. A., 31(55), *37*
Freiser, H., 315(87,88), *456*
Fried, I., 292(2), 304(2), *454*
Friedberg, S. J., 244(376), *283*
Friedman, S., 411(284a), *462*
Friedrichs, B., 231(292), *281*
Fritz, H., 178, *188*

Fritzsche, H., 441(361), *464*
Frohnert, P., 180(113), *189*
Frumkin, A. N., 316, 364(203), *457, 460*
Fujii, I., 60(152), 63(152), 65(152), *78*
Fujita, Y., 268(483), *286*
Fuller, R. C., 251(440), *285*
Fullerton, P. D., 254(459), *285*
Furlong, N. B., 204(66b), 206, 236(66b), *276*
Furmanski, P., 252(448), *285*

Gabay, S., 204(78), 205(78), 232(78), 233(78), *276*
Gabrielian, S. M., 54(93c), *76*
Gailiusis, J., 199(15), *274*
Gallo, R. C., 231(288), 249(414), *281, 284*
Gallwitz, D., 227(250), *280*
Galton, D. J., 254(460), *285*
Ganoza, M. C., 25(35a), *36*
Ganther, H. E., 41(12), *74*
Garland, P. B., 35(66), *37*
Garrison, A. K., 343(169), 345(169), *459*
Gatt, S., 244(378), *283*
Geiduschek, E. P., 202(48), 246(48), *275*
Geissner-Prettre, C., 317(101), 435(101), *457*
Gellrich, M., 336(158), 374(158), 375(158), 376(158), *458*
Gentner, N., 241(352), *283*
George, H., 204(78), 205(78), 232(78), 233(78), *276*
Gerber, G. B., 249(415), *284*
Gerstenfeld, S., 169(45), *187*
Getchell, G., 169(47), 147(47), *187*
Geyer, R. P., 208(100), *276*
Gholson, R. K., 231(296), *281*
Ghosh, S. B., *36*
Giacobini, E., 218(166), 226(166), 259 (166), 266(166), *278*
Gibbons, D., 58(119), 65(119), *77*
Gibert, L. I., 244(372), *283*
Gilbert, C. S., 40(5,6), *74*
Gilbert, L. I., 244(369), *283*
Gileadi, E., 316, *457*
Gill, D. M., 201(45), *275*
Gillette, J. R., 30(53), *37*
Gillis, J., 50, *75*
Gilsdorf, J. R., 31(56), *37*
Ginsburg, V., 230(275,284), 241(351), *281, 283*

Girardi, F., 60(153), 61, 63(153,165), 64(153), 66(179), *78*
Gissel-Nielson, G., 40, *74*
Giudicelli, H., 244(370), *283*
Glaser, L., 178(97), *188,* 228(254), 246 (387), *280, 283*
Gleicher, G. J., 317, *457*
Gleicher, M. K., 317, *457*
Gleit, C. E., 42, 62(16), *74*
Glick, D., 171, *188*
Glicksman, J., 403, *461*
Gnarino, A. M., 30(53), *37*
Goddard, J., 224(216), *279*
Godefroy, T., 199(13), 213(139), 237(139), 239(139), *274, 277*
Gold, A. H., 228(260), *280*
Gold, M., 224(223,226), *279, 280*
Goldberg, D. M., 245(382), *283*
Goldenberg, J., 8(14b), 29, *36*
Goldman, M., 66, *78*
Goldman, S. S., 33(58), *37*
Goldsby, R. A., 220(181), *278*
Goldstein, G., 202, 203, *275*
Goldstein, M., 223, *279*
Golewski, S., 428(316), 443(316), *463*
Gollnick, F. A., 295(8), *454*
Goodwin, J. F., 169(59), *187*
Gordon, A., 379(234), *461*
Gordon, H. L., 234(309), *282*
Gordon, M. P., 418(296), *462*
Gorlich, M., 20(26), *36*
Gorman, J. G., 57, *76*
Gorskaya, I. A., 368(206), *460*
Gorsuch, T. T., 43(32), 44, *74*
Goto, K., 203(56), 245(56), *275*
Gots, J. S., 411(284a), *462*
Gottlieb, A., 268(483), *286*
Gould, E. S., 46, *75*
Goulding, F. S., 65(174), *78*
Goulian, M., 220(178), *278*
Gounaris, A. D., 209(114), 251(438), *277, 285*
Goutier, R., 203(55), *275*
Graham, A. B., 28(40), *36*
Grahn, B., 209(112), 250(112), 251(112), *277*
Gram, T. E., 28(42,43), 29(43), 30(53), *36, 37*
Grancedo, C., 238(332), *282*
Grant, A. B., 44(40), 46(40), *74*

Grant, C. A., 43, 53(30), 58(30), *74*
Grav, H. J., 234(310), 236(310), 237(310), *282*
Greco, A. M., 232(302), *281*
Green, J. G., 80(5,6), 89(5), 95(6), 116(5,6), 134(5), 136(6), *152*
Green, R. H., 267(482), 268(482), *286*
Greenberg, E., 204(76), 206(76), 240(76), 241(352), *276, 283*
Greenberg, G. R., 225(240,242,244), *280*
Greenberg, L. J., 171, *188*
Greene, R. C., 204, 244(376), *276, 283*
Greene, R. F., *284*
Greengard, P., 234(313), *282*
Greenspan, M. D., 255(470), *285*
Gregory, F. J., 386(350), *463*
Griffin, D. A., 55, *76*
Griffiths, W. J., 169(39), *187*
Grodzka, P. G., 313(75,76), *456*
Grollman, A. P., 230(275), *281*
Gross, S., 428(321), *463*
Grunberg-Manago, M., 199(13), 213(139), 237(139), 239(139,340), *274, 277, 282*
Grzeszkiewicz, A., 398(255), 427(255), *461*
Guest, J. R., 225(238), *280*
Guidotti, G., 169(49), *187*
Guilbault, G. G., 173–175, 181, *188, 189,* 195(7), 197(7), *274*
Guinn, V. P., 65(169), *78*
Gulbinsky, J. S., 234(307), *281*
Gumaa, K. A., 251(436), *285*
Gunstone, F. D., 34(62), *37*
Gupta, P. L., 44, 53(49), *75*
Guroff, G., 212(128), 213(140), 214(140), 222(201–203), 223(203), 224(202), 239(140), *277, 279*
Gut, J., 385(269), 386(269), 401(268,269), 402(269), 426(268), 434(269), 436(269), *461*
Gutenmann, W. H., 42, *74*
Gutteridge, J. M. C., 170(74), *188*
Guzzi, G., 63(165), 66(179), *78*

Haas, R. G., 350(191), *459*
Haberman, V., 432(334), *463*
Hachadurian, A. K., 220(180), *278*
Hadjiioannou, S. I., 169(61), *187*
Hager, L. P., 209(114), 251(438), *277, 285*
Hager, S. E., 235(320), *282*
Hagerman, D. D., 197(9), *274*

Hagihara, B., 157(12), 160, *186, 187*
Hagopian, A., 239(276), *281*
Haight, D. E., 169(50), 178(50), *187*
Haines, J. A., 200(40), 213(40), 253(40), *275*
Haines, M. E., 239(342), *282*
Hakanson, R., 221(183,192), 224(215), 230(183), 251(183,192,426), *278, 279, 284*
Hall, C., 230(275), *281*
Hall, D. A., 160, *187*
Hall, J. W., 169(60), *187*
Hall, R. J., 44, 53(49), *75*
Hall, T. C., 204(72), 205(72), 221(198), 225(72,198), *276, 279*
Hall, W. K., 252(451), *285*
Haller, W. A., 58(137), 61, 62(137), 63 (137), 65(170), 66(137), 77, 78
Hallinan, T., 25, *36*
Halverson, A. W., 46(53), *75*
Hambly, A. N., 54, *76*
Hamer, D., 384(277), 385(277), 388(277), 405(277), 409(277), 411(277), 429(277), *462*
Hamilton, P., 95(37), *153*
Hamilton, P. B., 80(15), 87, 88, 94(21), *152*
Hamilton, R. G., 299(46), *455*
Hamilton, R. L., 30(50), *37*
Hammouya, G., 395(248), *461*
Hampel, A., 202(52), *275*
Hanahan, D. J., 20(31), 21(31), *36*
Handbook of Biochemistry, 380(245), *461*
Handley, R., 44(42), 47, 51, 66, *74, 78*
Hansen, A. B., 29(43), *36*
Hanson, R. W., 210(119),219(172), 251(119, 172,439), 255(119), *277, 278, 285*
Hanson, T. L., 31(56,57), *37*
Hanus, V., 312(71), *456*
Harada, T., 244(367), *283*
Hardman, J. G., 243, *283*
Harris, P. N., 380(242), *461*
Harrison, W. W., 57, *76*
Hart, P., 34(63), *37*
Hartly, A. M., 447(401), 448(401), 450(401), *465*
Hartree, E. F., 156(3), 158, 159, *186, 187*
Harvey, L. G., 50, 51, 53, 54, *75*
Harvey, R. A., 199(13), 213(139), 237(139), 239(139), *274, 277*

Harvis, S. A., 46(50), *75*
Haschemeyer, A. E. V., 331(131), *458*
Haskell, C. M., 215(148), *278*
Hassid, W. Z., 200(39), 213(132), 224(222), 230(132,271,272,277), 252(454), *275, 277, 279, 281, 285*
Hastings, A. B., 442(363), 443(363), *464*
Hatch, M. D., 209(113), 210(121), 219(113), 232(113), 251(113), *277*
Hauber, J., 20(29), *36*
Hawkins, T. R., 315(88), *456*
Hayaishi, O., 232(299), 239(341), 241(353), 248(341), *281–283*
Hayashi, S., 217(161,162), *278*
Haynes, G. R., 204(73), 240(345), *276, 282*
Hearn, V. M., 213(133), *277*
Heath, J., 388(280), 409(280), 428(280), *462*
Hechter, O., 212, 243(359), 246(391), *277, 283*
Heffernan, M. L., 420(300), *462*
Heirwegh, K. P. M., 10(16), 12, *36*
Heise, E., 20(26), *36*
Helfferich, F., 81(14), *152*
Hellstrom, H., 335(147,148), *458*
Hemmerich, P., 441(358), 442, 444(358), *464*
Hempstead, K. W., 231(291), 248(291), *281*
Henry, R. J., 171(82), *188*
Herman, H., 443(374), 445(374), *464*
Herman, H. B., 445(394), *464*
Herman, P. J., 316(138), *458*
Hers, H. G., 17(22), *36*, 200(38), 228(38, 258), 229, *275, 280*
Heyrovsky, J., 295(10), 347(187), *454, 459*
Heytler, P. G., 220(181), *278*
Hicks, G. P., 161, 169(41), *187*
Hiddleston, J. N., 312(72), *456*
Hidiroglou, M., 44(44), 53(44), *75*
Hill, D. L., 231(297), *281*
Hill, E. E., 2(8), *35*
Hill, J. B., 165, 170(34,77), 173, *187, 188*, 231(287), *281*
Hillcoat, B. L., 221(194), *279*
Hillebrand, H. W., 56, *76*
Hinse, C. M., 208(108), 209(108), 250(108), *277*
Hipple, J. A., 57(108), *76*
Hirata, M., 241(353), *283*

Ho, P. P. K., 248(412), *284*
Hoban, N., 266(479), *286*
Hodgman, J. E., 234(306), *281*
Hoffman, G. J., 224(215), *279*
Hoffman, H. N., 215(150), 229(150), *278*
Hoffman, I., 44(44), 53(44), *75*
Hoffman, M. A., 451(421), *465*
Hoffman, W. A., Jr., 400(261), *461*
Hofman, K., 405(275), *462*
Hogeboom, G. H., 2(2), *35*
Hogue, D. E., 58(136), 65(136), *77*
Hohmann, B., 180(113), *189*
Holis, D. P., 331(130), *458*
Holland, W. D., 42, 62(16), *74*
Holloway, P. W., 232(300), 252(300), 253 (457), *281, 285*
Hollowell, J. G., 252(451), *285*
Holzer, H., 237(330), 238(332), *282*
Homer, R. F., 345(183,184), *459*
Homolka, J., 296, *455*
Honjo, T., 239(341), 248(341), *282*
Horn, G., 401(265), 419(265), 427(265), 434(265), *461*
Horn, M. J., 49, *75*
Horne, R. N., 31(56), *37*
Horvath, C., 94(31), 95(40,41), 96(40), 98(44,45), 99(44), 102(45), 116(40,41), 118–125, 129–131, 133(40), 135, 136(45), 137, 143(41), *152, 153*
Hoshino, M., 170(67), *188*
Hoste, J., 50, 51(77), 59(148), *75, 77*
Hruska, F. E., 331(132), *458*
Hsia, D.Y-Y., 247(402), *284*
Hsia, Y-Y., 240(349), *282*
Hsie, A., 203(58), *275*
Hudgin, R. L., 28(41), *36*
Huggett, A. St. G., 169(44), *187*
Hughes, S. D., 203(58), *275*
Hulbert, M. H., 299(43), *455*
Humlova, A., 401(266), 411(266), 432(266), 433(266), *461*
Hunt, J. M., 221(184), 251(184), *278*
Hurlbert, R. B., 116(59,60), *153*
Hurtado, J. G., 337(163), 445(393), *459, 464*
Hurwitz, J., 224(223,226), 246(388), *279, 280, 283*
Husovsky, A. A., 364(204), *460*
Hussey, W. W., 319(104), 322(104), *457*
Hutton, J. J., 224(219), *279*

Hutton, R. F., 335(152), 336(152), *458*
Hutton, R. S., 49(62), *75*
Huyck, C. L., 451(415), *465*
Hyatt, E. A., 200(31), 240(31), *275*

Iacob, A., 379(227), *460*
Ibuki, F., 428(318), *463*
Icha, F., 387(312), 426(309), 427(312), 443(312), *462*
Ignarro, L. J., 221(188), 225(188), *279*
Ikawa, S., 170(65), *187*
Ikeda, M., 232(299), *281*
Ikram, M., 379(235), *461*
Imaizumi, R., 221(190), *279*
Imoto, E., 344(172a, 180c), *459*
Ingraham, L. L., 336(161), 442(372), 443 (372), 447(372), *459, 464*
Inoue, T., 163(32), *187*
Inouye, T., 60(152), 63(152), 65(152), *78*, 240(349), *282*
Inscoe, J. K., 216(151–153), 224(152,153), 225(230), *278, 280*
International Union of Biochemistry, 194, 195(6), 198, *274*
Ireland, D. M., 214(143), *278*
Irreverre, F., 252(450), *285*
Irvine, D., 493(23), *514*
Irwin, W. E., 230(279), *281*
Isbell, H. S., 156(1,2), 157(2), 166(2), 169(2), *186*
Isenhour, T. L., 60(151), 64, 65(151,175), *78*
Ishimoto, N., 211(127), 228(127), *277*
Ismail, I. A., 34(62), *37*
Ito, A., 20(27), *36*
Ito, E., 255(462,463), *285*
IUPAC-IUB Commission on Biochemical Nomenclature, *154*
Ives, D. H., 204(67b), 206(67b), 207(67b), 234(67b,311), 236(67b), 245, 246(67b), *276, 282*
Izatt, R. M., 296, *455*
Izumi, F., 221(190), *279*

Jabbal, I., 28(41), *36*
Jambor, B., 446(398), 451(398), *465*
Jamieson, D. R., 328(116), *457*
Janecek, J., 203(59), 246(394), *275, 284*
Janik, B., 295, 311(69,70), 313(74), 315 (69,70), 320(69), 323(69,70), 324(69),

330(70,129b,129c), 333(14),356(70), 380(14), 381(14), 385(251), 388, 389 (69), 390(69), 392(14,70), 393(251,252), 397(251,252,252a), 397(252a), 398, 399, 401(14,251), 407(70,74), 408(74), 409 (69,281–284), 410(74), 412(284), 421 (251), 422(69,70), 423(70), 425(70,74), 426(251), 428(70,74), 429(74), 432(14, 281,282), 435(69,70), 437, 439(70), 442(14), *454, 456, 458, 461, 462*
Janne, J., 250(431), *284*
J. Biol. Chem. 241,527 *(1966)*, 380(244), *461*
Jeauier, E., 217(158), 224(158), *278*
Jellinek, H. H. G., 379(234), *461*
Jemmali, M., 161, *187*
Jenkins, A. E., 473(3), 474(4), *514*
Jequier, E., 224(214), *279*
Jirku, H., 230(282), *281*
Johansen, H., 336(161), *459*
Johns, P., 54(90b), *76*
Johnson, C. M., 40(7), 44(42), 47, 51, *74*
Johnson, H., 43, 53(24), *74*
Johnson, J. E., 169(40), *187*
Johnson, M. J., 44(43), 51, *75*
Johnston, I. R., 239(342), *282*
Jolley, R. L., 94, *152*
Jones, E., 51(73), *75*
Jones, E. J., 57(108), *76*
Jones, J. G., 243(365), 244(365), *283*
Jones, L., 248(412), *284*
Jones, M. E., 209(117), 210(117), 235(117, 320), *277, 282*
Jost, J-P., 203(58), *275*
Jourdian, G. W., 252(455), *285*
Juneau, M., 174(89), *188*
Jungas, R. L., 204(82), 205(82), *276*
Justice, P., 247(402), *284*

Kadish, A. H., 160, 161, *187*
Kahle, K., 170(73), *188*
Kahlson, G., 250(425), 451(425), *284*
Kahn, H. L., 54(92,93), *76*
Kaihara, S., 208(97), 209(109), 249(97), 250(109), *276, 277*
Kajihara, T., 160, *187*
Kakac, B., 411(287), 412(287), *462*
Kakimoto, Y., 491(22), 492(22), *514*
Kalab, D., 399(264,326), 401(264), 409 (264), 432(264,325–328,330–333), 433(264), 439(328), *461, 463*

Kalbhen, D. A., 218(169), 244(169), *278*
Kalinsky, H., 226(247), *280*
Kalman, S. M., 209(118), 210(118), 235 (118), *277*
Kalvoda, R., 300(54), 432(335), *456, 463*
Kamel, M. Y., 185(121), *189*
Kamemoto, Y., 66(177), *78*
Kammen, H. O., 225(243), *280*
Kanfer, J. N., 247(399), *284*
Kanno, T., 169(52), *187*
Kaplan, A., 244(368), 255(368), *283*
Kaplan, N. O., 333(139), 343(170), 376(139), *458, 459*
Kapphahn, J. I., 180(115), *189*
Karrer, P., 441(361), *464*
Karrer, W., 374(347), *463*
Karzel, K., 218(169), 244(169), *278*
Kasinski, H. E., 428(319), *463*
Kastening, B., 299(42), *455*
Kates, M., 22(35), *36*
Kato, S., 344(172a,180c), *459*
Katz, S., 116(57), *153*
Katzen, R., 22(34), *36*
Kaufman, R., 204(74), 206(74), *276*
Kauss, H., 224(222), *279*
Kawàda, M., 34(64), *37*
Kawashima, T., 52, *76*
Kawerau, E., 170(69), 171(69), *188*
Kay, J. E., 200(35), 254(35), *275*
Kaye, R. C., 355(346), 443(346), 444(346, 389), 445(389), 451(389), 454(389), *463, 464*
Ke, B., 354(196), 355(196), 357(196), 358 (196), 359(196), 363(196), 368(196,208, 211), 369(196,208), 447(402), 449(402), *460, 465*
Kearney, E. B., 20(28), *36*
Keech, B., 209(115), 210(115), 255, *277*
Keenan, T. W., 230(274), *281*
Keilin, D., 156(3), 158, 159, *186, 187*
Keiser, H., 221(189), *279*
Kelleher, D. E., 57(116), *76*
Kelleher, W. J., 44(43), 51, *75*
Keller, D. M., 178, *188*
Kellerhohn, C., 58(127), 62(127), 65(127), 66(127), 77
Kelmers, A. D., 149(79), *154*
Kemp, R. G., 246(395), *284*
Kemula, W., 379(221), *460*
Kendall, R., 169(46), 171(46), *187*

Kenna, B. T., 58(143), 63, 65(158), 77, 78
Kennedy, E. P., 216(156), 227(156,249), 278, 280
Kennedy, W. P., 133, 139(68), 153
Kerber, J. D., 54(93a), 76
Kern, W., 413, 462
Kessel, D., 204(72), 205(72), 225(72), 276
Kessler, G., 170(77), 188
Keston, A. S., 156(4), 158, 165, 169, 172, 186, 187
Kevei, E., 451(418), 465
Khairallah, E. A., 208(103), 276
Khmelnitskaya, E. Yu.,444(390,391), 445 (390,391), 446(390), 464
Khorana, H. G., 200(30), 238(30), 275
Kilburn, D. M., 169(53), 171(53), 187
Killick, R. A., 59(145), 61(145), 63(145), 65(145), 77
Kim, K. H., 228(263,280), 280
Kim, S., 224(221), 279
Kimble, D. O., 450(411), 465
Kimura, H., 252(449), 285
King, J., 197(12), 274
King, L., 335(154), 336(154), 344(154), 371(154), 458
Kingsley, G. R., 169(47), 178(47), 187
Kiniwa, M., 51(76), 75
Kinoshita, T., 170(67), 188
Kinsey, S. J., 354(197a), 380(197a), 460
Kirkbright, G. F., 52, 54(90a), 76
Kirkland, J. J., 80(9a), 95(42,43), 97, 99(42), 116(43), 132, 133, 152, 153
Kirsten, E., 87, 152
Kirsten, R., 87, 152
Kisliuk, R. L., 218(165), 226(165), 278
Kitamura, M., 156, 186
Kittler, L., 385(267), 386(267), 390(267), 393(267), 401, 402(267), 410(267), 421(267), 425(267), 432(329), 461, 463
Kizzel, M., 451(418), 465
Klarwein, M., 170(73), 188
Klein, A. K., 42, 43, 49, 54, 56, 74
Klemperer, H. G., 204(73), 249(345), 276, 282
Klink, F., 213(135), 249(422), 255(135), 277, 284
Kloppstech, K., 213(135), 249(422), 255 (135), 277, 284
Knobloch, E., 337(162), 358(162), 443 (375), 444(375), 446(399), 451(375,399, 425), 459, 464, 465

Knobloch, K., 379(231), 460
Knoblock, E. C., 41(13), 44(41), 55(13,41, 97,98), 74, 76
Knox, W. E., 235(319), 282
Kobata, A., 230(284), 281
Kobayashi, Y., 200(20), 220(20), 221(20, 185,193), 274, 279
Kocent, A., 451(427,428), 465
Koch, R. C., 58(117), 65, 76
Kodiri, S., 61(154), 65(154), 78
Koelling, W., 48, 75
Koenigsbuch, M., 403(272), 404(272), 427(272), 435(272), 437(272), 461
Koh, C. K., 80(8), 94(8), 95(8), 116(8), 130(8), 142(8), 143(8), 152
Kohl, D. M., 284
Koletnikova, A. V., 368(207), 460
Kolthoff, I. M., 296(24), 325(24), 379(24), 449(24), 455
Komaki, T., 169(58), 170(58), 187
Komano, T., 170(68), 188
Konno, K., 160, 187
Kono, T., 368(209,212), 369(212,214), 270(209,212), 372(209), 460
Konobu, K., 169(58), 170(58), 187
Konupcik, M., 398(253,254), 427(253,254), 461
Kopecky, M., 379(230), 460
Kopp, L. E., 231(291), 248(291), 281
Kornberg, A., 200(22,23), 213(138), 214 (138), 236(138,326), 237(138,326), 239 (22,23), 246(23,138), 274, 277, 282
Korner, A., 200(35), 254(35), 275
Kornfeld, R., 241(351), 283
Kornfeld, S., 241(351), 283
Korosi, A., 176(96), 188
Korshunov, L. I., 451(429,431), 465
Koshland, D. E., Jr., 8(14a), 36
Kosower, E. M., 333(140), 343(140), 377 (140), 438(140), 458
Kotelnikova, A. V., 368(206), 460
Koukol, J., 252(452), 285
Koutecky, J., 312(71), 399, 456, 463
Kozak, L. P., 230(279), 281
Krakow, J. S., 200(26), 213(26), 237(26), 275
Kramer, D. N., 175(94), 181, 188, 189
Kratzing, C. C., 368(210), 372(210), 460
Kraus, P., 250(428), 284
Kravitz, E. A., 250(429), 284

Krawczyk, A., 443(388), *464*
Krebs, E. G., 246(395), *284*
Krishna, G., 242, 243(357), *283*
Krupicka, J., 385(269), 386(269), 401(268, 269), 402(269), 426(268,311), 427(311), 434(268), 435(269), 436(269), *461, 462*
Kruze, I. E., 451(416), *465*
Kubac, V., 379(229), *460*
Kubota, J., 40(8), *74*
Kuff, E. L., 27(39), 28(39), *36*
Kuik, M., 402(270), *461*
Kuhn, N. J., 230(273), *281*
Kuhn, R., 441(360), 442(364,365), 443 (364,365), 447(364), *464*
Kuo, J. F., 234(313), *282*
Kupelian, J., 221(193), *279*
Kurochkina, O. A., 379(228), 451(228), *460*
Kurtz, J., 224(219), *279*
Kusai, K., 157, 158, *186*
Kusnetzowa, L. A., 379(220), *460*
Kuto, J., 347(187), *459*
Kuthan, J., 344(180b), *459*
Kwan, T. H., 209(118), 210(118),235(118), *277*

Labow, R., 204(75), 206(75), 225(75), 236(328), 249(75,328), *276, 282*
Ladd, F. C., Sr., 80(5), 89(5), 116(5), 134(5), *152*
Laduron, P., 215(146), 221(146), 224(146), 251(441), *278, 285*
Laitinen, H. A., 351(192), 445(395), 446 (395), 450(192), *459, 464*
Lakin, H. W., 43, *74*
Lakshmannan, T. K., 112, *153*
Laland, S., 213(142), *278*
Lamb, B., 388(279), 389(279), 392(279), 409, *462*
Lambert, J. F. P., 58(128,129), 63(129), 65(128,129), 66, *77*
Land, E. J., 336(205), 347(205), 365(205), 368(205), 374(205), 375(205), *460*
Lands, W. E., 2(7,8), 26(7), *35*
Lands, W. E. M., 34(62,63), *37*
Lane, J. C., 44(37), 48(37), 53(37), *74*
Lansford, E. M., 200(41), 253(41), *275*
Lansing, A. I., 247(401), *284*
Lapedes, S. L., 159, *186*
Lapp, D., 229(269), *281*

Lardy, H. A., *285*
Larin, F., 233(304), *281*
Larner, J., 178, 188, 228(264), 229(264), *280*
Laster, L., 252(450), *285*
Lau, H. K. Y., 54, *76*
Lauber, J. K., 225(229), *280*
Lawford, H. G., 147(76), *153*
Layne, D. S., 230(282),*281*
Leach, S. J., 344(175), 345(175), 347(175), 348(175), *459*
Leanza, W. J., 386(350), *463*
Leddicotte, G. W., 58(141), 63(163), 64, *77, 78*
Lee, G. R., 208(104), *277*
Lee, J. C., 133, 139(68), *153*
Lee, N., 217(163), *278*
Lee, N. E., 141(73), *153*
Lee, N. M., 200(41), 253(41), *275*
Lee, S. M., 450(411), *465*
Lehman, I. R., 213(138), 214(138), 236 (138), 237(138), 246(138,386), *277, 283*
Leibetseder, J., 63(160), 65(160), *78*
Leinweber, F. J., 251(434), *284*
Leliaert, G., 59, *77*
Leng, M., 408(262a), *461*
Lenihan, J. M. A., 60(150), 62(150), 65(150), 66(150), *78*
Lenk, H-P., 247(403), *284*
Lenz, P. H., 169(57), *187*
Leone, J. T., 439(342), *463*
LeQuire, V. S., 30(50), *37*
Lerner, J., 116(63), *153*
Levander, O., 58(128), 65(128), 66(128), *77*
Leveille, G. A., 210(119), 251(119), 255 (119), *277*
Levey, G. S., 246(393), *283*
Levine, H., 51(72), *75*
Levine, R. J., 208(105), 251(433), 257(105), *277, 284*
Levit, A. S., 238(331), *282*
Levitt, M., 215(144), 222(144,205), 224 (205), *278, 279*
Levitzki, A., 8(14a), *36*
Levvy, G. A., 9, *36*
Lewis, M., 208(104), *277*
Leyko, W., 428(315,317,320,321), *463*
Li, Lu-Ku, 159(18), *186*
Liberti, A., 400(259), *461*

Lieberman, I., 247(401), *284*
Lieberman, S., 112, *153*
Lieser, K. H., 94, *152*
Like, A. A., 183, *189*
Lilley, R. McC., 231(298), *281*
Lin, E. C. C., 217(161,162), *278*
Lin, H-S., 225(245), *280*
Lindsay, R. H., 231(287), *281*
Lingane, J. J., 296(24), 325(24), 379(24), 443(381), 444(381), 449(24), 451(381), *455, 464*
Lingane, P. J., 446(396,397), 450(397), *464, 465*
Linn, S., 246(386), *283*
Linn, T. C., 219(176), *278*
Lipmann, F., 254(465), *284, 285*
Lippman, F., 200(36), 254(36), *275*
Lipsky, S. R., 95(40,41), 96(40), 98(44), 99(44), 116(40,41), 118–122, 124, 125, 129–131, 133(40), 137(40), 143(41), *153*
Lisanti, L. E., 66, *78*
Lisk, D. J., 42, *74*
Liss, M., 224(220), *279*
Litle, R. L., 161(28), *187*
Litwack, G., 233, *281*
Liu, C., 229(269), *281*
Lloyd, J. B., 169(54), *187*
Lloyd, P. H., 200(28), 238(28), *275*
Lockwood, D. H., 251(437), *285*
Loeb, L. A., 239(335), *282*
Loerch, J. D., 240(346), *282*
Logan, G. F., 22(33), *36*
Logan, J. E., 169(50), 178(50), *187*
Loken, M. K., 248(408), *284*
Lomax, M. I. S., 225(242,244), *280*
Lombard, S. M., 65(175), *78*
Loosli, J. K., 58(136), 65(136), *77*
Lopes, E. S., 379(222), *460*
Lopez, J. A., 18, *36*
Lorenz, W., 316(90), *456*
Lott, P. F., 43(21), 48, 51–54, 57, *74–76*
Loutfi, G. I., 197(9), *274*
Lovecchio, K. K., 344(180a), *459*
Lovenberg, W., 217(158), 221(189), 224 (158,214), *278, 279*
Lowenstein, J. M., 255(470), *285*
Lowry, O. H., 18, *36*, 180, *189*
Lowy, B. A., 384(258), 385(258), 387(258), 399(258), 400(258), 401(258), 405(258), 420(258), 421(258), *461*

Lubbe, W. V., 51(73), *75*
Lucas-Lenard, J., 199(13), 213(139), 237 (139), 239(139), *274, 277*
Lucena-Conde, F., 56(104), *76*
Lucier, G. W., 14, 15(18), 33(61), *36, 37*
Lueders, K. K., 27(39), 28(39), *36*
Luke, C. L., 48, 51(57), *75*
Lund, H., 307(65), *456*
Lundell, G. E. F., 56, *76*
Lundgren, R. A., 66, *78*
Lupien, P. J., 208(108), 209(108), 250(108), *277*
Luthy, N. G., 388(279), 389(279), 392(279), 409, *462*
Lynch, W. E., 247(401), *284*
Lynen, F., 35(67), *37*, 253(456), *285*
Lyons, W. S., Jr., 58(120), 59(120), 64(120), *77*

Maag, D. D., 44(46,47), 51(46,47), *75*
Maag, G. W., 44(47), 51(47), *75*
Macchia, V., 22(34), *36*
MacDonald, F. R., 95(38), 100(38), 116(38), 134(38), 140(38), 141(38), 146(38), *153*
MacDonald, J. R., 316, *457*
MacFarlane, M. G., 22(32), *36*
Madievskaya, R. G., 379(228), 451(228), *460*
Maddock, R. S., 63(162), *78*
Maddox, W. L., 202(54), *275*
Maeda, K., 169(87), 171(87), *188*
Magee, C. W., 57(115), *76*
Mager, M., 169(51), 179, *187–189*
Maggiora, G., 336(161), *459*
Magin, J. B., Jr., 51(72), *75*
Maguinary, A., 451(412), *465*
Mahler, H. R., 336(160), *459*
Mahley, R. W., 30(50), *37*
Maidlova, E., 432(334), *463*
Mairanovskii, S. G., 314(86), 316, 348(95), *456, 457*
Majer, J. R., 471(1), 473(2,3), 474(4), 476(6), 477(7), 478(8), 480(9), 481(10), 482(12), 484(12,13), 485(13), 491(18,19,21), 493 (23), 494(18,19,21), 498(28), 503(29), *514*
Majlat, P., 315(89), *456*
Makela, P. H., 240(348), *282*
Makin, H. L. J., 161, *187*
Makino, Y., 160, *187*

Maley, F., 116(66), *153,* 204(75), 206(75), 225(75), 236(328), 249(328,417), *276, 282, 284*
Maley, G. F., 204(75), 206(75), 225(75), 249(75,417), *276, 284*
Maley, G. J., 116(66), *153*
Malmstadt, H. V., 169(61), 175, *187, 188*
Malt, R. A., 201(44), *275*
Mandel, D., 147(77), *154*
Manian, A. A., 216(154), 224(154), *278*
Manning, D. C., 54(93b), *76*
Manousek, O., 398(253,254), 427(253,254, 313), 434(313), 436(313), *461, 463*
Mans, R. J., 200(32), 254(32), *275*
Marbach, E. P., 169(62), *187*
Marciszewski, H., 398(255), 427(255), *461*
Maresh, C. G., 209(118), 210(118), 235 (118), *277*
Marglin, A., 224(219), *279*
Marino, G., 232(302), *281*
Markland, F. S., 236(325), *282*
Markley, E., 251(435), *284*
Markman, A. L., 451(419), *465*
Markowitz, J. M., *463*
Marmur, J., 224(227), *280*
Marsh, J. M., 242(355), 243(355), *283*
Marshall, J. R., 397(250), *461*
Martin, J. L., 44(46,47), 51(46,47), *75*
Martin, S. M., 219(173), *278*
Martinek, R. G., 156, 171(84), *186, 188*
Martinez, A. M., 200(27), 238(27), *275*
Martonosi, A., 2(7), 26(7), *35*
Marvin, J. F., 248(408), *284*
Marx, W., 218(168), 244(168), 247(168), *278*
Massey, L. M., 208(106), *277*
Mastragostino, M., 299(47), *455*
Mathias, A. P., 239(342), *282*
Matsuhashi, S., 219(174), *278*
Matsushita, S., 428(318), *463*
Matta, G., 379(222), *460*
Matthaei, H., 238(333), 239(333), *282*
Matthews, H. B., 14(18), 15(18), *36*
Mattice, J. J., 47, 51, *75*
Mattock, P., 243(365), 244(365), *283*
Matula, T. I., 219(173), *278*
Matzuk, A. R., 386(350), *463*
Maudsley, D. V., 221(193), *279*
Mautner, H. G., 58(141), *77*
Maxam, A. M., 224(220), *279*

Maxwell, I. H., 202(53), *275*
Mayweg, V. P., 329(127), *458*
Maziere, B., 58(127), 62, 63(127), 65(127), 66, *77*
McCaman, M. W., 221(184), 251(184), *278*
McCaman, R. E., 193(3), 199, 208(101), 218(166), 221, 226(166), 251(184), 256, 259(166), 266, *274, 276, 278*
McClemens, D. J., 343(169), 345(169), *459*
McComb, R. B., 170, 171(78), *188*
McConnell, K. P., 58(138,141), 63(138), 65(138), *77*
McCord, T. G., 299, *456*
McCrosky, C. R., 56, *76*
McDaniel, O. S., 14(18), 15(18), 33(61), *36, 37*
McDonald, A. J., 51(74), *75*
McDonald, I. J., 219(173), *278*
McFarlane, E. S., 224(225), *280*
McGinn, F. A., 388(351), 389(351), 390 (351), 391(351), 392(351), *463*
McGuire, E. J., 28(41), *36,* 230(283), *281*
McKinley, W. P., 345(185), *459*
McKinney, P. S., 296(30), *455*
McLean, P., 251(436), *285*
McNab, J. M., 230(280), *281*
McNulty, J. S., 44(45), *75*
McPherson, S., 482(11), *514*
Medes, F., 233(303), *281*
Meeker, J., 482(11), *514*
Mees, G. C., 345(183), *459*
Mehlman, M. A., 210(122), *277*
Meinke, W. W., 63(162), *78*
Meissner, G., 26(38), *36*
Meissner, L., 335(153), 336(153), 371(153), 375(153), *458*
Meister, A., 235(321), *282*
Meites, L., 296(25,26), 300(60,61), 301(25), 302(25), 325(25,26,61), 353(25), *455, 456*
Melik-Gaikazyan, V. I., 364(203), *460*
Mellon, M. G., 439(344), *463*
Melo, A., 246(387), *283*
Merkel, J. R., 443(382), 444(382), 445(382), 451(382), 454(382), *464*
Merlin, L. M., 230(274), *281*
Merrick, J. M., 255(471), *285*
Merritt, L. L., 376(219), *460*
Mersmann, H. J., 228(262), *280*
Mertelsmann, R., 238(333), 239(333), *282*

Mesman, B. D., 52, *76*
Metge, W. R., 215(150), 229(150), *278*
Meyer, W. L., 336(160), *459*
Miana, G. A., 379(235), *461*
Micarelli, E., 379(226), *460*
Michaelis, L., 442(366–368), 443(366–368), *464*
Middleton, J. E., 169(39,43), *187*
Miech, R. P., 220(179), 231(291), 248(291), *278, 281*
Mihelson, S. H., 58(133), 59(133), 77
Mihelsons, H., 62(156), 63(156), 65(156), *78*
Milhorat, A. T., 43(22), *74*
Miller, H. K., 204(80), 248(80), *276*
Miller, J. W., 328, *457*
Miller, S. L., 116(62), *153*
Miller, W. L., 201(44), *275*
Mills, D. C. B., 214(143), *278*
Mills, L. W., 451(423), *465*
Mitoma, C., 223(210), *279*
Mittermayer, C., 234(308), *282*
Miwa, I., 159, 161(27), 162(27), 163(32), 166, 167, 169(35,87), 171, *186–188*
Miyahsita, H., 493(23), *514*
Miyasaka, K., 379(233), *461*
Miyoshi, Y., 234(316), *282*
Mizuno, D., *284*
Mnovcek, K., 451(425), *465*
Moellering, H., 184, *189*
Mohilner, D. M., 316, *457*
Mohnke, M., 87(20), *152*
Molemans, F., 240(344), *282*
Molinoff, P. B., 250(429), *284,* 494(24), *514*
Mollenhauer, H. H., 30(50), *37*
Mollica, R. F., 221(191), *279*
Molloy, J. P., 53(88), *76*
Molnar, G. D., 169(56), *187*
Monard, D., 203(59), 246(394), *275, 284*
Monro, R., 254(465), *285*
Montreuil, J., 115, *153*
vanMontagu, M., 240(344), *282*
Montgomery, G. A., 369(215), 275(215), *460*
Montgomery, P. O., 200(29), *275*
Moore, J. M., 379(232a), *460*
Moore, S., 80, *152*
Moore, W. M., 451(430), *465*
Moret, V., 337(165), 344(177,178), 348

(177,178), 359(201,202), 369(202), 447 (403), *459, 460, 465*
Mori, Y., 441(357a), 443(383,384), 444 (383,384), 451(383,384), *464*
Moriber, G., 48(59), 51(59), 53(59), 57(59), *75*
Morre, D. J., 30, *37,* 230(274), *281*
Morris, D. F. C., 59(145), 61(145), 63(145), 65(145), 77
Morrison, G. H., 57, 60(151), 64, 65(151), *76, 78*
Morrison, J. F., 204(70), 205(70), 235(70), *276*
Moruzzi, G., 442(365), 443(365), *464*
Moscarello, M. A., 179(111), *189*
Moses, A. J., 58(121), 77
Moshier, R. W., 476, *514*
Mosk. Inst. Biol. Med. Khim., 170(76), 171(76), *188*
Mosser, D. G., 248(408), *284*
Motegi, K., 170(66), *188*
Moule, Y., 29, *36*
Moxon, A. L., 40(3,4), 42(3), 47(3), 57(3), *74*
Moyer, R. W., 249(421), *284*
Mozingo, R., 46(50), *75*
Mudd, S. H., 225(237), 252(450), *280, 285*
Mueller, R. A., 224(217), *279*
Muench, K. H., 202(50), *275*
Mulford, C. E., 42, *74*
Müller, O. H., 295, 296(27), 325(27), *455*
Mullins, W. T., 66(178), *78*
Munk, N. M., 95(38), 100(38), 116(38), 134(38), 140(38), 141(38), 146(38), *153*
Munske, K., 243(362), *283*
Murad, F., 246(392), *283*
Murali, D. K., 251(445), *285*
Murao, S., 169(58), 170(58), *187*
Murata, K., 441(357a), 443(383,384), 444 (383,384), 451(383,384), *464*
Murea, L., 379(227), *460*
Murray, A. W., 199(17), 204(79), 231(79, 290,292,294,298), 234(17), *274, 276, 281*
Muto, H., 60(152), 63(152), 65(152), *78*
Myers, R. L., 299(41), *455*
Myers, V. B., 58(132), 77

Nabikawa, T., 379(233), *461*
Nadjo, L., 299(47), *455*
Nadkarni, R. A., 58(135), 59(147), 61(135, 147), 63, 65(135,147), 66, 77

Nadler, H. L., 240(349), *282*
Nagai, S., 227(251), *280*
Nagatsu, T., 215(144,145), 222, 224(145, 205), *278, 279*
Naidu, P. P., 52, 56(105), *76*
Nakagawa, H., 252(449), *285*
Nakai, M., 157(12), *186*
Nakamura, S., 232(299), *281,* 368(209), 370(209), 372(209), *460*
Nakanishi, Y., 244(367), *283*
Nakashima, S., 55, *76*
Nakaya, J., 344(172a,180c), 359(200), *459, 460*
Nakazawa, K., 239(341), 248(341), *282*
Nargolwalla, S. S., 66, *78*
Nasrallah, S., 220(180), *278*
Nathenson, S. G., 255(463), *285*
Naylor, A. W., 209(116), 210(116), 235 (116), *277*
Nazar, R. N., 147(76), *153*
Neidhart, F. C., 213(141), 253(141), *277*
Neithling, L. P., 58(124), *77*
Netter, H., 213(135), 249(422), 255(135), *277, 284*
Neufeld, E. F., 200(39), 252(454), *275, 285*
Newberry, C. L., 44(38), 49(38), 57(38), *74*
Newsholme, E. A., 204(68), 206, 234(68, 312), *276, 282*
Ng, W. G., 204(77), 205(77), 234(306), 240(77), *276, 281*
Nichol, S. L., 319(103), 322(103), *457*
Nicholls, P. B., 231(290), *281*
Nicholson, B. H., 200(28), 238(28), *275*
Nicholson, R. S., 299, *455*
Nickerson, W. J., 443(382), 444(382), 445(382), 451(382), 454(382), *464*
Nikaido, H., 240(348), *282*
Nikaido, K., 240(348), *282*
Nishihara, M., 249(420), *284*
Nishizawa, Y., 210(124), 244(124), *277*
Nishizuka, Y., 200(36), 232(299), 239(341), 248(341), 254(36), *275, 281, 282, 284*
Nixon, D. A., 169(44), *187*
Nixon, G. S., 58(132), *77*
Nixon, P. F., 221(194), *279*
Noll, H., 247(405), *284*
Nordlie, R. C., 2(9), 16(19–21), 19(19,21), 31(56,57), *35–37,* 235(317), *282*

Nordschow, C. D., 170(71), *188*
Norris, J. F., 56, *76*
Norum, K. R., 227(253), *280*
Novacek, E. J., 45(49a), *75*
Novacic, L., 379(226), *460*
Novelli, G. D., 200(32), 254(32), *275*
Novikov, E. G., 337(167), *459*
Nunley, C. E., 80(6), 95(6), 116(6), 136(6), *152*

Obara, T., 170(65), *187*
Obinata, M., *284*
O'Connor, C. M., 336(157), *458*
Oda, K., 224(227), *280*
Offner, H. G., 55(99), *76*
Ogawa, H., 34(64), *37*
Oka, M., 221(190), *279*
Okada, M., 65(171), *78*
Okamoto, K., 449(409,433), *465*
Okamura, K., 169(58), 170(58), *187*
Okazaki, Y., 379(233), 426(307,308), *461, 462*
Okuda, G., 161(26,27), 162, 167, 168(26), 169(87), 171(87), *187, 188*
Okuda, J., 161, 162, 163(32), 166, 167, 168(26), 169(35,87), 171(87), *187, 188*
Okunuki, K., 157(12), *186*
Okuyama, T., 170(68), *188*
Oldham, K. G., 193(1), 201(46,47), 204(87), 205(46,87), 208(92), 209(46), 210(46), 214(46), 222(46), 223(46), 225(47), 234(87), 235(46), 239(46,47), 245(381), 250(46), 252(46,47), 257(46,47), 258(46), 260(46), 261(46), 262(92), 263(46,47), 264(92), 267(92), *274–276, 283*
Oliverio, V. T., 265(478), *286*
Olmstead, M. L., 299(37,38,46), *455*
Olsen, E. C., 66, *78*
Olsen, I., 255(471), *285*
Olson, C., 176, *188*
Olson, D. C., 329, *458*
Olson, O. E., 43, 44(31), 45(31,49a), 46(53), 53, *74, 75*
Ondera, K., 60(152), 63(152), 65(152), *78*
O'Neal, D., 209(116), 210(116), 235(116), *277*
O'Neill, R. C., 386(350), *463*
Ono, S., 400(260), *461*
O'Reilly, J. E., 298, 308(68), 309(68), 314 (68), 395(68,249), 405(277a), 421(249), 422(68), 423(68,277a), *456 461, 462*

Orengo, A., 204(71), 234(71,315), 246(71), 276, 282
Orlando, J., 247(405), 284
Orr, G., 285
Osborn, M. J., 230(281), 281
Osborn, R. A., 43(26–28), 56, 74
Osburn, R. L., 52(85), 76
Ostashever, A. S., 226(247), 280
Oster, M. O., 218(164), 278
Ostrowski, Z., 443(388), 464
Otsuka, S., 200(20), 220(20), 221(20), 274
Otsuki, T., 379(233), 461
Otvinovski, V., 58(139), 77
Owen, C. A., 215(150), 229(150), 278
Owman, C., 221(183), 230(183), 251(183), 278
Ozaki, T., 236(326), 237(326), 282
Ozawa, H., 30(51), 37

Pace, G. F., 415(294), 418(294), 430(294), 462
Pace, S. J., 405(277a), 423(277a), 462
Paetkau, V., 285
Paigen, K., 33(60), 37
Paik, W. K., 224(221), 279
Painter, E. P., 40(2), 74
Paiss, Y., 348(190), 374(190), 459
Palade, G. E., 29(48), 30(48), 36
Palecek, E., 295, 296, 300(55–58), 380, 385(251), 392(262), 393(251,252,262, 356), 397(251,252), 398, 399, 401(251, 262–264), 408(262), 409(264,281–283), 421(251), 426(251), 432(263,264,281, 282,322,323), 433(263,264), 454, 456, 461–464
Palmer, I. S., 45(49a), 75
Pan, F., 232(301), 281
Panny, S. R., 249(419), 284
Panusz, H., 428(315–317), 443(316), 463
Pardue, H. L., 175, 188
Parizkova, H., 426(310), 462
Parkanyi, C., 319, 323, 457
Parker, C. A., 50, 51, 53, 54, 75
Parks, L. W., 225(239), 280
Parrish, J. R., 94(32), 152
Parsons, R., 316, 457
Partridge, S. M., 94(28), 152
Passannante, A. J., 169(57), 187
Passonneau, J. V., 180(114), 189
Pastan, I., 22(34), 36

Pastor, T., 379(236), 461
Patek, P., 59(146), 65(146), 66(146), 77
Patterson, B. W., 229(269), 281
Patterson, M. K., 285
Patterson, M. S., 204, 276
Paul, J. S., 200(29), 275
Paul, R., 246(387), 283
Paulis, J. W., 116(56), 153
Paulsen, P. J., 57(116), 76
Pauly, J., 63(165), 66(179), 78
Paysant, M., 26(37), 36, 244(377), 283
Peacocke, A. R., 200(28), 238(28), 275
Pearlman, R. E., 239(338), 246(338), 282
Pearson, W. N., 53(87), 76
Peaud-Lenoël, G., 241(350), 283
Pegg, A. E., 251(437), 285
Peingor, E., 337(166), 459
Pelekis, L., 58(133), 59(133), 62(156), 63 (156), 65(156), 77, 78
Pelekis, Z., 58(133), 59(133), 62(156), 63 (156), 65(156), 77, 78
Pence, D. T., 299(39), 455
Peng, L., 209(117), 210(117), 235(117), 277
Peniston, Q. P., 382(247), 461
Pentchev, P. G., 156(5), 158(14), 159(17), 186
Pepper, K. W., 152
Perdijon, J., 59(144), 63, 65(144), 77
Peres, K., 181(116), 189
Perkins, R. W., 58(123), 63(123), 65(123), 66(123), 77
Perrin, C. L., 301(62), 306, 320, 324, 329, 456
Perrin, D. D., 341(168a), 459
Perry, R., 476(6), 478(8), 480(9), 481(10), 482(12), 484(12,13), 485(13), 498(28), 514
Perry, S., 231(288), 249(414), 281, 284
Peterkofsky, B., 217(160), 221(160), 268 (483), 278, 286
Peterson, G. E., 54(93), 76
Peterson, J. I., 179, 188
Pfaff, K. J., 169(56), 187
Pfleiderer, G., 344(171), 369(171), 459
Pfister, K., 386(350), 463
Pfrogner, N., 246(389), 283
Philips, S., 495(27), 514
Phillips, R., 358(199), 359(199), 382, 409, 460

Pigman, W., 156(1), *186*
Piha, R. S., 253(458), 254(458), *285*
Pinamonti, P., 447(403), *465*
Pinteric, L., 28(41), *36*
Piras, R., 228(259), *280*
Pitt, W., Jr., 106(46), *153*
Pleticha, R., 379(232), *460*
Pleticha-Lansky, R., 439(339), *463*
Poels, C. L. M., 236(324), 237(324), *282*
Polcyn, D. S., 299(40), *455*
Polonovski, J., 26(37), *36,* 244(377), *283*
Poni, M., 379(224), *460*
Pope, A. L., 44(48), 48(48), 53(48), *75*
Popjak, G., 232(300), 252(300), 253(457), *281, 285*
Popp, F. D., 333, *458*
Porter, W. L., 266(479), *286*
Portillo, R., 451(413,420), *465*
Portnoy, M., 442(371), *464*
Posgay, E., 315(89), *456*
Potter, L. T., 245(379), *283*
Potter, V. R., 116(59,60), *153*
Pourian, S., 54(93c), *76*
Powers, J. J., 451(423), *465*
Powning, R. F., 368(210), 372(210), *460*
Pozdeeva, G. A., 337(167), *459*
Prager, M. D., 211(126), 215(147), 225(126, 246), *277, 278, 280*
Pratt, D. E., 451(423), *465*
Preiss, B., 95(40), 96(40), 116(40), 124(60), 133(40), 137(40), *153*
Preiss, J., 204(76), 206(76), 240(76), 241 (352), *276, 283*
Prescott, D. M., 262(474), *285*
Prochoroff, N., 223(207), *279*
Proctor, J. F., 58(136), 65(136), *77*
Puck, T. T., 224(216), *279*
Pullman, A., 293, 296, 317(101), 321, 323 (3), 435(3,101), *454, 455, 457*
Pullman, B., 293, 296, 307(66), 321, 323(3), 435(3), *454–455*
Pullman, M. E., 335(146), *458*
Purnell, H., 81(11), *152*
Purdy, W. C., 41(13), 44(41), 55(13,41,98), *74, 76,* 325(112), *457*
Pysh, E. S., 319(102), 322, *457*

Quan, L., 491(17,20), 494(17,20), *514*

Raabo, E., 171(85), *188*

Radhakrishnaa, A. N., 251(445), *285*
Radin, N. S., 247(400), *284*
Rafter, G. W., 335(150), 374(150), *458*
Rahn, K. A., 63(159), 65(159), 66(159), *78*
Raina, A., 250(431), *284*
Rainnie, D. J., 231(286), *281*
Ralea, R., 379(224), *460*
Ramakrishna, T. V., 52, *76*
Ramaley, R. F., 249(421), *284*
Rancitelli, L. A., 58(123,137), 61(137), 62(137), 63(123,137), 65(123,137,170), 66, *77, 78*
Randerath, E., 90(22), 149(22,78), *152, 154*
Randerath, K., 90, 94, 116(55), 149, *152–154*, 205(89), *276*
Rann, C. S., 54, *76*
Rao, G. G., 52, 56(105), *76*
Rao, G. S., 229(270), *281*
Rapaport, L. I., 426(308b), *462*
Ratliff, R. L., 200(27), 238(27), *275*
Raunio, R., 199(16), 233(16), *274*
Reade, M. J. A., 471(1), 474(4), 484(13), 485(13), 503(29), *514*
Reed, D. J., 193(2), 203(56), 245(56), *274, 275*
Reeves, H. C., 252(447,448), *285*
Reeves, J. Y., 248(413), *284*
Reichard, H., 208(93), 226(93), *276*
Reilley, C. N., 312, 316, *456, 457*
Reilly, M. A., 218(171), 251(171), *278*
Reinmuth, W. H., 313, *456*
Reitz, R. C., 34(62), *37*
Renold, A. E., 208(95), *276*
Renschler, H. E., 178, *188*
Renson, J., 224(213), *279*
Rettig, S., 51(72), *75*
Reynaud, J. A., 408(262a), *461*
Reynolds, R. C., 200(29), *275*
Reynolds, S. A., 63(163), 64, 66(178), *78*
Rhian, M., 40(4), *74*
Rhoads, C. A., 213(140), 214(140), 231(140), 239(140), *277*
Ricci, E., 64(166), 65, *78*
Rich, A., 331(131), *458*
Richardson, C. C., 200(22), 239(22), *274*
Richardson, W. J., 57(112), *76*
Rickenberg, H. V., 203(59), 246(394), *275, 284*
Riddle, D. C., 61(155a), *78*
Ridge, D., 239(342), *282*

Riley, W. D., 246(395), *284*
Rinne, R. W., 199(15), *274*
Robbins, J. A., 63(159), 65(159), 66(159), *78*
Roberts, D., 204(72), 205(72), 221(197, 198), 225(72,198), *276, 279*
Roberts, N. R., 180(115), *189*
Roberts, R. M., 206, 207, 212(129), 250 (129), *276, 277*
Robins, R. K., 439(343), *463*
Robinson, D. S., 221(189), 224(214), *279*
Robinson, J., 204(68), 206(68), 234(68, 312), *276, 282*
Robinson, R. A., 357(348), 443(348), *463*
Robinson, W. O., 42, 43, 56(14), 57(15), 72, *74*
Roboz, J., 57, *76*
Rock, M. K., 180(114), *189*
Rodriguez-kabana, R., 161, *187*
Rogers, E. F., 386(350), *463*
Rogers, L. A., 28(42), *36*
Romine, C. J., 231(287), *281*
Rosalki, S. B., 197(11), *274*
Roseman, S., 28(41), *36*, 230(283), 252 (455), *281, 285*
Rosen, O. M., 246(390), *283*
Rosen, S. M., 246(390), *283*
Rosenfeld, I., 41(10), 43(33), 47, *74*
Rosengren, E., 209(112), 250(112,425), 251(112,425), *277, 284*
Rosenkranz, H. S., 269(484), *286*
Rosenthal, I., 319(106), *457, 463*
Rosevear, J. W., 169(56), *187*
Roth, E. S., 221(191), *279*
Roth, J. S., 236(329), *282*
Roth, R., 229(267), *280*
Rothenberg, S. P., 221(195,196), *279*
Rothman, L. B., 228(259), *280*
Rothschild, J., *36*
Rouiller, C., 29(45), *36*
Roy, D. N., 208(99), *276*
Roy, J. E., 231(289), *281*
Rubin, I. B., 202(54), *275*
Rudy, H., 441(360), *464*
Rulfs, C. L., 296(28), 325(28), *455*
Runyan, W. S., 208(100), *276*
Russell, B. G., 51(73), *75*
Russell, D. H., 250(432), *284*
Russell, D. W., 223(311), *279*
Russell, J. A., 169(48), 171(48), *187*

Russell, P. J., Jr., 411(285), *462*
Russell, V., 243(361), *283*
Ryan, C., 247(402), *284*
Ryan, L., 203(58), *275*
Rytting, J. H., 296, *455*

Saavedra, J. M., 494(25), *514*
Saccoccia, P. P., 220(179), *278*
Sadar, M. H., 161(116), *189*
Saifer, A., 169(45), *187*
Saito, K., 20(31), 21(31), *36*
Saito, Y., 449(433), *465*
Sakae, Y., 169(52), *187*
Sakai, A., 170(67), *188*
Salach, J. I., 20(29,30), 21(30), *36*
Salmeron, P., 337(163), 445(393), *459, 464*
Salomon, H., 441(361), *464*
Salomon, L. L., 169(40), *187*
Samsahl, K., 62, *78*
Samuelson, O., 81(12), *152*
Sand, T., 213(142), *278*
Sancho, J., 337(163), 445(393), *459, 464*
Sann, E., 344(171), 369(171), *459*
San Pietro, A., 335(146), *458*
Santhanam, K. S. V., 446(400), *465*
Santos, J. I., 235(321), *282*
Santos, J. N., 231(291), 248(291), *281*
Sargent, M., 54(90a), *76*
Sartori, G., 400(259), *461*
Sasaki, N., 57, *76*
Sastry, K. S., 418(296), *462*
Satake, K., 170(68), *188*
Sato, R., 20(27), *36*
Saukkonen, J. J., 115, *153*
Saunders, G. F., 231(285), *281*
Saunders, M., 335(151), 336(151), *458*
Saunders, P. P., 231(285), *281*
Saveant, J. M., 299(45,47), *455*
Sawada, M., 34(64), *37*
Sawicki, E., 51(78), *75*, 482, *514*
Scamahorn, J. O., 228(263), *280*
Scardi, V., 232(302), *281*
Schachter, H., 28(41), *36*, 179(111), *189*
Schallis, J. E., 54(92,93), *76*
Schauder, H., 416(293), 443(293), *462*
Schayer, R. W., 218(171), 251(171), *278*
Schepartz, A. I., 116(63), *153*
Schersten, B., 178, 180, *188, 189*
Schertel, M. E., 379(232b), *461*
Schildkraut, C. L., 200(22), 239(22), *274*

Schimke, R. T., 208(96), 255(96), *276*
Schipper, E. S., 405(276), *462*
Schlender, K. K., 228(264), 229(264), *280*
Schlittler, E., 441(361), *464*
Schmakel, C. O., 337(168), 338(168), 339 (168), 340(168), 341(168), 343(168), 344(168), 345(168), 347(168), 348(168), 349(168), 350(168), 351(168), 354(168), 355(168), 357(168), 358(168), 359(168), 360(168), 362(168), 363(168), 364(168), 365(168), 366(168), 367(168), 368(168), 369(168), 370(168), 371(168), 373(168), 374(168), 375(168), 376(168), 379(168), *459*
Schmid, R., 13(17), *36*
Schmidt, C. L., 337(164), 338(164), 344 (164), 345(164), 347(164), 348(164), 358(164), *459*
Schmidt, F. H., 179, *188, 189*
Schmitz, H., 116(59,60), *153*
Schmuckler, H. W., 128, 134(69), 135, 138, *153*
Schmunk, R., 87(20), *152*
Schneider, M., 248(409), *284*
Schneider, W. C., 2(2), *35*
Schöpp, K., 441(361), *464*
Schoreder, D. H., 30(53), *37*
Schramm, G., 239(337), 246(337), *282*
Schrauzer, G. N., 329(127), *458*
Schrier, B. K., 227(248), *280*
Schubert, M. D., 442(368), 443(368), *464*
Schuberth, J., 211(125), 226, *277*
Schulte, B. M., 46(53), *75*
Schultz, G., 243(362), *283*
Schultz, H. P., 333, *458*
Schuly, H., 336(155, 156), 343(156), *458*
Schulz, D. W., 180(114), *189*
Schulze, W., 58(118), *77*
Schumann, H., 48, *75*
Schutte, E., 247(403), *284*
Schutz, H., 330, *458*
Schutzbach, J. S., 251(444), *285*
Schutze, H., 87(20), *152*
Schwabe, A. D., 209(110), *277*
Schwabe, K., 443(385), 451(385), *464*
Schwaer, L., 55, *76*
Schwarz, J. H., 204(81), 248(81,413), *276, 284*
Schwarz, K., 58(134,140,142), *77*
Schwarzenbach, G., 442(367), 443(367), *464*

Schweikert, E. A., 61(155a), *78*
Schweizer, M. P., 331(130), 408(357), *458, 464*
Scocca, J. J., 249(419), *284*
Scott, C. D., 80(7), 94, 95(7,39), 106(46), 116(39), 141(73), *152, 153*
Scott, J. F., 202(49), *275*
Scott, J. M., 247(398), *284*
Scrutton, M. C., 210(120), 255(120), *277*
Seagers, W. J., 451(414), *465*
Sease, J. W., 319(103), 322(103), *457*
Sebek, O., 451(422), *465*
Segal, H. L., 31(54), *37*, 228(262), *280*
Segal, S., 217(157), *278*
Segal, I. W., 228(256), *280*
Seim, H. J., 42(20), 53(20), *74*
Sekine, T., 384(302), 386(302), 387(302), 403(302), 421(302), 439(302), *462*
Sekuzu, I., 157(12), *186*
Semal, M., 203(55), *275*
Semerano, G., 328(121), *457*
Semonsky, M., 411(287), 412(287), *462*
Senda, M., 447(404), 448(404,406), 449 (404), *465*
Seng, R., 20(29,30), 21(30), *36*
Seo, E. T., 354(197a), 380(197a), *460*
Service, F. J., 169(56), *187*
Shackman, G. D., 240(343), *282*
Shain, I., 299, 347(33), *455*
Shamaa, M. M., 220(180), *278*
Shapiro, D., 247(399), *284*
Shapiro, D. M., 225(240), *280*
Shapiro, S. K., 225(235), *280*
Shaw, W. V., 227(252), *280*
Sheid, B., 224(228), *280*
Shell, J. W., 66, *78*
Shen, L., 241(352), *283*
Shendrikar, A. D., 52(85), *76*
Shepherd, D., 35(66), *37*, 228(256), *280*
Shepherd, D. M., 251(446), *285*
Sheppard, A. J., 379(232b), *461*
Sheppard, H. C., 267(480), *286*
Sherman, J. R., 204(66), 206, 234(66), *276*
Shideman, F. E., 221(188), 225(188), *279*
Shimada, K., 213(136), *277*
Shimizu, S., 244(367), *283*
Shimoishi, Y., 55(95), *76*
Shirai, K., 384(302), 386(302), 387(302), 403(302),421(302),439(302), *462*

Shive, W., 200(41), 253(41), *275*
Shmuckler, H. W., 115(48), *153*
Shōji, K., 170(66), *188*
Sholts, Kh. F., 368(206), *460*
Shrago, E., 255(469), *285*
Shugar, D., 246(383), *283*
Shuman, M. S., 299(44), *455*
Shuster, L., 218(165), 226(165), 227(248), *278, 280*
Siebke, J. C., 213(142), *278*
Siegel, J. M., 369(215), 375(215), *460*
Siekevitz, P., 29(48), 30(48), *36*
Sieraakowska, H., 246(383), *283*
Sievers, R. E., 476, *514*
Sih, C. J., 223(209), *279*
Silverman, D. A., 223(208), *279*
Silverman, H. P., 354(197a), 380(197a),*460*
Simek, M., 451(418), *465*
Simmonds, S., 248(407), *284*
Simms, E. S., 213(138), 214(138), 236(138), 237(138), 246(138), *277*
Simonek, V., 344(180b), *459*
Simpson, R. E., 58(128,129), 63(129), 65 (128,129), 66(128), *77*
Singer, T. P., 20(29,30), 21(30), *36*
Sinsheimer, J. E., 328(117), *457*
Sirlin, S., 223(207), *279*
Sjoerdsma, A., 217(158), 221(189), 224 (158,214), *278, 279*
Skafi, M., 94, *152*
Skoda, J., 426(311), 427(311), *462*
Skold, O., 234(314), *282*
Skulachev, V. P., 355(278), 388(278), 392 (278), 393(278), 407(278), *462*
Slack, C. R., 209(113), 210(121), 219(113), 229(266), 232(113), 251(113), *277, 280*
Slein, M. W., 22(33), *36,* 156, *186,* 178, *188*
Sluyters, J. H., 299(51,52), *455*
Sluyters-Rehbach, M., 299(51,52), *455*
Smellie, R. M. S., 234(310), 236(310), 237 (310), *282*
Smith, A. D., 244(373,374), *283*
Smith, D. A., 200(27), 238(27), *275*
Smith, D. E., 299, 316, 365(435), *455, 456, 465*
Smith, D. L., 295(13), 308(67), 309(133), 327(113), 328(116), 333(133), 367(133), 380(13), 381(13), 383(67), 384(67,113), 385(67,113), 388(113,133), 395(67),

397–399, 405(113,133), 407(133), 409 (133), 411(67,113,133), 413(113), 414 (13), 415(113,352), 416 (13), 418(113), 420(67,133), 421(67,113,133), 425(67, 133), 426(113), 428(133), 429(113,133), 431(113), 433(113), 439(67), *454, 456– 458, 463*
Smith, E., 328(117), *457*
Smith, I., 224(223), *279*
Smith, L. D., 231(296), *281*
Smith, M. S., 221(184), 251(184), *278*
Smith, N., 182(118), *189*
Smith, Z. G., 213(133), *277*
Smrt, J., 426(311), 427(311), *462*
Smythe, S. V., 442(368), 443(368), *464*
Snyder, S. H., 250(432), 251(442), *284, 285*
Soler, C., 244(377), *283*
Solga, J., 48(59), 51(59), 53(59), 57(59),*75*
Solomatina, V. V., 368(207), *460*
Solotorovsky, M., 386(350), *463*
Solymosi, F., 56(106), *76*
Somerford, H. W., 57(112), *76*
Sommes, R. G., 330(129b,129c), *458*
Somogyi, M., 171, 172(81), *188*
Sonneborn, D., 228(255), 229(255), *280*
Sorantin, H., 59(146), 65(146), 66(146),*77*
Sorm, F., 251(443), *285,* 344(176), 345(176), 346(176), 426(311), 427(311), *459, 462*
Sormova, Z., 344(176), 345(176), 346(176), *459*
Southgate, J., 221(187), *279*
Spanner, J. A., 267(481,482), 268(482),*286*
Spear, F. E., 169(46), 171(46), *187*
Spencer, B., 247(398), *284*
Spiegel, A. M., 204(84), 205(84), *276*
Sprecher, M., 403(272), 404(272), 427(272), 435(272), 437(272), *461*
Squires, J. M., 233, *281*
Srinivason, P. R., 224(228), *280*
von Stackelberg, M., 439(341), *463*
Staehelin, M., 224(224), *279*
Stalman, W., 228(258), *280*
Stanford, F. G., 267(481), *286*
Stanley, R. G., 229(265), *280*
Stanley, T. W., 482(11), *514*
Stanton, R. E., 51(74), *75*
Starchek, L. P., 61(154), 65(154), *78*
Stare, F. J., 442(369), 443(369), *464*
Steele, T. W., 51(73), *75*
Steeper, J. R., 203, 220(60), *275*

Stein, G., 336(217), 348(190), 374(190), 376(217), *459, 460*
Stein, G. A., 386(350), *463*
Stein, W. H., 80, *152*
Stemmle, D. W., 225(234), *280*
Stephen, W. I., 471(1), 476(6), 477(7), 481(10), *514*
Stern, K. G., 442(370), 443(370), *464*
Sternberg, J. C., 161(28), *187*
Stesney, J., 58(140), *77*
Stetten, M. R., 2(10), 17(23,24), 18(23,24), 19(23,24), *35, 36,* 235(318), *282*
Steuart, C. D., 203, 220(60), 239(336), *275, 282*
Stock, A., 344(171), 369(171), *459*
Stock, J. T., 328, *457*
Stonehill, H. I., 355(346), 443(346), 444 (346), *463*
Storey, I. D. E., 9, *36*
Stork, H., 179, *189*
Straat, P. A., 238(334), *282*
Stracke, W., 439(341), *463*
Strausbauch, P. H., 250(430), *284*
Strehlow, H., 329, *457*
Streitwieser, A., 321, 322(108), *457*
Strocchi, P. M., 344(174), 345(174), 346 (174), 347(174), 348(174), *459*
Stromberg, A. G., 436(355), *463*
Strominger, J. L., 211(127), 219(174), 228(127,255), 229(255), 255(462,463), *277, 278, 280, 285*
Struck, W. A., 294(5), 295(13), 301(5), 329(5), 330(129), 354(5), 377(129), 380(13), 381(13), 387(129), 400(129), 413(5), 414(13), 415(5), 416, 430(5), *454, 458*
Stubbs, C. S., 224(218), *279*
Stulberg, M. P., 200(42), 253(42), *275*
Stumm, W., 316, *457*
Stumpf, P. K., 333(144), *458*
Sturman, J. A., 220(177), *278*
Suchy, D., 55, *76*
Suekane, M., 369(214), *460*
Sugino, K., 384(302), 386(302), 387(302), 403(302), 421(302), 439(302), *462*
Sugino, Y., 213(136), 234(316), *277, 282*
Sund, H., 333(141,142), 336(142), 377(142), 378(142), 379(142), 438(14,142), *458*
Sunderman, F. W., 170(64), *187*
Sunderman, F. W., Jr., 170(64), *187*

Suntzeff, V., 344(180), 346(180), 355(180), 380(180,240–242), *459,. 461*
Sussenbach, J. S., 418(295,297), *462*
Sussman, M., 229(267), 240(347), *280, 282*
Sutherland, E. W., 243, *283*
Suzuki, M., 313(77,78), *456*
Suzuki, S., 244(367), *283*
Swallow, A. J., 336(205,217,218,434), 347 (205), 365(205), 368(205), 374(206), 375(205), 376(217,218), *460, 465*
Swan, S., 333, *458*
Symposium on Operational Amplifiers, 296 (29), *455*
Szenberg, A., 246(383), *283*

Tachi, I., 447(404), 448(404), 449(404), *465*
Tachi, T., 448(405), *465*
Taft, H. L., 17(23), 18(23), 19(23), *36*
Tagg, J., 223(210), *279*
Tai, M., 226(247), *280*
Takats, S. T., 239(339), *282*
Takazi, M., 400(260), *461*
Takeda, Y., 217(159), 226(159), 251(159), *278*
Takemori, Y., 448(406,407), 450(407,410), *465*
Takeuchi, K., 215(145), 224(145), *278*
Talalay, P., 223(208), *279*
Tammes, A. R., 170(71), *188*
Tanabe, M., 223(210), *279*
Tanaka, M., 52, *76*
Tani, A., 60(152), 63(152), 65(152), *78*
Tanner, J. T., 58(129), 63(129), 65(129), *77*
Tarr, H. L. A., 231(289), *281*
Tarver, H., 232(301), *281*
Tasaka, Y., 248(411), *284*
Taswell, H. F., 246(384), *283*
Tatwawadi, S. V., 351(193), 443(193,374), 445(193,374), 446(193,400), *459, 464, 465*
Taure, I., 58(133), 59(133), 62(156), 63 (156), *77, 78*
Tuassky, H. H., 43, *74*
Tay, B. S., 231(298), *281*
Taylor, D., 58(122), *77*
Taylor, J. D., 51(73), *75*
Taylor, K., 204(68), 206(68), 234(68), *276*
Taylor, P. M., 169(53), 171(53), *187*

Taylor, R. J., 224(218), 251(435), *279, 284*

Taylor, R. T., 225(236), *280*

Tech, J., 355(239), 380(239), *461*

Tedoraze, G. A., 444(390,391), 445(390, 391), 446(390), *464*

Teller, J. D., 169, *187*

Tener, G. M., 202(53), 248(410), *275, 284*

Terkildsew, T. C., 171(85), *188*

Terrill, K. D., 248(408), *284*

Thaber, W., 374(347), *463*

Thacker, L. H., 106(46), *153*

Thafvelin, B., 43(30), 53(30), 58(30), *74*

Thatcher, L. L., 51(72), *75*

Thevenot, D., 344(178a), 354(197b), 395 (248), *459–461*

Thevnos, T. G., 252(451), *285*

Thoenen, H., 224(217), *279*

Thomas, D. des S., 229(265), *280*

Thomas, J. A., 228, 229(264), *280*

Thomas, L. J., 203(57), 242(356), *275, 283*

Thomas, T. D., 257(472), *285*

Thompson, J. C., 209(110), *277*

Thompson, R. H., 169(42), *187*

Thompson, S. J., 60(150), 62(150), 65(150), 66(150), *78*

Thompson, U. B., 204(69), *276*

Thomson, I. J., 477(7), *514*

Thomson, J. F., 225(235), *280*

Throop, L. J., 46, 51(52), *75*

Thunberg, R., 250(425), 251(425), *284*

Tibbling, G., 180, *189*

von Tigerstrom, M., 248(410), *284*

Tikhomirova, G. P., 379(228), 426(308a), 451(228), *460, 462*

Timmer, B., 299, *455*

Timms, B. G., 170(70), 171(70), *188*

Tisdale, H., 20(30), 21(30), *36*

Toei, K., 51(76), 55, *75, 76*

Tomchick, R., 225(231), *280*

Tomlinson, R. H., 58(126), 61(126), *77*

Tomlinson, T. E., 345(183,184), *459*

Tompkins, P. C., 337(164), 338(164), 344(164), 345(164), 347(164), 348(164), 358(164), *459*

Tono, T., 382(246), *461*

Toto, G. T., 63(161), *78*

Tourian, A., 224(216), *279*

Tovey, K. C., 206, 207, 212(129), 250(129), *276, 277*

Trinder, P., 170(72,75), 171(75), *188*

Ts'o, P. O. P., 238(334), *282,* 331(130), 408(357), *458, 464*

Tsuji, A., 170(67), *188*

Tsuji, H., 232(299), *281*

Tsukamoto, T., 382(246), *461*

Tsunemi, T., 217(159), 226(159), 251(159), *278*

Tucker, D. M., 169(60), *187*

Tucker, V. S., 204(67b), 206(67b), 207(67b), 234(67b), 236(67b), 245(67b), 246(67b), *276*

Turini, P., 20(29), *36*

Turner, J. C., 204(64,65,88), 229(64,65,88), 267(480), 272(64,65,88), *276, 286*

Turner, J. J., 178(100), *188*

Turner, W. R., 292(2), 304(2), 307(64), *454, 456*

Tyson, B. C., Jr., 175(94), *188*

Uden, P. C., 477(7), *514*

Udenfriend, S., 215(144), 222, 223(203), 224(205,213,219), 268(483), *278, 279, 286*

Ueda, K., 239(341), 248(341), *282*

Ueno, K., 51(71), *75*

Ultmann, J. E., 208(98), 220(98), *276*

Underwood, A. L., 296, 343(169), 344(173), 345(169,173), 347(173,188), 348(173), 349(173), 350(173,188), 354(195,197), 355(195,197), 357(195,197), 358(195, 197), 359(195,197), 360(195), 363(195), 366(188), 368(195), 369(195,213), 370 (195), 371(195,197), 372(197), 373(213), 375(213), 376(213), *455, 459, 460*

Ungar, F., 344(172), 369(172), *459*

Updike, S. J., 161, *187*

Urabe, N., 439(340), *463*

Urks, M., 426(310), *462*

Utter, M. F., 210(120), 255(120), *277*

Uziel, M., 80(8), 94, 95(8), 116(8), 130, 142, 143(8), *152*

Vachek, J., 411(286–288), 412(287), 419 (286,288), 429(286,288), 431(288), *462*

Vainer, H., 228(261), *280*

Vajgand, V., 379(236), *461*

Valentinuzzi, M., 442(371), *464*

Van der Heiden, D. A., 171(79), *188*

Van Lancker, J. L., 235(323), 237(323), *282*

Van Roy, F. P., 10(16), 12, *36*

Varela, G., 451(413,420), *465*
Vaughan, M., 246(392), *283*
Veeger, C., 441(358), 442(358), 444(358), *464*
Vennesland, B., 335(145), *458*
Ventrice, M., 248(409), *284*
Veres, A., 61(155), *78*
Verzina, H. V., 426(308b), *462*
Vesely, J., 251(443), *285*
Vessey, D. A., 2(5), 4(13), 5(5,14), 8(14, 14b), 10(14), 21(5), 24, 27(5), 28(5), 29, *35, 36*
Vetter, K. J., 314, 347(82), *456*
Vetterl, V., 295, 316, 299(18,19,256,257), 408(256), 422(256), 423(18,19,256,257, 305,306), 439(306), *454,461,462*
Veveris, O., 58(133), 59(133), 62, 63(156), 65(156), *37, 38*
Vianello, E., 299(45), *455*
Vidre, J. D., 240(346), *282*
Villemez, C. L., 230(280), *281*
Virkola, P., 116(65), *153*
Viteri, F., 53(87), *76*
Vityaeva, S. I., 451(419), *465*
Volke, J., 295, 345(181,182), 379(238), 401, 443(16), *454, 459, 461*
Volkova, V., 345(181), 379(238), *459, 461*
Volveo, V., 344(180b), *459*
Voronova, K. R., 436(355), *463*

Wade, A., 87(16), 116(16), *152*
Wadkins, C. L., 236(325), *282*
Wagner, H. N., 208(97), 209(109,111), 249(97), 250(109), *276, 277*
Wagner-Jauregg, T., 441(360), *464*
Wahlgren, M. A., 65(173), *78*
Wainerdi, R. E., 65(168), *78*
Wald, R., 26(37), *36*, 244(377), *283*
Waldron, D. M., 384(277), 385(277), 388 (277), 405(277), 409(277), 411(277), 429 (277), *462*
Walker, J. J., 397(250), *461*
Wallenfels, K., 333(141), 336(155–159), 343(156,158), 374(158), 375(158), 376 (158), 438(141), *458*
Walls, J. S., 116(56), *153*
Walstenholme, W. A., 57, *76*
Walter, P., *285*
Walton, H. F., 81(13), *152*
Walzcyk, J., 43(21), 48(21), 51(21), 53(21), *74*

Wang, C. H., 203(56), 245(56), *275*
Wang, J. H., 335(151), 336(151), *458*
Wanka, F., 236(324), 237(324), *282*
Ware, A. G., 169(62), *187*
Warner, R. G., 58(136), 65(136), 77
Warren, P. J., 161, *187*
Wasa, T., 400(260), *461*
Washington, A., 43(22), *74*
Washko, M. E., 31(54), *37*
Watanabe, E., 57, *76*
Waterfield, W. R., 267(481,482), 268(482), *286*
Watkins, W. M., 213(133,134), 230(134), *277*
Watkins, W. W., 230(271), *281*
Watkinson, J. H., 40, 43, 44(1,23,39), 45(1), 46(23), 48, 53, 54, 67, *74*
Watson, J. G., 335(154), 336(154), 344(154), 371(154), *458*
Watts, D. E., 208(105), 257(105), *277*
Watts, R. W. E., 231(293), *281*
Wawzonek, S., 333, *458*
Weast, R. C., 63(164), *78*
Webb, E. C., 196, 197(8), 211(8), *274*
Webb, J. W., 327, 330(115), 393(252a), 397(252a), 399(252a), 423(115a), 439 (115), *457, 461*
Wegener, W. S., 252(447), *285*
Weicher, H., 178(103), *188*
Weiner, B., 248(409), *284*
Weiss, B., 218(167), 241(354), 242(358), 243(167), *278, 283*
Weiss, D. E., 94, *152*
Weiss, J. F., 149(79), *154*
Weiss, J. J., 439(339), *463*
Weiss, L., 170(73), *188*
Weissbach, H., 224(213), 225(236), *279, 280*
Weistein, A. W., 233(303), *281*
Weisz, H., 49, *75*
Weliky, N., 354(197a), 380(197a), *460*
Wellner, V. P., 235(321), *282*
Wempen, I., 413(291), *462*
Wenig, K., 379(230), *460*
Wenzel, M., 247(403), *284*
West, P. W., 49(65), 52, *75, 76*
West, T. S., 51(81), 52, 54(90a), *75, 76*
Westerby, R. J., 44(44), 53(44), *75*
Westergaard, O., 239(338), 246(338), *282*

Westheimer, F. H., 335(145,152), 336(152), *458*
Westwick, W. J., 231(293), *281*
Wever, G., 239(339), *282*
Whelan, W. J., 169(54), *187*
Whitehead, E. I., 45(49a), 46(53), *75*
Whitlock, J. P., 204(74), 206(74), *276*
Wiberg, J. S., 236(327), 237(327), *282*
Wieland, O., 170(73), *188*
Wilhelm, J. M., 200(33), 254(33,466), *275, 285*
Wilken, D., 255(469), *285*
Wilkinson, J. H., 197(11), *274*
Wilkinson, N. T., 43(34), 49, 57(34), *74*
Willard, H. H., 376(219), *460*
Williams, A. K., 208(95), *276*
Williams, D. L., 176, *188,* 200(27), 232(27), *275*
Williams, K. T., 42(15), 43, 57(15), 72(15), *74*
Williams-Ashman, H. G., 251(437), *285*
Wilson, A., 51(73), *75*
Wilson, A. M., 351(194), 365(194), 366 (194), *459*
Wilson, B. A., 231(285), *281*
Wilson, G. S., 447(401), 448(401), 450(401), *465*
Wilson, J. M., 299(37), *455*
Wilson, R. H., 204(86), 206(86,91), 232(86), 251(86), *276*
Wilson, T. H., 378(219b), *460*
Wimmer, E., 202(53), *275*
Winchester, J. W., 63(159), 65(159), 66(159), *78*
Winer, A. D., 369(216), *460*
Wing, J., 65(173), *78*
Winkler, H., 244(373–375), *283*
Winteringham, F. P. W., 210(123), 244(123), *277*
Wintrobe, M. M., 208(104), *277*
Wiseman, C., 251(426), *284*
Wiseman, G. E., 46, *75*
Witkop, B., 222(201,203), 223(203), 224 (213,219), 268(483), *279, 286*
Wituck, E. F., 55(99), *76*
Woeller, F. H., 208(107), 209(107), *277*
Wolf, D., 238(332), *282*
Wolf, D. E., 46(50), *75*
Wolf, G., 208(103), 238(331), *276, 282*
Wolstenholme, G. E. W., 336(157), *458*

Wong, J. T. F., 147(76), *153*
Wong, M-Y., 231(287), *281*
Wong, P. C. L., 231(292), *281*
Woo, J., 158(14), *186*
Wood, G. C., 28(40), *36*
Wood, H. C. S., 441(358), 442(358), 444 (358), *464*
Wood, N. P., 218(164), *278*
Wood, R. P., 53(87), *76*
Woodcock, B. G., 251(446), *285*
Woodhouse, D. L., 384(277), 385(277), 388(277), 405(277), 409(277), 411(277), 429(277), *462*
Woods, D. D., 225(238), *280*
Woodside, A. M., 239(335), *282*
Worrel, L. F., 328(117), *457*
Wright, E. B., 170(74), *188*
Wrona, M., 387(265a), 394(265a), 401(265a), *461*
Wu, P. H., 494(26), 495(27), *514*
Wurtman, R. J., 221(186), 233(304), *279, 281*
Wustrack, K. O., 251(433), *284*

Yagi, K., 441(359), *464*
Yahiro, A. T., 450(411), *465*
Yakeley, W. L., 58(131), 63(131), *77*
Yakolev, I. V., 58(139), *77*
Yamagishi, S., 66(177), *78*
Yamamoto, M., 231(289), *281*
Yamauchi, S., 157(12), *186*
Yang, N. C., 319(102), 322, *457*
Yarmolinsky, M. B., 335(149), 368(149), *458*
Yasuda, D., 223(210), *279*
Yasukochi, K., 439(340), *463*
Yates, D. W., 35(66), *37*
Yip, M. C. M., 235(319), *282*
Yoe, J. H., 52, *76*
Yoshida, H., 221(190), *279*
Yoshihara, K., 239(341), 248(341), *282*
Young, D. S., 179, *188*
Young, J. E., 211(126), 225(126,246), *277, 280*
Yuen, R., 179(111), *189*
Yule, H. P., 59, 63(149), 65, 66, *77, 78*
Yurachek, J. P., 57, *76*
Yushok, W. D., 170, 171(78), *188*

Zahler, W. L., 34(65), 35(65), *37*

Zahradnik, R., 319, *323*

Zakim, D., 2(4,5), 4(13), 5(5,14), 8(14, 14b), 10(14), 19(4), 21(5), 23(4), 24, 25, 27(5), 28(5), 29, *35, 36*

Zamecnik, P. C., 200(40), 213(40), 253(40), *275*

Zenk, M. H., 223(212), *279*

Zicha, B., 249(415), *284*

Zimmer, A. J., 451(415), *465*

Zimmer, M., 173(88), *188*

Zito, R., 232(302), *281*

Zlatkis, A., 81(10), *152*

Zolotovitskii, Ya. M., 444(390,391), 445 (390,391), 446(390), 451(429,431), *464, 465*

Zwiebel, R., 180, *189*

Zubillaga, E., 43(22), *74*

Zuman, P., 295, 300(59), 320, 325, 328, 329, 333, 337(20), 344(20), 355(20), 401(265), 402(270), 419(265), 427(265, 313), 434(265,313), 436(313,337,338), 442, 451, *454–458, 461, 463*

SUBJECT INDEX

Absorbance, 114
  at the peak maximum, 89
Activation, 60
Acylphosphate: D-glucose 6-phosphotrans-
    ferase, 184-185
Adenine, electrochemical analysis of, 428,
    430, 433
  electrochemical oxidation of, 414
  electrochemical reduction of, 388, 407
  nucleosides, electrochemical reduction of,
    392, 407-409
  nucleotides, electrochemical reduction of,
    392, 407-409
Adenines, electrochemical properties of,
    435, 437
  substituted electrochemical reduction of,
    407
Adjusted retention volume, 118
Adsorption, of purines and pyrimidines on
    electrodes, 421-424
  on electrodes, 314
    adsorption sites, 315
    measurement of adsorption, 315-316
  phenomena, 314-316, 350-353
    FAD, 448
    FMN, 448
    NAD⁺, 363
    NMN⁺, 363
    riboflavin, 444-446
Alkyl-substituted pyrimidines, 420
Allacil, 426
Allantoin, 416
Alloxan, 416
  electrochemical properties of, 400
  electrochemical reduction of, 387
Alloxantin, electrochemical properties of,
    400
Alternate pathway, 32
Alternating-current polarography, 299, 339
  FAD, 448
  FMN, 448

1-methyl-3-carbamoylpyridinium ion, 351
  NAD⁺, 363
  nicotinamide, 339
  NMN⁺, 363
  riboflavin, 442, 443
6-Amino purines, 436
Aminopyrimidines, electrochemical oxida-
    tion of, 403
  electrochemical reduction of, 384
2-Aminopyrimidines, 425
AMP, 135
Amphipathic and nonpolar substrates, 34-35
  carrier-bound substrate, 35
  micelles, 34
  nonspecific binding of substrates, 35
  substrate-induced inactivation, 34
Analysis, quantitative, by polarography, 325
Anion-exchange chromatography, 116
ATP, 137, 139, 392
8-Azaadenine, electrochemical reduction of,
    410
Azacytosine, electrochemical reduction of,
    402
Azapurines, electrochemical reduction of,
    390, 410-411
Azapyrimidines, electrochemical reduction
    of, 385, 401-402
Azauracil, electrochemical reduction of, 402
6-Azauracil, 434
  electrochemical analysis of, 426
  electrochemical properties of, 436
Azauracils, electrochemical reduction of,
    386
6-Azauridine, 427

Backflushing, 149
Band broadening, 84
Base line, 114
Bias voltage supply unit, 161-162
Biological applications of quantitative mass
    spectrometry, 468

543

Biological redox behavior, pyridine nucleotides, 377
Biological reduction mechanism, 377-379

Calibration, 93
Capacity factor, 82
Catalase, inhibition for, sodium azide, 162
  liberation of oxygen, from H_2O_2 by, 158
  presence of, in β-D-glucose oxidase, 162, 172
    in test material, 162
Cation-exchange chromatography, 116
Cellulose ion exchange, 116
Cell volume, 105
Characterization of chemical reactions, polarographic, 330
Chemical applications of quantitative mass spectrometry, 468
Chloropyrimidines, electrochemical properties of, 402-403
Chromatogram, 84
  evaluation of, 89-92
  of a single component, 81
  of a solute pair, 85
Chromatograph, liquid, 100-115
Chromatographic liquid velocity, 86
Chromatographic run, duration of, 113
Chromatography, ion-exchange, 115
  reverse-phase, 149
Chronopotentiometry, 351
  FAD, 448
  FMN, 448
  riboflavin, 442, 443
Closed-system combustion, 42-43, 71
  Parr bomb, 42, 46
  Schöniger flask, 42, 46
Column, 100
  conditioning, 110
  efficiency, 117, 124
  loading capacity, 117-118
  packing, 109, 110, 125
    dynamic method, 141
  regeneration, 126
  reusebility, 133
Competing enzymes, 33
Complex biological extracts, 490-498
  β-phenylethylamine in some tissues of the rat, 495
  p-tyramine in rat brain, 494
  p-tyramine in urine, 491

Concave gradient, 112
Concentration, 123
  distribution ratio, 83
Controlled-potential electrolysis, 340
  and coulometry, 300-301
  1-methyl-3-carbamoylpyridium ion, 348
  NAD^+, 368-371
  $NADP^+$, 371
  nicotinamide, 340
  NMN^+, 371
Conventional resins, 140, 144
Convex gradient, 112
Correlations involving potentials, 316-325
Cross linking, 144
Cross-sectional area of the column, 86
Cyclic AMP, 137
Cyclic voltammetry, 297, 339, 347
  FAD, 448
  FMN, 448
  1-methyl-3-carbamoylpyridinium ion, 350
  NAD^+, 366
  $NADP^+$, 366
  nicotinamide, 339
  NMN, 366
Cytosine, electrochemical analysis of, 426, 433
  electrochemical reduction of, 385, 397-399
  nucleosides, electrochemical properties of, 399
    electrochemical reduction of, 393
  nucleotides, electrochemical properties of, 399
    electrochemical reduction of, 393, 398

Dead volumes, 109
Deoxynucleotides, 134
Destructive analysis, 41
Detector, 87, 100, 102, 105
  sensitivity, 87, 142
  volume, 89
Detergents, 487-490
Detergent treatment of microsomes, 26
Dialuric acid, electrochemical properties of, 400
3,3'-Diaminobenzidine, 49, 50
2,3'-Diaminonaphthalene, 51, 53, 57
Dinucleotides, 140
Direct-current polarography, 337
  FAD, 447

FMN, 477
methylcarbamoylpyridinium ion, 347
NAD⁺, 358
NADP⁺, 358
nicotinamide, 337
NMN⁺, 358
pyridine nucleotides, 357
riboflavin, 442, 443
Distribution ratio, 82
Dry ashing, 42, 62

$E_{1/2}$ correlations, with Hammett constants,
    436
  with linear free-energy parameters, 435-
    436
  with MO parameters, 437-438
  with pK_a, 435
Electroactive sites, 309-310
  effect of substituents, 310-312
Electrochemical analysis, of FAD, 451
  of nicotinamide, 379
  of riboflavin, 451
  of 1-substituted 3-carbamoylpyridinium
    ions, 379
Electrochemical processes, literature, 295-
    296
  relation to biological processes, 293-295
Electrochemical reduction mechanism, 341-
    344
  1-methyl-3-carbamoylpyridinium ion, 353
  NAD⁺, 373
  NADP⁺, 373
  nicotinamide, 341
  NMN⁺, 373
Electrochemical symbols and acronyms,
    452-453
Electrochemical synthesis, 332-333
Electrode reaction, mechanisms, 306-316
  reversibility of, 305
Electrosynthesis, 439
Eluent flow, 100
Eluent velocity, 109
Enzymatic reduction, 334-336
  flavins, 441
  NAD⁺, 334, 335
    models, 336
  NADP⁺, 334
  redox behavior, 334-336
Enzyme activity, unit of, 194-195
Enzyme assay, radiometric, advantages of,

    249, 272-273
  automation of, 202, 209
  blank readings, 203, 205, 210, 256-258
  coupled enzyme assay methods, 215-217
  design of experiments, 255-272
  electrophoretic methods, 213
  general principles, 194-198
  involving the precipitation of macromole-
    cules, 200-202
  ion-exchange methods, 202-207
  isotope effects, 263-264
  methods, 193-274
    depending on the release or uptake of
      activity in a volatile form, 207-211
    involving adsorption on charcoal or
      alumina, 213-215
    involving precipitation of derivatives,
      217-218
    involving reverse isotope dilution assay,
      218
    of Enzymes, of Class 1, 219-224
      of Class 2, 224-244
      of Class 3, 244-249
      of Class 4, 249-252
      of Class 5, 252-253
      of Class 6, 253-255
    overcoming interference by other enzymes,
      233, 237, 240-242
    paper and thin-layer chromatographic
      methods, 211-212
    selection of optimum conditions, 258-260
    sensitivity of, 234, 256-258
    solvent extraction methods, 199-200
    special effects with tritium-labeled sub-
      strates, 266-270
    techniques, 198-218
Enzymes, classification of, 198
Enzymic cycling method, 180
"Enzymic peak shifts," 139, 141
Etimidine, 426

Faradaic current, 303
Flavin nucleotides, 440-452
  electrochemical reduction of, 447
Flavins, electrochemical reduction of, 442
Flow cell, 105
Flowmeters, 102, 108
Flow pulsation, 103
Flow rate, 124
Flow velocity, 124, 126

effect on column effeciency, 124
Fluorescence, of 2,2'-dihydroxybipheny-5,5'-diacetic acid, 174
of NADPH, 156, 180
of resorufin, 181
Formazan, 182
Free radicals, 341, 353, 366, 373

Gas chromatography, 81
Gaussian peaks, 84, 89, 90
Ghost peaks, 105, 115, 127
D-Glucose, determination of, 156-186
    with acylphosphate: D-glucose 6-phos-photransferase, 184-185
    with β-D-glucose oxidase, colorimetric, 169-173
        electrochemical, 175-176
        fluorometric, 173-175
        using oxygen electrode, 159-169
    with hexokinase, colorimetric, 182-183
        fluorometric, 180-181
        radioisotopic, 183-184
        spectrophotometric, 178-180
    equilibrium, 156, 162
    in blood (plasma, serum), 159-162, 164-165, 169-171, 178-179
    in urine, 161, 178-180
D-Glucose anomers, determination of, bromine method, 157
    gas chromatography, 157
    in blood, 165-167
    NMR spectrometry, 157
    polarimetry, 157, 169
    ratio of, 166-167
    with β-D-glucose oxidase, colorimetric, 172-173
    using oxygen electrode, 165-169
β-D-Glucose oxidase, 157-176
    from Aspergillus niger, 157
    from Penicillium amagasakiense, 157-158, 162
    from Penicillium notatum, 157-158
    -peroxidase-o-dianisidine system, 173
    properties of, 157
Glucose-6-phosphatase, 16-19
    mechanism of reaction, 16
    pH optimum, 17-19
    phosphate donors for phosphotransferase reaction, 19
    phosphohydrolase activity, with glucose-

6-P as substrate, 16
    with pyrophosphate as substrate, 17
    phosphotransferase activity, 18
    treatment with phospholipase, 25
D-Glucose-6-phosphate dehydrogenase, 156
    properties of, 178
β-Glucuronidase, 16
    inhibition of, by saccharic acid-1,4-lactone, 16
UDP-Glucuronyltransferase, 3-16
    activation, 5, 26, 27
    o-aminobenzoate as aglycone, 11
    o-aminophenol as aglycone, 9
    p-aminophenol as aglycone, 11
    bilirubin-albumin complex as substrate, 13
    bilirubin as aglycone, 12
    bilirubin-solubility as limiting factor, 13, 14
    choice of aglycone, 4
    cooperativity, 6, 8, 10
    determination of kinetic parameters, 5, 6, 10
    dieldrin as aglycone, 14
    effect of, metal ions, 10, 11
        metal ions (p-nitrophenol as aglycone), 6
        salt concentration, 5
    linearity of assay with o-aminophenol as aglycone, 9
    linearity of assay with p-nitrophenol, 5
    multiplicity, 4
    naphthol as aglycone, 14
    p-nitrophenol as aglycone, 4
    pH-activity curves, 5
    phenolphthalein as aglycone, 14
    radiochemical assay, 14
    reverse reaction, 15
    testosterone as aglycone, 14
    treatment with phospholipase, 23
Glutamic dehydrogenase, 180
Gradient elution, 83, 86, 105, 110, 112, 114-116, 119, 126, 133, 147
    calculation of concentration profile, 112
    concentration profiles, 113, 114
    delayed, 114
Guanine, adsorption of on electrodes, 422
    electrochemical analysis of, 430, 433
    electrochemical behavior of, 409
    electrochemical oxidation of, 416-418
    nucleosides and nucleotides,

electrochemical behavior of, 409

Half-wave potential, 301-303
  correlations involving, 316-325
  factors involved, 318-319
  linear free-energy correlations, 319-321
  miscellaneous types, 323-324
  quantum-mechanically calculated elec-
    tronic indices, 321
Heated graphite atomizer, 54
Hexokinase, 156, 177-184
  properties of, 177
High-performance liquid chromatography,
  80
High pressure, 80, 102, 144
Homogenization, 27-28
Homovanillic acid, 174
Hydrogenolysis, 46
p-Hydroxyphenylacetic acid, 174
Hydroxypurines, electrochemical analysis
    of, 430
  electrochemical oxidation of, 416-418
  electrochemical reduction of, 411
Hydroxypyrimidines, electrochemical analy-
    sis of, 426
  electrochemical oxidation of, 403
  electrochemical properties of, 399
  electrochemical reduction of, 385
2-Hydroxypyrimidine, 425
  electrochemical properties of, 399
4-Hydroxypyrimidine, electrochemical
    properties of, 400
Hypoxanthine, 388
  electrochemical analysis of, 429
  electrochemical oxidation of, 418
  electrochemical reduction of, 411

Injection port, 104, 106
Instrumental sensitivity, 471-472
  instrumental factors, 472
  sample factors, 471-472
Integrated ion current technique, 468-514
Integrator, 91
Interferences, 48-50, 52, 53, 55, 61-65
Interfering enzymes, 31, 34
  acyl-CoA ligase, 33
  β-glucuronidase, 33
  NADH-cytochrome b_5 reductase, 34
  NADPH-cytochrome c reductase, 34
  nonspecific phosphatases, 17

nucleotide pyrophosphatase, 32
Intermolecular association, of purines and
    pyrimidines on electrodes, 423
Internal standard, 93-94
Ion-exchange, capacities, 98
  chromatography, 80, 81
  resins, 95, 115-116
    conventional, 95
    pellicular, 94, 99
    superficial, 94
Ionic strength, 125
Iron(III) potentiated, 52
Isocratic elution, 83, 110, 114, 125, 131,
    133, 142, 144, 147
Isocytosine, electrochemical properties of,
    399
Isoguanine, electrochemical analysis of, 430
  electrochemical reduction of, 410
Isomerism, 498-505
  cis-trans isomers, 500-501
  conformational isomers, 502
  dioximes, 499
  isomeric polycyclic hydrocarbons, 501-
    502
  metal chelates, 503
  structural isomers, 499-500
Isotope dilution, 51, 53, 57
Isotope effects, in enzyme assays, 262-264

Kinetic factors, 312
Kinetic phenomena, heterogeneous electrode
    kinetics, 313-314
  kinetics near the electrode, 312-313

Lambert-Beer law, 88
Lead, in the atmosphere and in petroleum,
    487
  estimation of, 487
Linear gradients, 112
Liquid chromatograph, 100-115, 127
  high pressure, 101, 106
Liquid chromatography, high-performance,
    149
Liquid scintillation counting, difficulties
    with bicarbonate-^{14}C, 210
  of aqueous solutions, 204
  precipitation during counting, 229
  problems, 265
  quenching by organic solvents, 200
  self-absorption losses with soft beta

emitters, 201, 203, 204, 212-214,
    234, 238, 246, 269-270

Macroscale electrolysis, 439
Mass distribution ratio, 83
Methemoglobin, 160-161
2-Methyl-4-aminopyrimidine, 426
Methylcarbamoylpyridinium ion, electro-
    chemical reduction of, 347
1-Methyl-3-carbamoylpyridinium ions, 347
    alternating-current polarography, 351
    controlled-potential electrolysis, 348
    cyclic voltammetry, 347, 350
    direct-current polarography, 347
    electrochemical reduction mechanism,
        353
Methylnicotinamide, electrochemical reduc-
    tion of, 347
Microsomal enzymes, 1-35
    variability in properties, 30-31
        developmental changes, 30-31
Microsomal subfractions, 28-30
    differences between rough and smooth
        microsomes, 29
    isolation, 29
    uneven distribution of enzyme activity, 29
Microsomes, preparation of, 26-31
Millipore filters, 49, 73
Mixed chambers, 102, 106, 108, 112, 113
Molar absorptivities of RNA constituents,
    89
Mole ratios, 93
Mutarotase, effects of, determination of
        D-glucose and its anomers, 167, 171
    preparation of, from hog kidney, 159
    properties of, 158

Neutron activation analysis, 58-66
    error, 65, 66
    factors affecting detection limits, 63-66
    methods of measurement, 58-59
    modes of activation, 59-61
    sample preparation, 61-63
Nicotinamide, 337
    alternating-current polarography, 339
    controlled-potential electrolysis, 340
    cyclic voltammetry, 339
    derivatives, electrochemical reduction of,
        344
    direct-current polarography, 337

electrochemical reduction mechanism,
    341
electrochemical reduction of, 337
Noise level, 105
Nomenclature, 333-334
    flavin-adenine dinucleotide (FAD), 441
    flavin mononucleotide (FMN), 441
    nicotinamide-adenine dinucleotide (NAD$^+$),
        333
    nicotinamide-adenine dinucleotide phos-
        phate (NADP$^+$), 333
    nicotinamide mononucleotide (NMN$^+$),
        333
    purine and pyrimidine derivatives, 380
    purine and pyrimidine nucleosides, 381
    pyridine nucleotides, 333
    riboflavin, 441

Oligonucleotides, 131, 137
Optimization of separation efficiency, 125
Organic electrode reaction mechanisms, 306-
    316
Organic selenium compounds, 46, 47, 72
Orotic acid, electrochemical analysis of, 427
    electrochemical reduction of, 387
Oscillopolarography, 299
N-Oxides, electrochemical reduction of, 391
Oxygen electrode, 159-169
    Clark-type, 159-162
    Galvani-type, 162
    polarographic, 159-162, 166, 169
    rotating platinum, 160

Packing porosity, 88
Peak, areas, 90
    height, 90
    splitting, 119
    volumes, 109
    width, 84
Pellicular anion-exchange, 134
    resins, 97, 123
Pellicular cation-exchange, 122
    resins, 97
Pellicular ion-exchange resins, 117-140
Pellicular resins, 95, 96, 98, 109, 118, 119,
        121, 122, 124-127, 129, 133, 137,
        144
Peroxidase, 169-175
pH, effect, on chromatographic separation,
    118, 119

on peak shape and retention, 120
Phenazine methosulfate, 181-182
Phlorizin, 159
Phosphate ion, 156, 167, 178
6-Phospho-D-gluconate dehydrogenase, 180
Phospholipase A, 20-21, 24-26
  isoenzymes, 20
  metal requirements, 21
  properties, 21
  purification, 20
Phospholipase C, 21-22, 25
  isoenzymes, 22
  metal requirements, 22
Phospholipase D, 22
Phospholipase treatment of microsomes, 22-23
  activation of UDP-glucuronyltransferase, 23
  inactivation of enzyme, 24
  properties of phospholipase A-treated enzymes, 23
  rate of hydrolysis of microsomal phospholipids, 23
  stability of phospholipase treated enzymes, 24, 25
Phospholipid-protein interactions, 19-26
γ-Photon, 60
Piazselenol, 50, 53
Plate, height, 84, 85, 124
  number, 84, 86, 118
  per second, 86
  per atmosphere values, 144
Polarographic, analysis, 325
  amperometric titration, 328
  coulometric titration, 328
  coulometry, 328
  determination of concentration, 326-327
  titration, 327
  behavior, of organic compounds, interpretation, 301-333
  determination, chemical and physical properties, 329
  identification and characterization, 329
  measurement of reaction rates, 328-329
  observation, association (conformation and orientation), 330
  techniques, 296-301
    alternating-current polarography, 299
    controlled-potential electrolysis and

coulometry, 300-301
    cyclic voltammetry, 297-299
    oscillopolarography, 299-300
Polarography, 296
Polycyclic hydrocarbons, in extracts from solid samples, 485-487
  detection and analysis of, 485-487
  in the atmosphere, 481-485
  detection and analysis of, 481-485
Polyphosphates, 115
Precolumn, 103, 148
Pressure gauge, 102, 106
Prompt γ-ray emission, 60
Pulsing technique, 60
Pump, 102, 106
Pumps, 125
Purine, bases, 129
  electrochemical analysis of, 427
  electrochemical reduction of, 388, 405
  pyrimidine interaction, 332
Purines, adsorption of on electrodes, 421
  and pyrimidines, electrochemical analysis of, 424-434
    effects of substituents on electrochemical behavior of, 420
    electrochemical analysis of mixtures of, 433
    electrosynthesis of, 439
    oscillopolarographic behavior of, 432
  electrochemical analysis of, 427, 430-431
  electrochemical behavior of, 380
  electrochemical oxidation of, 413-420
  electrochemical properties of, 435, 437
  electrochemical reduction of, 405
  table for polarographic reduction data for, 388-391
  table for electrochemical oxidation data for, 415
Purine-6-sulfinic acid, electrochemical analysis of, 429, 432
  electrochemical oxidation of, 419
  electrochemical reduction of, 412-413
Purine-6-sulfonamide, electrochemical analysis of, 430
Purine-6-sulfonic acid, electrochemical analysis of, 430
  electrochemical reduction of, 413
Pyridine nucleotides, electrochemical reduction of, 357
  model compounds (1-substituted

3-carbamoylpyridinium ions), 344-354
  electrochemical reduction of, 344
Pyrimidine, bases, 129
  electrochemical analysis of, 425
  electrochemical reduction of, 383, 384
    mechanisms, 308-309
  purine, reduction mechanism, 395
Pyrimidines, electrochemical analysis of,
    425
  electrochemical behavior of, 380
  electrochemical oxidation of, 403-404
  electrochemical reduction of, 382
  polarographic reduction data for, table of,
    384-387
  voltammetric oxidation data for, table of,
    404
Pyrimidinyl-substituted sulfanilamides, 425

Quantitative analysis, 89-94, 325-329
Quantitative mass spectrometry, 468-514

Radiation decomposition, 293
  products, interference in enzyme assays
    by, 201, 239
Radioactivity monitor, 135
Raney nickel, 46
Reaction rates, polarographic, 312
  measurement, 328
Recorder, 105, 127
Reduction mechanism, 341-344
Reference column, 115
Refractometer, 87
Resazurin, 181
Resolution, 85
Resorufin, 181
Response factor, 89
Retardation ratio, 82
Retention, relative, 84
  time, 81, 123
    adjusted, 82
  volume, 82, 84
    adjusted, 82
Riboflavin, electrochemical reduction of,
    442
  polarographic determination, 451
Ribonucleosides, 127, 129, 130, 133, 142,
    143, 144
Ribonucleotides, 111, 122
Ribose, electrochemical reduction of, 382
Ring oven, 49

RNA constituents, 90

Salt concentration, 125
  effect on chromatographic separation, 122
"Salting in," 115
Sample, amount, 118
  injection, 103
  introduction, 103
  preparation, 147
    and storage, 40-41, 61
  volume, 103
"Sampling boat" technique, 54
Sampling cup technique, 54
Schlieren effect, 105
Selenium, analysis, methods, 67-73
  catalytic effect of, 52
  isolation of, 47-48
    arsenic coprecipitation, 48, 53, 70, 71
    distillation, 47, 56, 63, 72
    extraction, 50, 51, 53, 55, 69, 70
    gravimetric, 47
    ion-exchange treatment, 48
    neutron activation analysis, 53
    toluene-3,4-dithiol complex, 47-48, 53
  losses of, 40-44, 46, 51, 62
    sampling errors, 41
    volatile compounds, 41
  measurement of, 48-57
    atomic absorption spectrophotometry,
      54-55
    colorimetry, 49-50
    fluorometry, 52-54, 69
    gas chromatography, 55
    gravimetric, 56-57, 72
    neutron activation analysis, 58
    polarography, 57
    spark source mass spectrometry, 57
    spectrophotometry, 50
    titration, 56
    x-ray fluorescence analysis, 66
  reduction of, 46, 48, 49, 56, 57, 69, 73
Sensitivity of analysis, 87, 127, 149
Separation factor, 84
Sequence analysis of RNA, 149
Small-bore columns, 87, 88
Sodium azide, inhibition of catalase, 161-
    162
Solute concentration, maximum, in the ef-
    fluent peak, 87
Solute diffusion, 94

Sorption isotherm, 118
Standard deviation, 84, 92
Surface decontamination, 487-490
Synthesis, electrochemical, 332
Syringe, 108
  needle, 103

Temperature, 126
  effect of on, chromatographic separation,
    119, 120, 123
  the efficiency of separation, 122
  programming, 83, 122
Tetrazolium salt, 182
Theoretical plates, number of, 84
6-Thiopurine, electrochemical analysis of,
    429, 431
  electrochemical oxidation of, 419
  electrochemical reduction of, 411
Thiopurines, electrochemical analysis of,
    429-430, 431
  electrochemical oxidation of, 418-419
  electrochemical reduction of, 389, 411-
    412
6-Thiopurines, 434
Thiopyrimidines, electrochemical analysis
    of, 427
  electrochemical properties of, 402
Thymines, electrochemical oxidation of,
    404
  electrochemical properties of, 401
Time of analysis, 99
Trace metals, 473-481
  diketones analysis of, 476-480
    thermal reactions of, 480-481

  mercury analysis of, 473
  nickel analysis of, 473-474
  various oxinates analysis of, 474-475
    thermal stability of, 475

Uracil, electrochemical analysis of, 426
  electrochemical properties of, 401, 436
Uracils, electrochemical oxidation of, 404
  electrochemical properties of, 401
Uric acid, electrochemical analysis of, 431
  electrochemical oxidation of, 416
Ultraviolet absorption, NAD^+, 336, 371
  NAD dimer, 374
  NADH, 336, 371
  NMN dimer, 374
  NMN^+ dimer, 371
  NMNH, 371
  pyridine derivatives, 336
Ultraviolet spectra, 336
UV detector, 89, 91, 105, 106, 115, 135,
    148
  flow cell of, 107
UV photometer, 87

Voltammetry, 296

Wet digestion, 43-46, 56, 57, 62, 68
  mixtures containing perchloric acid, 44,
    53
  sulfuric acid-hydrogen peroxide, 44
  sulfuric and nitric acids, 43

Xanthine, electrochemical oxidation of,
    418

Methods of Biochemical Analysis

CUMULATIVE INDEX, VOLUMES 1–21 AND SUPPLEMENT

Author Index

	VOL.	PAGE
Ackerman, C. J., see *Engle, R. W.*		
Albertsson, Per-Ake, Partition Methods for Fractionation of Cell Particles and Macromolecules	10	229
Alcock, Nancy W., and *MacIntyre, Iain,* Methods for Estimating Magnesium in Biological Materials	14	1
Amador, Elias, and *Wacker, Warren E. C.,* Enzymatic Methods Used for Diagnosis	13	265
Ames, Stanley R., see *Embree, Norris D.*		
Andersen, C. A., An Introduction to the Electron Probe Microanalyzer and Its Application to Biochemistry	15	147
Anderson, N. G., Preparative Zonal Centrifugation	15	271
Andrews, P., Estimation of Molecular Size and Molecular Weights of Biological Compounds by Gel Filtration	18	1
Aspen, Anita J., and *Meister, Alton,* Determination of Transaminase	6	131
Augustinsson, Klas-Bertil, Assay Methods for Cholinesterases	5	1
Determination of Cholinesterases	Supp.	217
Awdeh, Z. L., see *McLaren, D. S.*		
Baker, S. A., Bourne, E. J., and *Whiffen, D. H.,* Use of Infrared Analysis in the Determination of Carbohydrate Structure	3	213
Balis, M. Earl, Determination of Glutamic and Aspartic Acids and Their Amides	20	103
Bauld, W. S., and *Greenway, R. M.,* Chemical Determination of Estrogens in Human Urine	5	337
Bell, Helen H., see *Jaques, Louis B.*		
Benesch, Reinhold, and *Benesch, Ruth E.,* Determination of—SH Groups in Proteins	10	43
Benesch, Ruth E., see *Benesch, Reinhold*		
Benson, E. M., see *Storvick, C. A.*		
Bentley, J. A., Analysis of Plant Hormones	9	75
Benzinger, T. H., see *Kitzinger, Charlotte*		
Berg, Marie H., see *Schwartz, Samuel*		
Bergmann, Felix, and *Dikstein, Shabtay,* New Methods for Purification and Separation of Purines	6	79
Berson, Solomon A., see *Yalow, Rosalyn S.*		
Bhatti, Tarig, see *Clamp, J. R.*		
Bickoff, E. M., Determination of Carotene	4	1
Bishop, C. T., Separation of Carbohydrate Derivatives by Gas-Liquid Partition Chromatography	10	1
Blackburn, S., The Determination of Amino Acids by High-Voltage Paper Electrophoresis	13	1
Blow, D. M., see *Holmes, K. C.*		

	VOL.	PAGE
Bodansky, Oscar, see Schwartz, Morton K.		
Bossenmaier, Irene, see Schwartz, Samuel		
Boulton, Alan A., The Automated Analysis of Absorbent and Fluorescent Substances Separated on Paper Strips	16	327
Boulton, A. A., see Majer, J. R.		
Bourne, E. J., see Baker, S. A.		
Brantmark, B. L., see Lindh, N. O.		
Bray, H. G., and *Thorpe, W. V.*, Analysis of Phenolic Compounds of Interest in Metabolism	1	27
Brierley, G. P., see Lessler, M. A.		
Brodersen, R., and *Jacobsen, J.*, Separation and Determination of Bile Pigments	17	31
Brodie, Bernard B., see Udenfriend, Sidney		
Bush, I. E., Advances in Direct Scanning of Paper Chromatograms for Quantitative Estimations	11	149
Bush, I. E., Applications of the R_M Treatment in Chromatographic Analysis	13	357
Erratum	14	497
Carstensen, H., Analysis of Adrenal Steroid in Blood by Countercurrent Distribution	9	127
Caster, W. O., A Critical Evaluation of the Gas Chromatographic Technique for Identification and Determination of Fatty Acid Esters, with Particular Reference to the Use of Analog and Digital Computer Methods	17	135
Chambers, Robin E., see Clamp, J. R.		
Chance, Britton, see Maehly, A. C.		
Chase, Aurin M., The Measurement of Luciferin and Luciferase	8	61
Chinard, Francis P., and *Hellerman, Leslie*, Determination of Sulfhydryl Groups in Certain Biological Substrates	1	1
Clamp, John R., and *Bhatti, T.*, and *Chambers, R. E.*, The Determination of Carbohydrate in Biological Materials by Gas-Liquid Chromatography	19	229
Clark, Stanley J., see Wotiz, Herbert H.		
Cleary, E. G., see Jackson, D. S.		
Code, Charles F., and *McIntyre, Floyd C.*, Quantitative Determination of Histamine	3	49
Cohn, Waldo E., see Volkin, Elliot		
Cotlove, Ernest, Determination of Chloride in Biological Materials	12	277
Craig, Lyman C., and *King, Te Piao*, Dialysis	10	175
see also *King, Te Piao*		
Crane, F. L., and *Dilley, R. A.*, Determination of Coenzyme Q (Ubiquinone)	11	279
Creech, B. G., see Horning, E. C.		
Creveling, C. R. and *Daly, J. W.*, Assay of Enzymes of Catechol Amines	Supp.	153
Curry, A. S., The Analysis of Basic Nitrogenous Compounds of Toxicological Importance	7	39
Daly, J. W., see Creveling, C. R.		
Davidson, Harold M., see Fishman, William H.		
Davis, Neil C., and *Smith, Emil L.*, Assay of Proteolytic Enzymes	2	215
Davis, R. J., see Stokstad, E. L. R.		

	VOL.	PAGE
Davis, Robert P., The Measurement of Carbonic Anhydrase Activity	11	307
Dean, H. G., see *Whitehead, J. K.*		
Dikstein, Shabtay, see *Bergmann, Felix*		
Dilley, R. A., see *Crane, F. L.*		
Dinsmore, Howard, see *Schwartz, Samuel*		
Dische, Zacharias, New Color Reactions for the Determination of Sugars in Polysaccharides	2	313
Dodgson, K. S., and *Spencer, B.*, Assay of Sulfatases	4	211
Dyer, John R., Use of Periodate Oxidations in Biochemical Analysis	3	111
Edwards, M. A., see *Storvick, C. A.*		
Elving, P. J., *O'Reilly, J. E.*, and *Schmakel, C. O.*, Polarography and Voltammetry of Nucleosides and Nucleotides and Their Parent Bases as an Analytical and Investigative Tool	21	287
Embree, Norris D., *Ames, Stanley R.*, *Lehman, Robert W.*, and *Harris, Philip L.*, Determination of Vitamin A	4	43
Engel, Lewis L., The Assay of Urinary Neutral 17-Ketosteroids	1	479
Engel, R. W., *Salmon, W. D.*, and *Ackerman, C. J.*, Chemical Estimation of Choline	1	265
Engelman, Karl, see *Lovenberg, S. Walter*		
Ernster, Lars, see *Lindberg, Olov*		
Fink, Frederick S., see *Kersey, Roger C.*		
Fishman, William H., Determination of β-Glucuronidases	15	77
Fishman, William H., and *Davidson, Harold M.*, Determination of Serum Acid Phosphatases	4	257
Fleck, A., see *Munro, H. N.*		
Fraenkel-Conrat, H., *Harris, J. Ieuan*, and *Levy, A. L.*, Recent Developments in Techniques for Terminal and Sequence Studies in Peptides and Proteins	2	359
Friedman, Sydney M., Measurement of Sodium and Potassium by Glass Electrodes	10	71
Frisell, Wilhelm R., and *Mackenzie, Cosmo G.*, Determination of Formaldehyde and Serine in Biological Systems	6	63
Gale, Ernest F., Determination of Amino Acids by Use of Bacterial Amino Acid Decarboxylases	4	285
Gardell, Sven, Determination of Hexosamines	6	289
Gofman, John W., see *Lalla, Oliver F. de*		
Goldberg, Nelson D. and *O'Toole, Ann G.*, Analysis of Cyclic 3′,5′-Adenosine Monophosphate and Cyclic 3′,5′-Guanosine Monophosphate	20	1
Grabar, Pierre, Immunoelectrophoretic Analysis	7	1
Greenway, R. M., see *Bauld, W. S.*		
Gross, D., see *Whalley, H. C. S. de*		
Haglund, Herman, Isoelectric Focusing in pH Gradients—A Technique for Fractionation and Characterization of Ampholytes	19	1
Haines, William J., and *Karnemaat, John N.*, Chromatographic Separation of the Steroids of the Adrenal Gland	1	171
Hanessians, Stephen, Mass Spectrometry in the Determination of Structure of Certain Natural Products Containing Sugars	19	105
Harris, J. Ieuan, see *Fraenkel-Conrat, H.*		
Harris, Philip L., see *Embree, Norris D.*		

VOL. PAGE

Hellerman, Leslie, see Chinard, Francis P.

Hermans, Jan, Jr., Methods for the Study of Reversible Denaturation of
Proteins and Interpretation of Data 13   81

Hjertén, S., see Porah, J.

Hjerten, Stellan, Free Zone Electrophoresis. Theory, Equipment and
Applications 18   55

Hoff-Jorgensen, E., Microbiological Assay of Vitamin B_{12} 1   81

Holman, Ralph T., Measurement of Lipoxidase Activity............................... 2   113

Measurement of Polyunsaturated Acids 4   99

Holmes, K. C., and Blow, D. M., The Use of X-ray Diffraction in the Study
of Protein and Nucleic Acid Structure 13   113

Homolka, Jiri, Polarography of Proteins, Analytical Principles and Appli-
cations in Biological and Clinical Chemistry.. 19   435

Horning, E. C., Vanden Heuvel, W. J. A., and Creech, B. G., Separation and
Determination of Steroids by Gas Chromatography...................................... 11   69

Horvath, C., High-Performance Ion-Exchange Chromatography with
Narrow-Bore Columns: Rapid Analysis of Nucleic Acid Constituents
at the Subnanomole Level 21   79

Hough, Leslie, Analysis of Mixtures of Sugars by Paper and Cellulose
Column Chromatography.. 1   205

Hughes, Thomas R., and Klotz, Irving M., Analysis of Metal–Protein
Complexes 3   265

Humphrey, J. H., Long, D. A., and Perry, W. L. M., Biological Standards in
Biochemical Analysis 5   65

Hutner, S. H., see Stokstad, E. L. R.

Jackson, D. S., and Cleary, E. G., The Determination of Collagen and
Elastin 15   25

Jacobs, S., The Determination of Nitrogen in Biological Materials Determination of Amino Acids by Ion Exchange Chromatography 14   177

Jacobsen, C. F., Léonis, J., Linderstrom-Lang, K., and Ottesen, M., The
pH-Stat and Its Use in Biochemistry 4   171

Jacobsen, J., see Brodersen, R.

James, A. T., Qualitative and Quantitative Determination of the Fatty
Acids by Gas–Liquid Chromatography... 8   1

Jaques, Louis B., and Bell, Helen J., Determination of Heparin 7   253

Jardetzky, C., and Jardetzky, O., Biochemical Applications of Magnetic
Resonance... 9   235

Jardetzky, O., see Jardetzky, C.

Jenden, Donald J., Measurement of Choline Esters .... Supp.   183

Jones, Richard T., Automatic Peptide Chromatography............................... 18   205

Josefsson, L. I., and Lagerstedt, S., Characteristics of Ribonuclease and
Determination of Its Activity.. 9   39

Jukes, Thomas H., Assay of Compounds with Folic Acid Activity............... 2   121

Kabara, J. J., Determination and Localization of Cholesterol 10   263

Kalckar, Herman M., see Plesner, Paul

Kapeller-Adler, R., Determination of Amine Oxidases .... Supp.   35

Kaplan, A., The Determination of Urea, Ammonia, and Urease 17   311

Karnemaat, John N., see Haines, William J.

Kearney, Edna, B., see Singer, Thomas P.

	VOL.	PAGE

Keenan, Robert G., see *Saltzman, Bernard E.*

Kersey, Roger C., and *Fink, Frederick C.*, Microbiological Assay of Antibiotics 1   53

King, Te Piao, and *Craig, Lyman C.*, Countercurrent Distribution............... 10   201
see also *Craig, Lyman C.*

Kitzinger, Charlotte, and *Benzinger, T. H.*, Principle and Method of Heatburst Microcalorimetry and the Determination of Free Energy, Enthalpy, and Entropy Changes 8   309

Klotz, Irving M., see *Hughes, Thomas R.*

Kobayashi, Yutaka, and *Maudsley, David V.*, Practical Aspects of Liquid-Scintillation Counting 17   55

Kolin, Alexander, Rapid Electrophoresis in Density Gradients Combined with pH and/or Conductivity Gradients 6   259

Kopin, Irwin J., Estimation of Magnitudes of Alternative Metabolic Pathways 11   247

Korn, Edward D., The Assay of Lipoprotein Lipase *in Vivo* and *in Vitro*..... 7   145

Kuksis, A., Newer Developments in Determination of Bile Acids and Steroids by Gas Chromatography 14   325

Kunkel, Henry G., Zone Electrophoresis.. 1   141

Kurnick, N. B., Assay of Deoxyribonuclease Activity 9   1

Lagerstedt, S., see *Josefsson, L. I.*

Lalla, Oliver F. de, and *Gofman, John W.*, Ultracentrifugal Analysis of Serum Lipoproteins 1   459

Lazarow, Arnold, see *Patterson, J. W.*

Leddicotte, George W., Activation Analysis of the Biological Trace Elements 19   345

Lehman, Robert W., Determination of Vitamin E 2   153
See also *Embree, Norris D.*

Leloir, Luis F., see *Pontis, Horacio G.*

Léonis, J., see *Jacobsen, C. F.*

Le Pecq, Jean-Bernard, Use of Ethidium Bromide for Separation and Determination of Nucleic Acids of Various Conformational Forms and Measurement of Their Associated Enzymes 20   41

Lerner, Aaron B., and *Wright, M. Ruth*, *in vitro* Frog Skin Assay for Agents That Darken and Lighten Melanocytes 8   295

Lessler, M. A., and *Brierley, G. P.*, Oxygen Electrode Measurements in Biochemical Analysis 17   1

Levy, A. L., see *Fraenkel-Conrat, H.*

Levy, Hilton B., see *Webb, Junius M.*

Lindberg, Olov, and *Ernster, Lars*, Determination of Organic Phosphorus Compounds by Phosphate Analysis 3   1

Linderstrom-Lang, K., see *Jacobsen, C. F.*

Lindh, N. O., and *Brantmark, B. L.*, Preparation and Analysis of Basic Proteins 14   79

Lissitsky, Serge, see *Roche, Jean*

Long, D. A., see *Humphrey, J. H.*

Lovenberg, S. Walter, and *Engelman, Karl*, Serotonin: The Assay of Hydroxyindole Compounds and Their Biosynthetic Enzymes .... Supp.   1

Loveridge, B. A., and *Smales, A. A.*, Activation Analysis and Its

	VOL.	PAGE
Application in Biochemistry	5	225
Lumry, Rufus, see Yapel, Anthony F., Jr.		
Lundquist, Frank, The Determination of Ethyl Alcohol in Blood and Tissues	7	217
McCarthy, W. J., see Winefordner, J. D.		
McIntire, Floyd C., see Code, Charles F.		
MacIntyre, Iain, see Alcock, Nancy W.		
Mackenzie, Cosmo G., see Frisell, Wilhelm R.		
McKibbin, John M., The Determination of Inositol, Ethanolamine, and Serine in Lipides	7	111
McLaren, D. S., Read, W. W. C., Awdeh, Z. L., and Tchalian, M., Micro-determination of Vitamin A and Carotenoids in Blood and Tissue	15	1
Maehly, A. C., and Chance Britton, The Assay of Catalases and Peroxidases	1	357
Majer, J. R., and Boulton, A. A., Integrated Ion-Current(IIC) Technique of Quantitative Mass Spectrometric Analysis: Chemical and Biological Applications	21	467
Malström, Bo G., Determination of Zinc in Biological Materials	3	327
Mangold, Helmut K., Schmid, Harald H. O., and Stahl, Egon, Thin-Layer Chromatography (TLC)	12	393
Margoshes, Marvin, and Vallee, Bert L., Flame Photometry and Spectro-metry: Principles and Applications	3	353
Maudsley, David V., see Kobayashi, Yutaka		
Meister, Alton, see Aspen, Anita J.		
Michel, Raymond, see Roche, Jean		
Mickelsen, Olaf, and Yamamoto, Richard S., Methods for the Determina-tion of Thiamine	6	191
Miller, Herbert K., Microbiological Assay of Nucleic Acids and Their Derivatives	6	31
Miwa, I., see Okuda, J.		
Montgomery, Rex, see Smith, Fred		
Müller, Otto H., Polarographic Analysis of Proteins, Amino Acids, and Other Compounds by Means of the Brdička Reaction	11	329
Munro, H. N., and Fleck, A., The Determination of Nucleic Acids	14	113
Natelson, Samuel, and Whitford, William R., Determination of Elements by X-Ray Emission Spectrometry	12	1
Neish, William J. P., α-Keto Acid Determinations	5	107
Novelli, G. David, Methods for Determination of Coenzyme A	2	189
Oberleas, Donald, The Determination of Phytate and Inositol Phosphates	20	87
Okuda, J., and Miwa, I., Newer Developments in Enzymic Determination of D-Glucose and Its Anomers	21	155
Oldham, K. G., Radiometric Methods of Enzyme Assay	21	191
Olson, O. E., Palmer, I. S., and Whitehead, E. I., Determination of Selenium in Biological Materials	21	39
O'Reilly, J. E., see Elving, P. J.		
O'Toole, Ann G., see Goldberg, Nelson D.		
Ottesen, Martin, Methods for Measurement of Hydrogen Isotope Exchange in Globular Proteins	20	135
Ottesen, M., see Jacobsen, C. F.		
Palmer, I. S., see Olson, O. E.		

	VOL.	PAGE
Patterson, J. W., and *Lazarow, Arnold*, Determination of Glutathione	2	259
Perry, W. L. M., see *Humphrey, J. H.*		
Persky, Harold, Chemical Determination of Adrenaline and Noradrenaline in Body Fluids and Tissues	2	57
Plesner, Paul, and *Kalckar, Herman, M.*, Enzymic Micro Determinations of Uric Acid, Hypoxanthine, Xanthine, Adenine, and Xanthopterine by Ultraviolet Spectrophotometry	3	97
Pontis, Horacio G., and *Leloir, Luis F.*, Measurement of UDP-Enzyme Systems	10	107
Porath, J., and *Hjertén, S.*, Some Recent Developments in Column Electrophoresis in Granular Media	9	193
Porter, Curt C., see *Silber, Robert H.*		
Poulik, M. D., Gel Electrophoresis in Buffers Containing Urea......................	14	455
Raaflaub, Jurg, Applications of Metal Buffers and Metal Indicators in Biochemistry	3	301
Radin, Norman S., Glycolipide Determination...	6	163
Ramwell, P. W., see *Shaw, Jane E.*		
Read, W. W. C., see *McLaren, D. S.*		
Robins, Eli, The Measurement of Phenylalanine and Tyrosine in Blood	17	287
Roche, Jean, Lissitzky, Serge, and *Michel Raymond*, Chromatographic Analysis of Radioactive Iodine Compounds from the Thyroid Gland and Body Fluids	1	243
Roche, Jean, Michel, Raymond, and *Lissitzky, Serge*, Analysis of Natural Radioactive Iodine Compounds by Chromatographic and Electrophoretic Methods	12	143
Roe, Joseph H., Chemical Determinations of Ascorbic, Dehydroascorbic, and Diketogulonic Acids	1	115
Rosenkrantz, Harris, Analysis of Steroids by Infrared Spectrometry	2	1
Infrared Analysis of Vitamins, Hormones, and Coenzymes	5	407
Roth, Marc, Fluorimetric Assay of Enzymes...	17	189
Salmon, W. D., see *Engel, R. W.*		
Saltzman, Bernard E., and *Keenan, Robert G.*, Microdetermination of Cobalt in Biological Materials	5	181
Schayer, Richard W., Determination of Histidine Decarboxylase Activity..	16	273
Determination of Histidine Decarboxylase...	Supp.	99
Schmakel, C. O., see *Elving P. J.*		
Schmid, Harold H. O., see *Mangold, Helmut K.*		
Schubert, Jack, Measurement of Complex Ion Stability by the Use of Ion Exchange Resins	3	247
Schuberth, Jan, see *Sörbo, S. Bo*		
Schwartz, Morton K., and *Bodansky, Oscar*, Automated Methods for Determination of Enzyme Activity	11	211
Schwartz, Morton K., and *Bodansky, Oscar*, Utilization of Automation for Studies of Enzyme Kinetics	16	183
Schwartz, Samuel, Berg, Marie H., Bossenmaier, Irene, and *Dinsmore, Howard*, Determination of Porphyrins in Biological Materials..................	8	221
Scott, J. E., Aliphatic Ammonium Salts in the Assay of Acidic Polysaccharides from Tissues	8	145

	VOL.	PAGE
Seaman, G. R., see *Stokstad, E. L. R.*		
Seiler, N., Use of the Dansyl Reaction in Biochemical Analysis	18	259
Shaw, Jane E., and *Ramwell, P. W.,* Separation, Identification, and Estimation of Prostaglandins	17	325
Shibata, Kazuo, Spectrophotometry of Opaque Biological Materials: Reflection Methods	9	217
Spectrophotometry of Translucent Biological Materials: Opal Glass Transmission Method	7	77
Shore, P. A., Determination of Histamine	Supp.	89
Silber, Robert H., and *Porter, Curt C.,* Determination of 17,21-Dihydroxy-20-Ketosteroids in Urine and Plasma	4	139
Silber, Robert H., Fluorimetric Analysis of Corticoids	14	63
Singer, Thomas P., and *Kearney, Edna B.,* Determination of Succinic Dehydrogenase Activity	4	307
Sjövall, Jan, Separation and Determination of Bile Acids	12	97
Skeggs, Helen R., Microbiological Assay of Vitamin B_{12}	14	53
Smales, A. A., see *Loveridge, B. A.*		
Smith, Emil L., see *Davis, Neil C.*		
Smith, Fred, and *Montgomery, Rex,* End Group Analysis of Polysaccharides	3	153
Smith, Lucile, Spectrophotometric Assay of Cytochrome *c* Oxidase	2	427
Sörbo, S. Bo, and *Schuberth, Jan,* Measurements of Choline Acetylase	Supp.	275
Spencer, B., see *Dodgson, K. S.*		
Sperry, Warren M., Lipid Analysis	2	83
Stahl, Egon, see *Mangold, Helmut K.*		
St. John, P. A., see *Winefordner, J. D.*		
Stokstad, E. L. R., Seaman, G. R. Davis, R. J., and *Hunter, S. H.,* Assay of Thioctic Acid	3	23
Storvick, C. A., Benson, E. M., Edwards, M. A., and *Woodring, M. J.,* Chemical and Microbiological Determination of Vitamin B_6	12	183
Strehler, Bernard L., Bioluminescence Assay: Principles and Practice	16	99
Strehler, B. L., and *Totter, J. R.,* Determination of ATP and Related Compounds: Firefly Luminescence and Other Methods	1	341
Talalay, Paul, Enzymic Analysis of Steroid Hormones	8	119
Tchalian, M., see *McLaren, D. S.*		
Thiers, Ralph E., Contamination of Trace Element Analysis and Its Control	5	273
Thorpe, W. V., see *Bray, H. G.*		
Tinoco, Jr., Ignacio, Application of Optical Rotatory Dispersion and Circular Dichroism to the Study of Biopolymers	18	81
Tolksdorf, Sibylle, The *in vitro* Determination of Hyaluronidase	1	425
Totter, J. R., see *Strehler, B. L.*		
Treadwell, C. R., see *Vahouny, George V.*		
Udenfriend, Sidney, Weissbach, Herbert, and *Brodie, Bernard B.,* Assay of Serotonin and Related Metabolites, Enzymes, and Drugs	6	95
Vahouny, George V., and *Treadwell, C. R.,* Enzymatic Synthesis and Hydrolysis of Cholesterol Esters	16	219
Vallee, Bert L., see *Margoshes, Marvin*		
Vanden Heuvel, W. J. A., see *Horning, E. C.*		

	VOL.	PAGE
Van Pilsum, John F., Determination of Creatinine and Related Guanidinium Compounds	7	193
Vessey, D. A., see *Zakim, D.*		
Vestling, Carl S., Determination of Dissociation Constants for Two-Substrate Enzyme Systems	10	137
Volkin, Elliot, and *Cohn, Waldo E.*, Estimation of Nucleic Acids	1	287
Wacker, Warren E. C., see *Amador, Elias*		
Waldemann-Meyer, H., Mobility Determination by Zone Electrophoresis at Constant Current	13	47
Wang, C. H., Radiorespirometry	15	311
Webb, Junius M., and *Levy, Hilton B.*, New Developments in the Chemical Determination of Nucleic Acids	6	1
Weil-Malherbe, H., The Estimation of Total (Free + Conjugated) Catecholamines and Some Catecholamine Metabolites in Human Urine	16	293
Determination of Catechol Amines	Supp.	119
Weinstein, Boris, Separation and Determination of Amino Acids and Peptides by Gas–Liquid Chromatography	14	203
Weissbach, Herbert, see *Udenfriend, Sidney*		
Whalley, H. C. S. de, and *Gross, D.*, Determination of Raffinose and Kestose in Plant Products	1	307
Whiffen, D. H., see *Barker, S. A.*		
Whitehead, E. I., see *Olson, O. E.*		
Whitehead, J. K., and *Dean, H. G.*, The Isotope Derivative Method in Biochemical Analysis	16	1
Whitehouse, M. W., and *Zilliken, F.*, Isolation and Determination of Neuraminic (Sialic) Acids	8	199
Whitford, William R., see *Natelson, Samuel*		
Willis, J. B., Analysis of Biological Materials by Atomic Absorption Spectroscopy	11	1
Winefordner, J. D., McCarthy, W. J., and *St. John, P. A.*, Phosphorimetry as an Analytical Approach in Biochemistry	15	369
Winzler, Richard J., Determination of Serum Glycoproteins	2	279
Woodring, M. J., see *Storvick, C. A.*		
Wotiz, Herbert H., and *Clark, Stanley J.*, Newer Developments in the Analysis of Steroids by Gas-Chromatography	18	339
Wright, M. Ruth, see *Lerner, Aaron B.*		
Yagi, Kunio, Chemical Determination of Flavins	10	319
Yapel, Anthony F., Jr. and *Lumry, Rufus*, A Practical Guide to the Temperature-Jump Method for Measuring the Rate of Fast Reactions	20	169
Yalow, Rosalyn S., and *Berson, Solomon A.*, Immunoassay of Plasma Insulin	12	69
Yamamato, Richard S., see *Mickelsen, Olaf*		
Zakim, D., and *Vessey, D. A.*, Techniques for the Characterization of UDP-Glucuronyltransferase, Glucose-6-Phosphatase, and Other Tightly-Bound Microsomal Enzymes	21	1
Zilliken, F., see *Whitehouse, M. W.*		

CUMULATIVE INDEX, VOLUMES 1–21 AND SUPPLEMENT

Subject Index

	VOL.	PAGE
Absorbent and Fluorescent Substances, The Automated Analysis of, Separated on Paper Strips (Boulton)	16	327
Activation Analysis and Its Application in Biochemistry (Loveridge and Smales)	5	225
Activation Analysis of Biological Trace Elements (Leddicotte)	19	345
Adenine, Enzymic Micro Determination, by Ultraviolet Spectrophotometry (Plesner and Kalckar)	3	97
Adrenal Gland, Steroids of, Chromatographic Separation (Haines and Karnemaat)	1	171
Adrenal Steroids in Blood, Analysis of, by Countercurrent Distribution (Carstensen)	9	127
Adrenaline, Chemical Determination, in Body Fluids and Tissues (Persky)	2	57
Aliphatic Ammonium Salts in the Assay of Acidic Polysaccharides from Tissues (Scott)	8	145
Alternative Metabolic Pathways, Estimation of Magnitudes of (Kopin)	11	247
Amine Oxidases, Determination of (Kapeller-Adler)	Supp.	35
Amino Acids, Analysis by Means of Brdička Reaction (Müller)	11	329
Amino Acids, Determination by High-Voltage Paper Electrophoresis (Blackburn)	13	1
Amino Acids, Determination by Ion Exchange Chromatography (Jacobs)	14	177
Amino Acids, Determination by Use of Bacterial Amino Acid Decarboxylases (Gale)	4	285
Amino Acids, Separation and Determination by Gas–Liquid Chromatography (Weinstein)	14	203
Ammonium Salts, Aliphatic, in the Assay of Acidic Polysaccharides from Tissues (Scott)	8	145
Ampholytes, A Technique for Fractionation and Characterization through Isoelectric Focusing in—pH Gradients (Haglund)	19	1
Antibiotics, Microbiological Assay (Kersey and Fink)	1	153
Ascorbic Acid, Chemical Determination (Roe)	1	115
Atomic Absorption Spectroscopy, Analysis of Biological Materials by (Willis)	11	1
ATP, Determination of Firefly Luminescence (Strehler and Totter)	1	341
Bacterial Amino Acid Decarboxylases in Determination of Amino Acids (Gale)	4	285
Basic Proteins, Preparation and Analysis of (Lindh and Brantmark)	14	79
Bile Acids, Newer Developments in the Gas Chromatographic Determination of (Kuksis)	14	325

	VOL.	PAGE
Bile Acids, Separation and Determination of (Sjövall)	12	97
Bile Pigments, Separation and Determination of (Brodersen and Jacobsen)	17	31
Biochemical Applications of Magnetic Resonance (Jardetzky and Jardetzky)	9	235
Biological Materials, Analysis by Atomic Absorption Spectroscopy (Willis)	11	1
Biological Materials, Determination of Nitrogen in (Jacobs)	13	241
Biological Materials, Determination of Porphyrins in (Schwartz, Berg, Bossenmaier, and Dinsmore)	8	221
Biological Materials, Determination of Zinc in (Malmström)	3	327
Biological Materials, Methods for Estimating Magnesium in (Alcock and MacIntyre)	14	1
Biological Materials, Microdetermination of Cobalt in (Saltzman and Keenan)	5	181
Biological Materials, Opaque, Spectrophotometry of; Reflection Methods (Shibata)	9	217
Biological Materials, Translucent, Spectrophotometry of; Opal Glass Methods (Shibata)	7	77
Biological Standards in Biochemical Analysis (Humphrey, Long, and Perry)	5	65
Biological Systems, Determination of Serine in (Frisell and Mackenzie)	6	63
Biological Trace Elements, Activation Analysis of (Leddicotte)	19	345
Bioluminescence Assay: Principles and Practice (Strehler)	16	99
Blood, Analysis of Adrenal Steroids in, by Countercurrent Distribution (Cartensen)	9	127
Blood, Determination of Ethyl Alcohol in (Lundquist)	7	217
Body Fluids, Chemical Determination of Adrenaline and Noradrenaline in (Persky)	2	57
Body Fluids, Chromatographic Analysis of Radioactive Iodine Compounds from (Roche, Lissitzky, and Michel)	1	243
Body Tissues, Chemical Determination of Adrenaline and Noradrenaline in (Persky)	2	57
Buffers, Containing Urea, Gel Electrophoresis in (Poulik)	14	455
Carbohydrate Derivatives, Separation of, by Gas–Liquid Partition Chromatography (Bishop)	10	1
Carbohydrate Structure, Use of Infrared Analysis in Determination of (Baker, Bourne, and Whiffen)	3	213
Carbohydrate, The Determination of, in Biological Materials by Gas–Liquid Chromatography (Clamp, Bhatti, and Chambers)	19	229
Carbonic Anhydrose Activity, Measurement of (Davis)	11	307
Carolene, Determination of (Bickoff)	4	1
Catalases, Assay of (Maehly and Chance)	1	357
Catechol Amine Biosynthesis and Metabolism, Assay of Enzymes of (Creveling and Daly)	Supp.	153
Catecholamines and Catecholamine Metabolites, Estimation of Total (Free + Conjugated), in Human Urine (Weil-Malherbe)	16	293
Catechol Amines, Determination of (Weil-Malherbe)	Supp.	119
Cell Particles and Macromolecules, Partition Methods for		

	VOL.	PAGE
Fractionation of (Albertsson)..	10	229
Cellulose Column Chromatography, Analysis of Mixtures of Sugars by (Hough)...	1	205
Centrifugation, Preparative Zonal (Anderson)...	15	271
Chloride in Biological Materials, Determination of (Cotlove)	12	277
Cholesterol, Determination and Microscopic Localization of (Kabara)	10	263
Cholesterol Esters, Enzymatic Synthesis and Hydrolysis of (Vahouny and Treadwell)..	16	219
Choline Acetylase, Measurements of (Sörbo and Schuberth)	Supp.	275
Choline, Chemical Estimation of (Engel, Salmon, and Ackerman)	1	265
Choline Esters, Measurement of (Jenden)..	Supp.	183
Cholinesterases, Assay Methods for (Augustinsson)...................................	5	1
Cholinesterases, Determination of (Augustinsson).....................................	Supp.	217
Chromatographic Analysis, Applications of the R_M Treatment in (Bush)	13	357
Chromatographic Analysis, Applications of the R_M Treatment in, Erratum (Bush)	14	497
Chromatographic Analysis of Radioactive Iodine Compounds from the Thyroid Gland and Body Fluids (Roche, Lissitzky, and Michel)	1	243
Chromatographic and Electrophoretic Methods, Analysis of Natural Radioactive Iodine Compounds by (Roche, Michel, and Lissitzky).....................	12	143
Chromatographic Separation of Steroids of the Adrenal Gland (Haines and Karnemaat)	1	171
Chromatography, Gas, in Determination of Bile Acids and Steroids (Kuksis)	14	325
Chromatography, Gas, Separation and Determination of Steroids by (Horning, VandenHeuvel, and Creech)	11	69
Chromatography, Gas–Liquid, Determination of the Fatty Acids by (James)	8	1
Chromatography, Gas–Liquid, Separation and Determination of Amino Acids and Peptides by (Weinstein)..	14	203
Chromatography, Gas–Liquid Partition, Separation of Carbohydrate Derivatives by (Bishop)	10	1
Chromatography, Ion Exchange, Determination of Amino Acids by (Jacobs)	14	177
Chromatography, Paper and Cellulose Column, Analysis of Mixtures of Sugars by (Hough)...	1	205
Chromatography, Thin-Layer (TLC) (Mangold, Schmid, and Stahl).............	12	393
Cobalt, Microdetermination of, in Biological Materials (Saltzman and Keenan)	5	181
Coenzyme A, Methods for Determination of (Novelli)...............................	2	189
Coenzyme Q, Determination of (Crane and Dilley)....................................	11	279
Coenzymes, Infrared Analysis of (Rosenkrantz)..	5	407
Collagen and Elastin, The Determination of (Jackson and Cleary)	15	25
Color Reactions, New, for Determination of Sugars in Polysaccharides (Dische)...	2	313
Column Electrophoresis in Granular Media, Some Recent Developments (Porath and Hjertén)	9	193
Complexes, Metal-Protein, Analysis of (Hughes and Klotz)........................	3	265
Complex Ion Solubility, Measurement by Use of		

	VOL.	PAGE
Ion Exchange Resins (Schubert)	3	247
Contamination in Trace Element Analysis and Its Control (Thiers)	5	273
Corticoids, Fluorimetric Analysis of (Silber)	14	63
Countercurrent Distribution (King and Craig)	10	201
Countercurrent Distribution, Analysis of Adrenal Steroids in Blood by (Carstensen)	9	127
Creatinine and Related Guanidinium Compounds, Determination of (Van Pilsum)	7	193
Current, Constant, Mobility Determination by Zone Electrophoresis at (Waldmann-Meyer)	13	47
Cyclic 3',5'-Adenosine Monophosphate and Cyclic 3',5'-Guanosine Monophosphate, Analysis of (Goldberg and O'Toole)	20	1
Cyclochrome c Oxidase, Spectrophotometric Assay of (Smith)	2	427
Dansyl Reaction, Use of the, in Biochemical Analysis (Seiler)	18	259
Dehydroascorbic Acid, Chemical Determination of (Roe)	1	115
Denaturation, Reversible, of Proteins, Methods of Study and Interpretation of Data for (Hermans, Jr.)	13	81
Density Gradients, Rapid Electrophoresis in (Kolin)	6	259
Deoxyribonuclease Activity, Assay of (Kurnick)	9	1
Diagnosis, Enzymatic Methods of (Amador and Wacker)	13	265
Dialysis (Craig and King)	10	175
Diffraction, X-ray, in the Study of Protein and Nucleic Acid Structure (Holmes and Blow)	13	113
17,21-Dihydroxy-20-Ketosteroids, Determination in Urine and Plasma (Silber and Porter)	9	139
Diketogulonic Acid, Chemical Determination of (Roe)	1	115
Dissociation Constants, Determination of, for Two-Substrate Enzyme Systems (Vestling)	10	137
Electron Probe Microanalyzer, An Introduction to, and Its Application to Biochemistry (Andersen)	15	147
Electrophoresis, Free Zone, Theory, Equipment, and Applications (Hjerten)	18	55
Electrophoresis, Gel, in Buffers Containing Urea (Poulik)	14	455
Electrophoresis, Paper, Determination of Amino Acids at High-Voltage by (Blackburn)	13	1
Electrophoresis, Rapid, in Density Gradients Combined with pH and/or Conductivity Gradients (Kolin)	6	259
Electrophoresis, Zone (Kunkel)	1	141
Electrophoresis, Zone, Constant Current Mobility Determination by (Waldmann-Meyer)	13	47
Electrophoresis in Granular Media, Column, Some Recent Developments (Porath and Hjertén)	9	193
Electrophoretic Methods, Analysis of Natural Radioactive Iodine Compounds by (Roche, Michel, and Lissitzky)	12	143
Elements, Determination of, by X-Ray Emission Spectrometry (Natelson and Whitford)	12	1
Enthalpy and Entropy Changes, Determination by Heatburst Microcalorimetry (Kitzinger and Benzinger)	8	309
Enzymatic Methods, in Diagnosis (Amador and Wacker)	13	265

	VOL.	PAGE
Enzyme Activity, Automated Methods for Determination of (Schwartz and Bodansky)	11	211
Enzyme Assay, Radiometric Methods of (Oldham)	21	191
Enzyme Kinetics, Utilization of Automation for Studies of (Schwartz and Bodansky)	16	183
Enzymes, Assay of in Catechol Amine Biosynthesis and Metabolism (Creveling and Daly)	Supp.	153
Enzymes, Fluorimetric Assay of (Roth)	17	189
Enzymes, Proteolytic Assay of (Davis and Smith)	2	215
Enzymes, Related to Serotonin, Assay of (Udenfriend, Weissbach, and Brodie)	6	95
Enzyme Systems, Two Substrate, Determination of Dissociation Constants for (Vestling)	10	137
Enzymic Determination of D-Glucose and Its Anomers, New Developments in (Okuda and Miwa)	21	155
Enzymic Analysis of Steroid Hormones (Talalay)	8	119
Estrogens, Chemical Determination of, in Human Urine (Bauld and Greenway)	5	337
Ethanolamine, Determination of, in Lipids (McKibbin)	7	111
Fatty Acid Esters, A Critical Evaluation of the Gas Chromatographic Technique for Identification and Determination of, with Particular Reference to the Use of Analog and Digital Computer Methods (Caster)	17	135
Fatty Acids, Determination by Gas–Liquid Chromatography (James)	8	1
Firefly Luminescence, Determination of ATP by (Strehler and Totter)	1	341
Flame Photometry, Principles and Applications (Margoshes and Vallee)	3	353
Flavins, Chemical Determination of (Yagi)	10	319
Fluids, Body, Chemical Determination of Adrenaline and Noradrenaline in (Persky)	2	57
Fluids, Body, Chromatographic Analysis of Radioactive Iodine Compounds from (Roche, Lissitzky, and Michel)	1	243
Fluorimetric Analysis of Corticoids (Silber)	14	63
Folic Acid Activity, Assay of Compounds with (Jukes)	2	121
Formaldehyde, Determination of, in Biological Systems (Frisell and Mackenzie)	6	63
Fractionation of Cell Particles and Macromolecules, Partition Methods for (Albertson)	10	229
Free Energy Changes, Determination by Heatburst Microcalorimetry (Kitzinger and Benzinger)	8	309
Frog Skin Assay for Agents that Darken and Lighten Melanocytes (Lerner and Wright)	8	295
Gas-Liquid Chromatography, The Determination in Carbohydrates and Biological Materials (Clamp, Bhatti, and Chambers)	19	229
Gel Electrophoresis in Buffers Containing Urea (Poulik)	14	455
β-Glucuronidases, Determination of (Fishman)	15	77
UDP-Glucuronyltransferase, Glucose-6-Phosphatase, and Other Tightly-Bound Microsomal Enzymes, Techniques for the Characterization of (Zakin and Vessey)	21	1
Glutamic and Aspartic Acids and Their Amides, Determination of (Balis)	20	103

	VOL.	PAGE
Glutathione, Determination of (Patterson and Lazarow)	2	259
Glycolipid Determination (Radin)	6	163
Glycoproteins, Serum, Determination of (Winzler)	2	279
Gradients, Density, Rapid Electrophoresis in (Kolin)	6	259
Heatburst Microcalorimetry, Principle and Methods of, and Determination of Free Energy, Enthalpy, and Entropy Changes (Kitzinger and Benzinger)	8	309
Heparin, Determination of (Jaques and Bell)	7	253
Hexosamines, Determination of (Gardell)	6	289
High-Performance Ion-Exchange Chromatography with Narrow-Bore Columns: Rapid Analysis of Nucleic Acid Constituents at the Subnanomole Level (Horvath)	21	79
Histamine, Determination of (Shore)	Supp.	89
Histamine, Quantitative Determination of (Code and McIntire)	3	49
Histidine Decarboxylase, Determination of (Schayer)	Supp.	99
Histidine Decarboxylase Activity, Determination of (Schayer)	16	273
Hormones, Infrared Analysis of (Rosenkrantz)	5	407
Hormones, Plant, Analysis of (Bentley)	9	75
Hormones, Steroid, Enzymic Analysis of (Talalay)	8	119
Hyaluronidase, in vitro Determination (Tolksdorf)	1	425
Hydrogen Isotope Exchange in Globular Proteins, Methods for Measurement (Ottesen)	20	135
Hypoxanthine, Enzymic Micro Determination, by Ultraviolet Spectrophotometry (Plesner and Kalckar)	3	97
Immunoassay of Plasma Insulin (Yalow and Berson)	12	69
Immunoelectrophoretic Analysis (Garbar)	7	1
Infrared Analysis, Use of, in the Determination of Carbohydrate Structure (Baker, Bourne, and Whiffen)	3	213
Infrared Analysis of Vitamins, Hormones, and Coenzymes (Rosenkrantz)	5	407
Infrared Spectrometry, Analysis of Steroids by (Rosenkrantz)	2	1
Inositol Determination of, in Lipides (McKibbin)	7	111
Iodine Compounds, Natural Radioactive, Analysis by Chromatographic and Electrophoretic Methods (Roche, Michel, and Lissitzky)	12	143
Iodine Compounds, Radioactive, from Thyroid Gland and Body Fluids, Chromatographic Analysis (Roche, Lissitzky, and Michel)	1	243
Ion Exchange Resins, Measurement of Complex Ion Stability by Use of (Schubert)	3	247
Isotope Derivative Method in Biochemical Analysis, The (Whitehead and Dean)	16	1
Kestose, Determination, in Plant Products (de Whalley and Gross)	1	307
α-Keto Acid Determinations (Neish)	5	107
17-Ketosteroids, Urinary Neutral, Assay of (Engel)	1	459
Lipase, Lipoprotein, Assay of, in vivo and in vitro (Korn)	7	145
Lipide Analysis (Sperry)	2	83
Lipides, Determination of Inositol, Ethanolamine, and Serine in (McKibbin)	7	111
Lipoprotein Lipase, Assay of, in vivo and in vitro (Korn)	7	145
Lipoproteins, Serum, Ultracentrifugal Analysis (de Lalla and Gofman)	1	459
Lipoxidase Activity, Measurement of (Holman)	2	113

	VOL.	PAGE
Liquid-Scintillation Counting, Practical Aspects of (Kobayashi and Maudsley)	17	55
Luciferin and Luciferase, Measurement of (Chase)	8	61
Magnesium Estimation, in Biological Materials (Alcock and MacIntyre)	14	1
Magnetic Resonance, Biochemical Applications of (Jardetzky and Jardetzky)	9	235
Mass Spectrometry in the Determination of Structure of Certain Natural Products Containing Sugars (Hanessian)	19	105
Melanocytes, Darkening and Lightening, Frog Skin Assay for (Lerner and Wright)	8	295
Metabolic Pathways, Alternative, Estimation of Magnitudes of (Kopin)	11	247
Metabolism, Analysis of Phenolic Compounds of Interest in (Bray and Thorpe)	1	27
Metal Buffers, Applications, in Biochemistry (Raaflaub)	3	301
Metal Indicators, Applications, in Biochemistry (Raaflaub)	3	301
Metal-Protein Complexes, Analysis of (Hughes and Klotz)	3	265
Microbiological Assay of Antibiotics (Kersey and Fink)	1	53
Microbiological Assay of Vitamin B_{12} (Hoff-Jorgensen)	1	81
Microbiological Assay of Vitamin B_{12} (Skeggs)	14	53
Microbiological Determination of Vitamin B_6 (Storvick, Benson, Edwards, and Woodring)	12	183
Mobility, Determination by Zone Electrophoresis at Constant Current (Waldmann-Meyer)	13	47
Molecular Size, Estimation of, and Molecular Weights of Biological Compounds by Gel Filtration (Andrews)	18	1
Neuraminic (Sialic) Acids, Isolation and Determination of (Whitehouse and Zilliken)	8	199
Nitrogen, Determination in Biological Materials (Jacobs)	13	241
Nitrogenous Compounds, Basic, of Toxicological Importance, Analysis of (Curry)	7	39
Noradrenaline, Chemical Determination, in Body Fluids and Tissues (Persky)	2	57
Nucleic Acid, Structure, X-ray Diffraction in the Study of (Holmes and Blow)	13	113
Nucleic Acids, Chemical Determination of (Webb and Levy)	6	1
Nucleic Acids, the Determination of (Munro and Fleck)	14	113
Nucleic Acids, Estimation (Volkin and Cohn)	1	287
Nucleic Acids and Their Derivatives, Microbiological Assay of (Miller)	6	31
Nucleic Acids of Various Conformational Forms and Measurement of Their Associated Enzymes, use of Ethidium Bromide for Separation and Determination of (Le Pecq)	20	41
Nucleosides and Nucleotides and Their Parent Bases as an Analytical and Investigative Tool, Polarography and Voltammetry of (Elving, O'Reilly, and Schmakel)	21	287
Optical Rotatory Dispersion, Application of, and Circular Dichroism to the Study of Biopolymers (Tinoco, Jr.)	18	81
Organic Phosphorus Compounds, Determination of, by Phosphate Analysis (Lindberg and Ernster)	3	1
Oxidations, Periodate, Use of, in Biochemical Analysis (Dyer)	3	111

	VOL.	PAGE
Oxygen Electrode Measurements in Biochemical Analysis (Lessler and Brierley)	17	1
Paper Chromatograms, Direct Scanning of, for Quantitative Estimations (Bush)	11	149
Paper Chromatography, for Analysis of Mixtures of Sugars (Hough)	1	205
Partition Methods for Fractionation of Cell Particles and Macromolecules (Albertsson)	10	229
Peptide Chromatography, Automatic (Jones)	18	205
Peptides, Separation and Determination, by Gas–Liquid Chromatography (Weinstein)	14	203
Peptides, Terminal and Sequence Studies in, Recent Developments in Techniques for (Fraenkel-Conrat, Harris, and Levy)	2	359
Periodate Oxidations, Use of, in Biochemical Analysis (Dyer)	3	111
Peroxidases, Assay of (Maehly and Chance)	1	357
Phenolic Compounds of Interest in Metabolism (Bray and Thorpe)	1	27
Phenylalanine and Tyrosine in Blood, The Measurement of (Robins)	17	287
pH Gradients, Isoelectric Focusing in—A Technique for Fractionation and Characterization of Ampholytes (Haglund)	19	1
pH-Stat and Its Use in Biochemistry (Jacobson, Léonis, Linderstrøm-Lang, and Ottesen)	4	171
Phosphate Analysis, Determination of Organic Phosphorus Compounds by (Lindberg and Ernster)	3	1
Phosphorimetry, as an Analytical Approach in Biochemistry (Winefordner, McCarthy, and St. John)	15	369
Phosphorus Compounds, Organic, Determination of, by Phosphate Analyses (Lindberg and Ernster)	3	1
Photometry, Flame, Principles and Applications of (Margoshes and Vallee)	3	353
Phytate and Inositol Phosphates, the Determination of (Oberleas)	20	87
Plant Hormones, Analysis of (Bentley)	9	75
Plasma, Determination of 17,21-Dihydroxy-20-Ketosteroids in (Silber and Porter)	4	139
Plasma Insulin, Immunoassay of (Yalow and Berson)	12	69
Polarographic Analysis of Proteins, Amino Acids, and Other Compounds by Means of the Brdička Reaction (Müller)	11	329
Polysaccharides, Acidic, from Tissues, Aliphatic Ammonium Salts in the Assay of (Scott)	8	145
Polysaccharides, End Group Analysis of (Smith and Montgomery)	3	153
Polysaccharides, Sugars in, New Color Reactions for Determination of (Dische)	2	313
Polyunsaturated Fatty Acids, Measurement of (Holman)	4	99
Porphyrins in Biological Materials, Determination of (Schwartz, Berg, Bossenmaier, and Dinsmore)	8	221
Prostaglandins, Separation, Identification, and Estimation of (Shaw and Ramwell)	17	325
Protein, Structure, X-ray Diffraction in the Study of (Holmes and Blow)	13	113
Protein, Terminal and Sequence Studies in, Recent Developments in Techniques for (Fraenkel-Conrat, Harris, and Levy)	2	359
Proteins, Analysis by Means of Brdička Reaction (Müller)	11	329

	VOL.	PAGE
Proteins, Basic, Preparation and Analysis of (Lindh and Brantmark)	14	79
Proteins, Polarography of, Analytical Principles and Applications in Biological and Clinical Chemistry (Homolka)	19	435
Proteins, Reversible Denaturation of, Methods of Study and Interpretation of Data for (Hermans, Jr.)	13	81
Proteolytic Enzymes, Assay of (Davis and Smith)....................................	2	215
Purines, New Methods for Purification and Separation of (Bergmann and Dikstein)	6	79
Quantitative Mass Spectrometric Analysis: Chemical and Biological Applications, Integrated Ion-Current(IIC) Technique of (Majer and Boulton)	21	467
R_M Treatment, Applications in Chromatographic Analysis (Bush)	13	357
R_M Treatment, Applications in Chromatographic Analysis, Erratum (Bush)	14	497
Radioactive Iodine Compounds, from Thyroid Gland and Body Fluids, Chromatographic Analysis of (Roche, Lissitzky, and Michel)	1	243
Radiorespirometry (Wang)	15	311
Raffinose, Determination in Plant Products (de Whalley and Gross)	1	307
Resins, Ion Exchange, Measurement of Complex Ion Stability, by Use of (Schubert)...	3	247
Resonance, Magnetic, Biochemical Applications of (Jardetzky and Jardetzky)	9	235
Ribonuclease, Characterization of, and Determination of Its Activity (Josefsson and Lagerstedt)...	9	39
Selenium in Biological Materials, Determination of (Olson, Palmer, and Whitehead)...	21	39
Serine, Determination of, in Biological Systems (Frisell and Mackenzie)......	6	63
Serine, Determination of, in Lipides (McKibbin)....................................	7	111
Serotonin: The Assay of Hydroxyindole Compounds and Their Biosynthetic Enzymes (Lovenberg and Engelman)	Supp.	1
Serotonin and Related Metabolites, Enzymes, and Drugs, Assay of (Udenfriend, Weissbach, and Brodie)	6	95
Serum Acid Phosphatases, Determinations (Fishman and Davidson)............	4	257
Serum Glycoproteins, Determination of (Winzler)	2	279
Serum Lipoproteins, Ultracentrifugal Analysis of (de Lalla and Gofman)	1	459
—SH Groups in Proteins, Determination of (Benesch and Benesch).............	10	43
Sialic Acids, see Neuraminic Acids		
Sodium and Potassium, Measurement of, by Glass Electrodes (Friedman) ...	10	71
Spectrometry, Infrared, Analysis of Steroids by (Rosenkrantz)	2	1
Spectrometry, Principles and Applications (Margoshes and Vallee)..............	3	353
Spectrometry, X-Ray Emission, Determination of Elements by (Natelson and Whitford)	12	1
Spectrophotometric Assay of Cytochrome c Oxidase (Smith)	2	427
Spectrophotometry of Opaque Biological Materials; Reflection Methods (Shibata)	9	217
Spectrophotometry of Translucent Biological Materials; Opal Glass Method (Shibata)	7	77
Spectrophotometry, Ultraviolet, Enzymic Micro Determinations of Uric Acid, Hypoxanthine, Xanthine, Adenine, and		

	VOL.	PAGE
Xanthopterine by (Plesner and Kalckar)	6	97
Standards, Biological, in Biochemical Analysis (Humphrey, Long, and Perry)	5	65
Steroid Hormones, Enzymic Analysis of (Talalay)	8	119
Steroids, Adrenal, in Blood, Analysis by Countercurrent Distribution (Carstensen)	9	227
Steroids, Analysis by Infrared Spectrometry (Rosenkrantz)	2	1
Steroids of the Adrenal Gland, Chromatographic Separation (Haines and Karnemaat)	1	171
Steroids, Newer Developments in the Analysis of, by Gas-Chromatography (Wotiz and Clark)	18	339
Steroids, Newer Developments in the Gas Chromatographic Determination of (Kuksis)	14	325
Steroids, Separation and Determination, by Gas Chromatography (Horning, VandenHeuvel, and Creech)	11	69
Succinic Dehydrogenase Activity, Determination of (Singer and Kearney)	4	307
Sugars, Analysis of Mixtures, by Paper and Cellulose Column Chromatography (Hough)	1	205
Sugars, in Polysaccharides, Determination, New Color Reactions for (Dische)	2	313
Sugars, the Determination of Structure of Certain Natural Products Containing Sugars (Hanessian)	19	105
Sulfatases, Assay (Dodgson and Spencer)	4	211
Sulfhydryl Groups, Determination in Biological Substances (Chinard and Hellerman)	1	1
Temperature-Jump Method for Measuring the Rate of Fast Reactions, a Practical Guide to (Yapel and Lumry)	20	169
Thiamine, Methods for the Determination of (Mickelsen and Yamamoto)	6	191
Thioctic Acid, Assay of (Stokstad, Seaman, Davis, and Hunter)	3	23
Thyroid Gland, Chromatographic Analysis of Radioactive Iodine Compounds from (Roche, Lissitzky, and Michel)	1	243
Tissues, Aliphatic Ammonium Salts in the Assay of Acidic Polysaccharides from (Scott)	8	145
Tissues, Body, Chemical Determination of Adrenaline and Noradrenaline in (Persky)	2	57
Tissues, Determination of Ethyl Alcohol in (Lundquist)	7	217
Trace Element Analysis, Contamination in, and Its Control (Thiers)	5	273
Transaminase, Determination of (Aspen and Meister)	6	131
Ubiquinone, Determination of (Crane and Dilley)	11	279
UDP-Enzyme Systems, Measurements of (Pontis and Leloir)	10	107
Ultracentrifugal Analysis of Serum Lipoproteins (de Lalla and Gofman)	1	459
Ultraviolet Spectrophotometry, Enzymic Micro Determinations of Uric Acid, Hypoxanthine, Xanthine, Adenine, and Xanthopterine by (Plesner and Kalckar)	3	97
Urea, Ammonia, and Urease, The Determination of (Kaplan)	17	311
Urea, Gel Electrophoresis in Buffers Containing (Poulik)	14	455
Uric Acid, Enzymic Micro Determinations, by Ultraviolet Spectrophotometry (Plesner and Kalckar)	3	97
Urinary Neutral 17-Ketosteroids, Assay of (Engel)	1	479

	VOL.	PAGE
Urine, Determination of 17,21-Dihydroxy-20-Ketosteroids in (Silber and Porter)	4	139
Urine, Human, Chemical Determination of Estrogens in (Bauld and Greenway)	5	337
Vitamin A, Determination of (Embree, Ames, Lehman, and Harris)	4	43
Vitamin A and Carotenoids, in Blood and Tissue, Microdetermination of (McLaren, Read, Awdeh, and Tchalian)	15	1
Vitamin B_6, Chemical and Microbiological Determination of (Storvick, Benson, Edwards, and Woodring)	12	183
Vitamin B_{12}, Microbiological Assay of (Hoff-Jorgensen)	1	81
Vitamin B_{12}, Microbiological Assay of (Skeggs)	14	53
Vitamin E Determination (Lehman)	2	153
Vitamins, Infrared Analysis of (Rosenkrantz)	5	407
Xanthine, Enzymic Micro Determination, by Ultraviolet Spectrophotometry (Plesner and Kalckar)	3	97
Xanthopterine, Enzymic Micro Determinations, by Ultraviolet Spectrophotometry (Plesner and Kalckar)	3	97
X-Ray Diffraction, in the Study of Protein and Nucleic Acid Structure (Holmes and Blow)	13	113
X-Ray Emission Spectrometry, Determination of Elements by (Natelson and Whitford)	12	1
Zinc, Determination of, in Biological Materials (Malmström)	3	327
Zone Electrophoresis (Kunkel)	1	141
Zone Electrophoresis, at Constant Current, Mobility Determination by (Waldmann-Meyer)	13	47